RICE UNIVERSITY STUDIES

THE OFFSHORE ECOLOGY INVESTIGATION

EFFECTS OF OIL DRILLING AND PRODUCTION IN A COASTAL ENVIRONMENT

C. H. WARD, M. E. BENDER, AND D. J. REISH, Editors

PUBLISHED BY
WILLIAM MARSH RICE UNIVERSITY
HOUSTON, TEXAS

Vol. 65, Nos. 4 & 5 Fall 1979

US-ISSN-0035-4996

US-ISBN-0-89263-243-7

ACKNOWLEDGMENTS

The Offshore Ecology Investigation (OEI) results described in section II and presented individually in section IV of this monograph are the product of the 23 principal investigators and 30 graduate students who participated in the program. The OEI could not have been conducted successfully without the knowledge and experience of these principal investigators nor without their dedication and hard work under very difficult and demanding schedules and field conditions. Each of the component investigations they conducted contributed data and information that were essential to accomplishing the program objectives.

Special recognition is due the four members of the OEI Program Planning Council: Dr. James P. Morgan, Chairman, Dr. Robert J. Menzies, Chief Scientist, Dr. Carl H. Oppenheimer, and Dr. Sayed Z. El-Sayed. In undertaking the design and direction of a multi-university interdisciplinary ecological field study, the members of this Council were faced with developing an especially challenging rationale and program for its implementation. There were no precedents for such a program; the objectives were predictably very difficult to achieve; and the subject of the investigation was highly controversial. The OEI was conducted largely because of their confidence that the program could be successful and because of their conviction that programs of this nature are especially important to the public, as well as for scientific purposes. The efficiency and effectiveness of the OEI were due in large part to the knowledge and planning skills of these Council members.

As Chief Scientist of the OEI, Dr. Robert J. Menzies was not only a principal architect of the program, he was the scientific innovator and the practical designer of aspects of the field program that were vital to its successful completion. His manual, "A Handbook on Procedures and Methods Employed in the Offshore Ecology Investigation (OEI)," was the first such compilation prepared for interdisciplinary ecological field studies. It specified all procedures for the field measurement and sampling exercises and for the subsequent laboratory analyses and data processing. In addition to being essential to the program, it ensured that any component of the OEI could be accurately repeated should subsequent studies be conducted. Above all, his conviction, dedication, and exceptional scientific understanding of marine ecology as an interdisciplinary science were vital influences. It is notable that even the independent re-appraisal of the OEI, conducted four years after completion of the OEI, was possible because of Dr. Menzies's early perception that the evaluation and interpretation of ecological data would require the development of computer software designed specifically for this purpose. Dr. Menzies was instrumental in initiating the development of such a capability by the Gulf Universities Research Consortium in 1971, and he was a principal investigator in the early phases of the program leading to its establishment.

In the latter regard, special recognition is due also to Mr. Henry S. Fleming. Mr. Fleming's extensive work in the organization of taxonometric data, and especially in the development of computer software and methods for computer-assisted analysis of biological and related environmental data, were the foundation on which the methods used for the re-appraisal reported in section III were built. Working with Dr. S. G. Appan, Mr. Fleming guided the development, testing, and demonstration of the ecological data analysis methods and associated software used in the re-appraisal.

The methodologies for applying this data base analysis capability were developed in large part by Dr. M. E. Bender, Dr. D. J. Reish, and Dr. C. H. Ward. They contributed improved rationales, procedures, and interpretive methods for the synthesis and analysis of multidisciplinary ecological data bases, which not only provided much greater scientific insight into and understanding of the OEI data, but which also should be generally applicable to many other data bases that have been generated in scientific and ecological monitoring programs. Of special importance in the re-appraisal of the OEI were their scientific perspective and ability to apply broad ecological and statistical principles in this program.

In their evaluation of the OEI data and their development of methods for synthesis, interpretation, and documentation of their re-appraisal, they were assisted by Dr. S. G. Appan. Special recognition is due Dr. Appan for his efficient support of the re-appraisal team in providing the tabulations, statistical treatment, and displays required for the re-appraisal and for his scientific contributions in support of their efforts.

Broad authorization was granted for use of articles and figures relating to the OEI study that have been published and copyrighted in various journals. They are: The Gulf Association of Geological Societies, *Transactions* (G. M. Griffin, 1978, Water clarity near oil production platforms on the Louisiana continental shelf and the source of the turbid bottom water layer); the Offshore Technology Conference, *Proceedings* (M. E. Bender, J. M. Sharp, D. J. Reish, S. G. Appan, and C. H.

Ward, 1979, An independent appraisal of the Offshore Ecology Investigation); the American Chemical Society, *Analytical Chemistry* (R. Miget, H. Kator, C. Oppenheimer, J. L. Laseter, and E. J. Ledet, 1974, New sampling device for the recovery of petroleum hydrocarbons and fatty acids from aqueous films); Marcel Dekker, Inc., *Analytical Letters* (E. J. Ledet and J. L. Laseter, 1974, A comparison of two sampling devices for the recovery of organics from aqueous surface films); and the American Association for the Advancement of Science, *Science* (E. J. Ledet and J. L. Laseter, 1974, Alkanes at the air-sea interface from offshore Louisiana and Florida). We appreciate their generous permission to reprint.

A special acknowledgment must be given to Mrs. Glenda R. Sowle. Since we lacked a large editorial support staff, the organization, compilation, typing, and editing of this monograph have been a formidable task. Its publication would have required much more time and effort were it not for her excellent and dedicated performance. She had responsibility for all of these very demanding aspects of the publication and accomplished them thoroughly and independently.

Finally, we wish to acknowledge the support of the 83 offshore industry companies that funded the OEI and the 19 offshore companies that funded the re-appraisal. In particular, the authors and editors appreciate the complete freedom given to all investigators in the design, conduct, and interpretation of their work, and these companies' insistence from the beginning that all data must be in the public domain, that the results be made public regardless of their nature, and that all investigators be encouraged to publish their results in the open scientific literature.

The Trustees of the Gulf Universities Research Consortium wish formally to recognize and express their appreciation to all of the contributors to this monograph, to all those scientists and graduate students whose work is reported, and to the companies that sponsored the work.

GULF UNIVERSITIES RESEARCH CONSORTIUM

James E. Brooks	Southern Methodist University
Agustin Ayala-Castanares	Universidad Nacional Autonoma de Mexico
William Daniel	Mobil Oil Corporation
F. E. Ellis	Continental Oil Company
Frank L. Keller	Tulane University
George F. Kirby	Texas Eastern Corporation
Marion T. Loftin	Mississippi State University
John Margrave	Rice University
George P. Mitchell	Mitchell Energy & Development Corporation
Anthony Randazzo	University of Florida
Joseph M. Reynolds	Louisiana State University
John S. Shaw, Jr.	Southern Natural Resources, Inc.
Robert T. van Aller	University of Southern Mississippi
Joseph F. Volker	University of Alabama System
William R. Walton	Amoco International Oil Company

RICE UNIVERSITY STUDIES

Vol. 65, Nos. 4 & 5 Fall 1979

THE OFFSHORE ECOLOGY INVESTIGATION

EFFECTS OF OIL DRILLING AND PRODUCTION IN A COASTAL ENVIRONMENT

I

INTRODUCTION

THE CUMULATIVE EFFECTS OF PETROLEUM DRILLING AND PRODUCTION IN COASTAL AND NEAR-SHORE AREAS

by J. M. Sharp

In the late 1960s there developed, on a universal scale, great concern about the deleterious effects of man's activities upon the environment. One area that received much attention and special scrutiny was the oil industry. In particular, the exploration, drilling, and production activities of the industry were extremely suspect. In an attempt to determine the impact, past and present, that the industry was truly causing, the Offshore Ecology Investigation (OEI) was initiated. This investigation was a multidisciplinary field study program consisting of eight seasonal synoptic data acquisition efforts of some 10 days each, carried out in 1972-1974. Biological, chemical, geological, and physical measurements were made at locations and on schedules designed to support valid synthesis and correlation of all data. The program was to investigate the possible cumulative ecological effects of oil drilling and production on coastal and near-shore ecosystems. It was carried out in Timbalier Bay and adjacent near-shore environments in an area in south Louisiana that had been subjected to more than 25 years of intensive oil and gas drilling and production. Scientific measurements were designed to provide the basis for identifying any cumulative effects present and for discriminating any variability of habitat or inhabitants according to whether they resulted from natural events or from drilling and production activities.

Conduct of the OEI involved the participation of 23 principal investigators. The data and results obtained from these investigations were evaluated and interpreted by the OEI Council, composed of four scientists selected by the Gulf Universities Research Consortium (GURC) Trustees. Their summary and interpretation were presented in a report,

J. M. Sharp is President, Gulf Universities Research Consortium.

which included the 23 individual studies (Menzies et al. 1974). Total synthesis and interpretation of some 600,000 data items obtained were not possible with the computerized programs available at the time of the final report in 1974. Since the advent of an adequate computerized data base service in 1976, the OEI data have been re-examined. This re-examination was conducted by scientists who had not participated in the original OEI program, to evaluate the adequacy and validity of the data, to determine whether the original consensus conclusions of the OEI Council were valid, and to arrive at additional conclusions on the basis of this re-investigation of data.

The principal purpose of this volume, therefore, is to present the final reports of the 23 principal investigators, the consensus opinion by the OEI Council, and the conclusions drawn from the independent appraisal of the data made in 1978-79.

MULTIPLE USES OF THE NORTHWESTERN GULF OF MEXICO

The coastal and near-shore regions of the United States, including those adjacent to the Gulf of Mexico, have been subjected to conflicting multiple-use demands. To sustain renewable resources and maintain environmental quality while attempting to satisfy these increasing demands is a very complex and difficult problem.

Renewable resources of the Gulf region have national significance. The annual tonnage of commercial fish landings in the Gulf is the largest of any region in the nation (U.S. Department of Commerce 1979). Sport fishing, hunting, and other forms of coastal recreation are important elements of the regional and national economy (U.S. Department of Interior 1978). Approximately 40% of the Gulf States' population lives within the coastal zone (U.S. Department of Commerce 1972). Maintaining the productivity and quality of the coastal and continental shelf areas of the region is important both to the inhabitants and to others who use the area for recreational purposes.

If human activities constitute a real threat to ocean ecosystems, the Gulf should be among the first places in the western hemisphere to experience degradation of its natural resources. This should be particularly true of the western Gulf because it has only limited water exchange with the eastern Gulf (Kuenen 1950; Parr 1935) and is thus relatively isolated from the Gulf Stream's Loop Current, which dominates eastern Gulf circulation and is the primary exchange mechanism between the Gulf and the Atlantic Ocean (Sverdrup et al. 1942). The exchange that does occur involves surface waters primarily (Kuenen 1950; Dietrich 1939). Gulf sediments remain in the Gulf (Kuenen 1950), including the 2 million tons/day discharged by the Mississippi River (Lynch 1954). Net water mass movement on the continental shelf west of the mouth of the Mississippi River is predominantly westward, carrying with it the largest part of the river's discharge (Bates 1953). The Mississippi River system drains some two-thirds of the area in the lower 48 states (Moody 1967). This area includes both heavy industrial and agricultural land use and large municipalities. Thus the northwestern Gulf receives a tremendous amount of industrial, agricultural, and municipal waste. Major influence of the Mississippi discharge on coastal and continental shelf waters extends to the midpoint of the Texas coastline (Lynch 1954; Price 1954).

Because of numerous natural port locations and the availability of abundant and cheap energy for many decades, large areas of the northwestern Gulf have become heavily industrialized. As a result, large-scale population increases have occurred in these coastal areas. There have been proportional increases in municipal and industrial effluents and in the use of coastal areas for residential, resort, and recreational purposes.

Twelve ocean ports are located on the Gulf. Nine of these are located on or west of the Mississippi. The latter include New Orleans and Houston, which are the nation's second and third largest ports (U.S. Army Corps of Engineers 1977). The heavily used Intracoastal Waterway interconnects the complex of ports and extends from Florida to the U.S.-Mexico border. Numerous branch canals interconnect with the waterway. This is particularly true in Louisiana, which has some 5,000 miles of inland waterways (Weidie et al. 1969). This vast transportation network is one of the major uses made of the coastal and continental shelf area in the northwestern Gulf. In addition to being a source of effluents, coastal modifications for transportation and those designed for flood control purposes have caused large-scale changes to the coastal zone.

Agriculture in the northwestern Gulf coastal area is highly productive (Wright et al. 1976). Discharges of agricultural chemicals have increased as more land has been used for this purpose. Also, hydrological modifications associated with a large amount of irrigation, particularly in the "rice belt" of southeast Texas and southwest Louisiana, have resulted in significant local changes in fresh water influx. These local influences are superimposed on those resulting from the westward flow of the Mississippi River discharge.

The Texas and Louisiana coastal areas have been major producers of oil and gas throughout this century. In 1977, the offshore areas of the Gulf of Mexico produced 10.1% of U.S. domestic crude oil and 22.6% of our natural gas (American Petroleum Institute 1978). This region now contains the world's largest concentration of offshore oil and gas wells.

Through these massive multiple uses of the northwestern Gulf estuarine and continental shelf areas, a very strong regional economy has been developed. There are many incentives to maintain that economy. There are equally strong incentives to maintain the quality of the ecosystem and its abundant renewable resources. Despite the heavy municipal, industrial, and transportation usage and the exploitation of both renewable and nonrenewable resources, the northwestern region remains one of the most biologically productive areas in the Gulf (National Academy of Sciences 1975). One must be concerned, however, with the reduction in the muskrat population and shifts in the oyster populations (National Academy of Sciences 1975), the near disappearance of the brown pelican, and other evidence of ecological change to which petroleum operations may have contributed.

Effective and ecologically sound management of the northwestern Gulf ecosystem is of great importance to the region, to the nation, and to Mexico. Uninformed management decisions could result in irreversible or prolonged damage to the ecosystem. Similarly, unnecessarily restrictive decisions could produce severe penalties to the regional and national economies, and thus to the security and quality of life of the nation.

The Offshore Ecology Investigation (OEI) and the series of post-OEI studies reported herein have been a systematic effort to learn basic facts and to develop basic understanding relative to the management of coastal and continental shelf resources as they pertain to one major human activity: offshore petroleum drilling and production. The study of this practical problem has led to the development and preliminary evaluation of quantitative methods and principles applicable to its solution. The methods and principles developed appear to have broader application in improving the scientific evaluation process as applied to multiple-use management.

TRANSIENT EFFECTS OF PETROLEUM IN THE MARINE ENVIRONMENT

Concern about the effects of petroleum on coastal, continental shelf, and ocean ecosystems has resulted primarily from the highly visible effects of massive oil spills. Since the *Torrey Canyon* and the Santa Barbara spills in the late 1960s, the effects of large spills have been studied intensively. The literature on this aspect of petroleum in the marine environment is abundant. Oil spill conferences (American Petroleum Institute 1977; American Institute of Biological Sciences 1978) are now held regularly to report the findings of multi-year retrospective studies of major spills and related laboratory and microcosm studies.

As a result of this intensive interest, the transient effects of large concentrations of petroleum in the marine environment are reasonably well understood, at least for tropic and temperate climates. As would be expected, ecological damage from spills is generally a direct function of the concentration and types of hydrocarbons and of the duration of exposure. In general, then, transient effects from spills are most severe when the spill is contained within a limited ecosystem and conditions promote mixing of the spilled product into the water column and sediments. Extensive damage is usually associated with refined products or with the light fractions of crude oil (National Academy of Sciences 1975). Unless there is containment and rapid mixing, however, light fractions in a crude oil spill are usually rapidly dispersed (e.g., by evaporation) so that ecological impacts are reduced unless the spill is very large.

The OEI was concerned only with petroleum drilling and production activities as one of the multiple uses of coastal and continental shelf areas. It is pertinent, then, to note that most of the large spills have occurred during the ocean transportation of crude oil and refined products, not in drilling and production operations. For example, since offshore oil and gas operations began in the Gulf of Mexico, only one major blowout-related spill had occurred in U.S. waters (Danenberger 1976). With modern control technology, the probability of blowouts, fires, and other causes of major spills has been reduced significantly. The recent Ixtoc I blowout, however, dramatically demonstrates that drilling technology is not infallible.

CUMULATIVE EFFECTS OF PETROLEUM IN THE MARINE ENVIRONMENT

The transient effects of large oil spills can be investigated with a high degree of confidence even in the presence of the large natural variabilities that usually exist. Where the conditions do not result in the required concentration and duration of exposure, effects have been reported to be of little biological significance (Chan 1977; Lasday et al. 1976; Kuhnhold 1978). These results, however, do not obviate concern regarding the more subtle effects on organisms that might result from exposure for long time periods to small concentrations of crude oil and other effluents related to offshore drilling and production.

The cumulative effects of sublethal petroleum discharges from drilling and production appear to be insignificant when compared to other sources of contamination or disturbance. Natural seeps of petroleum and natural gas have been reported in the Gulf of Mexico since the beginning of this century (ZoBell 1954), and recent comprehensive surveys confirm the wide distribution of seeps in the northwestern Gulf (Geyer 1979). Despite this continuing natural discharge of crude oil and natural gas into the environment over eons of time, the modern Gulf continues to have a very high level of biological productivity. Similar situations exist in other locales where large seeps occur (Straughan 1977).

Another pertinent comparison regarding contamination of the marine environment by coastal and offshore petroleum operations is that reported by the National Academy of Sciences (1975). That study attributes only 0.33% of the total hydrocarbons entering the world's oceans to normal marine production operations. It attributes 0.66% to major oil spills associated with production operations, 26.2% to river runoff, 9.8% to natural seeps, and 34.9% to marine transportation, with the remaining 28.1% attributed to diverse sources. For the Texas-Louisiana region, the actual amount that would be attributed to normal drilling and production operations should be higher than for the world's oceans. But these percentages are probably representative of this region also, since river runoff, seeps, and marine transportation in this area are also higher than world averages.

Since the advent of modern pollution control technology, hydrocarbons entering the environment during normal drilling and production operations have been reduced significantly. Current regulations (Federal Register 1979) require that the hydrocarbon content of produced brines disposed into marine waters must be maintained at a monthly average of not more than 48 mg/l with a daily maximum of 72 mg/l. During the decade preceding this specification, the regulatory agencies and industry attempted to maintain an average daily concentration of 50 ppm (essentially 50 mg/l). The magnitude discharged into the Gulf by small accidental spills is indicated by a total of 51,421 barrels being reported as "spills" in the four-year period ending 1975 (Danenberger 1976). This amounts to a spillage rate of 0.0028% of total production during that period. Such quantities appear small by comparison with the discharge of hydrocarbons by the Mississippi River. At its average flow rate of 500,000 cfs (World Almanac 1979) and assuming a lower than actual hydrocarbon concentration of 1.0 ppm, the Mississippi River would discharge more than 2.8 million barrels of hydrocarbons into the Gulf annually. Thus the threat posed by either normal or accidental discharges from oil and gas drilling and production appears to be relatively insignificant as compared to river discharge. In light of the small fraction of crude oil that may be ecologically damaging (Hodson et al. 1977) and the biodegradability of crude oil components (ZoBell 1954), hydrocarbons from petroleum operations appear even less important.

Such perspective does not, however, preclude the possibility of cumulative ecological effects of hydrocarbon effluents from drilling and production. The possibility of gradual accumulation of hydrocarbons in the sediments to levels of biological significance must still be considered.

Questions regarding the possibility of bio-accumulation, bio-magnification, and bio-transformation remain largely unanswered. Biomagnification of hydrocarbons in marine organisms has not been demonstrated (National Academy of Sciences 1975). Bioaccumulation does occur through feeding in some organisms (National Academy of Sciences 1975; Lee 1977); where contamination levels are not maintained, however, depuration usually occurs (Teal 1977; McAuliffe 1977; Koons et al. 1977). Studies of biotransformation of hydrocarbons indicate that biotransformation can produce metabolic products of higher toxicity than the parent hydrocarbons (Malins 1977; Stegeman 1977).

Of concern at least equal to that regarding hydrocarbons is contamination by drilling muds, antifouling paints, sacrificial electrodes, etc., involved in drilling and production operations. These materials contain trace metals that are considered ecologically important. As is true of hydrocarbon discharges, both total discharge and concentrations are low as compared with other sources. Again, however, trace metals from these sources might accumulate in the sediments or via biological processes to levels of ecological significance.

It was not possible in 1972 (nor is it now) to address all questions regarding specific concerns relating to petroleum operations. Should the cumulative effects of normal drilling and production resulting from the various processes be ecologically significant, however, they should be present in southern Louisiana, where operations have been most intensive for the longest period of time. As of December 31, 1973, cumulative production of crude oil from the Gulf of Mexico (principally Louisiana and east Texas) was 3.8 billion barrels, of natural gas liquids 0.71 billion barrels, and of natural gas 28.0 trillion cubic feet (American Petroleum Institute 1974). Therefore, when the OEI began in 1972, the several estuaries and offshore areas in southern Louisiana should have represented the end product of a real "model" of a marine ecosystem subjected to several decades of maximum cumulative impact, of which the first 20 years or so predated the advent of modern environmental management and control technology. During the two decades preceding the OEI, drilling and production were very intensive. Should ecological damage under these conditions be undetectable, drilling and production may be ecologically insignificant over the normal 10-50 year life of a field, at least as compared with larger sources persisting for longer time periods.

STATE OF KNOWLEDGE REGARDING CUMULATIVE EFFECTS

Cumulative effects resulting from small discharges or disturbances to natural ecosystems are difficult to detect and measure quantitatively. If one defines such effects as those that might result from persistent exposure to sublethal concentrations or to gradual irreversible changes in the environment, one might expect these effects to appear as small progressive incremental changes in the biota. These subtle effects must be detected and quantified in the presence of generally much larger natural variations. Also, it must be recognized that many species will exhibit resistance, adaptability, or resilience (Odum 1971), which would tend to mask or delay small progressive changes for some period of time. An additional complexity is that temporary improvements in the habitat not only permit some species to depurate but may allow time for recruitment and recolonization, which would further mask an effect or confuse interpretation. Finally, nonlinear responses of the biota to both natural variables (Slobodkin 1961) and specific contaminants (Neff et al. 1976) have been reported, indicating that cumulative change may not in all cases consist of small progressive increments. Rather, the change might be precipitous following the habitat's reaching a saturation value for a single variable or synergistic combination of variables. These inescapable difficulties have made the positive detection, and especially the quantification, of cumulative effects exceedingly complex.

The importance of knowing the natural variabilities of an ecosystem has long been recognized (Odum 1971). But developing a quantitative definition of the spatial and temporal variabilities of a coastal or continental shelf ecosystem (both of which are highly variable as compared, for example, with the deep ocean) requires the acquisition of an extensive data base for the ecosystem of interest. The expense and time required to acquire adequate data for this purpose have been primary deterrents to the determination of natural variabilities. In recent years, massive data bases for coastal and offshore ecosystems have been developed. However, the synthesis of the data from large ecological programs has proven very difficult. Even more difficult is the correlation of all variables related in time, space, and function. Adding to these difficulties is the lack of any commonly accepted means for establishing the statistical adequacy of ecological data and for defining statistical quantities needed for much of the interpretation process. So far, these deterrents have precluded quantitative determinations of variabilities, and thus quantitative measurement, of ecological damage or enhancement.

Nonetheless, the need for quantitative measures of ecological effects remains the most pressing need for informed environmental management. In the absence of this type of information, managers are forced to substitute exceedingly time-consuming procedures. Such procedures divert larger numbers of scarce experienced people into environmental management roles than would be required were this information available. Furthermore, its lack forces managers to apply subjective judgment, which is rarely either conclusive or convincing. This can lead to costly or socially disastrous mistakes in ecological management.

The need for quantitative measurement of ecological impact has led to a great deal of research. Because of the difficulty of developing quantitative cause-and-effect relationships in natural ecosystems, however, particularly as they relate to cumulative effects, most of the research has been conducted in laboratories or in microcosms. A large part of this work deals with investigation of the effects of single contaminants on single species. Although there has been excellent research carried out in natural ecosystems (Diaz et al. 1973), these too have been confined generally to the study of the effects of single contaminants. Both types of single-contaminant investigations have been encouraged by the need to obtain government permits relating to specific products or activities.

In summary, a great deal of useful information has been developed, particularly of the single-contaminant type. Given quantitative knowledge of the natural variabilities of an ecosystem, much of this information could be put to sound scientific use. Natural variabilities dictate the configuration and schedule for monitoring an ecosystem. Laboratory data relating to specific contaminants and species can then guide the selection of these variables for inclusion in the monitoring system. Implementing such a system should then permit development of realistic environmental control requirements as determined in the natural ecosystem.

The research effort on ecosystems has lacked a concerted attack on problems that are pertinent to ecological management. In general, the research effort has been diffuse and spread over a large number of ecosystems, and divided among an almost limitless number of possible cause-and-effect relationships. How can such entities as spatial and temporal variability, ecological health, ecological stability, significant biological levels, or biological similarity be meaningfully defined in scientifically reproducible terms? Some of these critical questions have been addressed to a small degree in carrying out the post-OEI studies reported in section III of this report, "Re-examination of the Offshore Ecology Investigation." While the evidence is preliminary, the results support the concept that cumulative effects can be determined quantitatively once statistically adequate variabilities are established. Perhaps of equal importance, the results indicate that the amount of baseline and monitoring data required for useful (not necessarily optimum) implementation of the method is considerably less than was expected. Indeed, it appears that sufficient data are being generated or have been generated through baseline, impact, and monitoring efforts to design and establish useful assessment and monitoring criteria.

THE OEI CASE STUDY

The OEI Case Study is a composite of three successive phases of investigation covering the period from 1971 through 1979. The specific purpose of the study was to determine the possible cumulative ecological effects of prolonged and intensive offshore oil and gas drilling and production. An underlying scientific purpose of the study was the investigation of methods whereby such effects can be determined quantitatively, and the development and evaluation of these methods as a basis for sound multiple-use management.

The phases of investigation and their successive objectives were:

Phase 1. *Field Studies* (the Offshore Ecology Investigation). The objective of the field studies was to acquire and interpret the minimum set of correlatable interdisciplinary data considered necessary to detect and identify any cumulative ecological effects resulting from normal offshore drilling and production operations. The studies consisted of two reconnaissances followed by eight seasonal synoptic field investigations. The 23 principal investigators interpreted and reported their individual findings. The total data and results were reviewed and overall conclusions were drawn by a group of four scientists charged with the design, implementation, and interpretation of the field studies. This phase is described in the following section of this volume, "Design of the Offshore Ecology Investigation," and the individual reports are collected in section IV.

Phase 2. *Method Development and Evaluation.* The objective of this phase was to develop and evaluate computer-assisted methods by which (a) interdisciplinary environmental data bases can be efficiently and comprehensively synthesized and analyzed by groups of scientists to arrive at consensus conclusions, and (b) the results of such analyses can be documented and displayed for effective communication with environmental managers. Software developed during 1971-1976 was used for data synthesis, selective retrieval, correlative analysis, and display. These methods were then employed to support 10 specific studies of OEI field data as the basis for evaluation of the efficiency and utility of the methods. The hundreds of map displays, graphs, horizontal and vertical cross sections, histograms, etc., that were generated provided rapid orientation to data distribution and content for scientists participating in Phase 3.

Phase 3. *Retrospective Scientific Appraisal and Method Application.* The objective of this phase was to obtain an independent appraisal of the OEI data and of the original conclusions, and the use of the methods developed in Phase 2 to evaluate methods for quantitative ecological assessment and monitoring. The investigations included numerous statistical appraisals and comparisons essential to the reappraisal. Methods employed in these appraisals were then evaluated as to their applicability to quantitative assessment and monitoring. The results of these investigations are presented in the third section of this volume, "Reexamination of the Offshore Ecology Investigation."

REFERENCES CITED

American Institute of Biological Sciences. 1978. Proceedings of the Conference on Assessment of Ecological Impacts of Oil Spills. Washington, D.C.

American Petroleum Institute. 1974. Reserves of Crude Oil, Natural Gas Liquids, and Natural Gas in the United States and Canada and United States Production Capacity as of December 31, 1973. Vol. 28. Washington, D.C.

———. 1977. Proceedings of the 1977 Oil Spill Conference (Prevention, Behavior, Control, Cleanup). Publication No. 4284. Washington, D.C.

———. 1978. Reserves of Crude Oil, Natural Gas Liquids, and Natural Gas in the United States and Canada and United States Production Capacity as of December 31, 1977. Vol. 32. Washington, D.C.

Bates, C. C. 1953. Physical and Geological Processes of Delta Formation. Ph.D. dissertation, Texas A&M University.

Chan, G. L. 1977. The five-year recruitment of marine life after the 1971 San Francisco oil spill. *In* Proceedings of the 1977 Oil Spill Conference (Prevention, Behavior, Control, Cleanup). Publication No. 4284. Washington, D.C.: American Petroleum Institute, pp. 543-545.

Danenberger, E. P. 1976. Oil Spills, 1971-75, Gulf of Mexico Outer Continental Shelf. U.S. Geological Survey Circular No. 741.

Diaz, R., M. Bender, D. Boesch, and R. Jordan. 1973. Water quality models and aquatic ecosystems: status, problems and prospectives. *In* R. A. Deininger, ed., Models for Environmental Pollution Control. Ann Arbor, Michigan: Ann Arbor Science Publishers.

Dietrich, G. 1939. Das Amerikanisch Mittelmeer. Zeitschrift für Erdkunde **3**:108-130.

Federal Register. April 13, 1979. Effluent Guidelines and Standards, Oil and Gas Extraction, Point Source Category, 40 CFR, Part 435. Washington, D.C.: U.S. Government Printing Office **44**(73):22069-22078.

Geyer, R. A. 1979. Naturally Occurring Hydrocarbon Seeps in the Gulf of Mexico and Caribbean Sea. College Station, Texas: Texas A&M University Press.

Hodson, R. E., F. Azam, and R. F. Lee. 1977. Effects of four oils on marine bacterial populations: control ecosystem pollution experiment. Bulletin of Marine Science **27**(1):119-126.

Koons, C. B., C. D. McAuliffe, and F. T. Weiss. 1977. Environmental aspects of produced waters from oil and gas extraction operations in offshore and coastal waters. Journal of Petroleum Technology **29**:723-729.

Kuenen, P. H. 1950. Marine Geology. New York: John Wiley.

Kuhnhold, W. W. 1978. Impact of the Argo Merchant oil spill on macrobenthic and pelagic organisms. *In* Proceedings of the Conference on Assessment of Ecological Impacts of Oil Spills. Washington, D.C.: American Institute of Biological Sciences, pp. 153-179.

Lasday, A. H. and E. W. Mertens. 1976. Fate and effects of oil on marine life: progress report on research sponsored by the American Petroleum Institute. *In* Proceedings of the Eighth Annual Offshore Technology Conference. Washington, D.C.: American Petroleum Institute. Vol. 1, Paper No. 2449, pp. 275-281.

Lee, R. F. 1977. Accumulation and turnover of petroleum hydrocarbons in marine organisms. *In* D. A. Wolfe, ed., Fate and Effects of Petroleum Hydrocarbons in Marine Ecosystems and Organisms. New York: Pergamon Press, Inc., pp. 60-70.

Lynch, S. A. 1954. Geology of the Gulf of Mexico. Fisheries Bulletin **55**(89):67-86.

Malins, D. C. 1977. Biotransformation of petroleum hydrocarbons in marine organisms indigenous to the arctic and subarctic. *In* D. A. Wolfe, ed., Fate and Effects of Petroleum Hydrocarbons in Marine Ecosystems and Organisms. New York: Pergamon Press, Inc., pp. 47-59.

McAuliffe, C. D. 1977. Dispersal and alteration of oil discharged on a water surface. *In* D. A. Wolfe, ed., Fate and Effects of Petroleum Hydrocarbons in Marine Ecosystems and Organisms. New York: Pergamon Press, Inc., pp. 19-35.

Menzies, R. J., J. P. Morgan, S. Z. El-Sayed, and C. H. Oppenheimer. 1974. The Offshore Ecology Investigation, Final Project Planning Council Consensus Report. Gulf Universities Research Consortium. Report No. 138.

Moody, C. L. 1967. Gulf of Mexico distributive province. American Association of Petroleum Geologists Bulletin **51**(2):179-199.

National Academy of Sciences. 1975. Petroleum in the Marine Environment. Washington, D.C.

Neff, J. M., J. W. Anderson, B. A. Cox, R. B. Laughlin, Jr., S. S. Rossi, and H. E. Tatem. 1976. Effects of petroleum on survival, respiration and growth of marine animals. *In* Sources, Effects and Sinks of Hydrocarbons in the Aquatic Environment. Washington, D.C.: American Institute of Biological Sciences, pp. 515-539.

Odum, E. P. 1971. Fundamentals of Ecology. 3rd Edition. Philadelphia: W. B. Saunders Company.

Parr, A. E. 1935. Report on hydrographic observations in the Gulf of Mexico and the adjacent straits made during the Yale oceanographic expedition on the Manel Taylor in 1932. Bingham Oceanographic Foundation Bulletin **5**(1):1-93.

Price, W. A. 1954. Shorelines and coasts of the Gulf of Mexico. Fisheries Bulletin **55**(89):39-62.

Slobodkin, L. B. 1961. Growth and Regulation of Animal Populations. New York: Holt, Rinehart and Winston.

Stegeman, J. J. 1977. Fate and effects of oil in marine animals. Oceanus **20**(4):59-66.

Straughan, D. 1977. The sublethal effects of natural chronic exposure to petroleum on marine invertebrates. *In* Proceedings of the 1977 Oil Spill Conference (Prevention, Behavior, Control, Cleanup). Washington, D.C.: American Petroleum Institute Publication No. 4284, pp. 563-568.

Sverdrup, H. U., M. W. Johnson, and R. H. Fleming. 1942. The Oceans: Their Physics, Chemistry, and General Biology. New York: Prentice-Hall, Inc.

Teal, J. M. 1977. Food chain transfer of hydrocarbons. *In* D. A. Wolfe, ed, Fate and Effects of Petroleum Hydrocarbons in Marine Ecosystems and Organisms. New York: Pergamon Press, Inc., pp. 71-77.

U.S. Army Corps of Engineers. 1977. Water Borne Commerce of the United States, Calendar Year 1977. Part 2, Waterways and Harbors: Gulf Coast, Mississippi River Systems and Antilles. New Orleans, Louisiana.

U.S. Department of Commerce. 1972. Census of Population, 1970. Washington, D.C.: U.S. Government Printing Office.

———. 1979. Fisheries of the U.S., 1978. Current Fisheries Statistics No. 7800. Washington, D.C.: U.S. Government Printing Office.

U.S. Department of Interior, Fish and Wildlife Service. 1978. Recreation, hunting and fishing survey—1977. Washington, D.C.: U.S. Government Printing Office.

Weidie, A. E., J. A. Wolleben, W. C. Craig, and E. Otvos. 1969. Mississippi River delta estuarine system. *In* D. E. Feray, ed., Case Studies of Estuarine Sedimentation and its Relation to Pollution of the Estuarine Environment. Report to Federal Water Pollution Control Authority. Washington, D.C.

World Almanac and Book of Facts. 1979. New York: Newspaper Enterprise Association, Inc.

Wright, R. J., S. A. Arbingast, R. Chandler, and P. Green. 1976. Texas Sources—A Bibliography. Austin: University of Texas Bureau of Business Research.

ZoBell, C. E. 1954. Marine bacteria and fungi. Fisheries Bulletin **55**(89):217-222.

II

METHODS AND SUMMARY RESULTS

DESIGN OF THE OFFSHORE ECOLOGY INVESTIGATION

by R. J. Menzies, J. P. Morgan, C. H. Oppenheimer, S. Z. El-Sayed, and J. M. Sharp

The Offshore Ecology Investigation (OEI) was a field study directed to a single objective: To assess the cumulative ecological effects of normal oil and gas drilling and production operations on estuarine and near-shore ecosystems.

It was designed to study the effects of persistent exposure of an estuary and the adjacent offshore area to the low-level discharge of contaminants and the physical disturbance associated with prolonged and intensive drilling and production operations. It was not concerned with the visible and more intensively studied effects of oil spills.

DISCHARGES FROM PETROLEUM OPERATIONS

Small amounts of crude oil and other contaminants may enter the environment during normal petroleum operations. Some of this contamination may be described as chronic in that the effluents are continuously or frequently discharged. Other discharges or disturbances may be infrequent, but may contain certain elements that might be retained in the environment so that persistent exposure of the biota would result.

Petroleum hydrocarbons originating from crude oil were one type of contaminant of primary interest in the OEI. It is not feasible to remove all crude oil from the brine that is usually produced with the oil and gas in petroleum operations. Especially in continental shelf areas, the con-

R. J. Menzies (deceased) was Professor of Oceanography, Florida State University, Tallahassee, FL; J. P. Morgan (emeritus) was Professor of Geology, Louisiana State University, Baton Rouge, LA; C. H. Oppenheimer is Professor of Microbiology at the Institute of Marine Science, The University of Texas at Austin, Port Aransas, TX; S. Z. El-Sayed is Professor of Oceanography, Texas A&M University, College Station, TX; J. M. Sharp is President, Gulf Universities Research Consortium, Houston, TX.

struction of larger or separate platforms to reduce oil concentration in brine is tremendously expensive. Where platforms can accommodate separators that will reduce crude oil concentration to the allowable daily average of 48 mg/l (Federal Register 1979), the produced brine is usually discharged directly into the environment. Otherwise, it must be barged or piped ashore for final separation and brine disposal. At a typical production platform in coastal or near-shore waters, discharging 1,000 to 10,000 barrels at 48 mg/l, the crude oil fraction at this level of concentration would amount to approximately 1/20 to 1/2 barrel of crude oil per day.

Accidental spills of crude oil occur under normal operating conditions. Several hundred such spills are reported annually. During the period 1971-1975, the spill total for the approximately 2,000 platforms in the Gulf was 51,420 barrels (Danenberger 1976). Hence, the average discharge from normal spills is of the order of 5 barrels per platform per year. Of the 5,857 spills recorded, 99.5% were less than 50 barrels.

The transportation of produced oil, gas, and, for some operations, brine mixtures is normally accomplished by pipeline. Underwater pipeline leaks of oil or gas are usually detected, minimizing chronic contamination. Pipeline breaks or leaks are also reported as spills. Such spills have contributed a locally significant amount of oil entering the marine environment; however, there is very little decisive evidence of harmful ecological effects (Council on Environmental Quality 1974).

The support of occupied platform operations requires frequent transportation of workers and materials by helicopters and service vessels. The contamination of the environment from fuels and lubricants used in these operations is another very small, but frequent, source of hydrocarbons.

Microbial processes, oxidation, evaporation, and other natural processes continuously modify both the concentrations and the chemical nature of hydrocarbon discharges resulting from petroleum operations (Frey 1977; Karrick 1977; McAuliffe 1977; National Academy of Sciences 1975; Sanders 1977). Light hydrocarbon fractions evaporate quickly unless there is turbulent mixing and adsorption to suspended sediments. Higher molecular weight fractions are degraded to lower molecular weight, principally by microbial processes. The rate of microbial degradation is dependent upon variables such as nutrients, dissolved oxygen, and temperature (National Academy of Sciences 1975). Hydrocarbon degrading bacteria are ubiquitous in shallow water sediments in the Gulf of Mexico (ZoBell 1954).

The aromatic fraction of crude oil has been of the greatest concern because of the reported higher toxicity of this fraction. The aromatic content of crude oils varies but is a small percentage for Louisiana crude oil (Hodson et al. 1977). Pyrolytically formed aromatic hydrocarbons, such as benzopyrene (Philpot et al. 1976; Gibson 1976), have been of particular concern. However, river runoff, marine transportation, and atmospheric influx are the predominant sources of such compounds and outweigh the influx from offshore production (National Academy of Sciences 1975).

Normal production operations also result in trace elements or "heavy metals." Elements considered to be toxic in low concentrations that may be included are As, Cd, Hg, Zn, Cu, and Pb. The produced brine that is discharged continuously may contain trace elements at levels of a few parts per million. Antifouling paints and rust inhibitors used on platforms, pipelines, and service vessels continuously release traces of some of these elements. Sacrificial electrodes, used for corrosion protection, primarily contain Zn and Al, but other elements may be present in negligible amounts as impurities. Hg is used in certain special-purpose electrodes, but in quantities that are undetectable, by conventional means, in receiving waters.

The trace elements and hydrocarbons contained in drilling fluids may have cumulative effects if retained in the sediments. In general, oil base drilling muds have been of greatest concern. But oil base muds are much more expensive than water base muds and are used only when water base muds would alter reservoir properties so as to penalize production. Thus, they are a small percentage of total mud used offshore. In any event, no oil base mud can be discharged into the environment; it is hauled ashore for disposal. The accumulation of drill cuttings and other solids around platforms could have localized effects on the benthos. Highest concentrations would be expected in the vicinity of platforms, but mixing and transport processes generally disperse and dilute the mud components. Probably for this reason studies of the effects of drilling muds in the environment have not revealed adverse effects on the biota (Ray et al. 1978; Monaghan et al. 1976).

BACKGROUND

The question posed to Gulf Universities Research Consortium (GURC) in 1971 was whether field studies might be able to detect and identify cumulative ecological effects of oil drilling and production operations. At that time, the only field studies undertaken to determine cumulative effects of petroleum operations were those in Lake Maracaibo, Venezuela (Templeton et al. 1975). At the time of the initiation of the OEI program, laboratory and microcosm studies had not developed probable cause-and-effect relationships. Critical pathways and points of accumulation for petroleum hydrocarbons had not been determined. Generally accepted methodologies for a coordinated study of an ecosystem were not available and ecological baselines had not been developed for any major oil-producing area.

The first consideration was evaluating the feasibility and practicality of achieving the specific objective by means of field studies. The principal concern was whether the expected small changes that might result from petroleum operations could be discriminated from other causes of ecological change. Seasonal and spatial variabilities in coastal and continental shelf areas were known to be large. Natural long-term trends or cycles are present. The effects of contaminants from other sources could mask or interfere with a determination of the effects of petroleum operations.

The second concern was whether scientific understanding of estuarine and offshore ecosystems was sufficient to support valid field study design and the interpretation of the data. Because of the general lack of accepted models of ecosystems, the design of the field study of the OEI was dependent upon the results of prior diverse scientific studies and on the collective knowledge of the investigators. Prior studies of ecosystem behavior were either limited in disciplinary content or disciplinary components were not directly correlative in time and space.

A three-day workshop involving some 125 scientists and engineers in 1971 resulted in a preliminary conclusion that a well-designed and coordinated field study could be successful in assessing the chronic effects of petroleum hydrocarbons. A four-man OEI Program Council was appointed further to evaluate feasibility and to devise a program design for review by the GURC Trustees. The sections that follow discuss the factors and rationale involved in program design. Review of this design led to the conclusion that achieving the objective was feasible and that the field studies should produce valid, pertinent, and scientifically supportable conclusions.

FIELD STUDY AREA SELECTION

The primary criterion in selecting the area for the field studies was that it must have experienced prolonged and extensive drilling and production activities. This was essential to assure the maximum potential for cumulative effects and the probability of detecting them. The southern Louisiana region was an obvious choice to meet these requirements.

Another factor considered to be important was that the area be biologically productive and diverse and that it receive a minimum of effluents from other sources. Both an estuarine and an offshore area should be studied.

In particular, sources of contaminants similar to those produced in normal petroleum operations were to be avoided as much as possible. This requirement dictated the choice of an area having no ocean port and a minimum of industrial development. It was equally necessary that it be as remote from river discharge as possible. In southern Louisiana, this meant avoiding the direct outflow of the Mississippi and the Atchafalaya Rivers as much as possible. Figure 1 shows the locations of Barataria, Timbalier, and Vermilion Bays, locations that best meet this requirement within the general area of intensive coastal and offshore petroleum activities.

To maximize the probability of cumulative effects, it was considered desirable that the estuary selected should be shallow and turbid. Timbalier Bay was deemed a more suitable choice than Barataria and Vermilion Bays since it has no large river system. Its shores and limited drainage area contained very little industrial and municipal development and no ocean port. It is more remote than Barataria Bay from the Mississippi River discharge. Also, Timbalier Bay is not in the discharge pattern of the Atchafalaya River, whereas Vermilion Bay is. Finally, oil and gas operations in Timbalier Bay and the adjacent offshore area had been as prolonged and intensive as in any of the areas in southern Louisiana.

Another significant factor was the need to appraise the possible cumulative effects of offshore petroleum operations on adjacent estuaries. The probability of persistent exposure of an estuary to offshore effluents would be enhanced by onshore winds, shoreward tidal components, and moderate fresh water outflow. Timbalier Bay satisfied these requirements. Probably the most significant offshore-to-onshore drive mechanisms in southern Louisiana are the persistent spring-summer and frequent fall-winter onshore winds.

One factor arguing against the selection of Timbalier Bay as the field study site was its lack of a convenient and suitable logistic support base for the field studies. Lack of municipal and industrial development meant a similar lack of shore facilities. Barataria Bay was a better choice in this regard. Grand Isle, which is the barrier island between Barataria Bay and the offshore area, had long been established as a primary base for both offshore service operations and fishing. However, the proximity of Barataria Bay to the Mississippi River's Southwest Pass and its higher flushing rates resulted in a choice of Timbalier Bay despite the inconvenience and increase in cost. Grand Isle was used as the support base for the program.

The foregoing considerations led to the selection of Timbalier Bay and the adjacent offshore area as the most suitable area for the field studies. In designing the program, the Program Council recognized that

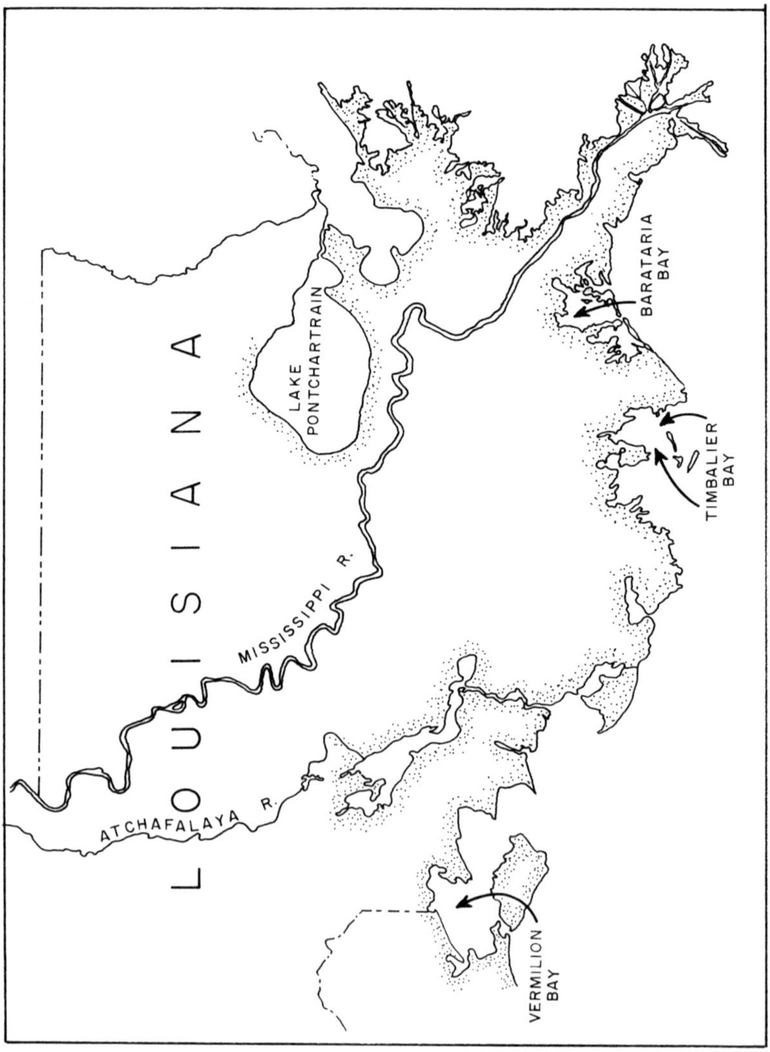

FIG. 1. LOCATIONS OF BARATARIA, TIMBALIER, AND VERMILION BAYS.

comparisons of Timbalier Bay with another bay having similar characteristics but not exposed directly to petroleum operations would be desirable. With the funding available, however, this could not be accomplished as part of the OEI. Therefore, comparisons of this nature were limited to those which could be made based on data from prior field studies and on the knowledge of the individual principal investigators.

VARIABILITY CONSIDERATIONS

Within Timbalier Bay, biological productivity, diversity, and community structure were expected to vary from the edge of the marshlands to the barrier island, with depth, salinity, nutrients, and bottom sediments being the controlling variables. The growth of subsurface flora was expected to be restricted by high turbidity in the bay.

Offshore spatial variabilities in biological communities were expected to be controlled by water depth and sediment properties. Very early in the OEI it was determined that turbidity was a variable having a major effect on biological distributions at depths greater than about 10 meters, and that during Mississippi River flood periods, salinity distributions contribute to major, though temporary, change.

Ecological seasons on the coast of Louisiana approximate the calendar seasons except for shortened winters and lengthened summers as compared with more temperate locales. Salinities and temperatures in the bay were expected to peak during the long summer, with a corresponding decrease in dissolved oxygen. The inverse was expected during fall and winter months. Smaller variations were expected offshore. Persistent, strong onshore winds occur during spring and early summer. Strong north winds followed by lighter onshore winds are common during late fall and winter. The windset accompanying these patterns produces higher than average water depths in the bay during the spring and summer and very low water periodically in the fall and winter. Windset is a primary driving force for water and sediment exchange between the bay and offshore areas, producing corresponding variabilities in temperature and salinity in the bay. The primary effect of the wind offshore is on the magnitude and direction of near-shore currents. These wind conditions cause persistent turbidity both in the bay and in near-shore areas.

The Louisiana coast is frequently subjected to natural episodic processes such as hurricanes, tropical storms, outbreaks of polar air ("northers"), and floods. These events usually cause major changes in both coastal and offshore areas. Physiographical and sedimentological changes accompanying these events can endure for long periods, possibly obscuring cumulative effects of petroleum operations.

Eight successive seasonal studies were recognized as being minimal to establish statistically valid results. Because of the more than 25 years of intensive petroleum activities in the region, it was expected that any significant cumulative effects would be detectable in this program.

PROGRAM DESIGN RATIONALE

Drilling and production activities could produce significant ecological effects if significant concentration gradients for petroleum hydrocarbons and petroleum-related trace elements exist near the platforms. High concentrations of contaminants and concentration gradients were considered most likely to exist in the sediments in the bay. A buildup of concentrations in the sediments was less likely offshore, where sediment transport is more rapid and pronounced.

Discrimination of the effects of petroleum activities from natural variability was the primary concern both in the bay and offshore. Discrimination of petroleum-related effects from natural variabilities required comparison of primary habitat variables and benthic communities at drilling and production sites with ecologically similar sites having no petroleum activity in the immediate vicinity. This would permit quantitative comparisons of differences in contamination and related biological effects. To minimize natural variability, the "control" or "ambient" sites should be located in similar habitats. Differences in the concentrations of petroleum hydrocarbons and trace elements, if present, could result in differences in biological community structure caused by cumulative effects.

The selection of adequate control sites offshore was less restrictive than selection of study sites because attention was focused on the chemistry of the water column. Water depth and sediment properties were the prime considerations. The control site should be sufficiently close to the primary site to experience similar and essentially concurrent variabilities but sufficiently remote from drilling and production activities to make comparison valid. It should be "upstream" from the platforms and not in the vicinity of other petroleum operations. Figure 2 shows the locations of the drilling and production platforms and the corresponding control sites selected in the bay and in the offshore area.

Concentrations of contaminants after 25 years of operations could be uniformly high, so that point source effects might be masked by widespread contamination and biological communities might have adapted to general contamination. This possibility dictated that conclusions not be based entirely on synoptic comparisons of contamination and biota at platforms and control sites. Hence it was necessary to sample the

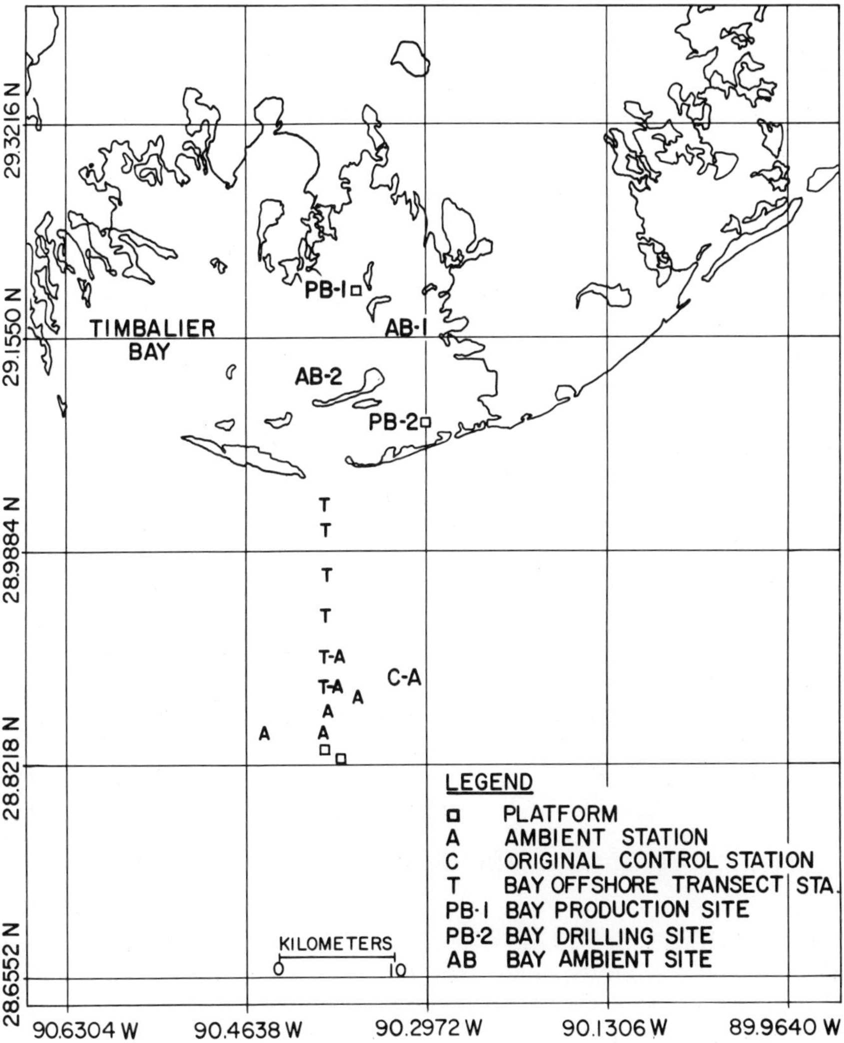

FIG. 2. THE OEI STUDY AREA SHOWING THE PLATFORMS AND CONTROL STATIONS USED.

ecosystem more generally rather than to limit measurements to platform and control sites. This need became more apparent when the first two seasonal field studies detected no significant differences between primary habitat variables or biota at the platforms and controls. Synoptic measurements at other sites and along transects (figure 3) were then included in the program. The four transects between the offshore sites

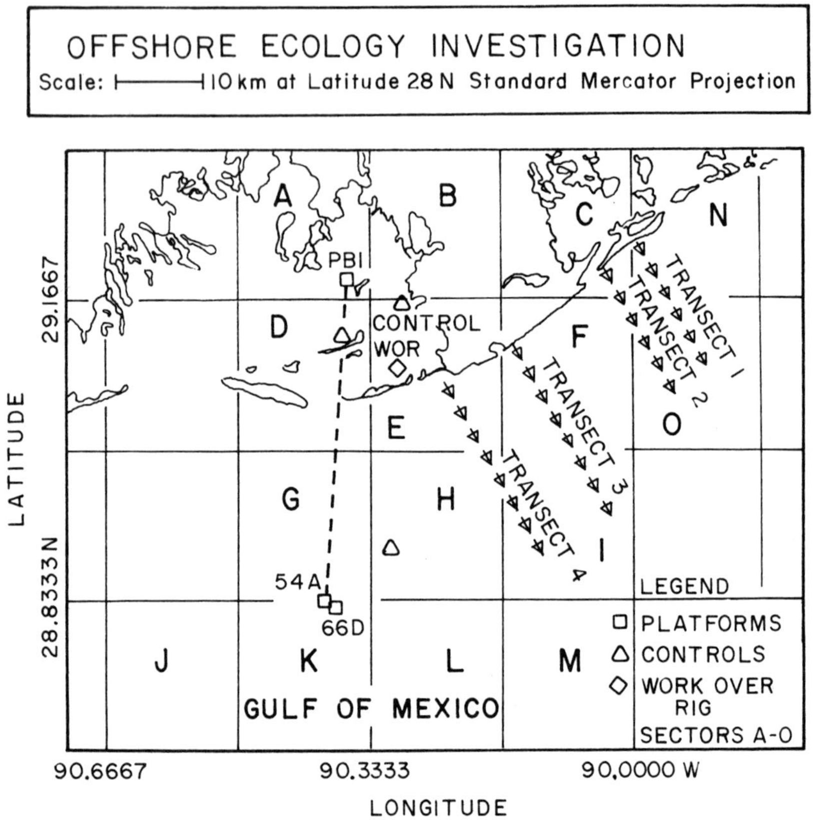

FIG. 3. THE REGION OF THE OFFSHORE ECOLOGY INVESTIGATION.

and the logistic support base at Grand Isle were selected because they (1) were upstream from the study area, (2) covered the same range of water depths as those in the study area, and (3) could be traversed conveniently and without considerable additional cost. The bay-to-offshore transect connects the bay and the offshore production platforms used in the study. It was selected for intensive study mainly because it follows the channel through which primary exchange occurs between the bay and the offshore area.

In the bay, individual investigators also conducted studies along transects connecting platform and control sites. Sampling stations along these transects were selected to enhance the discrimination of the effects of bottom sediment type and water depth from those that might result from contaminant concentrations.

In addition to platform-control comparisons, an overview was considered important in the assessment of cumulative effects. The suite of biological investigations had to permit a comprehensive assessment of the general ecological health of the study area. Therefore, the platform-control and other investigations collectively must include the study of major populations in each of the trophic levels. Where available, historical data were used to enhance interpretation of observed changes in populations and communities. Comparisons were also made with similar ecosystems.

A philosophy pervading the entire program design was that emphasis should be placed on detecting cumulative effects on the biological communities. While the determination of concentrations and distributions of petroleum-related contaminants was an essential part of the OEI, their presence would indicate environmental modification only, not ecological change in response to that modification.

Because of the large population of active production platforms in the OEI area and the adjacent areas, the possibility of a major oil spill occurring during the OEI had to be considered. Provision was made for a quick response team of OEI investigators, which would be activated immediately should such a spill occur in the study area. No major spills had occurred previously within the OEI study area, and none occurred during the 1972-1974 study period.

PROGRAM DESIGN

After selection of the study area, the sites to be used for point source investigations, and transects to observe general ambient conditions and exchange processes, it was necessary to determine a minimum set of disciplinary investigations. Of first consideration was the selection of those biological investigations that would collectively accomplish the platform-control comparisons, support a comprehensive appraisal of ecological health, contribute to the discrimination of natural and petroleum-related variations, and develop information directed to special concerns. It was then necessary to identify those physical, chemical, and geological investigations that would be essential in the interpretation of the biological data.

Twenty-three component disciplinary investigations were identified as being the minimum required. The objective, scope, and data requirements were defined for each investigation. Competitive proposals were requested from the university community in response to these specifications. The OEI Program Council, which was charged with program design, scientific direction, and interpretation, then selected the principal investigators based on the proposals submitted. Figure 4 shows the inves-

Major Aspect of Investigation			Principal Scientist	Region	Bay: Control Undisturbed	Bay: Drilling	Bay: Production	Offshore: Control Undisturbed	Offshore: Drilling	Offshore: Production
HABITAT										
Chemical	Hydrocarbon	Air/Sea Interface, Sediment, Beach	1. Laseter	1,2,3	●	●	●	●	●	●
Chemical	Hydrocarbon	Water Column, In Organisms	2. Oppenheimer	1,2	●	●	●	●	●	●
Chemical	Carbon & O	Water Column	3. Brent	✱	●	●	●	●	●	●
Chemical	Heavy Metal	Water Column	4. Montalvo	✱	●	●	●	●	●	●
Chemical	Nutrients	Water Column, Sediment	5. Burchfield	✱	●	●	●	●	●	●
Geological	Suspended Sediments	Water Column	6. Griffin	1,3,4				●	●	●
Geological	Sea Floor Sediments	Surficial	7. Oetking	1				●	●	●
Geological	Sea Floor Sediments	Bottom	8. Jones	1,3				●	●	●
Physical	Hydrography	Water Column	9. Oetking	✱	●	●	●	●	●	●
Physical	Currents	Water Column	10. Oetking	1,3,4				●	●	●
INHABITANTS										
Biological		Phytoplankton	11. El-Sayed	1				●		●
Biological		Phytobenthos	12. Humm	1,2	●	●	●			●
Biological		Zooplankton	13. Marum	1,2,3	●	●	●	●	●	●
Biological		Crustaceans & Molluscs	14. Farrell	1,2,3	●	●	●	●	●	●
Biological		Benthic Forams	15. Ostrom	2	●	●	●			
Biological		Benthic Polychaetes	16. Kritzler	2	●	●	●			
Biological		Beach Meiofauna	17. Fish	2	●	●	●			
Biological		Clams & Oysters	18. George	2	●	●	●			
Biological		Invertebrates	19. Waller	1				●	●	●
Biological		Fish	20. Perry	1,2	●	●	●	●	●	●
Biological		Fish & Shellfish	21. Thompson	1				●	●	●
Biological		Biofouling	22. George	1						●
Biological		Microbiology	23. Oppenheimer	1				●	●	●

1: OFFSHORE, 2: BAY, 3: BAY-OFFSHORE TRANSECT, 4: UPSTREAM TRANSECTS, ✱ ALL

FIG. 4. THE COMPONENT INVESTIGATIONS OF THE OEI STUDY.

tigations and their distribution in time and space, and identifies the principal investigator of each.

Where possible, methods and procedures for sampling, measurement, and analyses were standardized. A resulting "Handbook of Procedures and Standards" was prepared by Dr. Robert J. Menzies (1973), OEI Chief Scientist.

Practical considerations relating to standards and procedures, manpower, logistics, and funding had to be evaluated. For example, it was apparent that rigorous observance of common measurement sites and schedules by all investigators would be required. Yet the program had to be compatible with a multi-university effort. The shallow depths and limited shore facilities in the coastal oil-producing areas of interest would not permit the use of large marine research vessels. Therefore investigations involving several small vessels had to be scheduled and coordinated. Funding was limited to approximately $1.5 million, and no equipment purchases were authorized.

The logistic support plan was developed by Mr. Joe W. Tyson, GURC Coordinator for the OEI. Both the scientific and support plans were then tested by two reconnaissance field trips in the study area. Appropriate adjustments were made in schedules, sampling grids, and support vessels and equipment. After final plans for the initial seasonal field studies were made, Council members and the Coordinator conducted a dockside indoctrination of the entire OEI Project Team.

The field studies began in June 1972, and the eight seasonal studies were completed in February 1974. The component investigations of the OEI are presented in section IV, in papers individually written by the principal investigators. The consensus conclusions of the Council and an independent appraisal of the OEI are presented in the next section of this report.

REFERENCES CITED

Council on Environmental Quality. 1974. Outer Continental Shelf Oil and Gas: An Environmental Assessment. Vol. V. Washington, D.C.: U.S. Government Printing Office.

Danenberger, E. P. 1976. Oil Spills, 1971-75, Gulf of Mexico Outer Continental Shelf. U.S. Geological Survey, Circular No. 741.

Federal Register. April 13, 1979. Effluent Guidelines and Standards, Oil and Gas Extraction, Point Source Category, 40 CFR, Part 435. Washington, D.C.: U.S. Government Printing Office **44**(73):22069-22078.

Frey, M. G. 1977. Oil in marine waters. *In* Proceedings of the Ninth Annual Offshore Technology Conference. Vol. 1, paper no. 2752, pp. 227-230. Houston, Texas.

Gibson, D. J. 1976. Microbial degradation of carcinogenic hydrocarbons and related compounds. *In* Sources, Effects and Sinks of Hydrocarbons in the Aquatic Environment. Washington, D.C.: American Institute of Biological Sciences, pp. 224-238.

Hodson, R. E., F. Azam, and R. F. Lee. 1977. Effects of four oils on marine bacterial populations: controlled ecosystem pollution experiment. Bulletin of Marine Science **27**(1):119-126.

Karrick, N. L. 1977. Alterations in petroleum resulting from physico-chemical and microbiological factors. *In* D. C. Malins, ed., Effects of Petroleum on Arctic and Subarctic Marine Environments and Organisms. Vol. 1, Nature and Fate of Petroleum. New York: Academic Press, Inc., pp. 225-299.

McAuliffe, C. D. 1977. Dispersal and alteration of oil discharged on a water surface. *In* D. A. Wolfe, ed., Fate and Effects of Petroleum Hydrocarbons in Marine Ecosystems and Organisms. New York: Pergamon Press, Inc., pp. 19-35.

Menzies, R. J. 1973. Handbook of Procedures and Methods Employed in the Offshore Ecology Investigation (OEI). Gulf Universities Research Consortium. Report No. 121.

Monaghan, P. H., C. D. McAuliffe, and F. T. Weiss. 1976. Environmental Aspects of Drilling Muds and Cuttings from Oil and Gas Extraction Operations in Offshore and Coastal Waters. New Orleans: Offshore Operators Committee, Sheen Technical Subcommittee, P.O. Box 50751.

National Academy of Sciences. 1975. Petroleum in the Marine Environment. Washington, D.C.

Philpot, R. M., M. O. James, and J. R. Bend. 1976. Metabolism of benzo(a)pyrene and other xenobiotics by microsomal mixed function oxidases in marine species. *In* Sources, Effects and Sinks of Hydrocarbons in the Aquatic Environment. Washington, D.C.: The American Institute of Biological Sciences, pp. 184-199.

Ray, J. P. 1978. Drilling mud toxicity: laboratory and real world tests. Ocean Resources Engineering **12**:8-12.

Sanders, H. L. 1977. The West Falmouth Spill—Florida, 1969. Oceanus **20**(4):15-24.

Templeton, W. L., E. A. Sutton, R. M. Bean, R. C. Arnett, J. W. Blaylock, R. E. Wildung, and H. J. Moore. 1975. Oil pollution studies on Lake Maracaibo, Venezuela. *In* 1975 Conference on Prevention and Control of Oil Pollution. Washington, D.C.: American Petroleum Institute, pp. 489-496.

ZoBell, C. E. 1954. Marine bacteria and fungi. Fisheries Bulletin **55**(89):217-222.

III

INDEPENDENT APPRAISAL

RE-EXAMINATION OF THE OFFSHORE ECOLOGY INVESTIGATION

by M. E. Bender, D. J. Reish, and C. H. Ward

CRITIQUE OF THE OEI EXPERIMENTAL RATIONALE AND DESIGN

The objective of the OEI, according to Sharp and Appan (1978), was "to determine whether long-term and intensive petroleum drilling and production has resulted in harmful environmental or ecological effects in offshore and adjacent estuarine ecosystems." They also state that "because the objective emphasized *cumulative* effects, it [the OEI] was concerned principally with the effects of chronic low-level discharges of petroleum hydrocarbons, heavy metals, and the like, which might be attributed to drilling and production operations."

The experimental logic and design used to accomplish this objective included: (1) point source discharge (local impact) evaluations based on data sets taken synoptically at active platform "experimental" sites and at "control" sites, both in Timbalier Bay and offshore Louisiana; (2) dispersed source discharge (general ecosystem health) evaluations based on comparisons of OEI data with that from similar ecosystems and historical data considered applicable to the OEI study sites; and (3) system dynamics evaluations by (a) relating habitat parameters to principal driving forces such as tides, Mississippi River influences, etc., (b) sampling four roughly parallel synoptic transects generally perpendicular and "upstream" from the platform and control sites offshore and remote from intensive petroleum production activities, and (c) determination of exchange processes between bay and offshore areas by sampling a connecting 25-mile transect.

M. E. Bender is Assistant Director and Head of the Division of Environmental Sciences and Engineering, Virginia Institute of Marine Science, Gloucester Point, VA. D. J. Reish is Professor of Biology, California State University at Long Beach, CA. C. H. Ward is Professor of Biology and Environmental Science and Chairman, Department of Environmental Science and Engineering, Rice University.

Since they lacked comprehensive pre-drilling ecological baselines and had only general descriptive information regarding current systems, biomes, variabilities, etc., the 1972 logic and rationale employed for the OEI appeared, at the time, to be the soundest approach possible. In this re-examination of the OEI, however, it became apparent that either the point source studies based on platform-control site comparisons were an unproductive approach (for the objectives as specified), or the control sites, particularly offshore, were not appropriate to implement the approach effectively.

The experimental logic and design failed to include the concept of "qualification" of control sites based on specific parameters to be measured. In other words, the experimental design failed to validate the "criteria" used in selecting control sites. The logic was to compare platform sites (point sources) with control sites unaffected by oil drilling and production operations as a basis for judging impacts of "chronic low-level discharges of petroleum hydrocarbons" (Sharp and Appan 1978). However, the data collected by Oppenheimer et al. (1974) and by Laseter and Ledet (1974) strongly suggest that petroleum hydrocarbon levels were not significantly different at platform and control sites. Oppenheimer's group chose not to speculate on the origin of the petroleum-derived hydrocarbons found in the Gulf of Mexico, but stressed that "the OEI study areas do not differ appreciably in hydrocarbon concentrations or types relative to non-production 'Gulf' control areas several hundred miles from the study sites." They also state, "Based on all evidence it would appear that the nearshore Gulf of Mexico is uniformly exposed to hydrocarbons which probably are derived from petroleum and partially degraded remnants of the parent mixture can be found in seawater and associated with marine plankton and other organisms." The data of Laseter and Ledet do not contradict these conclusions. However, Brent et al. (1974) felt that their "offshore investigations revealed [total] organic carbon loading of the marine system from operating platforms and wells ranging from 2-4 mg/l above background levels. The effects usually extend some 800 to 1600 m from the platform." They also felt that the "offshore area near Grand Isle, Louisiana is one of the most organic rich ecosystems in the world of its type, quite atypical of an 'open' continental shelf region." Because of the naturally high organic load and its effects on oxygen resources, they warned that "vigilance should be maintained in minimizing effects from oil platform effluents and operational spills because of the instability of this ecosystem during spring and summer months." Brent's methods did not allow him to differentiate among petroleum hydrocarbons, other hydrocarbons, and total organic carbon.

The OEI data do not support a conclusion of "no effect" where the validity of the comparisons between platform and control sites is dependent on there being a difference in hydrocarbons present. Rather, it appears that most of the OEI data were collected from the same "population" and should be expected to be the same within statistical errors of sampling. In addition, great temporal and spatial variability of data on specific species, coupled with the probability of the Gulf being "completely mixed" with reference to hydrocarbon content, probably prohibits utilization of classical experimental versus control design concepts.

Several investigators firmly state that no evidence was found to indicate oil drilling and production effects. However, many directly or indirectly agree with Marum (1974) (who worked on zooplankton) that "variability of the results was high. Since the area is one of extreme environmental fluctuations, temporary pollution damage, if it occurred, would be difficult to detect." We believe that a conclusion of no effect must be followed by the question "Was the experimental design, quantity of data, and quality of data adequate to detect an effect if present?" The answer for much of the OEI is, we believe, no. However, we believe also that it is probably unreasonable to expect that the limited data-gathering effort of the OEI could generate the information required definitively to detect low-level effects of oil, which, by definition, must be minimal.

Other investigators appear to have found effects of oil drilling or production, regardless of whether they chose to discount them or felt that their findings did not indicate adverse effects on the general health of the Gulf. Perry (1974) found with reference to fish that "the offshore production site had the lowest biomass value and the lowest species diversity." He felt that this finding was "symptomatic of an adverse effect caused by the discharge of drilling mud solutions which form a hard crust on the bottom to the detriment of the infauna. . . . This lack of numbers of bottom fishes [near production platforms] is strongly believed to be influenced by the lack of bottom-dwelling, burrowing type invertebrates which these fishes feed on. . . . The invertebrate section of this study has indeed found this to be the case. This apparently is the result of the overboard discharge of large amounts of drilling mud solutions. . . . In lease tracts with heavy drilling pressure such a situation could become detrimental to the welfare of a considerable amount of viable offshore bottomland for a considerable time."

Montalvo and Brady (1974) found that "zinc, lead, and cadmium [in water samples] are higher near the rig sites than at areas remote from the rig sites." They considered the limited density of samples at platform and control sites and the lack of trace metal and sediment analyses serious deficiencies in experimental design. Their studies did not establish cause-effect relationships, but they felt that they had "obtained some data that indicates that there is at least a possibility of danger [from Zn, Pb, and Cd concentrations at the rigs]."

Fish et al. (1974), with reference to the meiofauna of Timbalier Bay, felt that "those stations located near active oil production sites consistently yielded a lower calculated species diversity index than did their corresponding control sites. As the only observable difference in each pair of stations was their proximity to active oil wells, the above data indicates that the environmental stress that was recorded for these environments may have been caused by the nearby active oil production sites." Without complete hydrocarbon analysis to aid in the interpretation of their data, they concluded that "active oil wells located within Timbalier Bay have had only a minimal effect on the distribution and abundance of Foraminifera inhabiting the area."

The findings or opinions of the OEI investigators noted above do not clearly indicate effect or no effect of oil drilling and production in the Gulf. Hence, we feel that the OEI Council conclusions (Morgan et al. 1974) should be reviewed and restated to indicate clearly the limitations of the experimental design and the data base for definitively answering questions on chronic effects of oil.

This re-examination of the OEI has shown that the natural variability inherent in the biological, physical, and chemical processes in the Gulf of Mexico precludes acquisition of adequate, representative, or valid baseline data in a period limited to two years. The quality of the OEI data is in some cases excellent, however, even though many of the investigators appear to have worked largely independently. Because their research was often conducted independently, the usefulness of the total OEI data base is limited for some scientific purposes. For instance, if total organic carbon, soluble organic carbon, particulate organic carbon, phytoplankton (or chlorophyll), primary productivity, zooplankton, hydrocarbons, and microbial flora had all been determined in the same rather than in different samples, the OEI data base would represent a unique resource for answering questions vital to our current understanding of carbon partitioning and flow in ocean ecosystems (Pomeroy 1974).

Regardless of the shortcomings in experimental design and density of data on specific parameters, the data amassed by the OEI investigators is invaluable to U.S. science, government, and industry, and at the time it was collected, it represented the best and most cost-effective data base in existence on any part of the Gulf of Mexico. The OEI data, in combination with more recent studies such as those sponsored by the Bureau of Land Management, represent the most comprehensive data base available on a salt water system. We believe that it is fortuitous that much of the OEI data appears to have come from the same "population."

Every effort should be made to find avenues to expand the Gulf environmental data base and to establish quantitative indicators of "acceptable limits of performance." But such indicators can only be identified with adequate knowledge of spatial and temporal variability of parameters important to the functioning of marine ecosystems. New avenues to, and philosophies for, experimental design for environmental assessment should be explored in order to develop approaches fully

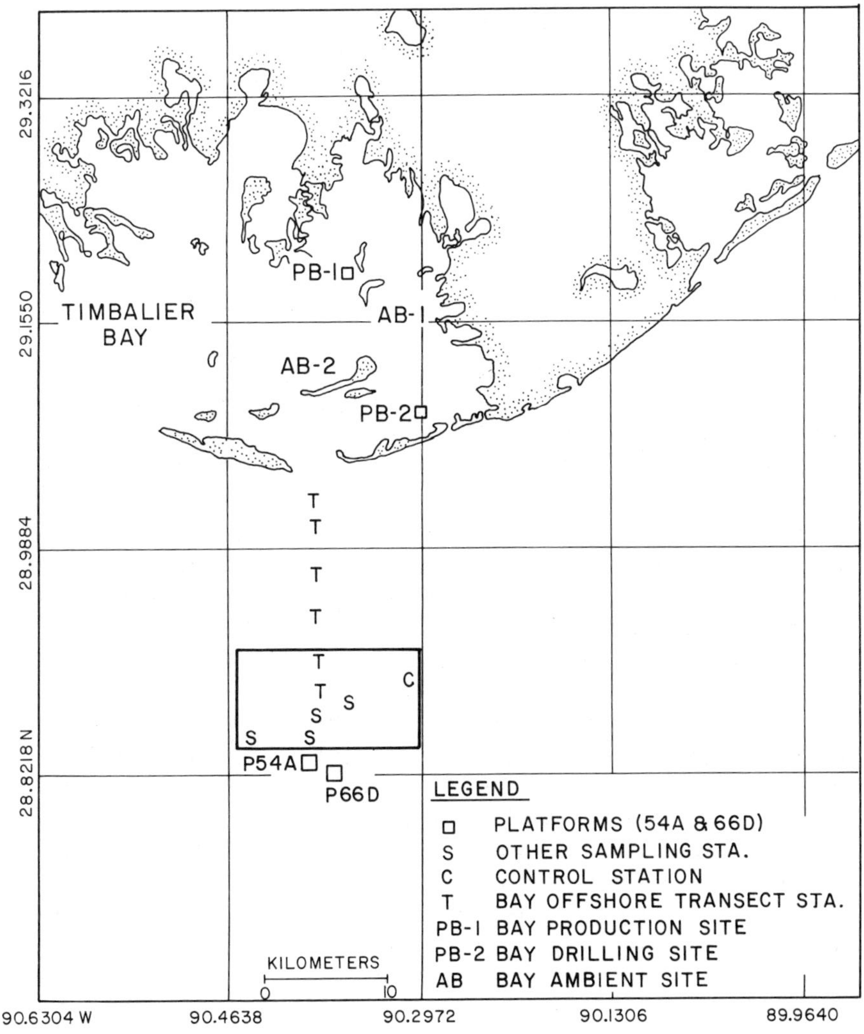

FIG. 1. LOCATIONS OF BAY AND OFFSHORE SAMPLING STATIONS used in the re-examination of the Offshore Ecology Investigation.

consistent with the scientific method or to provide acceptable alternatives. Consensus qualification of selected indicators as criteria for judging "good ecological health" might serve as a first approach.

RE-EXAMINATION AND ANALYSIS OF OEI DATA

Water column and sediment characteristics

Platform Versus Control

Minimum statistical treatment of data was evident in most of the OEI investigators' final reports, and conclusions reached were often limited to opinions based on qualitative review of the data. The density of the OEI data appeared to be the primary factor limiting quantitative comparisons.

Figure 1 illustrates the location of bay sampling sites, transect stations, platforms, other sampling stations, and the control, which was about 11 km east of the platforms.

To explore the degree to which the offshore control site was environmentally comparable to the platform site—except, of course, for the physical presence of the platform—we selected the four (of the eight) seasonal exercises that offered the most complete sets of data. Computer-generated displays were obtained of selected data taken at both the platform site (54A, also referred to as HOR-54A and Exxon 54A in other papers in this volume) and the control. Several sampling stations near the platform were chosen to illustrate the natural variability in the immediate platform area. Environmental parameters used in the comparisons were salinity, dissolved oxygen, temperature, optical transparency (transmittance), depth, and bottom type. Figures 2 through 5 are representative of the data that collectively indicate that the sites were comparable in these regards. Depths and bottom types were also similar.

Water quality parameters used to compare the platform and control sites were trace metals, nutrients, and carbon chemistry (organic carbon, inorganic carbon, and total carbon). Tables 1, 2, and 3 (p. 45) present representative data, most of which is insufficiently dense to permit statistical treatment. Site-related differences are not readily apparent; some seasonal and depth variations can be distinguished, however.

Taken as a whole, these data roughly show that the primary environmental difference between the two offshore sites was the physical presence of the platform in one and not the other, that the habitats varied essentially simultaneously in response to meteorological or seasonal events, and that prevailing current direction and speed (figure 6, p. 46) were such that materials originating at the platform would not normally reach the control site during most of the year.

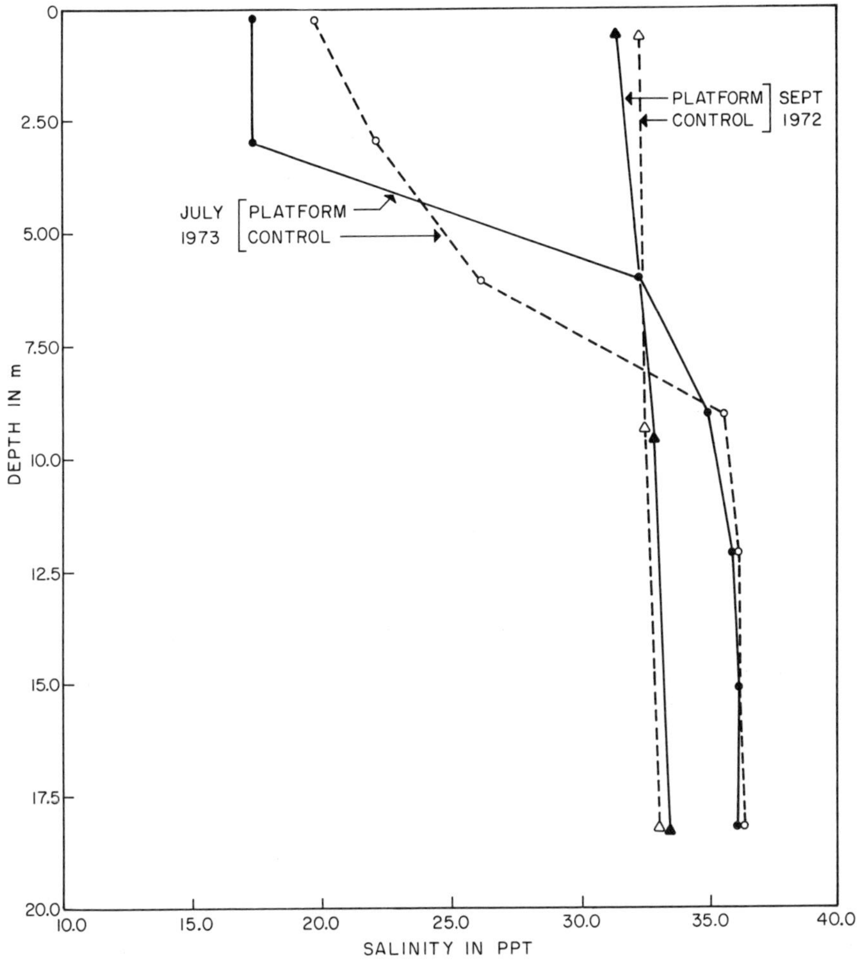

FIG 2. SALINITY WITH DEPTH DURING TWO SEASONAL SAMPLING EXERCISES.

Entrained material in the Mississippi River outflow, runoff from adjacent land, and platforms are probable sources of nutrients, trace metals, and organic carbon in the OEI study area. Water column measurements for arsenic, mercury, lead, cadmium, and zinc showed that arsenic was always below detectable levels. Trace metals, probably bound to suspended materials, were generally higher in water samples taken from the nepheloid layer. Figure 7 (p. 47) shows a non-statistical, but informative, treatment of the similarity between platform zinc concentrations and concentrations at stations 1,500 m or more away from the platform.

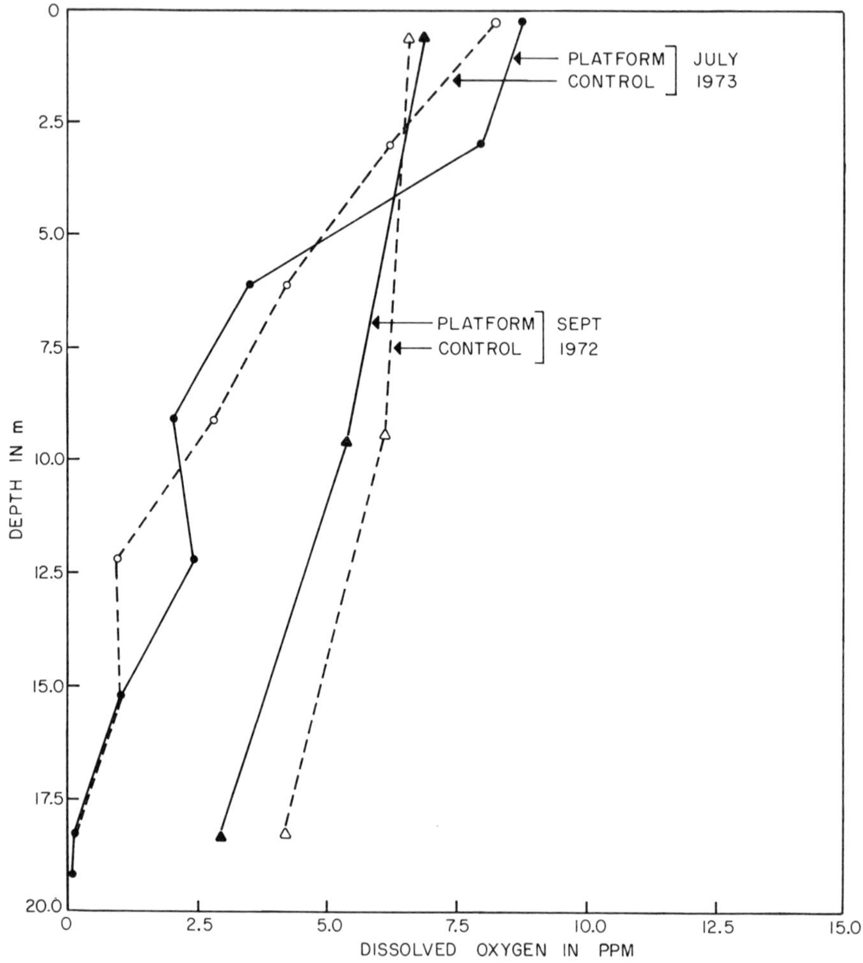

FIG. 3. DISSOLVED OXYGEN WITH DEPTH DURING TWO SEASONAL SAMPLING EXERCISES.

Dissolved oxygen concentrations varied more significantly with depth than with location. As would be expected, surface values were uniformly higher than those from deeper waters, and dissolved oxygen concentrations varied directly in relation to water temperature. During the summer of 1973, near-bottom values approached zero over the entire offshore region.

Except during the post-flood period, Biochemical Oxygen Demand (BOD) was usually slightly higher at the production platform offshore than at the drilling platform, and both were usually higher than at the

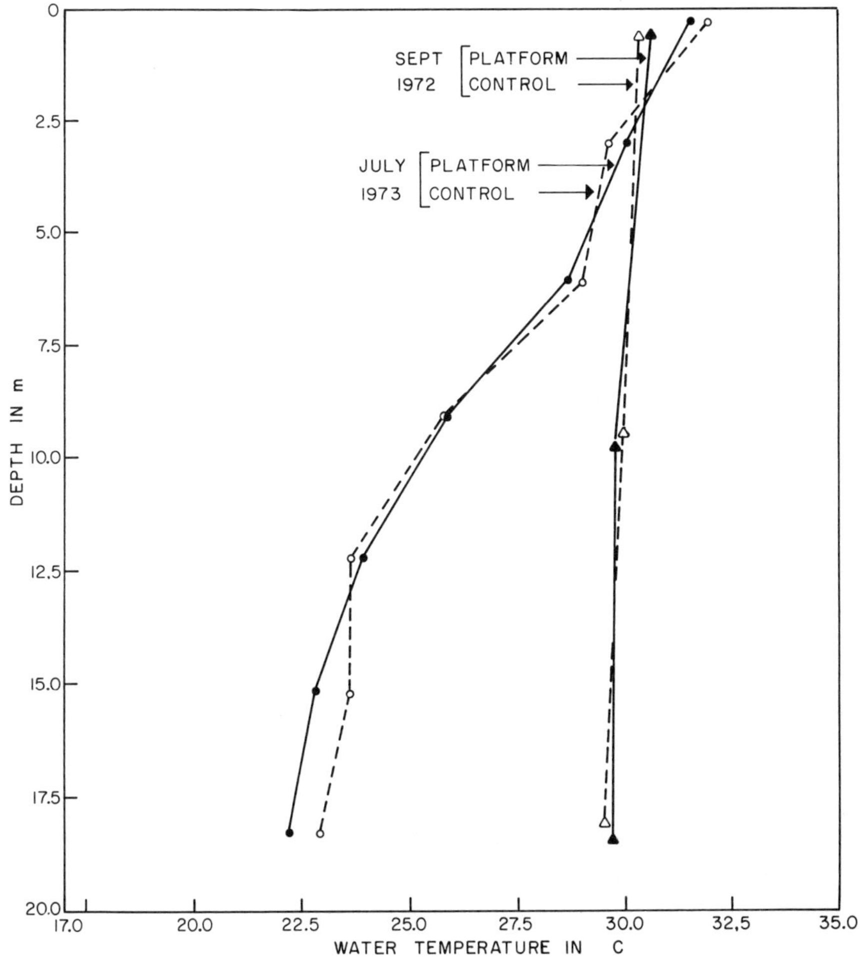

FIG. 4. WATER TEMPERATURE WITH DEPTH DURING TWO SEASONAL SAMPLING EXERCISES.

control site. The large fouling community that colonized the legs of the platforms may have contributed to the higher BOD values around the platforms. Differences between the production and drilling platforms, similarly, may have resulted from the smaller submerged surface area at the drill site. Surface values of BOD ranged from 0 to 2.7 mg/l at the platform and from 0 to 2.6 mg/l at the control over a period of four seasons. Near-bottom values during the same time ranged from 0.1 to 2.2 mg/l at the platform and from 0.7 to 0.9 at the control. Inadequate

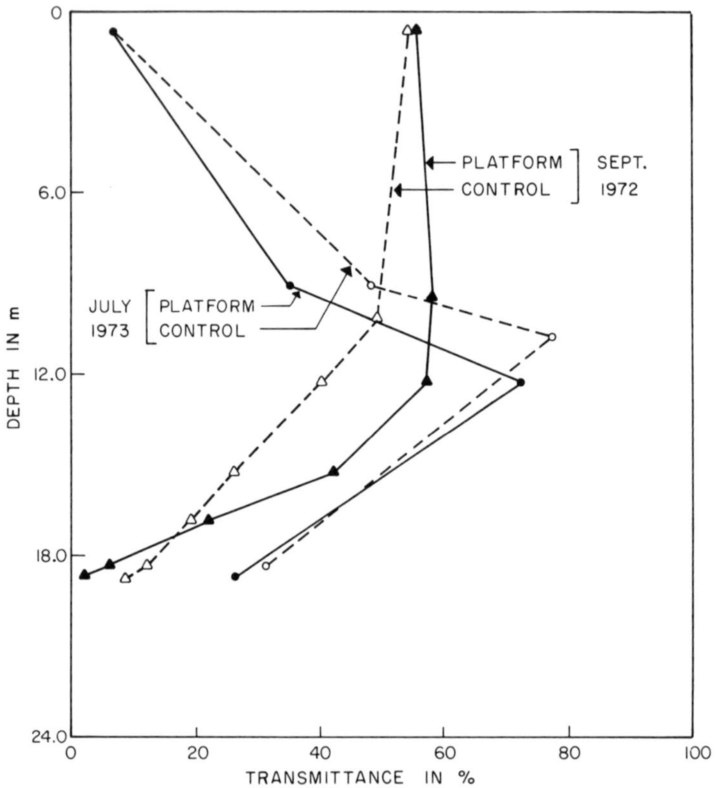

FIG. 5. TRANSMITTANCE WITH DEPTH DURING TWO SEASONAL SAMPLING EXERCISES.

BOD data were available to permit statistical treatment of possible differences between platform and control sites.

Slightly higher concentrations of total organic carbon (TOC) were often found around platforms (see table 3). Possible sources of organic carbon at platforms include kitchen and human wastes, oil, and contributions from the fouling communities. Data collected in the OEI do not permit differentiation of these sources and there was no obvious relationship between various hydrocarbon fractions and TOC.

Although only qualitative comparisons can be made between possible contaminants present at the OEI platform and control sites, it would appear that the effluents from drilling and production (1) are so small as to cause no significant concentrations of contaminants, (2) are rapidly diffused, and/or (3) are masked by distributed source effects in the intensely oil-active OEI area. In any event, such effects as may be

TABLE 1. Mercury concentrations (ppb) in the water column with depth at platform and control sites during four of eight OEI sampling exercises.

Date	Depth	Hg Range		Mean		No. of Observations	
		Platform	Control	Platform	Control	Platform	Control
Exercise 1	Top	0.4-3.0	0.4-0.8	0.9	0.6	6	2
1972 Sept.	Middle	0.3-0.6	0.4-0.5	0.4	0.4	6	2
	Bottom	0.3-2.9	0.3-0.4	0.8	0.3	6	2
Exercise 3	Top	0.3-0.3	0.3	0.3	0.3	2	1
1972 Nov.-Dec.	Middle	0.3-0.7	0.3	0.5	0.3	2	1
	Bottom	0.3-0.3	0.3	0.3	0.3	2	1
Exercise 5	Top	0.3	-	0.3	-	1	0
1973 Apr.	Middle	-	-	-	-	0	0
	Bottom	-	-	-	-	0	0
Exercise 6	Top	0.3-0.3	0.3-0.3	0.3	0.3	4	2
1973 July	Middle	0.3-0.3	0.3	0.3	0.3	2	1
	Bottom	0.3-0.5	0.3	0.3	0.3	7	1

Top, 0-5 m; Middle, 8-13 m; Bottom, 14 m - bottom.

TABLE 2. Nitrate plus nitrite nitrogen (ppm) in the water column with depth at platform and control sites during four of eight OEI sampling exercises.

Date	Depth	$NO_3 + NO_2$ Range		Mean		No. of Observations	
		Platform	Control	Platform	Control	Platform	Control
Exercise 1	Top	0 -0.45	0-0.45	0.06	0.22	7	2
1972 Sept.	Middle	0 -0.55	0-4.00	0.09	2.00	7	2
	Bottom	0 -0.70	0-0	0.13	0	7	2
Exercise 3	Top	4.34	2.34	4.34	2.34	1	1
1972 Nov.-Dec.	Middle	1.70	5.38	1.70	5.38	1	1
	Bottom	0	52.98	0	52.98	1	1
Exercise 5	Top	0.17	0.58	0.17	0.58	1	1
1973 Apr.	Middle	-	0.70	-	0.70	0	1
	Bottom	0.47-0.59	-	0.52	-	3	0
Exercise 6	Top	0.09-0.33	0.19-0.35	0.18	0.28	6	6
1973 July	Middle	0.14-0.04	0.04-0.32	0.19	0.22	2	3
	Bottom	0.05-0.29	0.05-0.30	0.15	0.19	6	3

Top, 0-5 m; Middle, 8-13 m; Bottom, 14 m - bottom.

TABLE 3. Total Organic Carbon (ppm) in the water column with depth at platform and control sites during four of eight sampling exercises.

Date	Depth	TOC Range		Mean		No. of Observations	
		Platform	Control	Platform	Control	Platform	Control
Exercise 1	Top	5.3-6.6	5.1-6.2	5.9	5.6	2	2
1972 Sept.	Middle	6.9-7.9	6.0-6.6	7.4	6.3	2	2
	Bottom	5.3-7.6	5.6-6.0	6.4	5.8	2	2
Exercise 3	Top	2.7	-	2.7	-	1	0
1972 Nov.-Dec.	Middle	-	-	-	-	0	0
	Bottom	-	-	-	-	0	0
Exercise 5	Top	8.2	7.2	8.2	7.2	1	1
1973 Apr.	Middle	-	2.8	-	2.8	0	1
	Bottom	-	3.3	-	3.3	0	1
Exercise 6	Top	4.4-11.7	6.8	7.3	6.8	5	1
1973 July	Middle	2.6-2.8	-	2.7	-	2	0
	Bottom	1.6-7.1	-	4.4	-	4	0

Top, 0-5 m; Middle, 8-13 m; Bottom, 14 m- bottom.

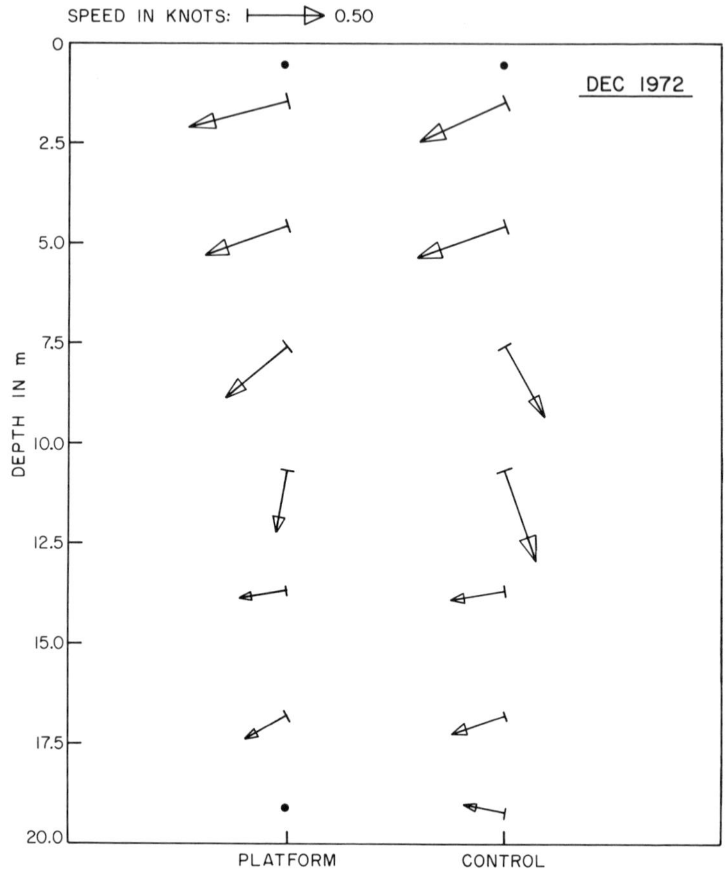

FIG. 6. CURRENT DIRECTION AND SPEED WITH DEPTH FOR PLATFORM AND CONTROL DURING ONE SEASONAL SAMPLING EXERCISE.

present appear to be small in comparison with seasonal variabilities. It should be noted that the observed variabilities were experienced at both platform and control sites at about the same time and magnitude, however, indicating that relative effects originating from oil drilling and production might be observed, if they occur, in the presence of large seasonal variations, provided that monitoring data are adequate to permit rigorous statistical treatment.

Since the density of the OEI data would not permit, for the most part, meaningful statistical comparisons between platform and control sites, the OEI as originally designed provided little evidence for or against cumulative point source effects.

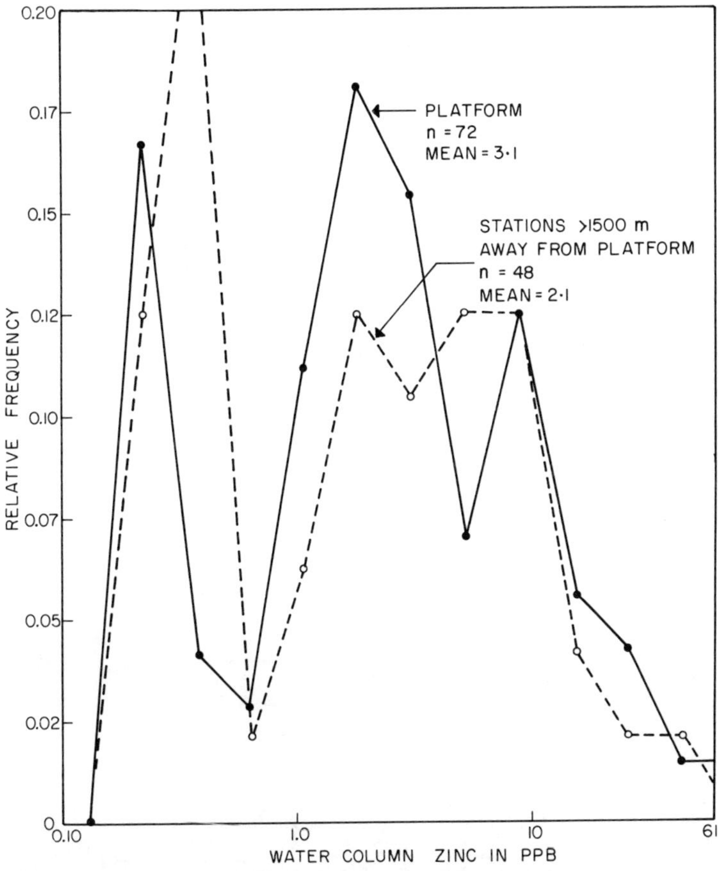

FIG. 7. FREQUENCY DISTRIBUTION OF WATER COLUMN ZINC CONCENTRATIONS for Platform 54A and stations greater than 1500 m from the platform.

Platform Versus Ambient

Since the OEI sampling design, primarily because of lack of replication, precluded statistical treatment of the data, it was decided to explore the possibility of expanding the data base to provide additional stations against which the platform stations might be compared. It was necessary, therefore, to explore the degree to which additional "control" sites might be environmentally comparable to the platform sites, except, of course, for the physical presence of the platform itself.

To accomplish these comparisons, the computerized data files were queried to provide plots displaying selected data taken at the platforms

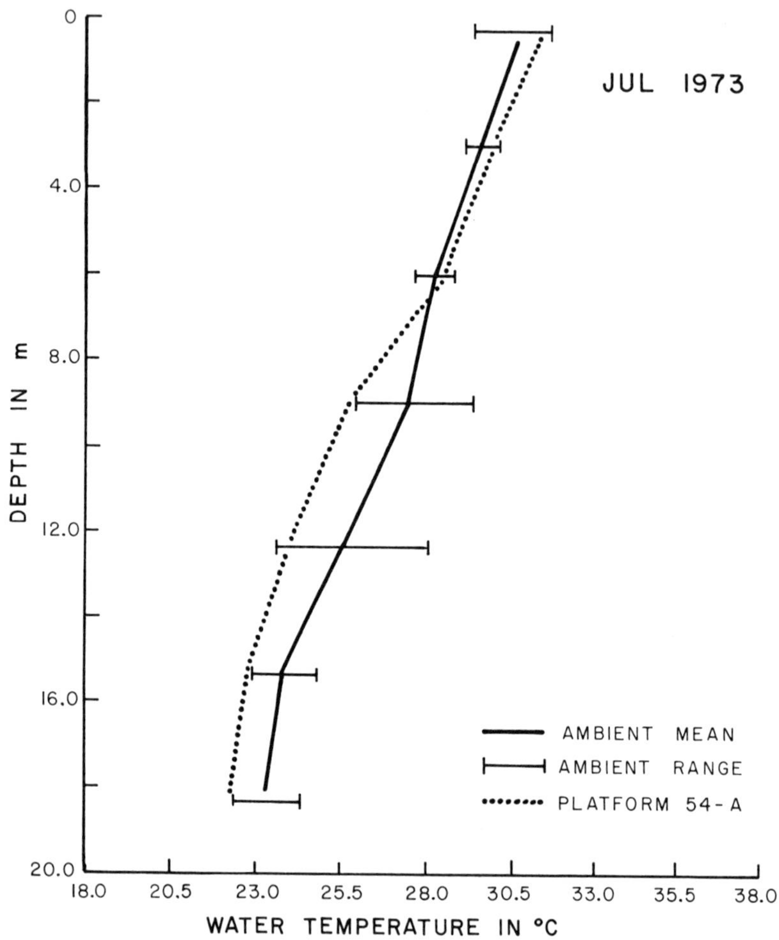

FIG. 8. WATER TEMPERATURE WITH DEPTH FOR AMBIENT AND PLATFORM STATIONS.

and potential additional controls. Environmental parameters used in the comparisons were salinity, dissolved oxygen, temperature, optical transparency, current, depth, and bottom type. For a station to qualify as an additional control in the near-shore region, the criteria were that it must: (1) be located at least 1,000 meters from the platform, (2) be sampled the same day as the platform site, (3) have similar salinities and temperatures, (4) have similar bottom substrates, (5) have depths within 5 meters of those at the platform with which comparisons were made, and (6) not receive transported materials originating at the platform during the sampling period.

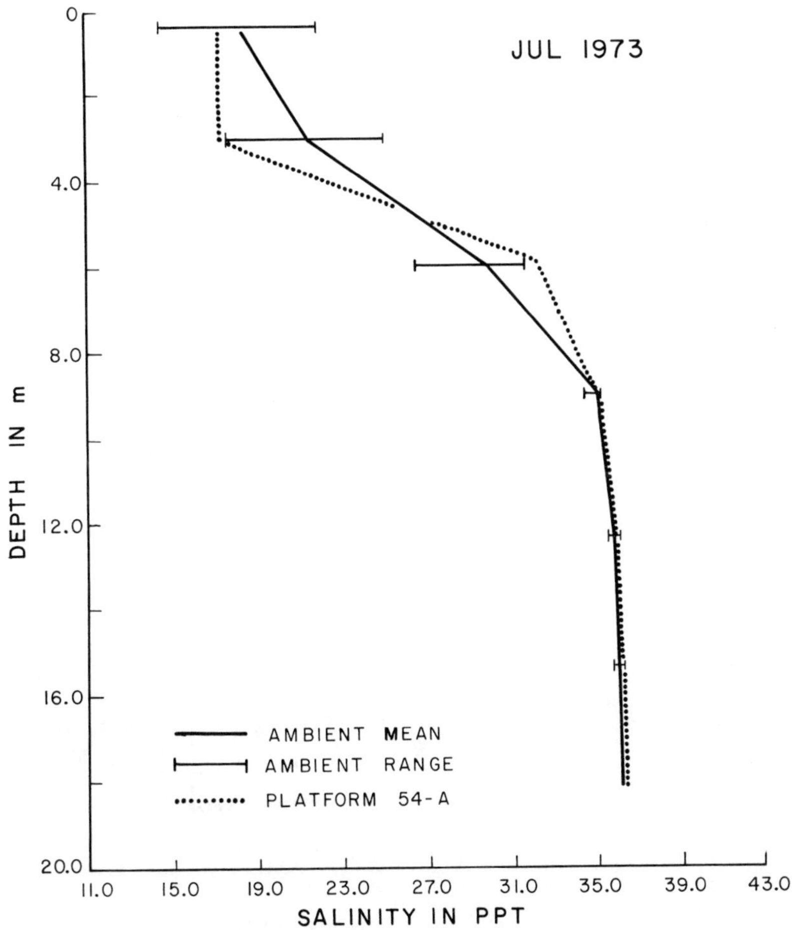

FIG. 9. SALINITY WITH DEPTH FOR AMBIENT AND PLATFORM STATIONS.

Figures 8, 9, and 10 are representative of the data that supported the inclusion of additional stations for statistical comparisons. A review of the OEI data showed that six stations, in addition to the original control, qualified as ambient stations (figure 1, stations included in rectangle).

Each seasonal field exercise was screened to determine whether a sufficient number of comparable ambient stations had been sampled to yield at least three samples for comparison with the platform. If so, the normality of the data was tested by comparing means and standard deviations with "corner tests" as described by Olmstead and Tukey (1947).

In the offshore region, the data indicated that normal statistics could

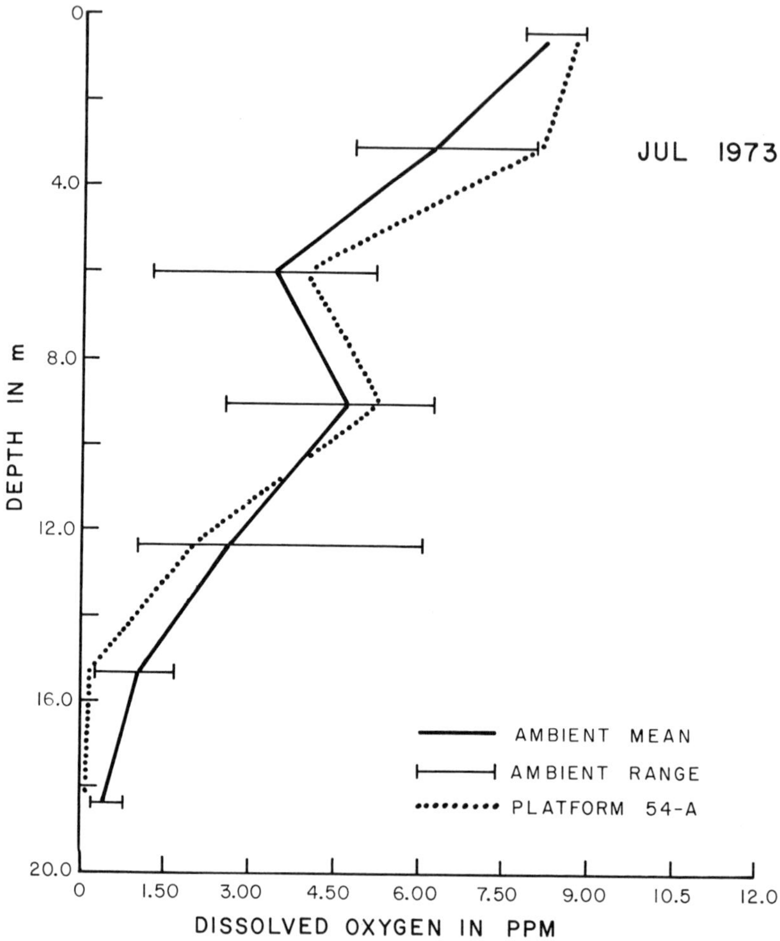

FIG. 10. DISSOLVED OXYGEN WITH DEPTH FOR AMBIENT AND PLATFORM STATIONS.

be used to test differences between platform and ambient stations for the parameters of lead, ammonia, and nitrate-nitrite nitrogen. Logarithmic transformations of the data not indicated as normal were performed before the data were used for "t" tests. In Timbalier Bay, square root transformations of nutrient and trace metal data were necessary before normal statistics could be applied.

Offshore—Table 4 shows the results of "t" test comparisons of eight parameters between platform 54A and the newly defined offshore ambient stations. Measurements were paired by depth at each station for

TABLE 4. Paired 't' test comparisons of water quality parameters between platform 54-A and offshore ambient stations.

Date	TOC 1)	Zn 2)	Pb	Cd 2)	Dissolved Silica 2)	NH_3-N	NO_3+NO_2	O-PO_4 2)
Exercise 1 1972 Sept.	(6)	(12)	(12)	(12)	(12)	(12)	(12)	(12)
Exercise 2 1972 Oct.	IS (1)	(3)	(3)	(3)	(3)	(3)	(3)	(3)
Exercise 3 1972 Nov.-Dec.	IS (0)	(6)	(6)	(6)	(3)	(3)	(3)	(3)
Exercise 4 1973 Jan.-Feb.	IS (1)	IS (0)	IS (0)	IS (0)	IS (1)	IS (0)	IS (0)	IS (1)
Exercise 5 1973 Apr.	IS (1)	IS (0)	IS (0)	IS (0)	IS (1)	IS (1)	IS (1)	IS (1)
Exercise 6 1973 July	IS (0)	(7)	(7)	(7)	(18)	(18)	(18)	(16)
Exercise 7 1973 Oct.	IS (2)	(8)	(8)	(8)	(12)* Amb.High	(12)* 54A High	(12)	(12)* Amb.High
Exercise 8 1974 Jan.	(9)	(8)	(8)	(8)	(12)	(12)	(12)* 54A High	(12)

1) data not normally distributed even when log transformed. Chi-square used to test significance;
2) log transformed; ( ) number of paired observations; * significant difference at 0.05 level;
IS inadequate sample size, n = <3 pairs.

The water column was divided into 3 depth zones: Top, 0.00 to 5.00 m; Mid, 5.01 to 13.00 m; Bottom, 13.01 to bottom. For each cruise a Top value of 54-A was paired with a Top value of Ambient. The pairing was done using random numbers. When a matching value was not available for pairing that measurement was not included in this 't' test.

AMB	n=	6	2			1	2		4	9
66D	n= 16						2	15	1	27
54A	n=	6	2	2		6	2	14	2	23
ALL	n=22	42	28	9		16	19	53	12	99

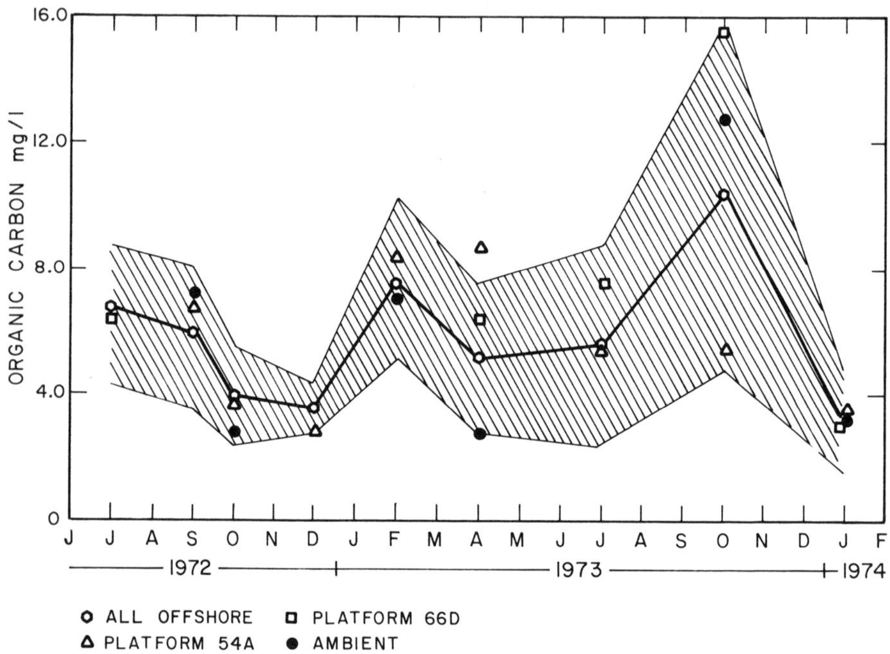

FIG. 11. SEASONAL VARIABILITY OF ORGANIC CARBON IN THE WATER COLUMN OFFSHORE.

each of the eight sampling exercises. Comparisons were not made when there were fewer than three pairable observations between the platform and one of the ambient stations. Single instances occurred where dissolved silica, ammonia nitrogen, nitrate-nitrite nitrogen, and orthophosphate were significantly higher or lower at platform 54A. Since these were single instances there is little basis for interpretation. It is important to note, however, that the analysis gave no indication of differences between platform and ambient in organic carbon or metals in the water column.

Further analysis of the TOC and BOD data of Brent et al. (1974) with season (figures 11 and 12) does not clearly demonstrate the presence of platforms 54A and 66D based on these parameters. The dashed area in each figure is one standard deviation above and below the mean of all

AMB	n=	6	2		1	2		4	9
66D	n= 16					2	15	1	27
54A	n=	6	2	2	6	2	14	2	23
ALL	n=22	42	28	9	16	19	53	12	99

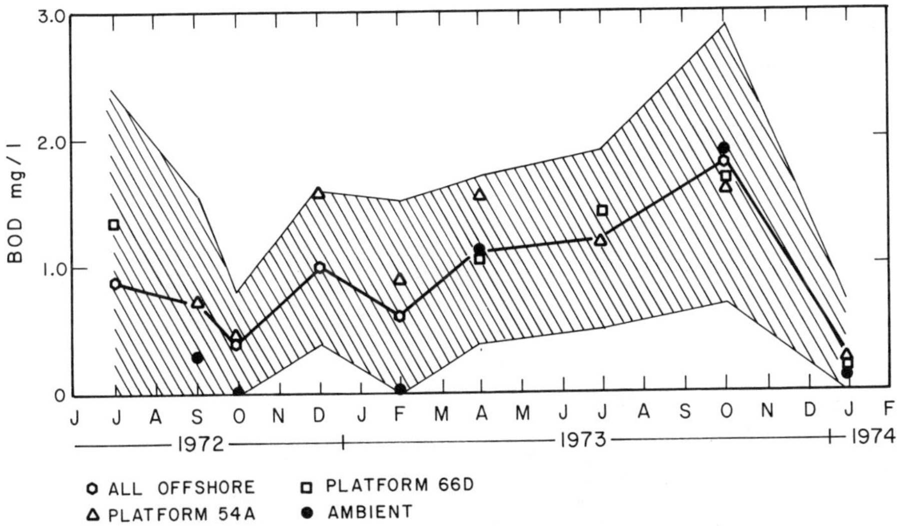

FIG. 12. SEASONAL VARIABILITY OF BIOCHEMICAL OXYGEN DEMAND IN THE WATER COLUMN OFFSHORE.

offshore data. Platforms 54A and 66D are the means of all data within a 750 m × 750 m area around each platform. "Ambient" is the mean of all data from all stations falling in the rectangular area shown in figure 1. Poor correlation exists between BOD and TOC concentrations at platform and ambient stations (figure 13).

Figures 14 and 15 present the BOD and TOC data with distance from platform 54A. The horizontal line connects the mean of all the data within each 1,000 m segment of the range from platform 54A, spanning an azimuth of 360 degrees. The number of observations in each mean is indicated at the top of each figure. The thick vertical line represents the 95% confidence interval of the mean and the thin vertical line is one standard deviation. Platform 54A is located at 0 on the ordinate; platform 66D at about 2,000 m. Both the BOD and TOC decrease initially with distance from 54A, but the number of observations also decreases with distance and variability of the data increases. The presence of the platforms could not be demonstrated from the collection of data, nor are the platforms significant point sources of TOC or biodegradable carbon.

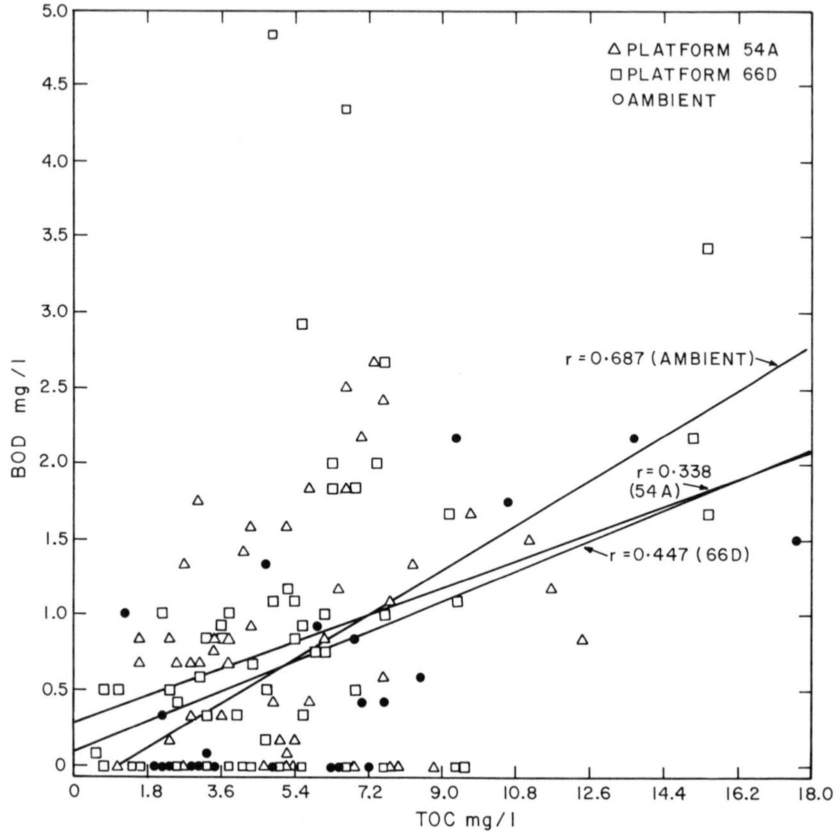

FIG. 13. RELATIONSHIP BETWEEN BOD AND TOC AT PLATFORM AND AMBIENT SAMPLING STATIONS OFFSHORE.

Laseter and Ledet (1974) analyzed sediment, beach, and air/sea interface samples for hydrocarbons and free fatty acids using gas chromatography and mass spectrometry. The air/sea and beach data were not sufficiently replicated to permit statistical comparisons; however, "t" test comparisons of the hexane, benzene, and methanol eluates of sediment samples from platform 54A and ambient stations showed no differences at the 5% confidence level (table 5, p. 56).

Figures 16, 17, 18, and 19 (pp. 57-58) present hydrocarbon data with distance from platform 54A. The horizontal line connects the mean of all data of all cruises within each 2000-meter segment of the range from 54A, spanning an azimuth of 360 degrees. The vertical line is one standard deviation. The data points are coded by cruise number. The hexane eluate contains the paraffins; the benzene eluate the aromatics, high mo-

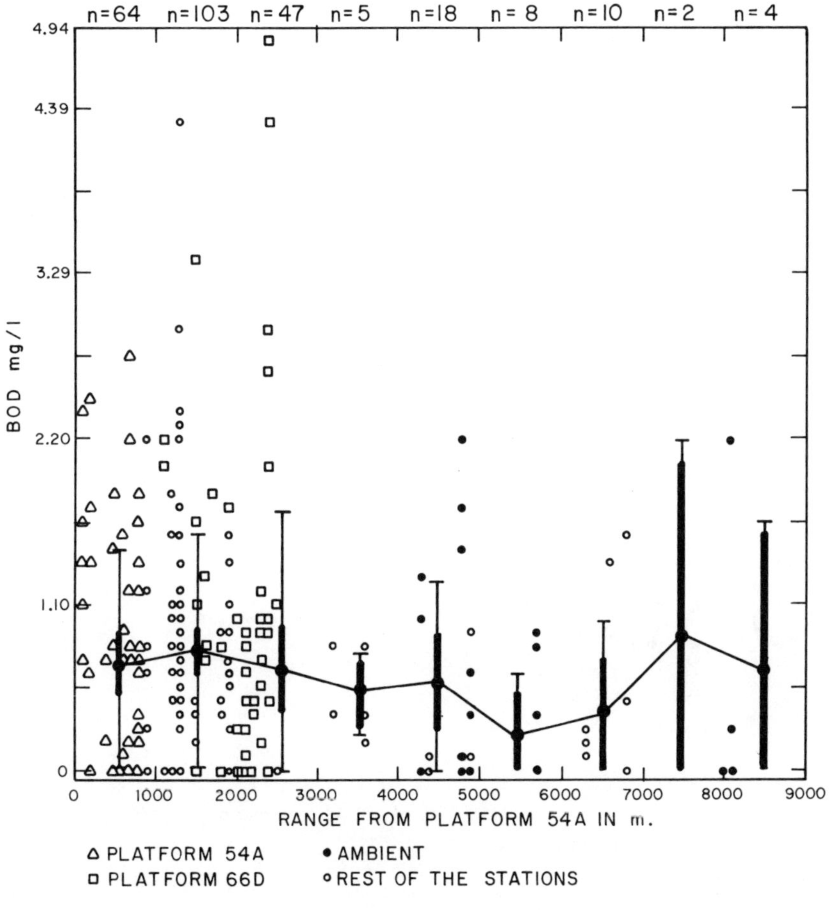

FIG. 14. VARIABILITY OF BOD IN THE WATER COLUMN WITH DISTANCE FROM PLATFORM 54A.

lecular weight ketones, and esters; the methanol eluate the free fatty acids and polar compounds. Since the data for each eluate represent mass per unit area and not percentage of the total hydrocarbons recovered, higher values would be expected at the platform if it were a significant source of petroleum hydrocarbons. This is clearly not the case but the low density of the data prevents rigorous statistical treatment.

Figures 20, 21, and 22 (pp. 59-61) present data for three hydrocarbon eluates with season. Inadequate benzene data were available for this type of analysis. The dashed area is one standard deviation above and below the mean of all offshore data. The points labeled "platform" are the mean of all data within a range of 0 to 2,500 m from 54A, spanning an azimuth of 360 degrees, which includes platform 66D. "Ambient" in

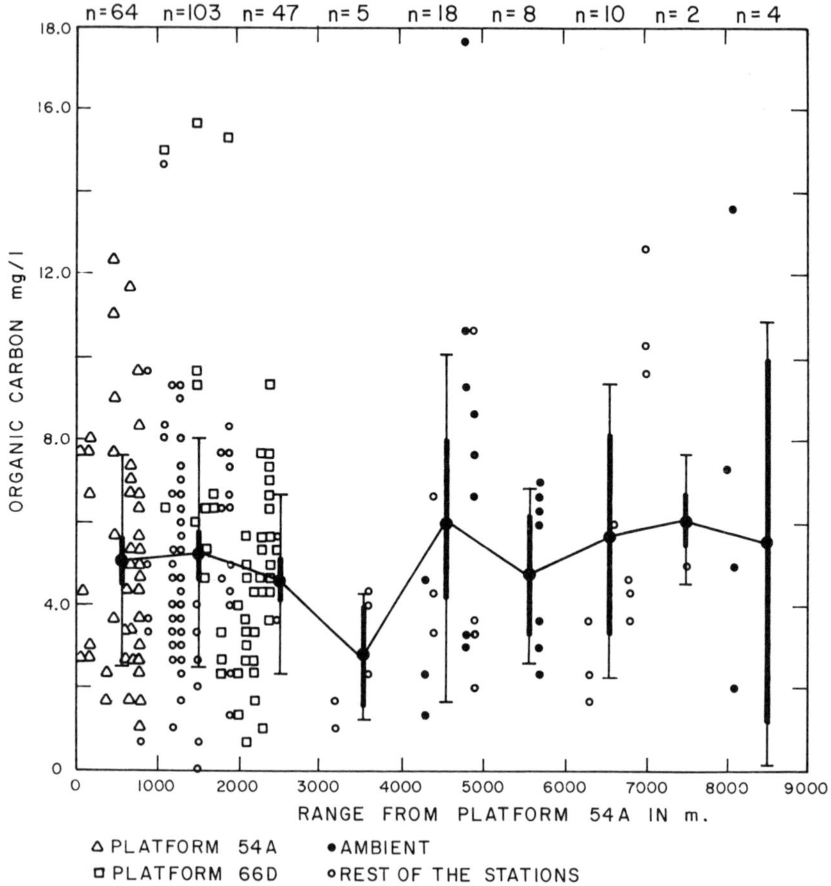

FIG. 15. VARIABILITY OF ORGANIC CARBON IN THE WATER COLUMN WITH DISTANCE FROM PLATORM 54A.

TABLE 5. Comparison between hydrocarbon fractions in offshore sediment samples collected during August, September, October, and December 1972.

Eluate	Range[1)]	n	Min	Max	$\bar{x}$	Std. error	s	Student's t	t.05
Hexane eluate mg/100 g	Platform	6	1.90	14.00	4.97	1.89	4.63	0.63	2.11
	Ambient	13	1.80	14.50	6.17	0.99	3.57		
Benzene eluate mg/100 g	Platform	6	0.30	18.20	5.73	2.72	6.65	0.44	2.11
	Ambient	13	1.00	23.30	7.21	1.96	7.06		
Methanol eluate mg/100 g	Platform	6	6.80	48.80	19.90	6.86	16.80	1.92	2.11
	Ambient	13	10.00	80.00	39.17	6.06	21.84		

1) Platform = 0 to 1,600 m range from platform 54A. This includes Platform 66D samples.
Ambient = 1,601 to 12,500 m range from platform 54A.

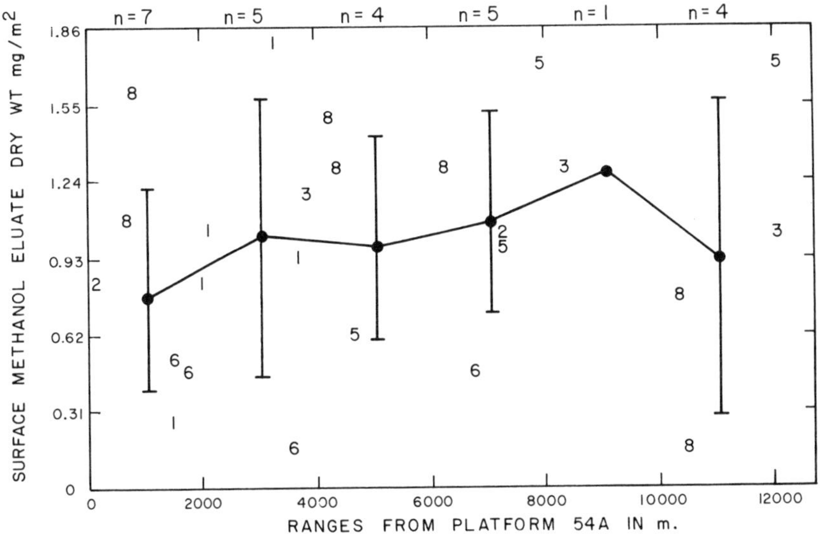

1 = SEP 72 3 = NOV-DEC 72 5 = APR 73 7 = OCT 73
2 = OCT 72 4 = JAN-FEB 73 6 = JUL 73 8 = JAN 74

FIG. 16. METHANOL ELUATE OF SURFACE WATER SAMPLES WITH DISTANCE FROM PLATFORM 54A.

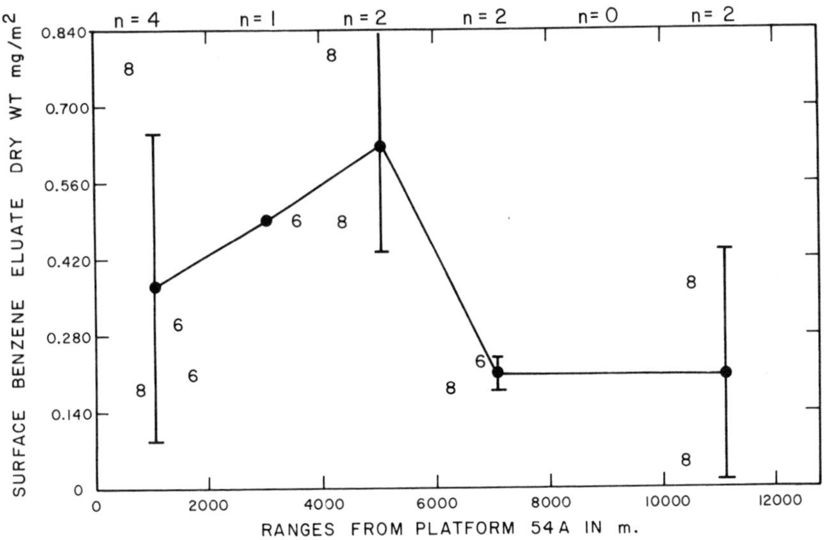

1 = SEP 72 3 = NOV-DEC 72 5 = APR 73 7 = OCT 73
2 = OCT 72 4 = JAN-FEB 73 6 = JUL 73 8 = JAN 74

FIG. 17. BENZENE ELUATE OF SURFACE WATER SAMPLES WITH DISTANCE FROM PLATFORM 54A.

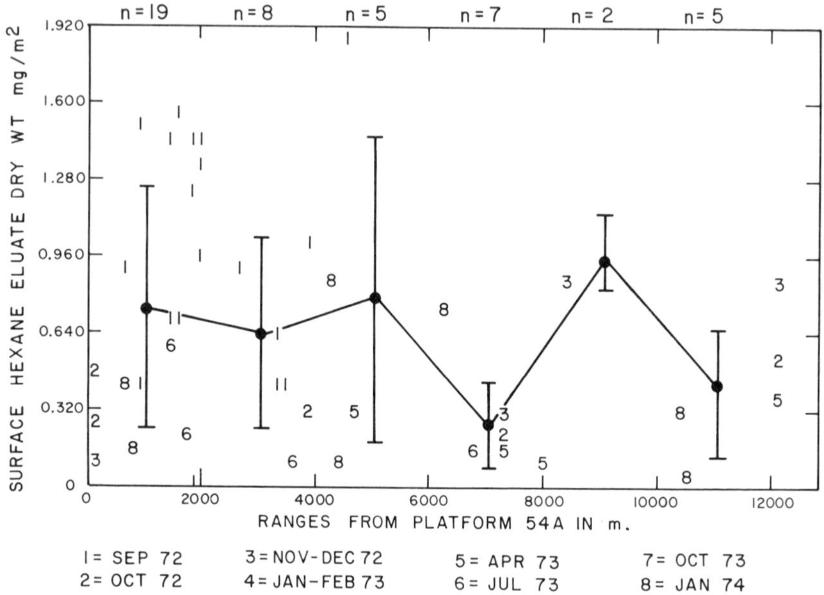

FIG. 18. HEXANE ELUATE OF SURFACE WATER SAMPLES WITH DISTANCE FROM PLATFORM 54A.

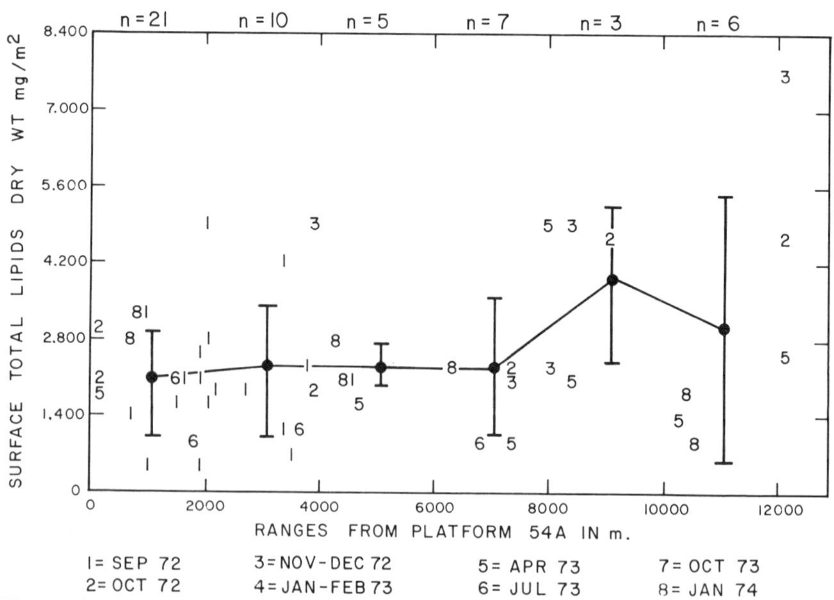

FIG. 19. TOTAL LIPIDS OF SURFACE WATER SAMPLES WITH DISTANCE FROM PLATFORM 54A.

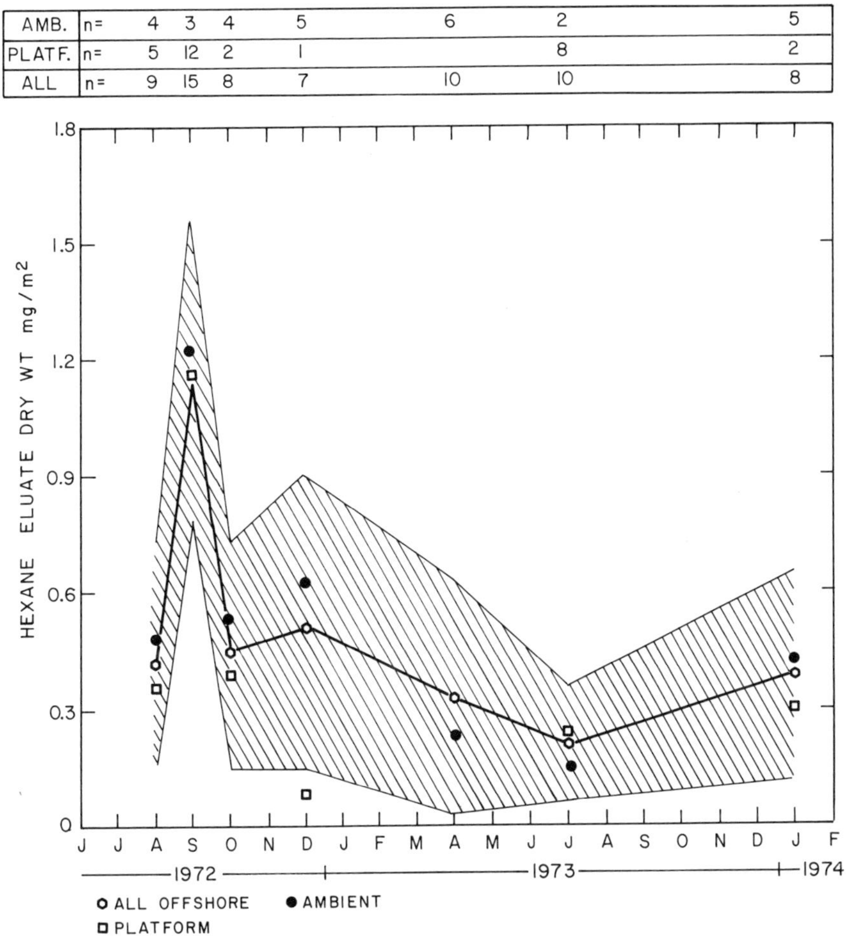

FIG. 20. SEASONAL VARIABILITY OF HEXANE ELUATE FROM SURFACE WATER SAMPLES OFFSHORE.

these figures is the mean of all data within a range of 2,501 to 12,500 m from 54A, spanning an azimuth of 360 degrees. The number of observations in each mean is given at the top of each figure. Again, the platform would be expected to have higher concentrations of the various hydrocarbon eluates if it were a significant point source of hydrocarbon-containing effluents. This does not seem to be the case, since the ambient means are frequently higher than the means for the platform.

Although the concentrations of TOC, BOD, and hydrocarbons do not appear to differ between platform and ambient sites, the ternary diagrams in figures 23 and 24 suggest that relative amounts of types of

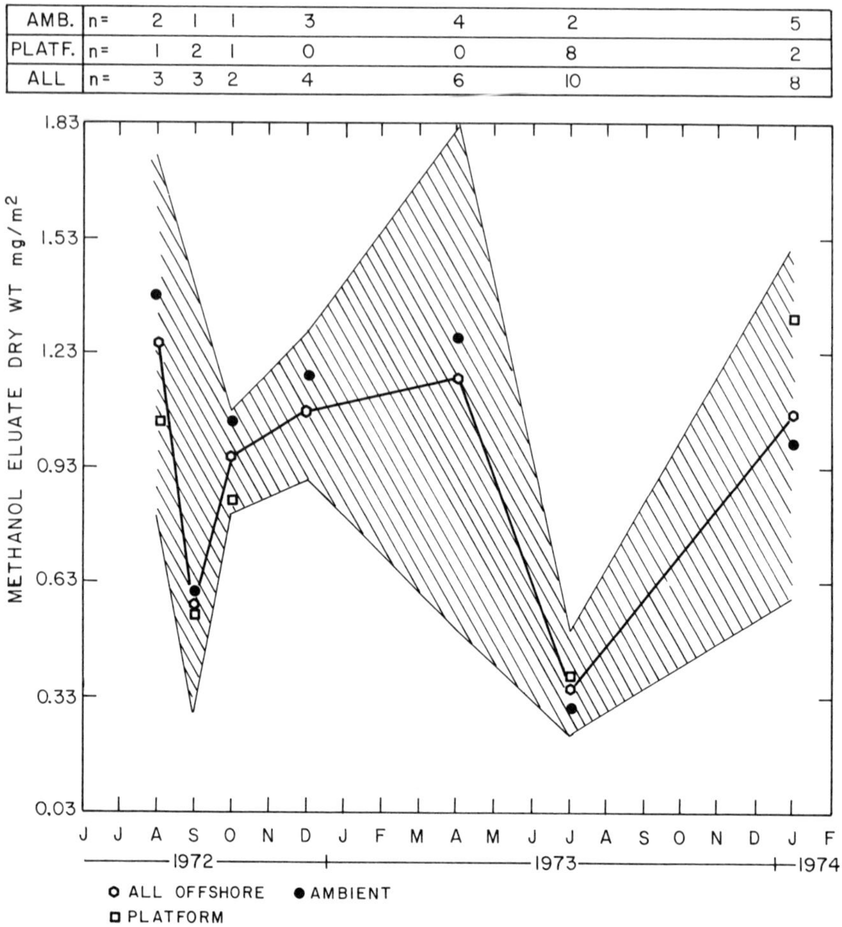

FIG. 21. SEASONAL VARIABILITY OF METHANOL ELUATE FROM SURFACE WATER SAMPLES OFFSHORE.

hydrocarbons in the air/sea interface and in sediments may differ between platform and ambient sites. The arrows in figure 23 (p. 62) indicate how to read a ternary diagram. "Platform" represents all data of all samples collected within a range of 0 to 2,500 m from 54A, spanning an azimuth of 360 degrees, which includes platform 66D. "Ambient" represents all data within a range of 2,501 to 12,500 m from 54A, spanning an azimuth of 360 degrees. The platform data tend to cluster and, with two notable exceptions, have the highest levels of benzene eluates (aromatics). Since aromatics usually disappear most rapidly with weathering of oil,

AMB.	n=	4	3	4	5	6	2	5
PLATF.	n=	4	11	2	1	1	8	2
ALL	n=	8	14	8	7	10	10	8

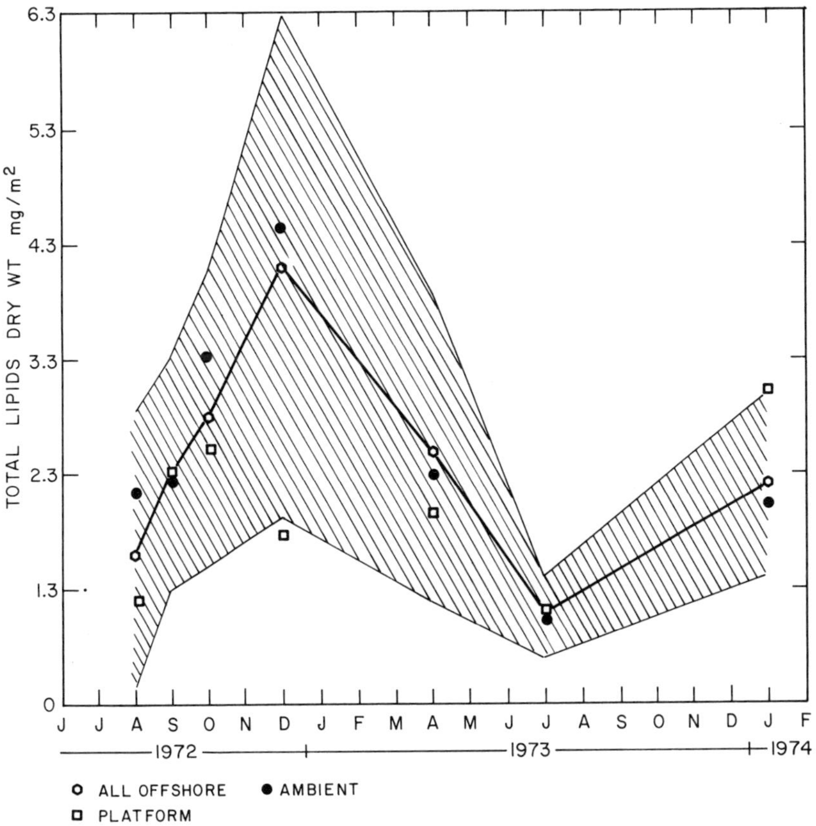

FIG. 22. SEASONAL VARIABILITY OF TOTAL LIPIDS FROM SURFACE WATER SAMPLES OFFSHORE.

this analysis suggests that the platform may be the source of fresh petroleum.

Figure 24 (p. 63) indicates that aromatic hydrocarbons do not tend to accumulate in sediments offshore, perhaps because they rapidly disappear during weathering of oil. The two samples within 750 m of 54A had the highest aromatic content. There were no samples taken within 750 m of platform 66D. All platform samples were taken within 2,500 m of 54A and ambient samples within the range of 2,501 to 12,500 m.

Data in figure 25 (p. 64) indicate that petroleum hydrocarbons present

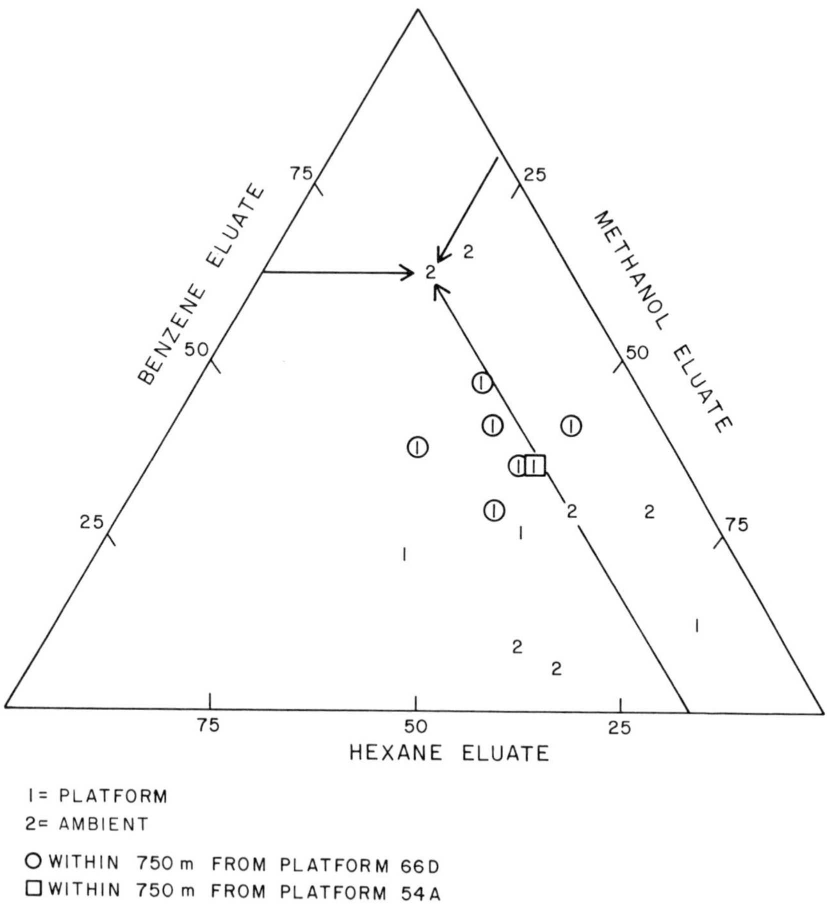

FIG. 23. TERNARY DIAGRAM OF HEXANE, BENZENE, AND METHANOL ELUATES OF SURFACE WATER SAMPLES AT PLATFORM AND AMBIENT STATIONS OFFSHORE.

in sediments are more weathered (have lower aromatic content) than those present in air/sea interface and beach samples. This diagram presents all the hydrocarbon data for all of the cruises.

In summary, we do not feel that the OEI data quantitatively demonstrate that platforms are significant sources of TOC, BOD, or hydrocarbons; on the other hand, display of the data in ternary diagrams suggests that platforms are sources of unweathered oil. The OEI data are inadequately replicated to support rigorous statistical treatment of platform versus ambient comparisons even with the redefinition of control sites.

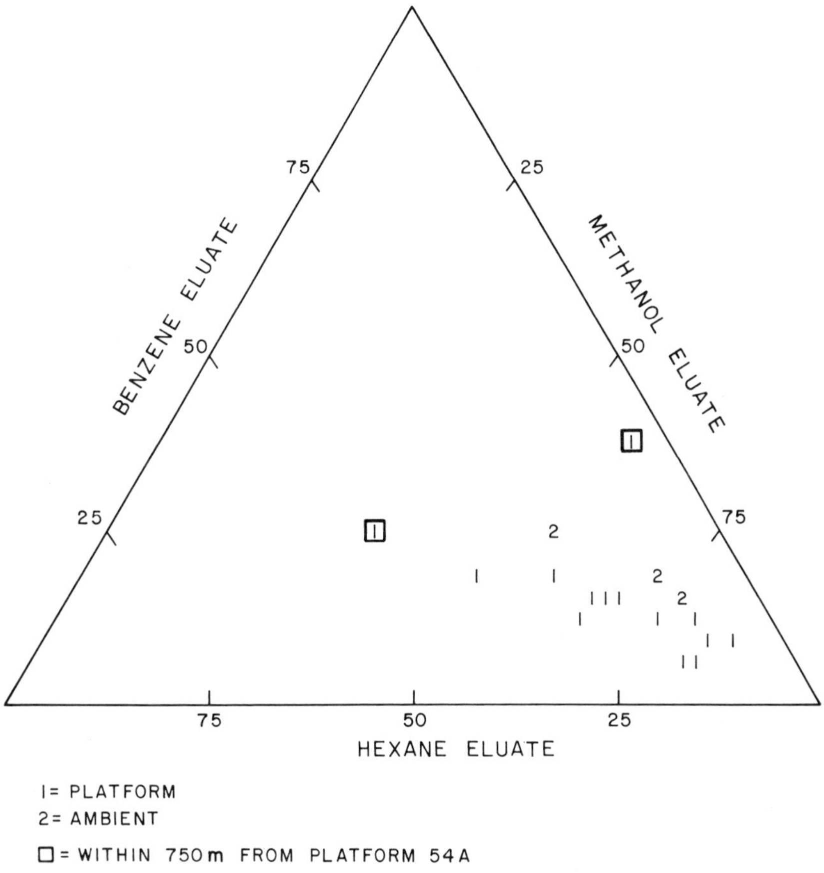

FIG. 24. TERNARY DIAGRAM OF HEXANE, BENZENE, AND METHANOL ELUATES OF SEDIMENT SAMPLES AT PLATFORM AND AMBIENT STATIONS OFFSHORE.

Timbalier Bay—Tables 6 and 7 present data for eight water quality parameters for production, workover, and ambient sites in Timbalier Bay. Data for zinc, lead, and cadmium were taken only during the first two sampling exercises. Significant differences between production or workover sites and ambient sites were determined by "t" tests.

The significant differences observed between PB1 and ambient sites and between PB2 and ambient sites are difficult to interpret without detailed analyses of flow patterns from inner to outer bay and influences from Mississippi River flow. This is illustrated for dissolved silica, orthophosphate, ammonia nitrogen, and nitrate-nitrite nitrogen during the

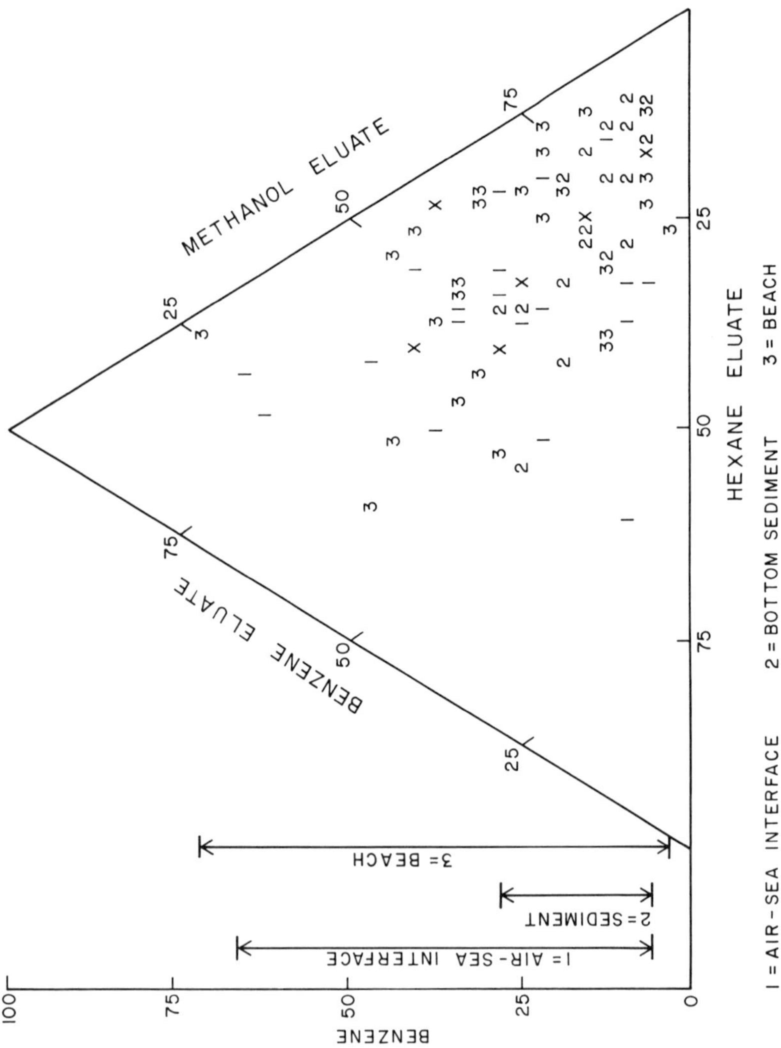

FIG. 25. TERNARY DIAGRAM OF HEXANE, BENZENE, AND METHANOL ELUATES OF ALL SAMPLES TAKEN FOR HYDROCARBON ANALYSIS.

TABLE 6. "t" test comparisons of metals concentrations in water samples taken at production, workover, and ambient sites in Timbalier Bay. (Values reported are before square root transformation.)

Date	Station	Zn (ppb)			Pb (ppb)			Cd (ppb)		
		x	s	n	x	s	n	x	s	n
Exercise 1	PB-1	8.8	14.1	15	2.1	1.9	15	0.2	0.0	15
1972 Sept.	PB-2	-	-	0	-	-	0	-	-	0
	Amb.	7.8	5.1	5	1.2	0.2	5	0.2	0.0	5
Exercise 2	PB-1	10.8	6.3	11	4.5	3.9	11	0.2	0.1	11
1972 Oct.	PB-2	10.3	6.2	17	6.9	9.8	17	0.5	0.5	17
	Amb.	11.1	7.3	30	6.6	18.5	30	0.8	1.8	30

Data were square root transformed before "t" test.

PB-1 = Production platform site: 750m x 750m area; PB-2 = Workover site: 750m x 750m area; Amb. = Ambient: rest of the Bay.

July sampling exercise, which coincided with the 100-year flood of the Mississippi River. When different, these variables were significantly lower in the inner bay from the generally distributed ambient stations. This pattern could indicate transport of nutrients from the Mississippi or enhanced flow of fresh water into the inner bay. Significant differences in organic carbon at PB1 or PB2 from the ambient stations generally reflect the gradient of organic carbon from inner to outer bay (see figure 30, p. 71). The significant difference indicated for nitrate-nitrite nitrogen during the January 1974 exercise appears impossible but resulted from the use of square root-transformed data.

Bay-Offshore Transect

The OEI included samples collected along a transect ranging from the inner part of Timbalier Bay to the offshore production and drilling platforms. Figure 26 illustrates the 13 sectors sampled during eight sampling exercises.

Figures 27 through 32 (pp. 68-73) present data for 14 water column and two sediment parameters. Water column data are means of all measurements made during all of the exercises averaged over depth. Sediment data for total phosphate and total nitrogen are averaged over all exercises. Shaded areas are one standard deviation from the means.

Salinity increases steadily from inner bay to offshore, transmissivity decreases markedly immediately outside the bay, perhaps because of littoral currents, and temperature appears to fluctuate more in the bay than offshore (figure 27). Sediment phosphate appears to be higher in the vicinity of the offshore platforms and total nitrogen decreases steadily from inner to outer bay (figure 28). Water column orthophosphate, ammonia nitrogen, and nitrate-nitrite nitrogen appear to have no definite pattern from inner bay to offshore (figure 29). Organic carbon and BOD decrease rapidly from inner bay to outer bay, but COD does not appear

TABLE 7. "t" test comparisons of water quality parameters at production, workover, and ambient sites in Timbalier Bay. (Values reported are before square root transformation.)

Date	Station	TOC[1] (mg/l)			Diss. Silica (mg/l)			O-PO_4 (µg/l)			NH_3-N (mg/l)			NO_3+NO_2 (mg/l)		
		$\bar{x}$	s	n	$\bar{x}$	s	n	$\bar{x}$	s	n	$\bar{x}$	s	n	$\bar{x}$	s	n
Exercise 1	PB-1	15.0	0.5	2	1.3	0.2	18	52.8	62.1	18	0.1	0.3	18	0.3	0.4	18
1972 Sept.	PB-2	-	-	0	-	-	0	-	-	0	-	-	0	-	-	0
	Amb.	12.5	2.5	10	1.2	0.2	8	36.0	36.8	8	0.0	0.0	8	0.6	1.4	8
Exercise 2	PB-1	-	-	0	2.0	0.5	18	169.7*	119.8	18	0.1	0.2	18	0.0	0.1	18
1972 Oct.	PB-2	7.0	4.3	12	1.4*	0.5	18	476.3	390.8	18	0.0**	0.0	18	0.0	0.1	18
	Amb.	7.3	4.0	2	2.0	0.8	14	400.3	305.8	14	0.1	0.1	14	0.1	0.1	14
Exercise 3	PB-1	14.0	1.4	2	0.1	0.0	18	133.3	137.4	18	1.2	2.3	18	5.8	8.0	18
1972 Nov.-Dec.	PB-2	7.4*	0.3	4	0.6	0.5	6	112.5	28.0	6	1.9	2.9	5	25.9	25.9	5
	Amb.	13.8	1.5	11	0.3	0.3	10	125.0	60.2	10	0.3	0.4	10	17.6	26.2	10
Exercise 4	PB-1	13.9*	4.4	13	0.2	0.2	24	111.2	28.4	24	5.9	11.1	24	10.1	19.9	24
1973 Jan.-Feb.	PB-2	-	-	0	-	-	0	-	-	0	-	-	0	-	-	0
	Amb.	9.0	5.9	13	0.1	0.1	22	109.5	43.7	22	6.6	11.8	22	7.7	14.9	22
Exercise 5	PB-1	21.5	0.0	1	0.7	0.3	18	102.8*	53.1	18	0.1*	0.0	18	0.1	0.1	18
1973 Apr.	PB-2	-	-	0	-	-	0	-	-	0	-	-	0	-	-	0
	Amb.	6.7	2.8	7	0.9	0.4	24	145.5	81.6	24	0.2	0.2	24	0.1	0.2	24
Exercise 6	PB-1	-	-	0	1.0**	0.3	28	24.4**	40.5	28	0.4**	0.1	28	0.3**	0.1	28
1973 July	PB-2	-	-	0	1.6**	0.1	8	92.9	23.3	8	0.7**	0.1	8	0.4	0.1	8
	Amb.	-	-	0	1.3	0.5	62	82.7	44.3	62	0.6	0.1	62	0.8	1.0	62
Exercise 7	PB-1	14.2	3.3	4	2.2*	0.4	30	89.4	52.0	30	0.3	0.1	30	0.1	0.0	30
1973 Oct.	PB-2	-	-	0	-	-	0	-	-	0	-	-	0	-	-	0
	Amb.	12.4	2.2	3	2.0	0.6	51	77.1	24.8	50	0.4	0.1	50	0.1	0.1	50
Exercise 8	PB-1	9.2	5.0	12	0.6	0.1	26	88.8	49.2	26	0.6**	0.2	26	0.1**	0.1	26
1974 Jan.	PB-2	5.5*	1.5	9	0.9**	0.2	18	98.9	37.9	18	0.5**	0.1	18	0.1	0.1	18
	Amb.	9.9	3.6	24	0.7	0.3	50	90.8	33.8	50	0.7	0.2	50	0.1	0.1	50

Data were square root transformed before "t" test.
* significant at 0.05 level.
** significant at 0.01 level.
1) data not normally distributed even when square root transformed. Chi-square used to test significance.
PB-1 = Production platform site: 750m x 750m area; PB-2 = Workover site: 750m x 750m area;
Amb. = Ambient: rest of the Bay.

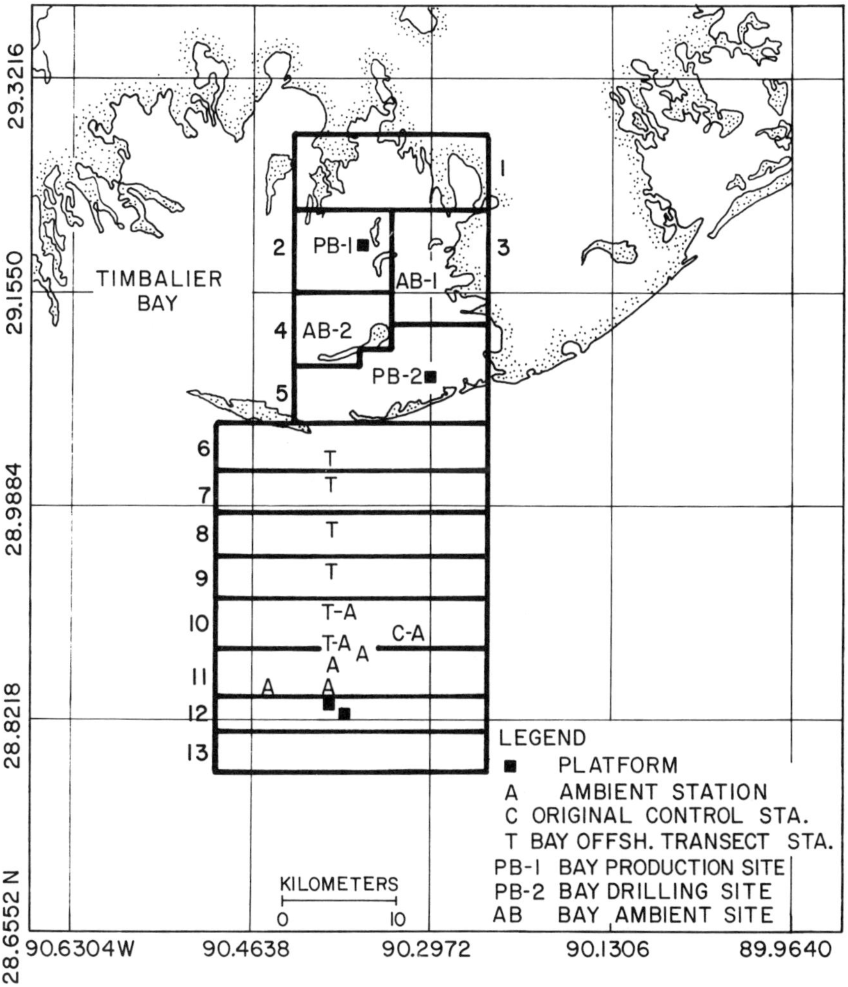

FIG. 26. SECTORS SAMPLED IN BAY-OFFSHORE TRANSECT (see figs. 27 through 32).

to be strongly related to position on the transect. Mercury appears to increase somewhat offshore, while zinc, cadmium, and lead decrease (figures 31 and 32). Lead concentrations are surprisingly high and variable at the mouth of the bay (figure 32).

With the possible exception of sediment phosphate near the offshore platform, the transect data do not demonstrate the platforms under study in the OEI to be possible sources of chronic low level discharges of pollutants.

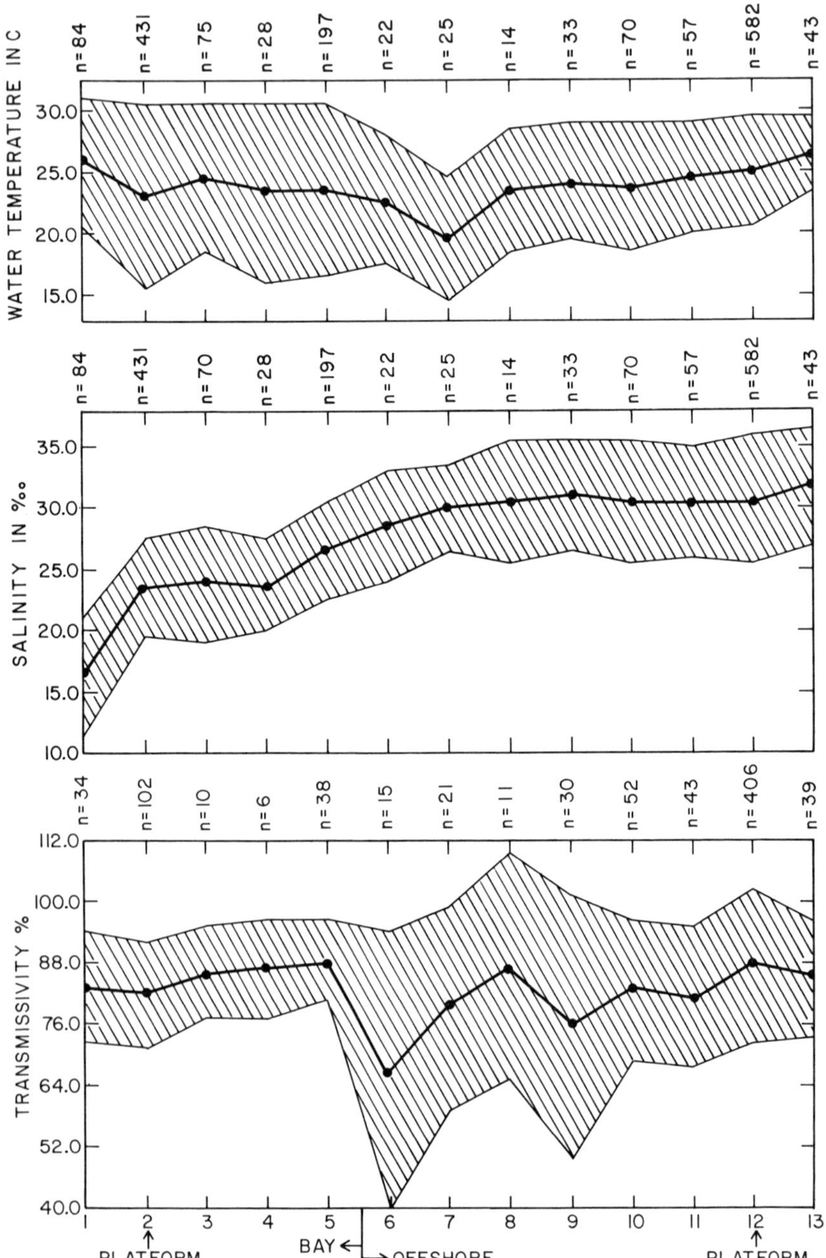

FIG. 27. VARIABILITY OF WATER TEMPERATURE, SALINITY, AND TRANSMISSIVITY FROM INNER BAY TO OFFSHORE.

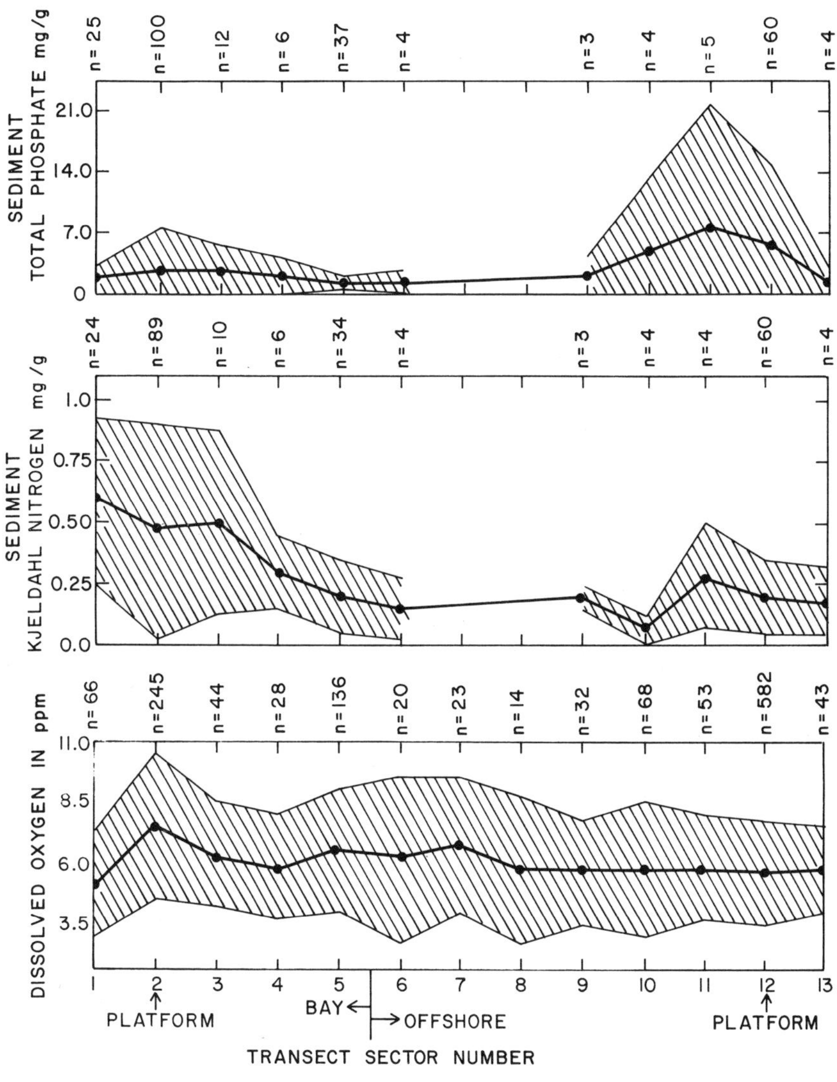

FIG. 28. VARIABILITY OF SEDIMENT PHOSPHATE AND NITROGEN AND DISSOLVED OXYGEN IN THE WATER COLUMN FROM INNER BAY TO OFFSHORE.

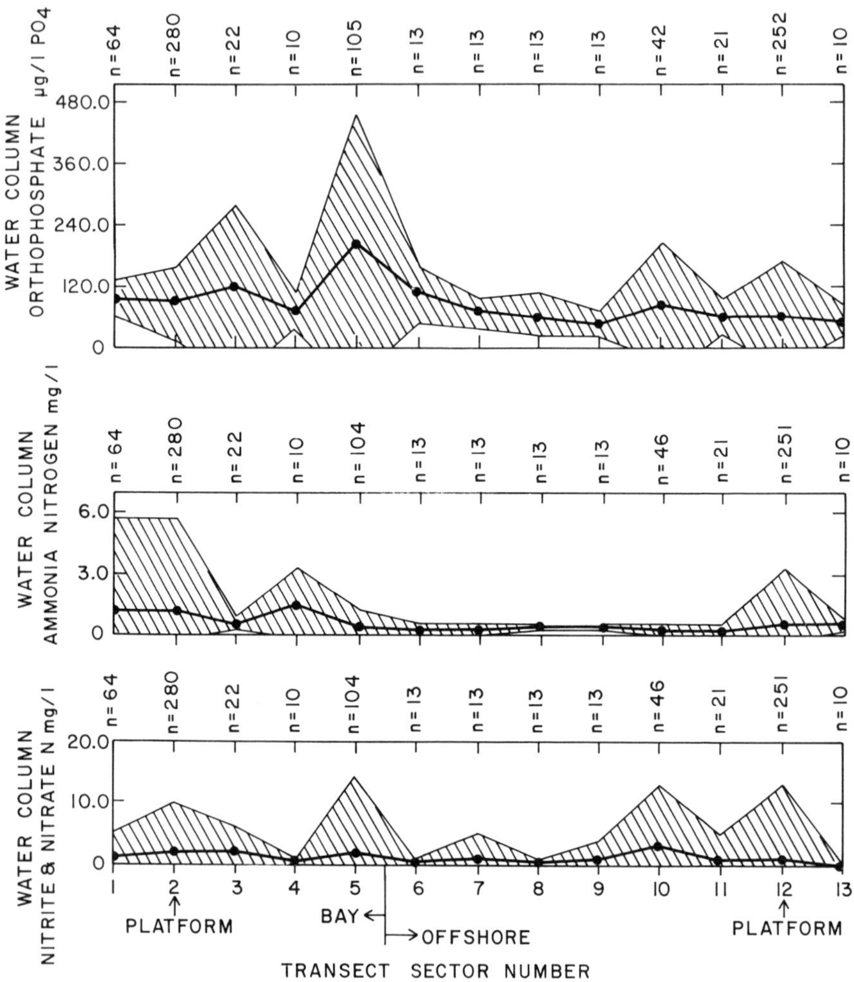

FIG. 29. VARIABILITY OF PHOSPHATE, AMMONIA NITROGEN, AND NITRITE-NITRATE NITROGEN IN THE WATER COLUMN FROM INNER BAY TO OFFSHORE.

Benthic biology

The benthic biology program of the OEI in 1972-1974 consisted of separate investigations in the following disciplines: Timbalier Bay —polychaetous annelids (Kritzler 1974; Fish et al. 1974), molluscs and crustaceans (Perry 1974; George 1974; Farrell 1974a), fish (Perry 1974), foraminifera (Ostrom 1974; Fish et al. 1974), and algae (Humm 1974).

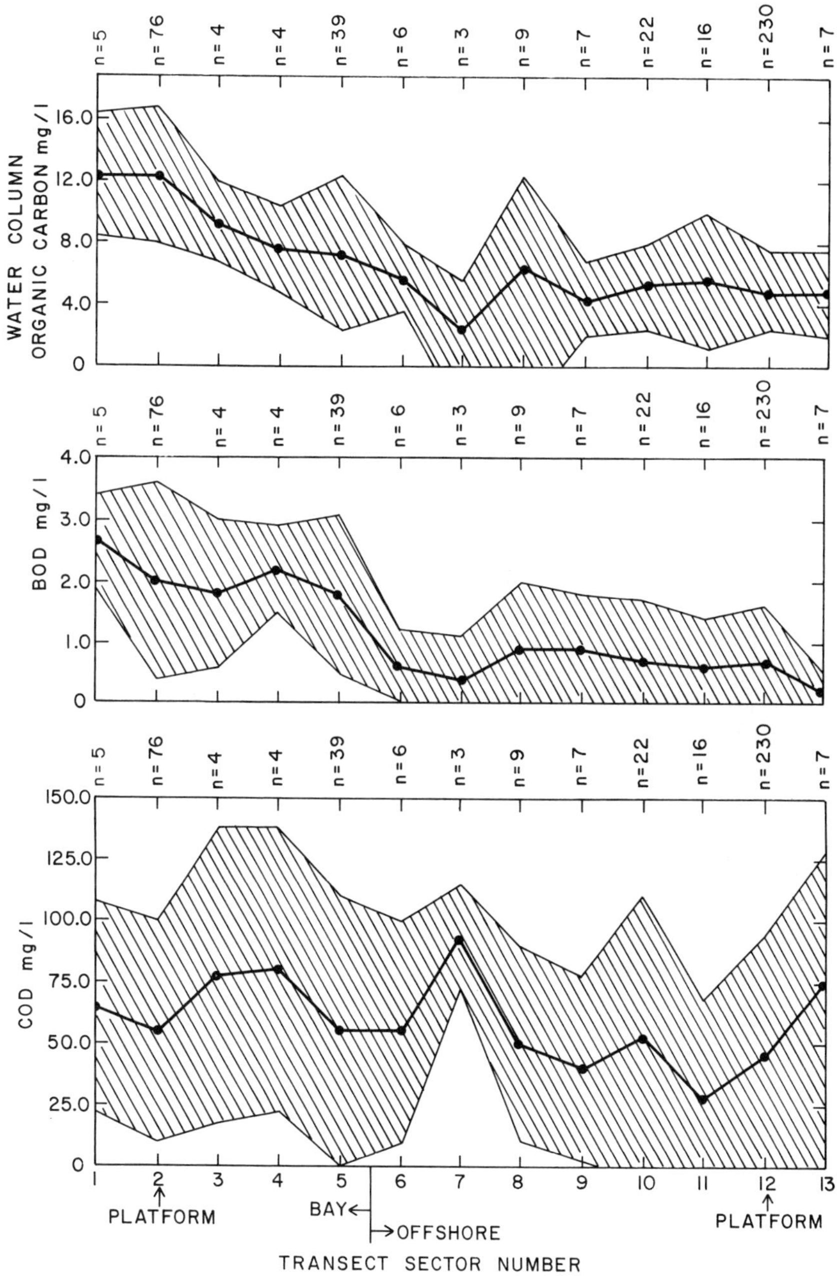

FIG. 30. VARIABILITY OF ORGANIC CARBON, BOD, AND COD IN THE WATER COLUMN FROM INNER BAY TO OFFSHORE.

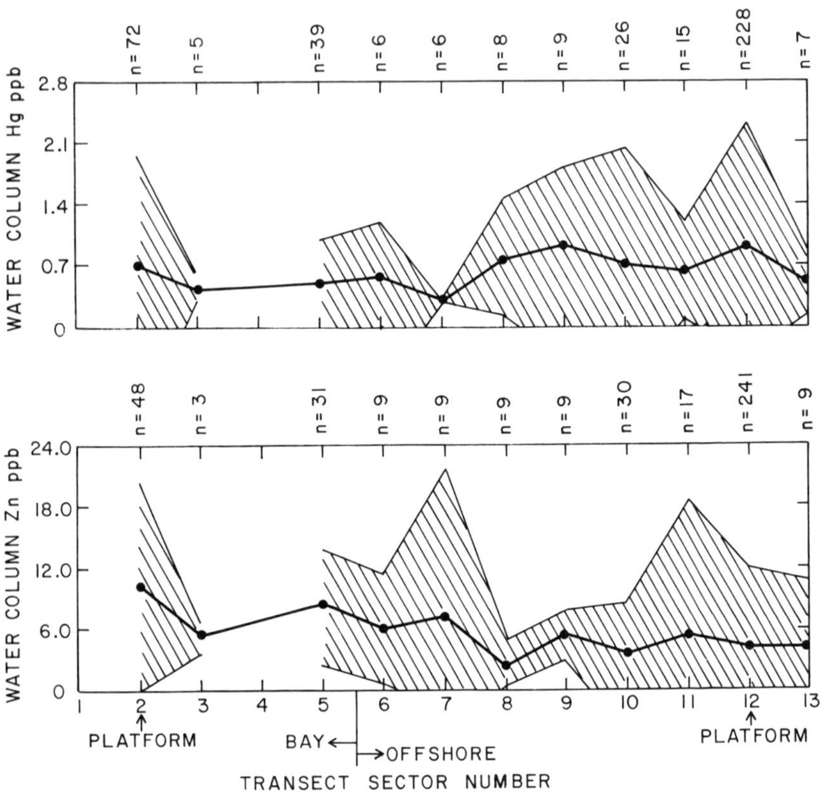

FIG. 31. VARIABILITY OF MERCURY AND ZINC IN THE WATER COLUMN FROM INNER BAY TO OFFSHORE.

Offshore biological data consisted of molluscs and crustaceans (Waller 1974; Farrell 1974b), fish (Perry 1974), algae (Humm 1974), and fouling organisms (George and Thomas 1974). Because of the varied methods of data collection used in these studies, it was not possible to analyze quantitatively all the data generated by these investigations. It was only possible to analyze the polychaete, mollusc, and crustacean data with computer-assisted programs. Because of the importance of certain other organisms to the understanding of environmental changes, however, some salient points dealing with the benthic foraminifera, algae, and fouling organisms will be discussed here.

Benthic organisms, especially infaunal species, have been considered useful in assessing environmental conditions because they have been

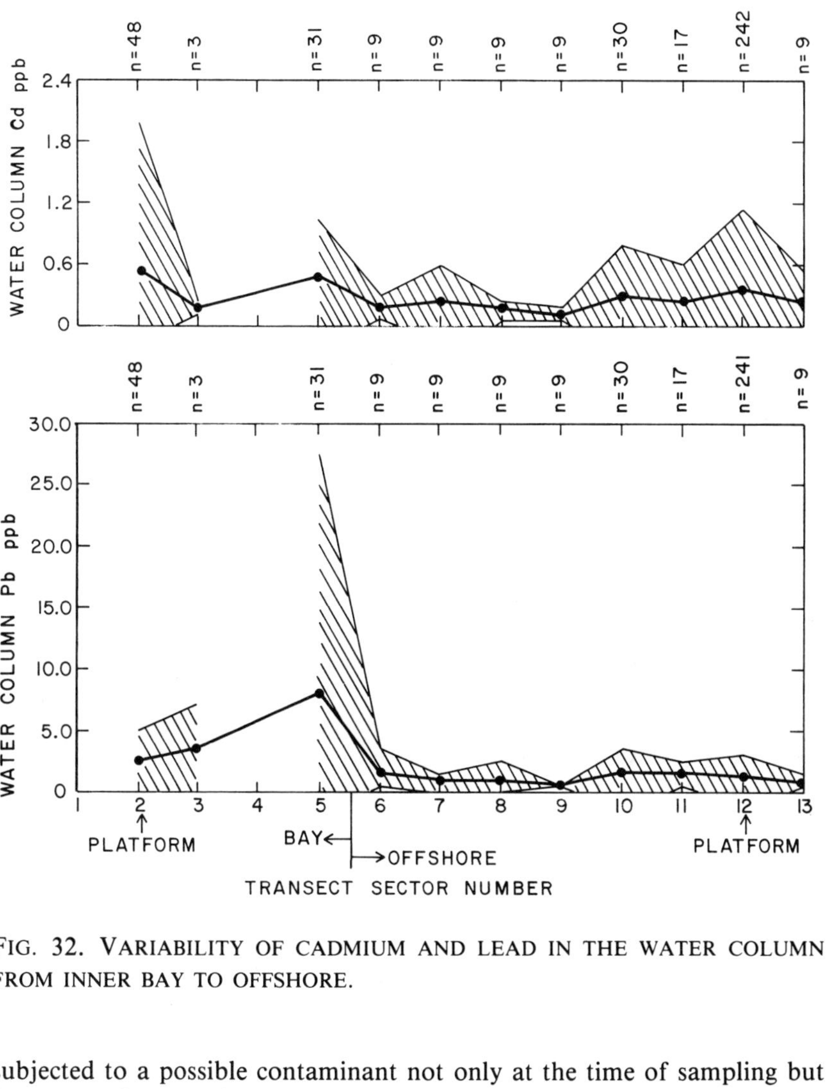

FIG. 32. VARIABILITY OF CADMIUM AND LEAD IN THE WATER COLUMN FROM INNER BAY TO OFFSHORE.

subjected to a possible contaminant not only at the time of sampling but for some time previously (Reish 1973). In contrast, pelagic organisms may have entered the particular area just prior to capture and may not have been previously exposed to the contaminant. Polychaetes, molluscs, and crustaceans are the most abundant groups of macroinvertebrates in the marine benthos; they constitute approximately 90% of the species and specimens present within a benthic sample (Jones 1969). It is therefore imperative to include the identification of these groups of species in the assessment of the environmental condition of any marine area. Since a large amount of data on these benthic groups in Timbalier Bay was generated but not synthesized, it was apparent that this appraisal should include a synthesis of the polychaete, mollusc, and crustacean occurrences and distributions. No previous studies dealing with the benthic polychaetes, molluscs, and crustaceans within the OEI study area had been undertaken.

Timbalier Bay Macroinvertebrates

Data Base—Polychaetes were collected five times at four stations during 1973. Samples were taken with a plastic tube, which sampled an area of 625 cm^2 to a depth of 10 m. The sampler was connected to an air hose, which lifted the material into the boat, where it was washed through a 0.5 mm mesh sieve; the material retained on the sieve was preserved (Kritzler et al. 1974). Sixteen replicates were taken at each station, giving a total surface area of 1 m^2. Two stations were located near oil operations (production PB1 and drilling PB2), and two were located as far as possible from such activity to represent natural environmental conditions (AB1 and AB2) (table 8, figure 33). The data from the winter, spring, and summer collections were coincident in time and space with mollusc and crustacean data.

Molluscs and crustaceans were taken with a Van Veen grab, which sampled an area of 2000 cm^2 to a depth of 10 cm at seven stations within Timbalier Bay. Collections were taken during winter, spring, and summer in 1973. Since one of the stations was located in the Little Lake area, which was far from any polychaete sampling station, the data generated at that locality were not included in these analyses (table 8, figure 33). Samples were washed through a 0.5 mm mesh sieve and preserved. Since the Van Veen sampled an area of 0.2 m^2, the data, for a number of specimens collected, was multiplied by five in order to coincide with the total surface area sampled for polychaetes.

Methods of Analysis—The combined data for polychaetes, molluscs, and crustaceans were grouped according to the two platforms and two ambient stations (table 8, figure 33) for the winter, spring, and summer collections. The data were analyzed for species diversity (Shannon and Weaver 1963) and faunal affinities (Morisita 1959; Ono 1961), and the stations were grouped by cluster analysis.

The cluster analysis program in this study was the Bray-Curtis similarity coefficient and group average sorting (Boesch 1977). The similarity (or dissimilarity) coefficient can be expressed as:

$$D_{jk} = \frac{i\,(x_{ij} - x_{ik})}{1\,(s_{ij} + x_{ik})}$$

TABLE 8. Corresponding station locations for analyses of polychaete, mollusc, and crustacean data, Timbalier Bay, 1973.

Sampling Site	Combined Station Number	Polychaete Stations			Mollusc-Crustacean Stations		
		Number	Latitude	Longitude	Number	Latitude	Longitude
Production Platform	PB-1	PBO3	29°11.90'	90°21.50'	7	29°11.47'	90°21.62'
					9	29°11.88'	90°21.43'
Ambient	AB-1	BCO2	29°09.50'	90°18.30'	12	29°09.70'	90°17.50'
Ambient	AB-2	BPO1	29°07.79'	90°24.16'	6	29°07.59'	90°23.96'
Drilling Platform	PB-2	GF13	29°04.93'	90°19.13'	3	29°03.72'	90°21.26'
					4	29°04.16'	90°23.35'

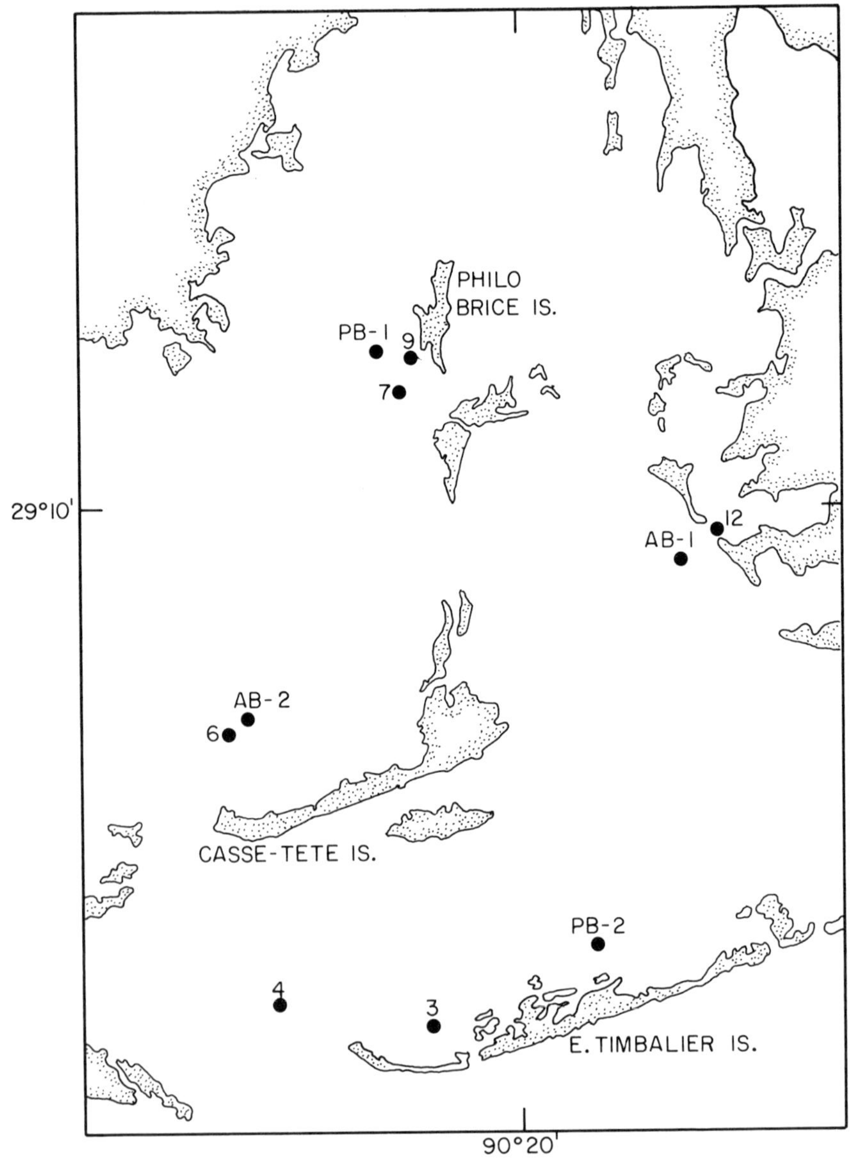

FIG. 33. BENTHIC SAMPLING STATIONS IN TIMBALIER BAY.

where D_{jk} is the distance between entities j and k; x_{ij} and x_{ik} are the values of attribute i in entities j and k, respectively. The group averaging sorting strategy was used to form the dendrogram for each of the four stations for the three collecting periods. In the clustering used herein, the numbered value ranges from 0 to 1 with the 0 indicating complete similarity and 1 indicating complete dissimilarity. The calculations are made with the aid of a computer-assisted program and are then summarized graphically in a dendrogram. Species that were encountered only five or fewer times in the entire study were excluded from this analysis.

Results—The data on benthic fauna were classified according to the Shannon-Weaver index of species deversity (H ′) and are summarized in table 9. Because of the dominance of the polychaete *Spiochaetopterus oculatus*, the indices are given with and without the data for this species. The H ′ is also given with the mollusc and crustacean data multiplied by five, with and without the *Spiochaetopterus* data. H ′ ranged from 0.79 to 3.94 when the data for *Spiochaetopterus* were included. Two groups of H ′ values were noted; one varied from 3.40 to 3.94 at the production station PB1 and ambient station PB2, and the other varied from 0.79 to 1.40 at the ambient station AB1 and the drilling station PB2. When the number of specimens of molluscs and crustaceans was multiplied by five, the H ′ index ranged from 0.80 to 4.12; the separation of the two groups of stations was evident but not as marked. Exclusion of the *Spiochaetopterus* data altered the H ′ index. It ranged from 3.17 to 4.42 when the mollusc-crustacean data were not multiplied by a factor of five and from 2.50 to 4.31 when they were. The dominance of *Spiochaetopterus* at these stations within Timbalier Bay is apparent from table 9. The H ′ values of ambient station AB1 and drilling site PB2 are influenced by the large quantity of this species at these two sites in comparison to the other two. No relationship between the H ′ value and the season of collection is apparent from these data.

The Morisita-Ono index of faunal similarity (λ) among the four groups of stations is summarized according to the period of collection and the composition of the data in figures 34 and 35. Indices of similarity were calculated utilizing four different data sets, as was done for the Shannon-Weaver index. The occurrence of the dominant species by station, by data, by total number of specimens, and by percentage of occurrence is included in table 10 (p. 82).

The index of similarity varies from 0.28 to 0.99 when the *Spiochaetopterus* data are included in the analyses, and from 0.13 to 0.87 when the data for this polychaete are excluded. The greatest similarity is noted for all three collecting periods between the ambient station AB1 and the drilling station PB2 when the *Spiochaetopterus* data

TABLE 9. Species Diversity of Polychaetes, Crustaceans, and Molluscs in Timbalier Bay, 1973.

(Shannon-Weaver H')

Station Number	Number of Species	Number of Specimens	No. of Specimens excluding *Spiochaetopterus*	H'	H' excluding *Spiochaetopterus*	H' X 5 Mollusc and Crustacea	H' excluding *Spiochaetopterus* and X 5 Mollusc and Crustacea
	Winter 1973						
Production Platform							
PB-1	30	361	248	3.45	3.72	2.99	2.83
Drilling Platform							
PB-2	40	21[illegible]1	259	1.04	4.20	1.46	3.28
Ambient							
AB-1	42	4670	460	0.79	3.39	0.80	3.42
AB-2	39	486	416	3.40	3.28	3.33	3.18

Spring 1973							
Station Number	Number of Species	Number of Specimens	No. of Specimens excluding *Spiochetopterus*	H'	H' excluding *Spiochaetopterus*	H' X 5 Mollusc and Crustacea	H' excluding *Spiochaetopterus* and X 5 Mollusc and Crustacea
Production Platform							
PB-1	34	709	501	3.58	3.83	3.65	3.85
Drilling Platform							
PB-2	56	3954	671	1.40	4.42	2.60	3.93
Ambient							
AB-1	35	4961	645	0.96	3.17	1.06	3.26
AB-2	39	344	288	3.74	3.71	2.68	2.50

Summer 1973							
Station Number	Number of Species	Number of Specimens	No. of Specimens excluding *Spiochaetopterus*	H'	H' excluding *Spiochaetopterus*	H' X 5 Mollusc and Crustacea	H' excluding *Spiochaetopterus* and X 5 Mollusc and Crustacea
Production Platform							
PB-1	25	515	401	3.47	3.47	3.53	3.54
Drilling Platform							
PB-2	28	1993	285	1.07	3.41	1.50	2.59
Ambient							
AB-1	36	3133	580	1.30	3.30	1.31	3.33
AB-2	41	333	254	3.94	4.13	4.12	4.31

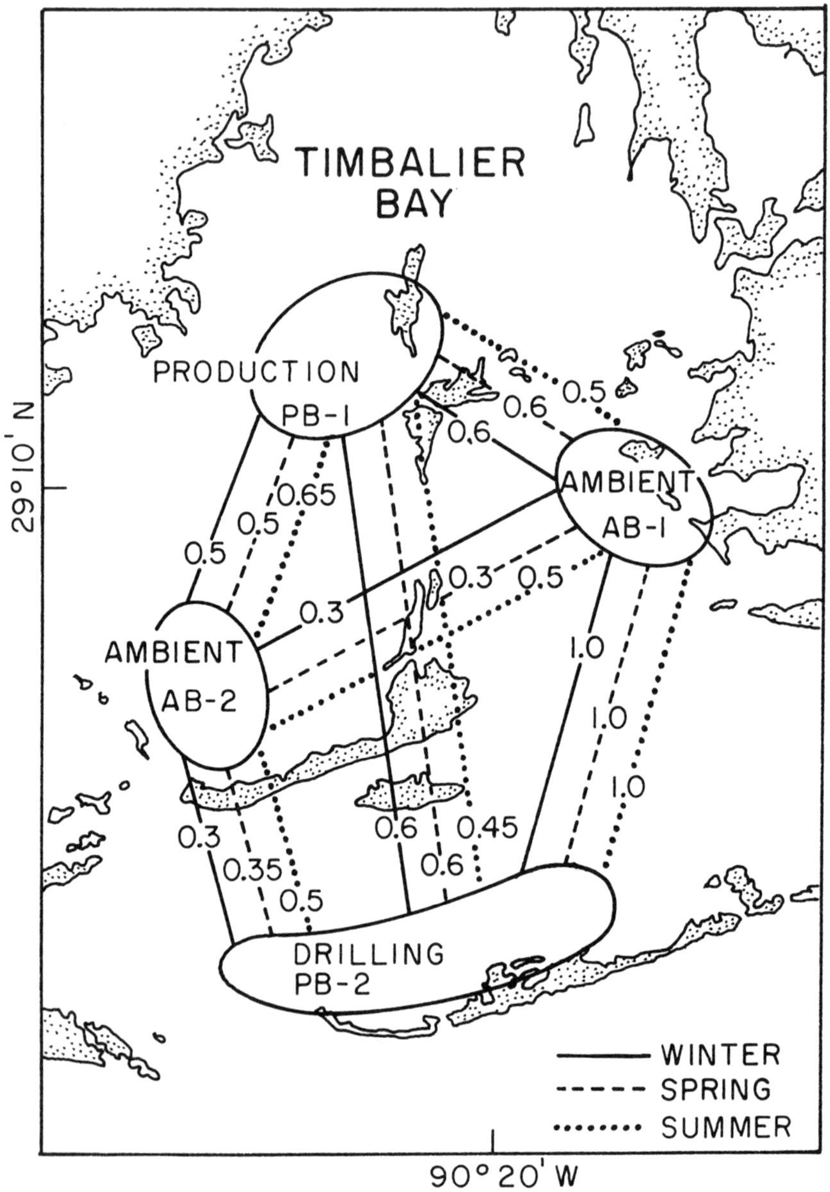

FIG. 34. DIAGRAMMATIC REPRESENTATION OF THE MORISITA-ONO INDEX OF FAUNAL SIMILARITY IN TIMBALIER BAY (from Bender et al. 1979; Copyright, Offshore Technology Conference, 1979; reproduced by permission).

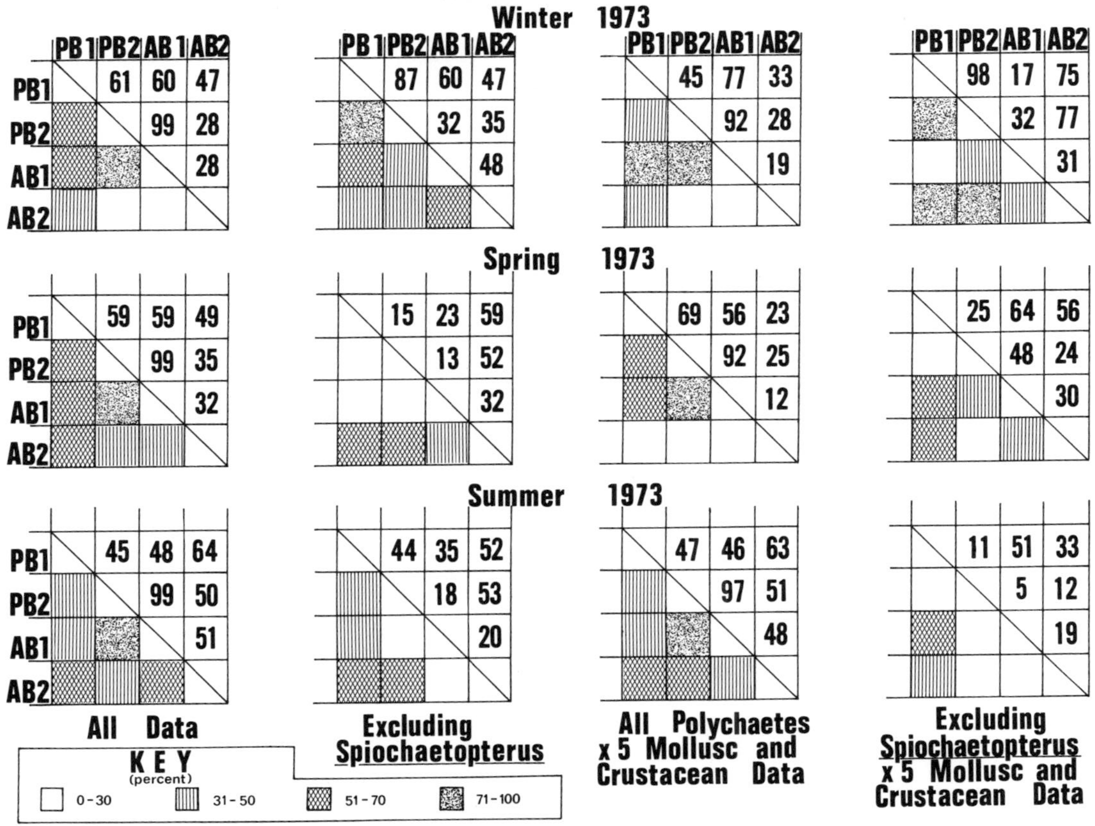

FIG. 35. TRELLIS REPRESENTATION OF THE MORISITA-ONO INDEX OF FAUNAL SIMILARITY (× 100) IN TIMBALIER BAY.

TABLE 10. Dominant species of benthic invertebrates as measured by percent occurrence, Timbalier Bay, 1973.

Species[1]	Station and Collection Period												Total Number of Specimens	Percent Occurrence
	PB-1			AB-1			AB-2			PB-2				
	W	Sp	Sum	W	Sp	Sum	W	Sp	Sum	W	Sp	Sum		
Spiochaetopterus oculatus	X	X	X	X	X	X	X	X	X	X	X	X	19,502	100
Paraeurythoe americana	X	X	X	X	X	X	X	X	X	X	X	X	604	100
Lumbrineris parvapedata	X	X	X	X	X	X	X	X	X	X	X	X	447	100
L. bassi	X	X	X	X	X	X	X	X	X	X	X	X	259	100
Glycinde polygnatha	X	X	X	X	X	X	X	X	X	X	X	X	153	100
Harmothoe trimaculata	X	X	X	X	X	X	X	X	X	X	X		670	92
Heteromastus filiformis	X	X	X	X	X	X		X	X	X	X	X	181	92
Notomastus latericeus	X	X		X	X	X	X	X	X	X	X	X	145	92
Magelona cincta		X	X	X	X	X	X	X	X	X	X	X	61	92
Sigambra tentaculata	X	X	X	X	X	X	X	X	X		X	X	43	92
Gyptis capensis	X	X	X	X	X		X	X	X	X	X	X	34	92
Prionospio sexoculata	X	X	X		X	X		X	X	X	X	X	243	83
Euclymene sp.	X	X	X	X	X	X	X		X	X	X		84	83
Polydora socialis	X	X		X	X	X	X	X	X		X	X	42	83
Myriochele bioculata	X	X	X	X	X		X	X	X	X			367	75
Mulinia lateralis	X[2]	X[2]			X		X	X	X	X[2]	X[2]	X[2]	299	75
Cossura hoodi	X	X	X	X	X	X	X	X	X				77	75
Aedicira pallida			X	X		X	X	X	X	X	X	X	44	75
Neanthes succinea			X	X		X	X	X	X	X	X	X	30	75
Paraprionospio pinnata	X	X	X	X	X	X			X			X	81	67
Ancistrosyllis hamata	X		X		X	X	X	X	X	X			15	67

1) All species are polychaetous annelids except for the pelecypod *Mulinia lateralis*.
2) Data included for two stations in the vicinity.

TABLE 11. Principal living foraminifera species of Timbalier Bay, 1973.

Species	Intertidal			Subtidal
	Mud	Sand	All Substrates	
Ammoastuta inepta	X			
Ammonia b. var. *parkinsoniana*			X	X
A. b. var. *tepida*			X	X
Ammotium dilatatum	X			
A. salsum	X			X
Arenoparella mexicana	X			
Bolivina striatula				X
Elphidium gunteri			X	X
E. matagordanum		X		X
E. poeyanum		X		X
E. vadescens				X
Haplophragmoides manilaensis	X			
H. wilberti	X			
Nonionella atlantica		X		
N. opima		X		
Palmerinella gardenislandensis				X
Quinqueloculina compta		X		
Q. funafutiensis		X		
Q. lamarckiana		X		
Q. rhodiensis		X		
Q. tenagos				X
Trochammina inflata	X			

are included. Regardless of which data are used in the analyses of index of similarity, the figure is less than 0.7 at least 75% of the time. While there were 11 species of polychaetes that were present in at least 90% of the collections (table 11), the dominance of *Spiochaetopterus* at all stations in all collections reduces the index values when data for this polychaete are excluded. For example, the high index of similarity between ambient station AB1 and drilling station PB2, which was essentially 1.0 at all three dates, dropped to between 0.18 and 0.32 when the data for *Spiochaetopterus* were excluded.

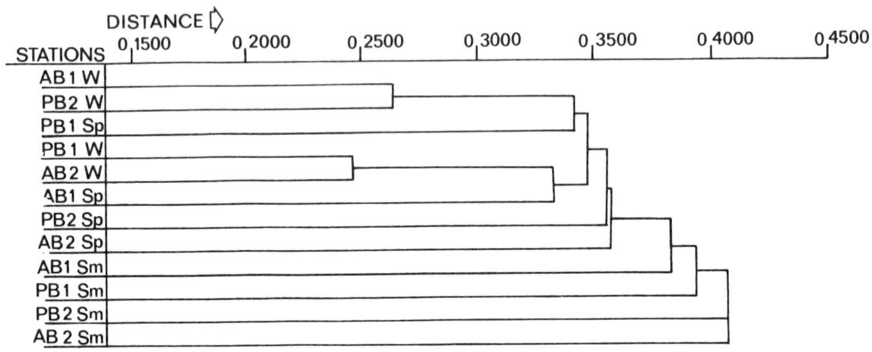

FIG. 36. DENDROGRAM FROM THE CLUSTER ANALYSIS OF BENTHIC BIOLOGICAL DATA FROM TIMBALIER BAY.

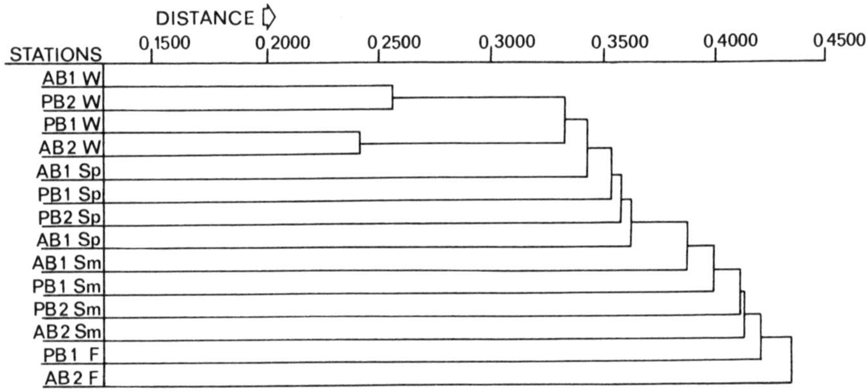

FIG. 37. DENDROGRAM FROM THE CLUSTER ANALYSIS OF THE POLYCHAETE BENTHIC DATA FROM TIMBALIER BAY.

The index of similarity does not seem to indicate any marked seasonal difference in the benthic populations. There is a slight indication of greater similarity between stations in summer than in winter. This can be noted particularly in the comparisons of ambient station AB1 with the other stations.

The benthic biological results in Timbalier Bay were subjected to cluster analysis using all polychaete, crustacean, and mollusc data (figure 36), and using all polychaete data (figure 37). The combined data included collections made in winter, spring, and summer 1973, and the polychaete data included collections made during the same seasons plus two samplings taken in fall 1973. With the combined data for the three animal groups, three distinct assemblages are noted, which coincided, in part, with the seasons (figure 36). The four summer stations fell out as a group. The four winter stations were grouped with two spring stations. The two remaining spring stations were together in one group. No benthic biological differences existed between the platform stations and the ambient stations when this method of analysis was used.

The dendrogram drawn from the polychaete data showed a separation of the stations according to the season (figure 37). The four winter collections were separated as a group with two subdivisions. The remaining 10 collections were grouped according to season but were added one-by-one in a step-wise cluster.

Sample Replication—In conducting his study of the polychaetes of Timbalier Bay, Kritzler (1974) made 16 replicates at each station during each collection period. Each replicate was processed separately, and the data were presented separately in individual reports. Although not all replicates and stations were analyzed, the data presented afford a unique opportunity to determine how many replicates one should take at a site to characterize that particular locality adequately. These data based on 14-16 replicates at 14 stations were plotted in figure 38 according to range, mean, and standard deviation for those polychaete species represented six or more times when the data from all replicates were combined. This latter exclusion of some data was based on the elimination of species with five or fewer specimens from cluster analysis (Boesch 1977).

It is apparent from both graphs in figure 38 that with each additional replicate, the ranges and standard deviations become less. It was necessary to take 10 replicates in order to gather 90% of the polychaete species present within an area, but when those species present fewer than six times were excluded, the 90% recovery was reached by the third replicate. Apart from the mathematical interpretation of these data, one must also consider the effort and money involved in sorting and identifying the organisms from the additional replicates. Each replicate requires essentially the same amount of time for identification and enumeration. Do we achieve, in this instance, over three times the amount of information for analyzing 10 samples rather than three? We think not. Since pelagic larvae are common in polychaetes, larval settlement can occur over wide areas, including those not necessarily within their optimum environment. Undoubtedly, we are encountering many such species within Timbalier Bay. Of the 82 species of polychaetes reported by Kritzler from Timbalier Bay, 39, or 48%, of the species encountered were represented by fewer than 6 specimens. Many of these species are more common offshore; but because of the absence of knowl-

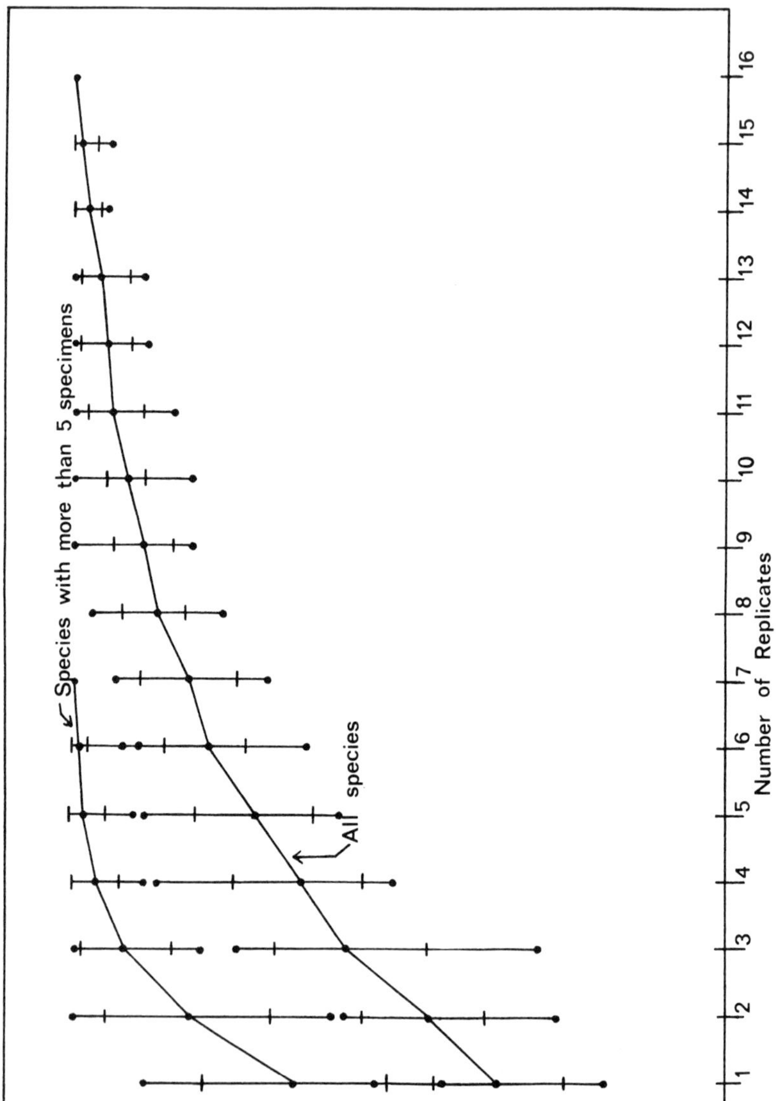

FIG. 38. NUMBER OF POLYCHAETE SPECIES PER REPLICATE EXPRESSED AS A PERCENTAGE OF THE TOTAL.

edge of the quantitative distribution of most of these species, the number of typically offshore inhabitants settled in Timbalier Bay is unknown.

Timbalier Bay Foraminifera

The intertidal benthic Foraminifera population in Timbalier Bay was determined by Fish et al., and the subtidal by Ostrom. A total of 59 living species was collected from the bay in these two studies (Appendix 1, p. 103). No additional analyses of these data have been carried out in connection with this re-examination; because Foraminifera are so important in oil exploration, however, and because these organisms have been utilized as indicators of stressed conditions in the environment (Shafer 1973), a generalized synthesis of these reports is presented here. In addition, Foraminifera are the only group of benthic organisms for which prior data for Timbalier Bay were available (Waldron 1957). Only the living species are discussed here.

The presence or absence of a particular species of Foraminifera in Timbalier Bay was found to be dependent upon the type of substrate, the salinity, and the tidal level. In general, the sediments became coarser in texture as one moved from the inner bay to the entrance. The salinity also increased towards the entrance of the bay.

The subtidal benthos in Timbalier Bay was dominated by 10 living species of Foraminifera, all of which were distributed throughout the bay. Generalized peaks in distribution were noted in some of these species. Among those species that reached population maxima in the inner reaches of the bay were *Ammonia beccarii* var. *parkinsoniana*; *Ammonia beccarii* var. *tepida*, which peaked near ambient station AB1; and *Elphidium vadescens*, which peaked near production station PB1. Middle bay primary inhabitants included *Elphidium matagordanum*, *E. poeyanum*, and *Quinqueloculina tenagos*. Principal inhabitants of lower bay areas included *Ammotium salsum, Elphidium gunteri,* and *Palmerinella gardenislandensis*. *Bolivina striatula* is an offshore inhabitant, but some living specimens were found throughout the bay.

Many of the principal intertidal species were also dominant members of the subtidal benthos (table 11). While these species were widely distributed throughout the intertidal reaches of the bay, the majority of the species were more commonly present in association with a particular substrate type. The intertidal mud association was dominated by seven species of Foraminifera, one of which, *Ammotium salsum,* was also an important species in the subtidal zone. Eight species were the principal components of the sand association, of which *Elphidium matagordanum* and *E. poeyanum* were important in the subtidal environment. Three species, *Ammonia beccarii* var. *parkinsoniana*, *A. b.* var. *tepida*, and

Elphidium gunteri, were present in all intertidal collections, as well as being important components of the subtidal foraminiferan fauna.

In summary, a knowledge of the living foraminiferan populations can be useful in indicating environmental conditions. It has been shown by others (e.g., Shafer 1973) that living species composition is a useful indicator of varying degrees of environmental stress near domestic sewage outfall discharges. While the relationship of foraminiferan species to oil spills or oil refinery discharges has not been established, it is reasonable to assume that some type of correlation can be expected. Comparisons of the Foraminifera data reported by Ostrom (1979) and Inabinet and Fish (1979) to the earlier work by Waldron indicate that there has been no appreciable change in the foraminiferan fauna in Timbalier Bay in the past two decades.

Marine Algae

Qualitative comparisons of the benthic algae of Timbalier Bay and offshore waters are indicated in Appendix 1. Numerically, there were 28 and 40 species of Chlorophyta, 26 and 22 species of Cyanophyta, 12 and 20 Phaeophyta, and 18 and 38 Rhodophyta in the bay and offshore area respectively. The reduction in the number of species of Chlorophyta, Phaeophyta, and Rhodophyta in Timbalier Bay is probably related to the lower salinities and higher turbidities (Humm and Bert 1979). The environmental difference noted in the offshore area is similar to one that shows the influence of the Mississippi River plume.

Twenty-three species of algae were present within the bay but not offshore, whereas 55 species were restricted to offshore habitats. Fewer species of algae, especially blue-greens and reds, were collected from those offshore platforms within or near the Mississippi River plume than in Timbalier Bay. The four distant offshore platforms studied were rich in greens, browns, and reds.

The use of benthic algae in assessing potential damage as a result of oil drilling and production activities is of limited value unless it is known what species were present before operations began. The use of algae as a means of assessing environmental damage as a result of an oil spill would be valuable in determining the latitudinal, longitudinal, and bathymetrical extent of any destruction. Because of the influence of such environmental parameters as salinity and turbidity, particular care should be exercised in making such comparisons; platforms must be within similar conditions in order to draw valid conclusions.

On the basis of data collected during the OEI program and algal studies previously conducted in Louisiana (Humm and Bert 1979), we have identified no significant adverse effect upon the algae of Timbalier Bay or offshore waters during the past 35 years of operations.

Offshore Oil Platform Fouling Organisms

The occurrence and bathymetric distribution of fouling organisms on offshore oil platforms were studied by the examination of existing populations on platform legs and by the suspension of test panels at various depths for a 60-day period. Knowledge of the existing population on the platform legs is particularly valuable in assessing the damage to marine organisms if there is an oil spill or other environmental stress. Because such organisms exist from the intertidal zone to the sediment line, it is also possible to measure the extent of damage due to such a stress. Since the larger and more conspicuous fouling organisms are attached species, they serve as indicators not only of present environmental conditions, but also of previous ones, much the same as benthic species serve as indicators of bottom environmental conditions.

Because the data collected from platform 54A are largely qualitative, it is not possible to analyze them with a computer-assisted program. The purpose of this supplemental analysis is to discuss the importance of this biological community as a means of monitoring the environmental conditions of the water mass surrounding an oil platform.

The dominant species present on platform 54A's legs are diagrammatically represented in figure 39. Beginning at the mean low tide level, the green alga *Enteromorpha* sp. dominates the upper 1 m of the leg, with the amount rapidly diminishing below the upper 5-10 cm of growth. The barnacles *Balanus reticulatus* and *B. improvisus* and the hydroid *Syncoryne* sp. are interspersed throughout the *Enteromorpha* zone. Below 2 m, *Enteromorpha* is replaced by the sea anemone *Aiptasia* sp., and there are larger and more frequent occurrences of the two species of barnacles. Between the depths of 7.5 and 12 m, the barnacles diminish in numbers and the hydroid *Bouganvillia tennella* dominates the association. *Balanus reticulatus* is the principal barnacle species, while *B. tintinnabulum* and *B. improvisus* are only infrequently present. Below 12 m, hydroids such as the *Bouganvillia* dominate the fouling material; a few serpulid polychaete worm tubes are also present.

Many additional species of marine invertebrates are present within the masses of dominating fouling organisms. Such organisms may either be free-living or attached forms, such as sponges, nematodes, polychaetes, crustaceans, and molluscs. Many may occur seasonally, so a knowledge of the seasonal occurrence of these smaller species is essential before such data can be utilized in environmental assessment.

Suspension of test panels for a sufficient length of time has been useful in assessing environmental conditions (Reish 1971). Because succession may occur, however, the species composition may not necessarily correspond with that present on a platform leg at a given depth. The chief value of the test panels is that they ease collection and

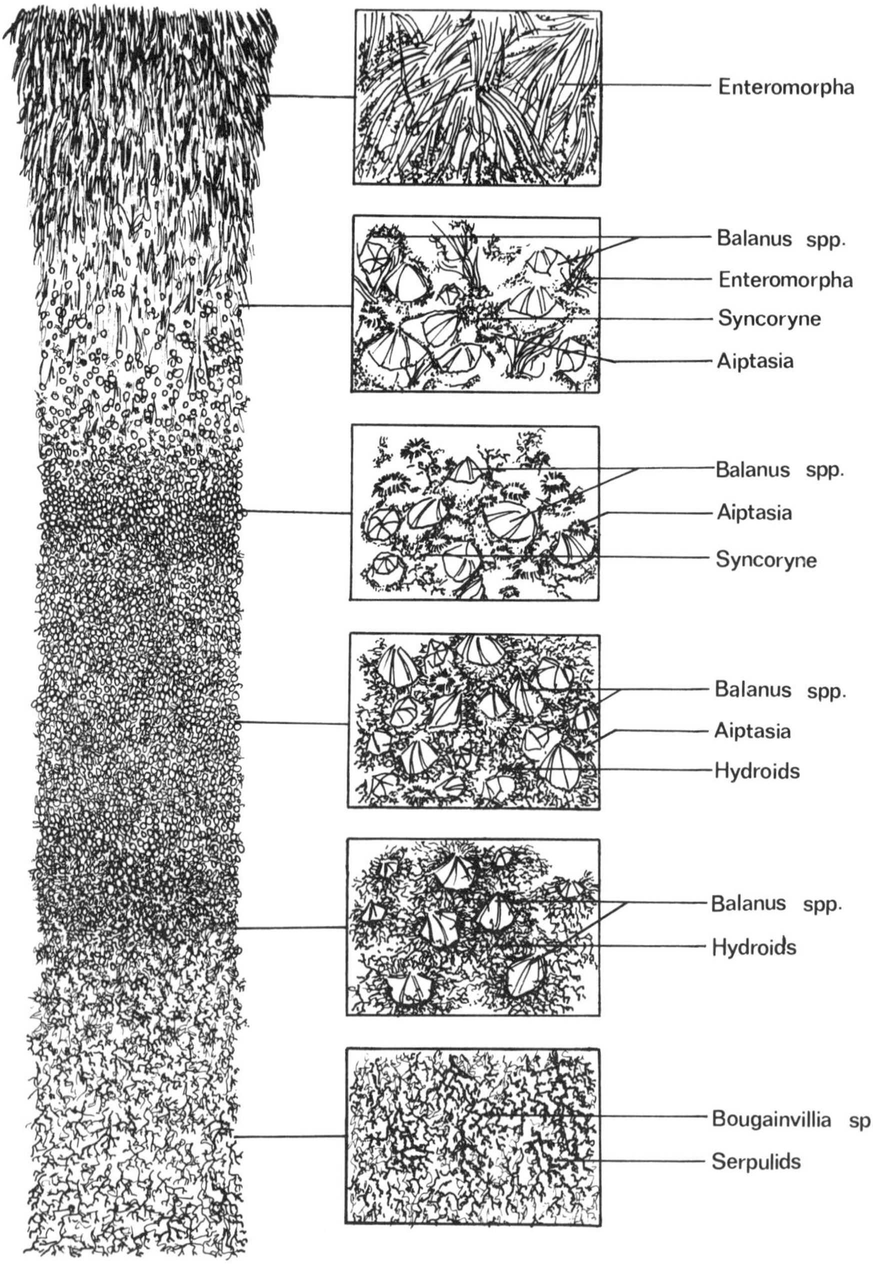

FIG. 39. DIAGRAMMATIC REPRESENTATION OF THE PRINCIPAL MARINE FOULING ORGANISMS ATTACHED TO PILINGS AT PLATFORM 54A.

biological analysis. As in the case of studying the fouling growth on platform legs, the larger the number of oil platforms used for suspension of test panels, the more useful the data.

In summary, an in-depth knowledge of the fouling organisms present on either platform legs or on test panels can be useful in measuring the effects of an oil spill or other environmental stress, if it should occur. Particular care must be exercised in identification of the organisms, and in noting the tidal level, the season, and the locality of the collection, in order to obtain meaningful data. There are limitations to the use of fouling organisms in evaluating the possible effects of oil platform activity in the OEI studies since there are no other offshore structures (not associated with oil activities) with which to make comparisons. As we stated above, fouling organisms are particularly useful as a monitoring device. In the event of an environmental stress or disaster at a given point, the biological communities on platform legs should be examined for destruction of marine life. Nearby platforms should be examined, in addition, in order to assess the extent of these effects. Test panels, if they have been utilized previously, can also be examined for possible bathymetric damage as well as latitudinal and longitudinal damage. The more thorough the background data on fouling organisms, the more useful these data become in assessing damage.

Offshore Macroinvertebrates

Benthic samples used for analyses of marine macroinvertebrates from the shallow offshore Louisiana waters are limited in both the number taken and in biological analyses made. It is possible to compare only the mollusc and crustacean data for the January and April 1973 periods from the production platform station 54A, and ambient stations 17A and 22A (located 2 and 12 miles, respectively, from station 54A). The sediments at all three stations were predominately sandy. An evaluation of the molluscan and crustacean data presented by Farrell (1979) does not give any indication of a clearly defined benthic community. All samples contained pelecypods, gastropods, amphipods, and cumaceans, but of the species in these groups, none was characteristic of all stations. The dwarf surf clam, *Mulinia lateralis*, was very abundant at the station 12 miles from the platform, but only one specimen was taken elsewhere. Other species were also found in abundance at one or two stations but not at other stations.

On the basis of the offshore benthic biological data, we found no indications that the benthic macroinvertebrate community structure has been altered as a result of the offshore oil production in Louisiana. It is unfortunate that the other groups of invertebrates, including coelenterates, nemerteans, sipunculoids, polychaetes, ophiuroids, and cephalochordates were not identified, because with such additional data it may have been possible to draw a more positive conclusion.

Discussion

The OEI benthic biological data from Timbalier Bay and the shallow offshore waters of Louisiana were re-examined with and without the assistance of computer analysis. Data consisted of the quantitative distribution of polychaetes, molluscs, crustaceans, and foraminiferans, as well as the qualitative occurrence of benthic algae in Timbalier Bay and offshore platforms, and also of fouling organisms attached to offshore platform legs.

With the exception of a previous study of the foraminiferans in Timbalier Bay, no data from previous quantitative studies of these organisms in the OEI study area were found. Ostrom occupied some of the same stations as Waldron. The data from these two studies were compared using the Morisita-Ono index of faunal similarity, which indicated no significant change during the past two decades.

In the absence of previous data on the occurrence and distribution of benthic fauna, data collected near platforms were compared to those collected some distance from any oil production or drilling activity. By and large, regardless of the type of analysis, the conclusions were similar—ambient station AB1 had greater affinity to drilling station PB2 than to the other stations. The close affinity of these two stations was largely the result of a large population of *Spiochaetopterus*, since the index of affinity was low when the *Spiochaetopterus* data were excluded (see figure 35). In one type of cluster analysis with the polychaete data (figure 37), the separation of stations was by season rather than by station.

Analyses of the species composition of bay and offshore benthic algae indicate that salinity and turbidity play an important role in the distribution of these species. Salinity reduction and increased water turbidity from outer to inner Timbalier Bay or offshore towards the Mississippi plume caused a reduction in the number of species of greens, browns, and reds.

If drilling and production operations had added appreciable amounts of hydrocarbons to Timbalier Bay during the past 25 years, then one would expect to find measurable effects in the benthic community. The faunal index would be reduced at the stressed stations, and the populations in the affected areas would separate from the ambient stations in a cluster analysis. Furthermore, if there had been a buildup of hydrocarbons or other pollutants, then one would expect a reduced benthic fauna and the appearance of pollution-tolerant species, such as *Capitella capitata*, in large numbers at the affected localities. No separate pollution-tolerant population of benthic organisms was found at any of the localities studied.

It is undetermined how the benthic fauna in the shallow Timbalier Bay area is affected by trawling for shrimp or by boating, but it is known that the wake of such boat activity is etched into the surface sediments. If such activity is of significance, then some macroinvertebrates, especially free-living epibenthic species, would be susceptible to exclusion, but others, such as *Spiochaetopterus*, would retreat downward in their tubes and later reconstruct the upper end, if it had been cut off. *Spiochaetopterus* constructs a cylindrical, annulated, translucent, flexible, horny tube, one end of which is extended into the substrate. This species is apparently able to tolerate sediment removal and deposition that may be the result of any natural or man-made cause. None of the dominant species listed in table 10 is truly epibenthic; *Paraeurythoe americana* and *Gyptis capensis* live in the upper 1 to 2 cm of sediment. As they would presumably be affected by boat activity, they may be stirred up with the sediment and transferred to another locality. The constant reworking of the sediment would, in the long run, lead to a similarity in sediment characteristics throughout the bay, which in turn would lead to a similar benthic community in the bay. This is essentially what the benthic biological conditions were in Timbalier Bay in 1973, when it was characterized by sandy silt or silty sand sediments and dominated by the polychaete *Spiochaetopterus*.

Conclusions

Analysis of the benthic biological data in Timbalier Bay and offshore waters does not indicate any environmental stress resulting from oil drilling and production activities. Variations in biota are the result of salinity or turbidity variability rather than of environmental changes resulting from petroleum-related operations. The presence of a rich and diversified polychaete, algae, and foraminiferan population at all areas sampled, together with the absence of any indication of a stressed biotic community at any locality, lead us to conclude that oil drilling and production activities have not adversely affected environments in the OEI study area of Louisiana.

CRITIQUE OF THE OEI "CONSENSUS CONCLUSIONS"

The 1974 consensus conclusions of the GURC OEI Program Council have been presented previously (Morgan et al. 1974). One objective of the re-examination was evaluation of the OEI data base using computerized and computer-assisted methods for synthesis and interpretation.

These evaluations have also been directed to determining the value of the experience and knowledge gained from the OEI as related to the original conclusions. We hoped to gain knowledge from retrospective evaluation of the program rationale and design to serve as the basis for formulation of recommendations for future studies. A restatement of the OEI conclusions and our critique of each conclusion follows:

> **Conclusion 1:** "Timbalier Bay has not undergone significant ecological change as a result of petroleum drilling and production since just prior to 1952 when other more limited baseline data were generated."

An intensive review of the data base and references to historical data that are pertinent to time-change comparison purposes leads us to the conclusion that, on the basis of historical comparison alone, this conclusion is not sufficiently supported by the evidence. Direct evidence acquired during the OEI, however, does support the conclusion.

In the logic and rationale of the OEI program, historical comparisons were only one of three elements that constituted the "substitute time trajectory." Comparisons of ecological health with comparable ecosystems and the exercise of individual judgment based on cumulative scientific experience were also used to arrive at the Council's consensus conclusions. In our judgment, the combination of these three criteria would suggest more scientifically supportable conclusions:

> **Existent historical ecological data, which mostly consist of knowledge of the Foraminifera, indicate no significant changes in the Timbalier Bay ecosystem as a result of petroleum drilling and production. Further, the ecological health (as judged against comparable ecosystems) and the low concentrations of both hydrocarbons and trace metals in the environment argue that the OEI study area probably has not suffered.**

While sufficient baseline data were not available to support Conclusion 1 fully, the existing data indicate that the conclusion is probably valid, but further confirmation is needed. The existence of the OEI data base, which is now a 5- to 7-year-old baseline, offers the opportunity for at least a partial, and important, confirmation or invalidation of this conclusion, by offering a means for determining whether changes subsequent to the OEI have occurred.

> **Conclusion 2:** "Every indication of good ecological health is present. The region of the sampling sites is a highly productive one

> from the biological standpoint, more so than other regions thus far studied in the eastern and open Gulf of Mexico.''

The OEI data support this conclusion. While our appraisals were confined to the OEI data *per se* and to the reports of the OEI Principal Investigators, those indices currently used by others to indicate ecological health support the conclusion, in our opinion. Furthermore, the in-depth examinations of the data—exercised primarily in seeking alternative rationales and experimental approaches—have developed statistical summaries that further support the conclusion, while generating no statistical evidence to the contrary.

More conclusive statistical evidence to validate or invalidate the conclusion might have been generated through a more effective basic rationale using different measurements, parameters, spatial and temporal density, and patterns.

Probably the evidence to support this conclusion is more convincing for the bay than for the offshore area, as a consequence of there being greater data density for important parameters (especially the benthos) for the bays.

Therefore, we concur with the conclusion, but with the reservation that it should be further confirmed by using the newly available baseline data for other Gulf areas as the means for conducting a more comprehensive comparison of habitats and communities to determine relative ecological health and productivity as a function of habitat parameters. The data of Fucik and El-Sayed (1979), Marum (1979), and Brent et al. (1979) adequately document the high biological productivity of the OEI study area.

> **Conclusion 3:** ''Concentration of all compounds which are in any way related to drilling or production are sufficiently low to present no known persistent biological hazard.''

The OEI data on hydrocarbon and trace metal contamination that might be related to oil drilling and production generally support this statement, with the exception that the OEI did not address ''all compounds'' in a comprehensive sense. The OEI data show concentrations within the bay and the offshore area as being generally below those considered significant for biological uptake and cumulative effects. However, the data regarding hydrocarbon compounds are inadequate to eliminate the possibility of subtle cumulative effects that may be due to specific hydrocarbon compounds or specific chemical combinations of trace metals in the environment. Again, in the absence of data to support a completely sound statistical treatment, it is necessary to view the rather extensive evidence that is contained in the OEI data base in the light of more broadly applicable evidence—i.e., that of very low concentrations of trace metals and petroleum fractions, of concentrations of fractions less than or comparable to those reported for the Gulf and for the world's oceans, and concentrations quite low as compared with regulatory limits—to draw the more supportable conclusion:

> **Concentrations of petroleum-related hydrocarbons and trace metals found in the OEI area are below those generally considered to have biological significance.**

There were insufficient sediment samples analyzed for hydrocarbons in the OEI adequately to assess buildup or possible effects on benthic organisms.

> **Conclusion 4:** "Natural phenomena such as seasonality, floods, upwellings, and turbid layers have much greater impact upon the ecosystem than do petroleum drilling and production activities."

The OEI provided abundant evidence to support this conclusion. Fortunately, the record Mississippi River flood occurred near the midpoint of the seasonal exercises, so that its effects throughout the OEI area were dramatically recorded and, importantly, could be compared with more stable pre-flood and post-flood periods. The turbid layer, its ecological effects, and its origin were defined, conclusively providing baseline information about ecological impacts of natural phenomena. The seasonal variabilities were of the character and magnitude to be expected and were conclusively demonstrated. Such variabilities now permit the development of more definitive judgments of the effects of either natural or human activity-related variations within the study area described by the OEI data.

The study area chosen for the OEI is heavily influenced by the adjacent land mass and receives significant hydrocarbon inputs from the Mississippi, which drains two-thirds of the continental United States. Petroleum-related hydrocabrons flowing into the Gulf from the Mississippi probably exceed all estimates of inputs attributable to oil drilling and production activities.

CONCLUSIONS BASED ON RE-EXAMINATION OF THE OEI DATA BASE

1. The platform versus control site experimental design used in the OEI would be valid only if the sites selected proved to be different in one or more specific parameters believed to be related to oil drilling and production. Data collected in the OEI did not substantiate that the sites were different. However, analyses for hydrocarbons, metals, and nutrients were mostly limited to the water column.

2. Methods and procedures used in the OEI were carefully recorded (Menzies 1973) to permit subsequent duplication of individual sampling and analytical methods and procedures, but use of standard procedures was not enforced. Hence, for example, benthic fauna in Timbalier Bay and offshore were collected by seven different investigators using different techniques, making synthesis and interpretation of benthic data a formidable task. The synthesis accomplished in this re-examination allows for only limited interpretation, an unfortunate circumstance considering the time and expense incurred in collecting the data.

3. The detection and interpretation of low-level, chronic disturbances of ecosystems requires careful attention to experimental design, including such things as selection of appropriate parameters for judging effects and methods of collection and statistical treatment of data. Based on our review of the OEI program, it is our opinion that most of the investigators believed that oil operations associated with platforms would produce large, easily detectable sources of contamination, and hence would be easily monitored by water column measurements, with little need for replication and statistical treatment of data. This did not prove to be the case.

4. Analyses of water column characteristics (physical, chemical, and biological) did not lead to definitive conclusions because of the highly variable nature of the data collected.

5. Our analyses of the benthic biological data in both Timbalier Bay and offshore did not reveal any indication of a stressed environment near or around oil drilling and production platforms.

6. We wish to re-emphasize conclusion number 4 of the OEI Council on the importance of recognizing the effects of natural phenomena on results obtained from environmental monitoring and analysis. For example, the data of Humm (1974) clearly show a direct relationship between the number of algal species on platforms and distance from the flow of the Mississippi River.

7. Programs designed to detect the presence or effects of petroleum hydrocarbons should have as a primary component of the experimental design the most advanced analytical support for quantifying hydrocarbons in the most likely sinks, especially in sediments and organisms.

8. The ability to distinguish environmental change at any specific time, based on monitoring programs, is greatly enhanced by the availability of historic baseline data. Unfortunately, with the exception of the Foraminifera, little meaningful historic data was available to the OEI. Hence, the OEI data base represents an invaluable resource for detection of changes in the Gulf ecosystem that may occur in the future.

RECOMMENDATIONS FOR FUTURE STUDIES

Existing data bases

The OEI was conducted with the single purpose of determining the cumulative ecological effects of normal petroleum drilling and production operations. Although the results of the investigation did not yield evidence of cumulative effects, certain aspects of the study design preclude a definitive statement to that effect. Since the OEI was conducted, however, numerous comprehensive environmental data bases have been generated for the northern continental shelf and coastal ecosystems of the Gulf of Mexico. These data bases contain much of the necessary information to assess the cumulative impacts of offshore development, but they have not been put to that use. Using investigative methods similar to those employed in this re-examination of the OEI in combination with a regional data base, which merges the data produced as a result of government and industrial programs in this geographical area, we recommend that: (1) the data be synthesized for interpretation, (2) the data base be analyzed for adequacy and validity to determine process variabilities, patterns of accumulation, etc., (3) relationships between contaminants and community structure be determined, and (4) the information be synthesized for management decision.

The data base so generated should also systematically accumulate data submitted to federal and state agencies in compliance with drilling permits and other petroleum operations. Such a program would continuously expand the data base, fill in gaps in time and space, and provide valuable quantitative information about point sources concerning contaminants and their concentrations.

Participation in a Gulf study of the nature described above should be national in experimental design and in data synthesis and interpretation.

Future studies

If a program with the objectives of the OEI were undertaken today, we would recommend an experimental design that would concentrate much less effort on measuring biological and chemical characteristics of the water column. Plankton populations are generally so spotty and have such high population resiliency that they make detection of chronic effects from field studies almost impossible. This same reasoning applies to measurements of petroleum hydrocarbons and metals in the water column. Intermittent releases at the platforms would result in patchiness in distribution, which would be compounded by variable current patterns. These factors are often further compounded by analytical problems, suggesting that efforts in measuring chemical contamination would be better spent in monitoring sediments and/or sessile biota.

An adequate prior knowledge of the physical environment including currents, sediment composition, and stability must be obtained prior to the selection of sampling stations. Once knowledge of these factors is available, the sampling design can be determined. We would suggest that any study designed to detect chronic effects include determinations of the following: (1) benthic community composition and structure, (2) petroleum hydrocarbon and trace metal contamination of sediments and sessile biota, (3) petroleum hydrocarbon and trace metal contamination and tainting of commercially important biota, and (4) provision to allow for special studies to investigate the significance of any unusual contamination found in the field monitoring. In addition, many habitats may present special cases when the further stratification of sampling efforts must be considered, e.g., benthic substrates, when more than one type are important in the area.

Whenever any large-scale marine ecological program is contemplated, it is imperative that all phases of sample collection and data analysis and interpretation be thoroughly planned and discussed with all scientific disciplines. The nature and amount of data collected must be adequate, within the limits of monetary constraints, to accomplish the objectives of the study. Collection of data without analysis is meaningless and a waste of resources. Constant vigilance and re-analysis is essential to avoid merely collecting data for data's sake; all data must be meaningful and useful to the study. More specific recommendations are:

1. All sampling stations must be coordinated so that the data from one discipline can be used by another.
2. Natural, as well as anthropogenic, environmental changes must be considered in the selection of stations and sampling techniques.
3. The number of samples taken must be carefully considered in order to assure valid statistical analysis.
4. If the purpose of the monitoring program is to ascertain whether or not a particular man-caused activity has caused damage, natural areas of sufficient distance, but similar characteristics, must be studied in order to draw valid conclusions. Naturally-caused environmental events, such as seasonal run-off from rivers, must be taken into consideration in assessing the environmental condition of the area.
5. Since the knowledge of what species of organisms are present in a particular environment is essential for making an accurate assessment of the state of the environment, it is important that the species be identified by competent scientists to the lowest possible taxon,

preferably to the level of species. Identification of all species of macroflora and fauna may be impossible in some instances because of the lack of reasonably good reference works. Therefore, we recommend that identification of marine organisms be limited to macroalgae, polychaetes, molluscs, crustaceans, echinoderms, and fish, with the proviso that such requirements be expanded to include other groups of organisms that may be of importance to a particular area, e.g., corals in tropical seas. Because of the importance of Foraminifera in oil exploration, we recommend that this group also be included in benthic analyses. Only by making such limitations on identification of organisms can there be hope to accomplish this difficult and time-consuming task with sufficient replication to minimize errors and to compensate adequately and correctly for natural variability.

6. All procedures for the collection and analysis of data should be standardized.

Exceptional caution should be observed in drawing conclusions without either a comprehensive synoptic interdisciplinary data base or a pre-established minimum set of monitoring data. Misinterpretation of ecological phenomena could easily occur if conclusions are based on single elements of the ecosystem or on a limited number of parameters. While it appears likely that the number of parameters required for ecological monitoring can be reduced, reduction in number can be done only after accounting for seasonal variabilities and parameter interactions.

REFERENCES CITED

Boesch, D. F. 1977. Application of Numerical Classification in Ecological Investigations of Water Pollution. Special Scientific Report No. 77. Gloucester Point: Virginia Institute of Marine Science.

Brent, C. R., H. P. Williams, W. A. Bergin, T. E. Myers, and J. L. Tyvoll. 1974. Organic carbon, inorganic carbon and related variables in offshore oil production areas of the northern Gulf of Mexico and in Timbalier Bay, Louisiana. Gulf Universities Research Consortium. Appendix to Report No. 138.

———, H. P. Williams, W. A. Bergin, J. L. Tyvoll, and T. E. Myers. 1979. Organic carbon, inorganic carbon, and related variables in offshore oil production areas of the northern Gulf of Mexico. Rice University Studies **65**(4):245-264.

Farrell, Douglas. 1974a. Benthic communities in the vicinity of producing oil wells in Timbalier Bay, Louisiana. Gulf Universities Research Consortium. Appendix to Report No. 138.

———. 1974b. Benthic communities in the vicinity of producing oil wells on the shallow Louisiana Continental Shelf. Gulf Universities Research Consortium. Appendix to Report No. 138.

______. 1979. Benthic molluscan and crustacean communities in Louisiana. Rice University Studies **65**(4):401-436.

Fish, A. G., L. L. Massey, J. R. Inabinet, and P. L. Lewis. 1974. A study of the effects of environmental factors upon the distribution of selected sandy beach organisms of Timbalier Bay, Louisiana. Gulf Universities Research Consortium. Appendix to Report No. 138.

Fucik, Kenneth W. and S. Z. El-Sayed. 1979. Effect of oil production and drilling operations on the ecology of phytoplankton in the OEI study area. Rice University Studies **65**(4):000-000.

George, R. Y. 1974. Structure and function of oyster reef and sandy shore community in the vicinity of oil platforms in Louisiana Coast. Gulf Universities Research Consortium. Appendix to Report No. 138.

______ and P. J. Thomas, 1974. Aspects of fouling on offshore oil platforms in Louisiana Shelf in relation to environmental impact. Gulf Universities Research Consortium. Appendix to Report No. 138.

Humm, H. J. 1974. The effect of the offshore oil and gas wells on the benthic marine plants of Louisiana—I. Timbalier Bay & II. Offshore. Gulf Universities Research Consortium. Appendix to Report No. 138.

______ and Theresa M. Bert. 1979. The benthic marine algae of Timbalier Bay, Louisiana. Rice University Studies **65**(4):379-399.

Inabinet, J. R. and A. G. Fish. 1979. Distribution and ecology of the littoral Foraminifera of Timbalier Bay, Louisiana. Rice University Studies **65**(4):491-509.

Jones, G. F. 1969. The Benthic Macrofauna of the Mainland Shelf of Southern California. Allan Hancock Foundation Monographs in Marine Biology, No. 4. University of Southern California.

Kritzler, H. R. 1974. Oil production and polychaetous annelids in a Louisiana estuary. Gulf Universities Research Consortium. Appendix to Report No. 138.

———, R. M. Hiskey, and P. J. Thomas, 1974. A system for the quantitative sampling of shallow water benthos. International Revue der gesamten Hydrobiologie **59**:621-627.

Laseter, J. L. and E. J. Ledet. 1974. Hydrocarbons and free fatty acids associated with the air/water interface, sediments, and beaches of the Timbalier Bay and offshore Louisiana area. Gulf Universities Research Consortium. Appendix to Report No. 138.

Marum, J. P. 1974. Effects of long-term oil drilling and production on the zooplankton of Louisiana coastal waters. Gulf Universities Research Consortium. Appendix to Report No. 138.

______. 1979. Significance of distribution patterns of planktonic copepods in Louisiana coastal waters and relationships to oil drilling and production. Rice University Studies **65**(4):355-377.

Menzies, R. J. 1973. A Handbook of Procedures and Methods Employed in the Offshore Ecology Investigation (OEI). Gulf Universities Research Consortium Report No. 121.

Montalvo, J. G. and D. V. Brady. 1974. Toxic metal determinations in offshore water samples, Part 2. Gulf Universities Research Consortium. Appendix to Report No. 138.

Morgan, J. P., R. J. Menzies, S. Z. El-Sayed, and C. H. Oppenheimer. 1974. The Offshore Ecology Investigation, Final Project Planning Council Consensus Report. Gulf Universities Research Consortium. Appendix to Report No. 138.

Morisita, M. 1959. Measuring interspecific association and similarity between communities. Memoirs, Faculty of Science, Kyushu University, Series E (Biology) **3**:65-80.

Olmstead, P. S. and J. W. Tukey. 1947. A corner test for association. Annual of Mathematical Statistics **18**:495-513.

Ono, Y. 1961. An ecological study of the brachyuran community on Tomiolca Bay, Amakusa, Kyushu. Records of Oceanographic Works in Japan. Special Report No. 5, pp. 199-210.

Oppenheimer, C. H., B. M. Miget. and H. Kator. 1974. Hydrocarbons in seawater and organisms and microbiological investigations. Gulf Universities Research Consortium. Appendix to Report No. 138.

Ostrom, Christopher. 1974. Relationships between the distribution of selected taxa of recent benthic Foraminifera and measured environmental parameters in Timbalier Bay, Louisiana. Gulf Universities Research Consortium. Appendix to Report No. 138.

______. 1979. The distribution of recent Foraminifera in Timbalier Bay, Louisiana. Rice University Studies **65**(4):447-471.

Perry, A. 1974. A bottom/surface trawl and bottom grab study of areas of oil production activity and areas of no activity in estuarine and offshore environments. Effect of platforms on biota (fishes) offshore. Gulf Universities Research Consortium. Appendix to Report No. 138.

Pomeroy, L. R. 1974. The ocean's food web, a changing paradigm. Bioscience **24**:499-504.

Reish, D. J. 1971. Seasonal settlement of polychaetous annelids on test panels in Los Angeles-Long Beach Harbors 1950-1951. Journal of the Fisheries Research Board of Canada **28**:1459-1467.

______. 1973. The use of benthic marine animals in monitoring the marine environment. Journal of Environmental Planning and Pollution Control **1**:32-38.

Shafer, C. T. 1973. Distribution of Foraminifera near pollution sources in Chaleur Bay. Water, Air, and Soil Pollution **2**:219.

Shannon, C. E. and W. Weaver. 1963. The Mathematical Theory of Communication. Urbana: University of Illinois Press.

Sharp, J. M. and S. G. Appan. 1978. Cumulative effects of oil drilling and production on estuarine and near-shore ecosystems. *In* M. L. Wiley, ed., Estuarine Interactions. New York: Academic Press, pp. 57-73.

Waldron, R. P. 1957. A seasonal ecological study of foraminifera from Timbalier Bay, Louisiana. *In* Technical Report No. 16, Part B, pp. 132-188, Coastal Studies Institute, Louisiana State University.

Waller, Richard S. 1974. Effects of platforms on biota (invertebrates). Gulf Universities Research Consortium. Appendix to Report No. 138.

NOTE: Individual papers of the OEI Principal Investigators for 1979 referred to in this section can be found in section IV of this book.

APPENDIX I

APPENDIX 1

Systematic List of the Marine Organisms Reported from the OEI Study Area

1972-1974

Compiled by Donald J. Reish and S.G. Appan

	Timbalier Bay	Offshore Waters
Phylum Chlorophyta		
Batophora oerstedii		X
Bryopsis hypoides		X
B. pennata		X
Chaetomorpha brachygona		X
C. gracilis		X
C. linum	X	X
C. minima	X	X
Cladophora albida	X	X
C. brachyclona		X
C. catenata		X
C. dalmatica	X	X
C. delicatula	X	
C. flexuosa		X
C. gracilis	X	X
C. repens		X
C. sericea		X
C. vagabunda		X
Cladophoropsis membranacea	X	X
Derbesia lamourouxii		X
D. vaucheriaeformis		X
Enteromorpha chaetomorphoides		X
Phylum Chlorophyta (continued)		
E. clathrata	X	X
E. compressa	X	X
E. erecta		X
Enteromorpha flexuosa	X	X
E. intestinalis	X	X
E. lingulata	X	X
E. plumosa	X	X
E. prolifera	X	X
E. salina		X
E. sp.		X
Entocladia viridis		X
Gomontia polyrhiza	X	X
Monostroma oxyspermum	X	
Ostreobium quekettii	X	X
Phaeophila dendroides		X
Protoderma marinum		X
Pseudoendoclonium marinum	X	
Pseudotetraspora antillarum	X	X
Rhizoclonium hookeri	X	X
R. kerneri	X	
R. kochianium	X	X

	Timbalier Bay	Offshore Waters
Phylum Chlorophyta (continued)		
R. riparium	X	X
R. tortuosum	X	
Stichococcus marinus	X	X
Ulothrix flacca	X	
Ulva curvata	X	
Ulvella lens	X	X
Phylum Pyrrhophyta		
Ceratium fusus		X
Exuviaella compressa		X
Gonyaulax diegensis		X
G. fragilis		X
G. monilata		X
G. polygramma		X
G. sp.		X
G. turbyne		X
Peridinium pellucidium		X
P. trochoideum		X
Prorocentrum gracile		X
Phylum Chryophyta		
Asterionella japonica		X
Bacteriastrum delicatum		X
Biddulphia mobiliensis		X
Cerataylina pelagica		X
Chaetoceros affinis		X
C. atlanticus		X
C. compressus		X
C. curvisetus		X
C. decipiens		X
C. didymus		X
C. laciniosus		X
C. lorenzianus		X
C. messanensis		X
C. pelagicus		X

	Timbalier Bay	Offshore Waters
Phylum Chryophyta (continued)		
C. pseudocurvisetus		X
C. sp.		X
Corethron criophilum		X
Coscinodiscus centralis		X
C. radiatus		X
Cyclotella striata		X
C. sp.		X
Denticula elegans		X
Dictyocha fibula		X
Eucampia cornuta		X
Guinardia flaccida		X
Hemiaulus sinensis		X
Leptocylindrus danicus		X
Melosira sp.		X
Nitzschia closterium		X
N. pungens		X
N. subfraudulenta		X
N. vitrea		X
N. sp.		X
Pleurosigma naviculaceum		X
Rhizosolenia alata		X
R. calcar-avis		X
R. delicatula		X
R. fragilissima		X
R. setigera		X
R. shrubsolii		X
R. stolterfothii		X
R. styliformis		X
Skeletonema costarum		X
Synedra acus?		X
Thalassionema nitzschioides		X
Thalassiosira aestivalis		X
T. decipiens		X
T. excentrica		X
T. oestrupii		X

APPENDIX 1 (continued)

	Timbalier Bay	Offshore Waters
Phylum Chryophyta (continued)		
T. rotula		X
T. sp.		X
Thallassiothrix mediterranea		X
Phylum Cyanophyta		
Agmenellum thermale	X	X
Anacystis aeruginosa	X	X
A. dimidiata	X	X
A. marina	X	
A. montana	X	X
Calothrix crustacea	X	X
Coccochloris elabens	X	X
C. stagnina	X	X
Entophysalis conferta	X	X
E. deusta	X	X
Gomphosphaeria aponina	X	X
Johnnesbaptista pellucida	X	
Mastigocoleus testarum	X	X
Microcoleus lyngbyaceus	X	X
M. vaginatus	X	X
Nostoc spumigena	X	
Oscillatoria erythraca	X	X
O. lutea	X	X
O. submembranacea	X	X
Porphyrosiphon kurzii	X	
P. notarisii	X	X
Schizothrix arenaria	X	X
S. calcicola	X	X
S. mexicana	X	X
S. tenerrima	X	X
Spirulina subsalsa	X	X

	Timbalier Bay	Offshore Waters
Phylum Phaeophyta		
Acinetospora crinita		X
Ascocyclus orbicularis		X
Bachelotia antillarum	X	X
Ectocarpus dasycarpus	X	
E. elachistaeformis		X
E. intermedius	X	X
E. variabilis	X	X
Giffordia conifera	X	X
G. indica	X	X
G. mitchellae	X	X
G. rallsiae	X	X
Herponema tortugense		X
Hummia onusta		X
Myriotrichia subcorymbosa	X	X
Phaeostroma pusillum		X
Pseudostictyosiphon onustus	X	
Sargassum fluitans	X	X
S. natans		X
S. sp.		X
Sphacelaria furcigera	X	X
S. tribuloides		X
Spongonema tomentosum		X
Phylum Rhodophyta		
Acrochaetium antillarum		X
A. crassipes	X	
A. dufourii		X
A. flexuosum		X
A. gracile	X	X
A. hoytii		X
A. sargaeanum	X	X
A. sancti-thomae		X

	Timbalier Bay	Offshore Waters
Phylum Rhodophyta (continued)		
A. seriatum		X
A. thureitii		X
A. trifilum		X
A. virgatulum	X	
Antithamion cruciatum		X
Asterocystis ramosa		X
Bangia fuscopurpurea	X	X
Bostrichia radicans	X	
B. r. forma *moniliforme*	X	
B. tenella	X	
Ceramium byssoideum		X
C. fastigiatum		X
C. f. forma *flaccida*		X
Erythrocladia recondita		X
E. subintegra		X
E. vagabunda		X
Erythrotrichia carnea	X	X
Fosliella atlantica		X
F. farinosa		X
F. f. var. *solmsiana*		X
Goniotrichum alsidii	X	X
Gracilaria foliifera	X	X
G. verrucosa	X	
Gracilariopsis sjoestedtii	X	
Herposiphonia secunda		X
Hypnea musciformis	X	
Jania adherens		X
Laurencia papillosa		X
L. poitei	X	
Lophosiphonia cristata		X
L. saccorhiza		X
L. subadunca	X	X
Polysiphonia echinata	X	
P. havanensis		X
P. hemisphaerica	X	X
P. howei		X
P. subtilissima	X	X
P. sp.		X
Pterocladia americana		X
Spermothamium investiens		X
Phylum Protozoa (Foraminifera)		
Ammoastuta inepta	X	
Ammobaculites crassus	X	
A. dilatatus	X	
A. exiguus	X	
A. subcatenulatus	X	
Ammonia beccarii var. *parkinsoniana*	X	
A. beccarii var. *tepida*	X	
A. beccarii var. A	X	
A. beccarii var. B	X	
A. pauciloculata	X	
A. rolshauseni	X	
Ammotium dilatatum	X	
A. fragile	X	
A. salsum	X	
Ammoscalarin pseudospiralis	X	
Arenoparella mexicana	X	
Bolivina lowmani	X	
B. striatula	X	
Bucella hannai	X	
Bulimenella bassendorfensis	X	
B. elegantissima	X	
Clavulina gracilis	X	
Cornuspira sp.	X	
Elphidium advena	X	
E. delicatulum	X	
E. discoidale	X	
E. gunteri	X	
E. limosum	X	

APPENDIX 1 (continued)

	Timbalier Bay	Offshore Waters
Phylum Protozoa (continued)		
E. matagordanum	X	
E. mexicana	X	
E. poeyanum	X	
E. vadescens	X	
Epistominella vitrea	X	
Gaudryina exilis	X	
Hanzawaia strattoni	X	
Haplophragmoides manilaensis	X	
H. wilberti	X	
Jadammina polystoma	X	
Miliammina fusca	X	
Nonionella atlantica	X	
N. opima	X	
Palmerinella gardenislandensis	X	
Proteonina lagenaria	X	
Quinqueloculina compta	X	
Q. funafutiensis	X	
Q. lamarckiana	X	
Q. poeyana	X	
Q. rhodiensis	X	
Q. seminulum	X	
Q. tenagos	X	
Q. wiesneri	X	
Recurvoides sp.	X	
Sulcophax sp.	X	
Triloculina sidebottomi	X	
Trochammina comprimata	X	
T. inflata	X	
T. laevigata	X	
T. macrescens	X	
Typhotrocha comprimata	X	

	Timbalier Bay	Offshore Waters
Phylum Porifera		
Cliona sp.		X
Lissodeudoryx sp.		X
Unidentified		X
Phylum Coelenterata		
Aiptasia sp.		X
Astrangia astreiformis		X
Bougainvillia tennella		X
Bundosoma cavernata		X
Calliactis tricolor		X
Chrysaura quinquecirrha		X
Pelagia noctiluca		X
Renilla mulleri		X
Syncoryne sp.		X
Anthozoa, unidentified		X
Phylum Platyhelminthes		
Leptoplana sp.		X
Phylum Nematoda		
Nematodes, unidentified	X	
Phylum Nemertea		
Cerebratulus lacteus		X
Phylum Annelida		
Class Polychaeta		
Aedicira belgicae	X	
Aglaophamus verrilli	X	
Ammotrypane aulogaster	X	
Amphinomidae	X	
Anaitides maculata	X	

	Tambalier Bay	Offshore Waters
Class Polychaeta (continued)		
Ancistrosyllis hamata	X	
Arabella iricolor	X	
Aricidea fauveli	X	
Armandia polyophthalma	X	
Asychis urceolata	X	
Autolytus faciatus	X	
Axiothella mucosa	X	
Bhawania goodei	X	
Branchioasychis americana	X	
Cabira brevicirrus	X	
Capitella capitata	X	
Cirratulus cirratus	X	
Cirriformia tentaculata	X	
Clymenella torquata calida	X	
Cossura delta	X	
Dasybranchis lumbricoides	X	
Diopatra cuprea spirobranchus	X	
D. c. cuprea	X	
Dorvillea rudolphi	X	
Drilonereis magna	X	
D. sp.	X	
Eteone heterpoda	X	
Euclymene sp.	X	
Eumida sanguinea	X	
Eunice subra	X	
Exogone dispar	X	
E. sp.	X	
Glycera americana	X	
G. rouxi	X	
Glycinde polygnatha	X	
Gyptis capensis	X	
Haploscoloplos robustus	X	
Harmothoe lunulata	X	
H. trimaculata	X	
Hesionid	X	

	Tambalier Bay	Offshore Waters
Class Polychaeta (continued)		
Heteromastus filiformis	X	
Laonice cirrata	X	
Loandalia americana	X	
Loimia viridis	X	
Lumbrineris bassi	X	
L. erecta	X	
L. parvapedata	X	
L. sp.	X	
Magelona cincta	X	
M. longicornis	X	
Malacoceros vanderhorsti	X	
Maldane sarsi	X	
Maldanid	X	
Megalomma bioculatum	X	
Megalona sp.	X	
Melinna maculata	X	
Melinnexis sp.	X	
Myriochele oculata	X	
Myriowenia sp.	X	
Mystides borealis	X	
Naineris laevigata	X	
N. setosa	X	
Neanthes arenaceodentata	X	
N. succinea	X	X
Nephtys bucera	X	
N. picta	X	
Nereiphylla fragilis	X	
N. parreti	X	
Nereis sp.		X
Nerine agilis	X	
Ninoe nigripes gracilis	X	
Notomastus hemipodus	X	
N. latericeus	X	
Onuphis eremita oculata	X	
Ophiodromus sp.	X	

APPENDIX 1 (continued)

	Timbalier Bay	Offshore Waters
Class Polychaeta (continued)		
Owenia fusiformis	X	
Paraeurythoe americana	X	
Paraonid	X	
Paraprionospio pinnata	X	
Pectinaria gouldii	X	
Pilargis verrucosa	X	
Pista mirabilis	X	
Poecilochaetus sp.	X	
Polydora socialis	X	
Pomatoceros sp.		X
Prionospio caspersi	X	
P. cirrifera	X	
P. sexoculata	X	
Sabella microphthalma	X	
Scoloplos rubra	X	
Semiodera roberti	X	
Sigambra tentaculata	X	
Sphaerosyllis sp. a	X	
S. sp. b	X	
Spiochaetopterus costarum	X	
Spiophanes bombyx	X	
S. soederstromi	X	
Sthenelais articulata	X	
S. boa	X	
Streblospio benedicti	X	
Syllidia armata	X	
Tharyx monilaris	X	
Phylum Sipunculoidea		
Sipunculids		X
Phylum Mollusca		

	Timbalier Bay	Offshore Waters
Class Gastropoda		
Anachis avara	X	X
A. obesa	X	
Busycon contrarium		X
Cantharus cancellarius		X
Crepidula fornicata	X	X
Distorsio clathrata		X
Epitonium krebsi		X
E. rupicola	X	
Mitrella sp.		X
Murex fulvescens		X
Nassarius acutus	X	X
N. vibex	X	X
Nudibranchia		X
Oliva sayana		X
Polinices duplicatus	X	X
Sinum perspectivum		X
Terebra dislocata		X
Thais haemastoma	X	X
Thais sp.		X
Turbonilla sp.		X
Class Pelecypoda		
Aequipecten irradians		X
Anadara ovalis	X	
A. transversa	X	X
Chione cancellata	X	X
C. grus		X
C. latilirata		X
Corbula sp.		X
Crassostrea virginica	X	X
Diplodonta punctata	X	X
Docinus discus	X	

	Timbalier Bay	Offshore Waters
Class Pelecypoda (continued)		
Donax texasianus	X	
Ensis minor	X	
Lithophaga aristata		X
Macoma mitchilli	X	
M. sp.		X
Mercenaria campechiensis	X	X
Mulinia lateralis	X	
Noetia ponderosa		X
Nucula proxima		X
Nuculana concentrica	X	X
Ostrea equestris		X
Solen viridis		X
Tellidora cristata	X	
Tellina alternata	X	
T. iris	X	X
Class Cephalopoda		
Dorytheuthis plei		X
Loligo pealei		X
Lolliguncula brevis		X
Octopus vulgaris		X
Phylum Arthropoda		
Class Crustacea		
Subclass Copepoda		
Order Caligoida		
Caligus sp. a	X	
C. sp. b	X	X
C. sp. c	X	
C. sp. d	X	
Lernaeocera sp.	X	
Order Calanoida		
Acartia danae		X
A. tonsa	X	X
Order Calanoida (continued)		
Acrocalanus longicornis		X
Anomalocera ornata		X
Calanopia americana		X
Calocalanus pavo		X
C. styliremis		X
Candacia curta		X
C. pachydactyla		X
Centropages furcatus	X	X
C. hamatus	X	X
C. velificatus	X	X
Clausocalanus furcatus		X
C. jobei		X
Eucalanus crassus		X
E. hyalinus		X
E. pileatus	X	X
E. sewelli		X
Euchaeta marina		X
Euchirella rostrata		X
Eurytemora hirundoides	X	
Labidocera acutifrons		X
L. aestiva	X	X
Lucicutia gaussae		X
Nannocalanus minor		X
Neocalanus gracilis		X
N. robustior		X
Paracalanus aculeatus		X
P. crassirostris	X	X
P. indicus	X	X
Paracandacia simplex		X
Pleuromamma abdominalis		X
P. piseki		X
Pontella securifer		X
Pontellopsis villosa		X
Pseudodiaptomus coronatus	X	X
Rhincalanus cornutus		X

APPENDIX 1 (continued)

	Timbalier Bay	Offshore Waters
Order Calanoida (continued)		
Scolecithrix danae		X
Temora stylifera	X	X
T. turbinata		X
Tortanus setacaudatus	X	
Undinula vulgaris		X
Order Cyclopoida		
Copilia mirabilis		X
Corycaeus amazonicus	X	X
C. clausi		X
C. giesbrechti		X
C. latus		X
C. lautus		X
C. limbatus		X
C. ovalis		X
C. speciosus		X
C. subulatus		X
Cyclopoida sp.	X	
Farranula carinata	X	X
F. gracilis	X	X
F. rostrata		X
Lubbockia squillimana		X
Oithona brevicornis	X	X
O. nana		X
O. plumifera	X	X
O. similis		X
O. simplex		X
Oncaea curta		X
O. mediterranea		X
O. venusta	X	X
Sapphirella sp. a	X	
S. sp. b		X

	Timbalier Bay	Offshore Waters
Order Cyclopoida (continued)		
Sapphirina nigromaculata		X
Order Harpacticoida		
Clytemnestra rostrata		X
C. scutellata		X
Euterpina acutifrons	X	X
Macrosetella gracilis	X	X
Microsetella norvegica	X	X
M. rosea	X	X
Miracia efferata		X
Harpacticoid sp. a	X	
Harpacticoid sp. b	X	
Harpacticoid sp. c		X
Harpacticoid sp. d	X	
Harpacticoid sp. e		X
Harpacticoid sp. f	X	
Harpacticoid sp. g		X
Harpacticoid sp. h	X	
Harpacticoid sp. i		X
Harpacticoid sp. j	X	
Harpacticoid sp. k		X
Subclass Cirripedia		
Balanus amphitrite	X	X
B. eberneus		X
B. improvisus		X
B. reticulatus		X
B. tintinnabulum		X
Lepas sp.		X
Subclass Malacostraca		
Order Stomatopoda		

	Timbalier Bay	Offshore Waters
Order Stomatopoda (continued)		
Squilla empusa	X	X
Order Mysidacea		
Mysidacea		X
Order Cumacea		
Cyclaspis sp.		X
Eudorella monodon	X	
Leucon sp.	X	
Oxyurostylis smithi	X	X
Order Tanaidacea		
Leptochela serratorbita		X
Tanaids		X
Order Isopoda		
Aegathoa oculata	X	
Ancinus depressus	X	X
Cymothoa excisa	X	
Edotea montosa	X	X
Idotea sp.	X	
Limnoria tripunctata		X
Lironeca ovalis	X	
L. texana		X
Munna sp.		X
Neroclia lanceolata		X
Sphaeroma quadridentatum		X
Order Amphipoda		
Acanthohaustorius sp.	X	
Ampelisca holmesi	X	X
A. vadorum	X	X
Amphilochus sp.		X
Batea catharinensis	X	X
Caprella sp.	X	X
Order Amphipoda (continued)		
Cerapus tubularis		X
Corophium acherusicum	X	X
C. sp.		X
Elasmopus rapax	X	
Erichthonius brasiliensis		X
Haustorius sp.	X	
Hemigena minuta		X
Hyperiidae		X
Lembos sp.		X
Listrella clymenella	X	X
Microprotopus raneyi	X	
Monoculoides nyei	X	X
Orchestia platensis	X	
Protohaustorius sp.	X	X
Stenothoe minuta	X	X
S. sp.		X
Synchelidium sp.	X	X
Order Decapoda		
Acetes americanus		X
Albunea pareti		X
Alpheus sp. a		X
A. sp. b		X
Calappa sulcata		X
Callinectes sapidus	X	X
C. similis	X	X
C. sp.	X	
Clibanarius vittatus	X	
Emerita portoricensis	X	
Euceramus praelongus		X
Eurypanopeus abbreviatus		X
Hepatus epheliticus		X
H. sp.		X
Hexapanopeus angustifrons	X	X
H. paulensis		X

APPENDIX 1 (continued)

	Timbalier Bay	Offshore Waters
Order Decapoda (continued)		
Isocheles wurdmanni	X	X
Labina dubia	X	
Latreutes fucorum		X
L. parvulus	X	X
Leander tenuicornis		X
Libinia dubia		X
Metoporhaphis calcarata		X
Neopanope texana		X
Ogyrides alphaerostris		X
O. limicola	X	X
Ovalipes guadulpensis	X	X
Paguridae	X	
Paguristes hummi		X
Pagurus annulipes	X	
P. longicarpus	X	
P. pollicaris	X	X
P. sp.		X
Palaemonetes vulgaris	X	
Panopeus herbstii	X	
Penaeus aztecus	X	X
P. duorarum	X	
P. fluviatilis	X	X
P. setiferus	X	
Persephona crinita		X
P. punctata		X
P. sp.		X
Petrochirus diogenes		X
P. bahamensis		X
Pilumus pannosus		X
Pinnixa cristata		X
P. sp.		X
Porcellanidae		X

	Timbalier Bay	Offshore Waters
Order Decapoda (continued)		
Porcellana sayana		X
Portunidae		X
Portunus gibbesi	X	X
P. spinimanus		X
Processa hemphilli		X
Sicyonia brevirostris		X
S. dorsalis		X
Synalphaus townsendi		X
Trachypenaeus constrictus	X	X
T. similis	X	X
Upogebia sp.		X
Xanthidae	X	
Xiphopeneus kroyeri	X	
Phylum Ectoprocta		
Acanthodesia sp.		X
Bugula neritina	X	X
Membranipora sp.		X
Phylum Echinodermata		
Class Asteroidea		
Astropecten duplicatus		X
Luidia clathrata		X
Class Echinoidea		
Arbacia punctulata		X
Class Ophiuroidea		
Ophiuroidea		X
Phylum Chordata		
Class Tunicata		

	Timbalier Bay	Offshore Waters
Class Tunicata (continued)		
Ascidians		X
Class Pisces		
Anchoa hepsetus	X	X
A. lyolepis	X	X
A. mitchilli		X
A. nasuta		X
A. sp.		X
Anquilliformes	X	X
Antennarius radiosus		X
Archosargus probatocephalus		X
Arius felis	X	
Bagre marinus	X	
Bairdiella chrysura		X
Balistes capriscus		X
Bothidae		X
Brevoortia patronus	X	
Caranx crysos		X
C. hippos	X	X
Carcharhinus leucas		X
Canthidermis sufflamen		X
Centropristis ocyurus		X
C. philadelphicus		X
Chaetodipterus faber	X	X
Chloroscombrus chrysurus	X	X
Citharichthys spilopterus		X
Clupeidae	X	
Cynoscion arenarius	X	X
C. nebulosus	X	
C. nothus	X	X
Dasyatis sabina	X	
Diplectrum bivittatum		X
Dorosoma cepedianum	X	
Etropus crossotus	X	X
Eucinostomus argenteus		X

	Timbalier Bay	Offshore Waters
Class Pisces (continued)		
E. gula		X
Halieutichthys aculeatus		X
Harengula pensacolae	X	X
Hypleurochilus geminatus		X
Lagodon rhomboides	X	X
Larimus fasciatus	X	X
Leiostomus xanthrus	X	X
Lutjanus campechanus		X
L. griseus		X
L. synagris		X
Membras martinica	X	
Menticirrhus americanus	X	X
Micropogon undulatus	X	X
Mugil curema	X	
Myrophis punctatus	X	
Ogcocephalus parvus		X
Ophidion welshi	X	
Opisthonema oglinum	X	X
Orthopristis chrysopterus	X	X
Paralichthys lethostigma		X
Peprilus alepidotus	X	X
P. burti	X	X
Polydactylus octonemus	X	X
Pomatomus saltatrix	X	X
Pontinus longispinis		X
Porichthys porosissimus		X
Prionotus pectoralis		X
P. roseus		X
P. rubio	X	X
P. salmonicolor		X
P. scitulus	X	
P. sp.	X	X
P. tribulus	X	X
Rhizoprionodon terraenovae	X	
Saurida brasiliensis		X

APPENDIX 1 (continued)

	Timbalier Bay	Offshore Waters
Class Pisces (continued)		
Sciaenidae	X	
Scomberomorus maculatus	X	
Selene vomer	X	X
Serranidae		X
Sphoeroides parvus	X	X
Stellifer lanceolatus	X	
Stenotomus caprinus		X
Syacium gunteri		X
Symphurus civitatus	X	X
S. plagiusa		X
Syngnathus louisianae	X	
Synodus foetens	X	X
Trachurus lathami		X
Trichiurus lepturus		X
Urophycis floridanus		X
Vomer setapinnis	X	X

IV

RESEARCH REPORTS BY PRINCIPAL INVESTIGATORS

PHYSICAL STUDIES OF THE NEAR-SHORE CONTINENTAL SHELF OF SOUTH CENTRAL LOUISIANA: CURRENTS AND HYDROGRAPHY

by P. Oetking, R. Back, R. Watson, and C. Merks

ABSTRACT

Measurements of currents and hydrography were made in the study area over an 18-month period in 1972-1974, as part of a multidisciplinary study to provide baseline information for evaluating the ecological effects of drilling and production of oil in offshore areas.

Both long and short duration current measurements indicated that there is a predominant drift eastward of the currents in summer and westward in winter and early spring, with tidal currents dominating during periods of transition. These findings support those reported by previous investigators.

Salinity, temperature, dissolved oxygen, and transmissivity were measured and analyzed for daily, seasonal, and areal trends. Large variations were shown to exist.

INTRODUCTION

Interpretation of the data obtained in this study is facilitated by first examining the general pattern of water circulation in the Gulf of Mexico (hereinafter referred to as the Gulf), as shown in figure 1. Using pilot charts, based upon ships' drift, Curray (1960) described the current patterns of the Gulf as follows:

At the time of this investigation, P. Oetking was Director and R. Back and C. Merks were Research Scientists at the Southwest Research Institute's Ocean Science and Engineering Laboratory at Corpus Christi, TX. R. Watson was at the Marine Science Institute, The University of Texas, Port Aransas, TX.

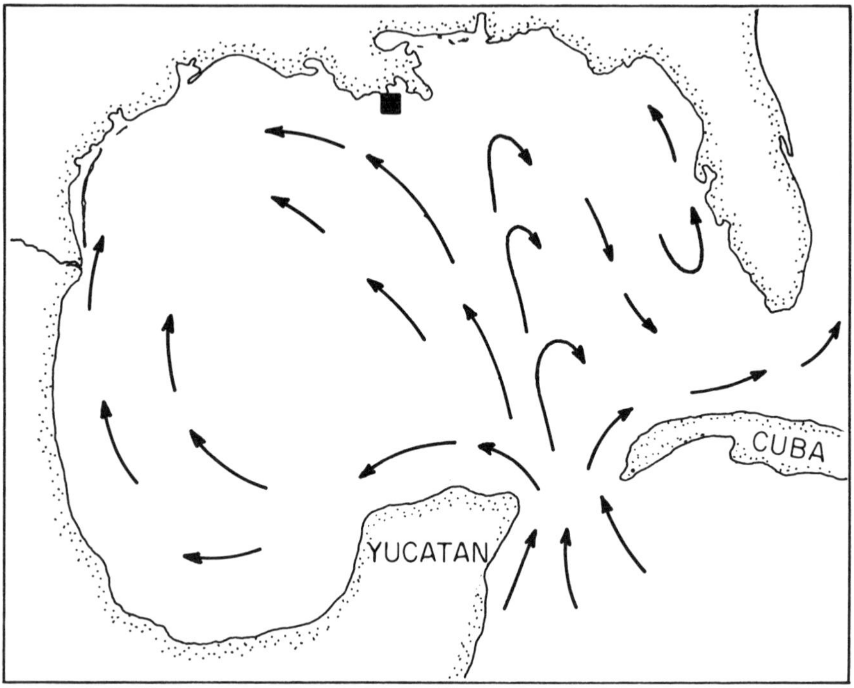

FIG. 1. GENERALIZED REGIONAL CURRENTS OF THE GULF OF MEXICO (U.S. Naval Oceanographic Office).

Water from the Caribbean entering through the Yucatan Channel southeast of the Mississippi Delta, divides approximately along a line connecting with the Mississippi Delta. Part of the eastern flow turns sharply to the east along the northern coast of Cuba into the Florida current and the Gulf Stream and part circulates clockwise in the northeastern Gulf. . . . The western cell also consists of two parts, one circulating clockwise in the southwestern Gulf of Mexico and the other part circulating counterclockwise in the northwestern Gulf.

During January, February, and March, there is a strong westward and southwestward flow across the shallow Louisiana Shelf, extending as far south as about latitude 28°. This flow converges near the coast at about latitude 27°30′ to 28° with the northward current flowing past the Rio Grande. In May, the winds are nearly southeast and the convergence starts to shift northward to approximately latitude 28° to 29°. The northeastward shift of the convergence resulting from the strong southeast and south-southeast winds continues through July. . . . The most sudden change of the year is in September when the winds shift abruptly to the east and the convergence moves in response to the wind. During the fall months the convergence occurs between latitudes 27° and 28°30′, apparently because of the northeast winds. The southwest nearshore current increases to a maximum in the winter at the time the winds have the greatest north component. The shelf currents off Louisiana are to

> the west . . . throughout the year. The winds apparently influence only the intensities and minor fluctuations in direction.

Drift bottle studies (Kimsey and Temple 1962; Kimsey 1963) on the northwest Gulf indicate that the currents moved east and southeast along the Louisiana Shelf and west, south, and east just west of the Mississippi River Birdfoot Delta (hereinafter referred to as the Delta). There appeared to be a large-scale eddy developed in the "embayment" west of the Delta when the easterly current was deflected seaward by the Delta.

Currents on the near-shore shelf are primarily driven by local and regional wind systems, passage of the diurnal tide, and impingement of regional Gulf currents onto the shelf. The resulting speed and direction of flow of the shelf currents at any time, location, and depth are dependent upon the vector sum of the various driving forces present and the constraint provided by the orientation of the shoreline and near-shore bottom topography. Winds are probably the principal force controlling currents. Curray (1960) described the wind patterns of the Gulf as follows:

> In January the winds are northeasterly along the eastern and the central Gulf shore and in the Gulf itself, while there is an onshore southeasterly wind along the western shore. During March, the northerly component diminishes, and by May the winds are generally east-southeast everywhere in the region. The southerly component is strongest and most consistent in June, July, and August, producing strong *resultant winds*. This is a difference in uniformity of direction rather than in average wind speed. The wind speed averaged without regard to differences in direction is, in fact, lower during the summer months than during winter months. The direction, however, is much more uniform during the summer, and the vector *resultant wind* is higher. An abrupt change to easterly winds occurs in September. During November, the effect of the local storms called "northers" can be seen by the increase in the northerly component in both the sea and land stations.

Wind data gathered at Ft. Livingston Light, Grand Isle, Louisiana, indicate that the prevailing winds (according to Barrett [1971]) are: September through February, N to NE; April through August, generally SE to SW; and March, variable N to S. Both the regional and local wind patterns tend to produce surface currents to the west and southwest during the winter months. Local southwest winds coupled with development of northeasterly drift on the upper Texas coast, driven by strong southeast winds during the summer months, probably generate the easterly currents on the Louisiana Shelf.

In summary, the Offshore Ecology Investigation (OEI) study area is located on the northeast corner of a counterclockwise circulation cell in the northwest Gulf. The circulation is modified by local winds, tidal currents, and possibly by eddies caused by the projection of the Delta. An examination of the data should show an enhancement of the westerly current during the winter months and retardation or even reversal of it during summer months. Any effects due to tides should appear throughout the year superimposed on the net drift of the regional currents.

Tides in the Delta region are dominantly diurnal and have a maximum tide range of about 2.5 feet (75 cm) when the sun reaches maximum declination. The minimum range occurs when the sun is over the equator. Thus, annually the maximum tide range and, hence, the maximum influence of the tide on the shelf currents occurs in December and June, the minimum influence in March and September. The largest monthly fluctuations occur when the moon's declination is greatest; the cycle is fortnightly (National Ocean Survey 1972 and 1973).

The studies reported on herein were conducted to provide information about near-shore water current and hydrographic characteristics to confirm the work of earlier investigators and to supplement the biological, chemical, and physical data collected by other investigators.

Although the current data were not sufficiently complete to provide seasonal or annual estimates of direction frequency, they were representative of the currents at the study site for the duration of each of the measurements. Further, the data identified the major current patterns on the Louisiana Shelf and indicated their probable driving force. The site of the study, approximately 50 miles (92 km) west of the Delta and only 5 to 15 miles (9 to 28 km) offshore in water depths of less than 30 meters, is an area where shoreline influence on the regional shelf currents may be severe.

METHODS AND MATERIALS—CURRENTS

Measurements of current speed and direction were made with the Marine Advisors Model Q-12. Four meters were used to make long duration measurements in which the data were recorded within the instrument in integral Rustrak recorders. A fifth meter was used for point sampling of the currents. This meter was adapted to provide a direct readout of the data aboard the vessel. The instruments consist of a savonius rotor and a magnetically referenced direction vane. Both configurations are capable of reading on either of two scales: 0.05-2.0 knots or 0.05-5.0 knots. Accuracy of the system is $\pm 2\%$ of the reading for speed and $\pm 10°$ for direction. Calibration of the systems was maintained by simulating the output of the savonius rotor at various speeds with a pulse generator and adjusting the readout to a standard established by the manufacturer.

The water depth at each sampling station was measured by an EDO Western, Model 353, piezoelectric transducer (34 kHz) and marked on a Gifft, Model 4000T, recorder. In addition, a metering wheel on the winch cable was used to establish water column sampling positions.

The R/V *Southwest Researcher I* was the principal sampling platform for the study. Cruises were made in August, September, October, and December 1972; January, February, April, July, and October 1973; and January 1974. Both long and short duration current measurements were made. Long duration (referred to as continuous) current measurements were made by suspending current meters from South Timbalier oilfield production platforms HOR-ST-66D and SH-ST-26C at the surface, mid-depth, and bottom, and date of emplacement, location, and duration were recorded. The current velocity was measured automatically every half hour for 3 minutes throughout each of these periods. Twenty of the 50 short duration measurements of the currents were made within a radius of 0.2 km of platform HOR-ST-54A (also called Exxon 54A in other papers in this collection) and 0.5 km of HOR-ST-66D and the GURC grid number of each station was recorded. Current measurements were made throughout the water column at each sampling station after the vessel was anchored or secured to a platform. A total of 4087 readings was made with short duration and continuous current meters.

RESULTS AND DISCUSSION—CURRENTS

Current direction modes

The continuous and short duration measurements indicate four major current direction modes. Figure 2 summarizes all continuous and short duration measurements taken from August 1972 to January 1974. Flood tidal currents are represented by the 315°-45° sector, and the ebb tidal currents occupy the 105°-225° sector. The winter westerly drift varies from southwest to northwest, occupying the 225°-315° sector, and the summer easterly drift (developed from strong southerly winds) occupies the 45°-105° sector. The respective mean speeds of flood and ebb tidal currents are the following: surface 0.45 knots, 0.35 knots; mid-depth 0.30 knots, 0.26 knots; and bottom 0.21 knots, 0.22 knots. The westerly drift flowed with an average speed of 0.82 knots at the surface, 0.26 knots at mid-depth, and 0.22 knots at the bottom. The easterly current flowed only slightly faster than the tidal currents at average speeds of 0.40 knots at the surface, 0.30 knots at mid-depth, and 0.23 knots at the bottom. The number of observations upon which these averages were based is tabulated and recorded. Since the current direction data are not continuous for an entire year, but represent the sum of a series of shorter-term measurements, the direction frequency may be biased. The frequency diagrams are intended to show which current directions are prominent.

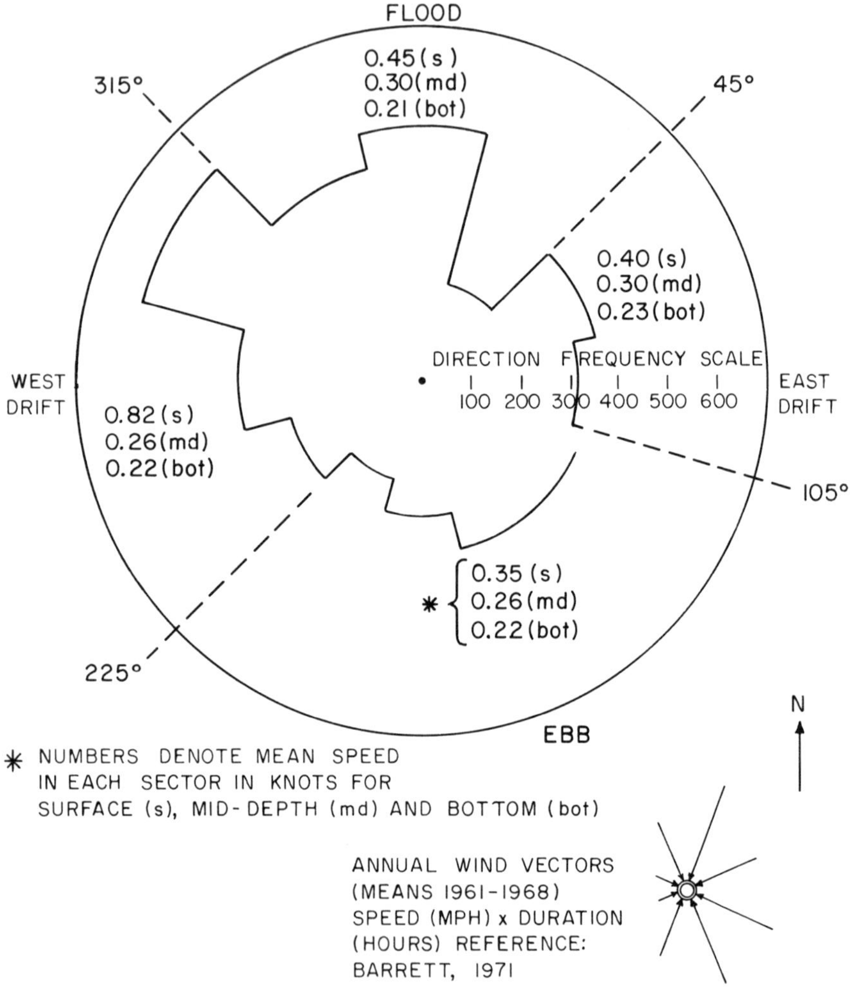

FIG. 2. CURRENT DIRECTION FREQUENCY DIAGRAM: All short duration and continuous measurements in total water column.

The effect of strong tidal currents on either an easterly or a westerly drift produces a pattern of loops with net movement in the dominant drift direction. The loops in the progressive current vector diagrams occur daily and are thought to be the result of Coriolis deflection of tidal current (MacMillan 1966). In the absence of a drift current, tidal forces produce a current pattern consisting of overlapping closed loops. During ebb tide the current was deflected to the west, and during floodtide the current was deflected to the east. The net result is depicted as an ellipse

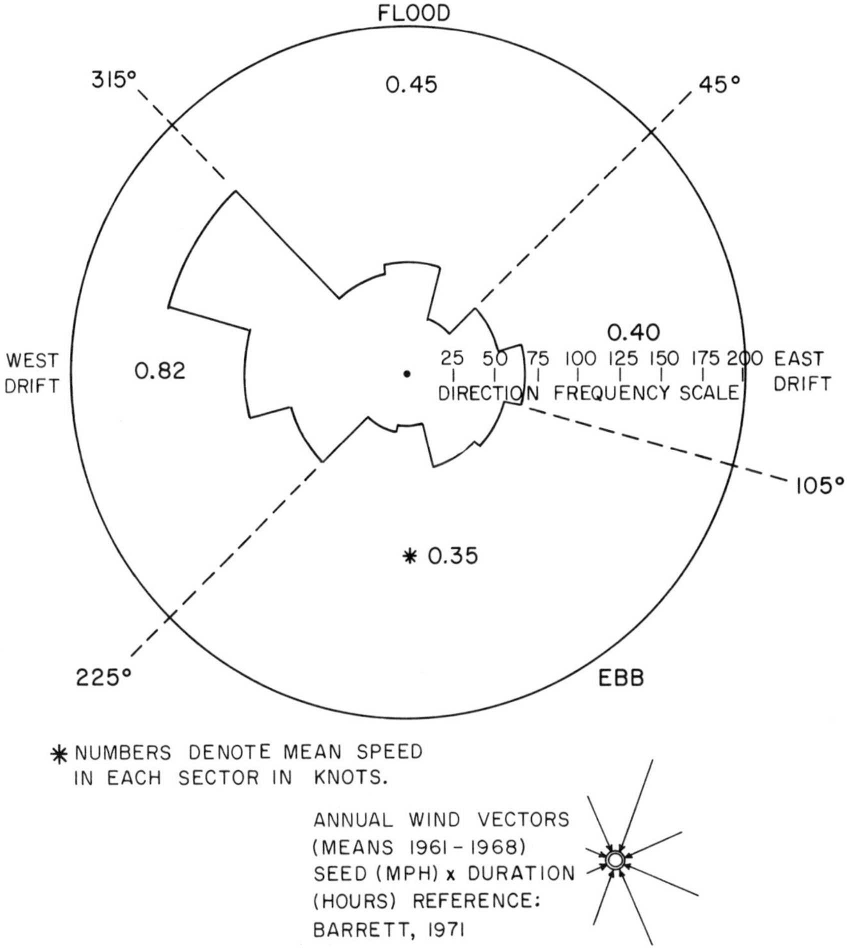

FIG. 3. SURFACE CURRENT DIRECTION FREQUENCY DIAGRAM: All short duration and continuous current meter measurements.

that traces the flow of the tidal current. When the strength of the easterly or westerly drift was increased, the daily loops were separated by a distance equivalent to the daily drift of the regional current. If the regional drift was fast relative to the velocity of the tidal currents, the loops no longer existed because the tidal velocities opposing the regional current never became fast enough to cause a current reversal.

Current direction frequencies

Direction frequency diagrams for each of the sampling depths (i.e.,

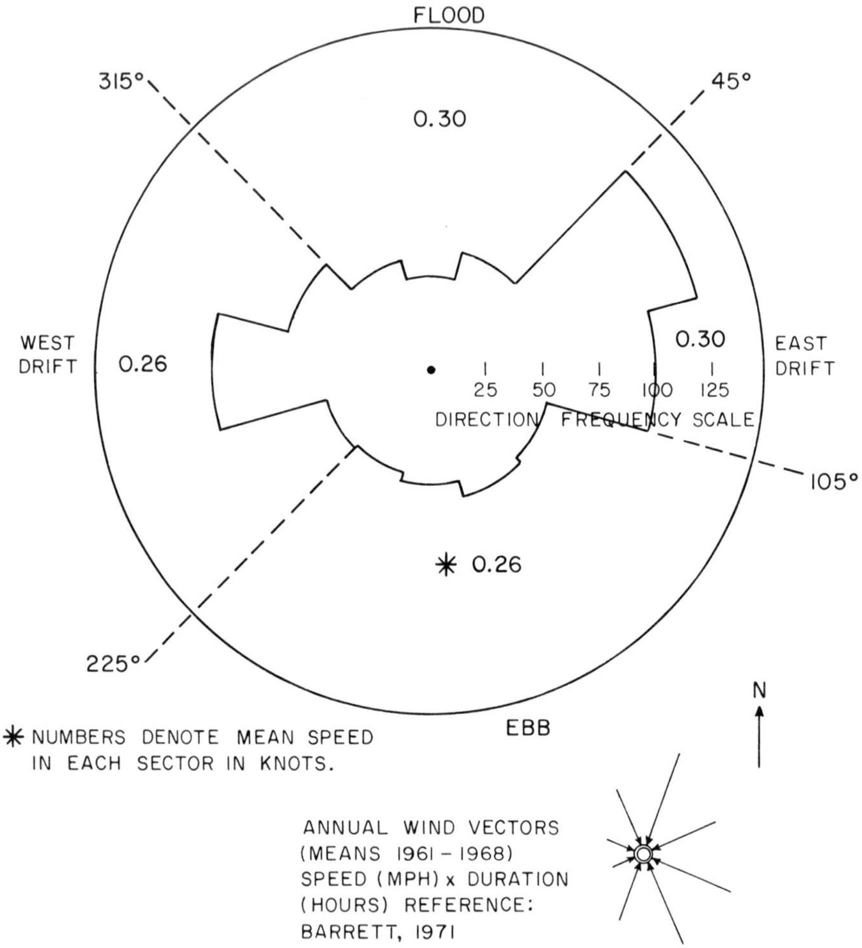

FIG. 4. MID-DEPTH CURRENT DIRECTION FREQUENCY DIAGRAM: All short duration and continuous meter measurements.

surface, mid-depth, and bottom) were constructed, based on both the short duration and continuous measurements (figures 3, 4, and 5). It can be seen from these diagrams that the wind-driven easterly and westerly drift currents were dominant at the surface and mid-depth, whereas onshore currents (flood tides) were dominant at the bottom of the water column. Ebb currents appeared the strongest at the surface and mid-depth, although either flood or ebb tides could dominate the current direction at any depth.

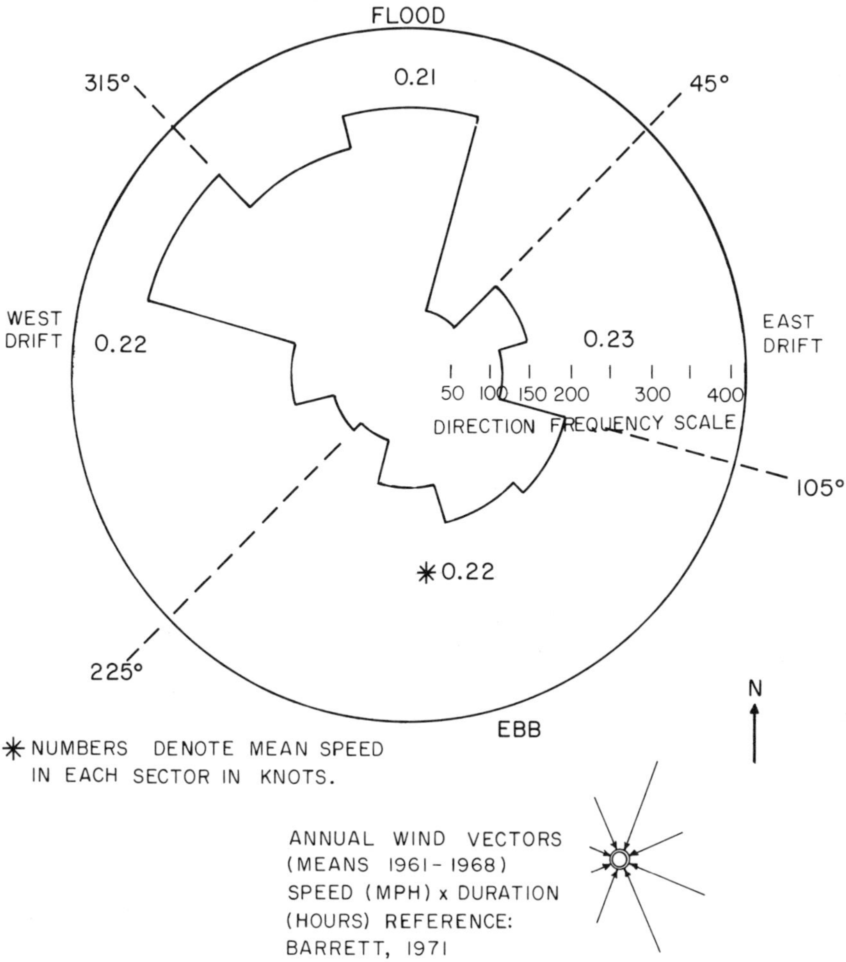

FIG. 5. BOTTOM CURRENT DIRECTION FREQUENCY DIAGRAM: All short duration and continuous current meter measurements.

Summer regime. The direction frequency diagrams for August 1972 depict surface currents to the east and northeast with some ebb and flood currents. Current flow in the ebb direction was the strongest at mid-depth, whereas the currents in the flood direction were most pronounced at the bottom. The direction frequency diagram for July 1973 also indicates that the easterly drift current was strong at the surface. At mid-depth the continuous readings indicated flow in the easterly, southerly (ebb tide), and northerly (flood tide) directions. The more variable currents at the bottom flowed predominantly to the east, west, and north. In general, the presence of strong easterly drift on the Louisiana Shelf during summer agreed with the findings of Kimsey and Temple (1962).

Winter regime. In December, the dominant surface currents flowed to the northwest. Mid-depth currents were primarily westerly and offshore, indicating the effect of ebb tidal currents, whereas the bottom currents were westerly but strongly influenced by the tides. A progressive vector diagram for December 5-8, 1972, shows a strong northwest surface drift with no tidal current loops. The effect of the tidal currents does not appear on the diagram because they were weak relative to the swift westerly surface current, which was driven by strong regional winds. Current data gathered in January, February, March, and April 1973 are also representative of the winter regime. An example of local wind modification of the currents was seen in January 1973. On the morning of January 23, the current was flowing to the west at 1.0 to 1.2 knots (50 to 60 cm/sec). At the same time the wind changed from westerly to northerly and became stronger. Near midnight the current had become southwesterly at 0.8 knots (40 cm/sec). The current became westerly again on January 26 after the north winds had subsided.

Transitional months. September and October current characteristics were transitional between the summer and winter regimes. The short duration current patterns for September 1972 depict the effect of the flood tides at all depths. However, the continuously recorded data from platform HOR-ST-66D show strong northeasterly drift in addition to significant flood and ebb currents. The progressive surface current vector diagram for September shows a strong easterly drift superimposed with pronounced tidal current loops on September 17. But by September 22, the loops disappear because the tidal currents became weaker as determined from the tide table. For several days preceding these current measurements, the winds were southerly, a phenomenon that may have given a northward component to the easterly drift current producing a net flow to the northeast. Apparently it was too early in the fall for the westerly drift to have developed.

The short duration current measurements for October 1972 were dominated by flood tide at the bottom and ebb tides at both mid-depth and the surface. The continuous surface measurements taken at platforms HOR-ST-66D and SH-ST-26C appear anomalous. The current at platform HOR-ST-66D was predominantly onshore and westward, while at SH-ST-26C (15 miles, or 24 km, to the northeast) the surface current for the same period was offshore and westward. Since October 1972 was characterized by neither an easterly nor westerly drift, the tidal currents appeared to have formed a large clockwise eddy just west of the Delta. With the development of either a strong easterly or westerly drift, such an eddy would probably soon dissipate. In October 1973, continuous current measurements showed well-defined westerly and southwesterly currents at surface and mid-depth, possibly indicating the establishment of the winter westerly drift as early as October.

Unfortunately, there are no current data for May or June. It is likely that the spring transitional period occurred then. Since the regional currents offshore were dominantly westerly and were enhanced by the winter north and northeast winds producing westerly surface drift with average speeds more than twice as great as easterly drift or tidal currents, the westerly drift was likely to persist until late in the spring, well after the south and southeast winds had begun on the Texas and Louisiana coasts. Drift bottle data from the Texas coast have indicated that tongues of southward-flowing waters were detectable within the general northward-flowing water mass well after the beginning of the strong southeastern wind season (Watson and Behrens 1970).

SUMMARY—CURRENTS

1. The net movement of the water column, which was driven primarily by regional winds, was easterly in the summer and westerly from winter through early spring.
2. Onshore and offshore currents prevailed in the fall and probably late spring.
3. Surface currents were generally easterly in the summer (0.40 knots) and westerly from winter through early spring (0.82 knots).
4. Mid-depth currents (0.26 knots) were easterly, onshore and offshore in the summer; westerly and offshore from winter through early spring.
5. Bottom currents (0.22 knots) were mostly onshore and easterly in the summer; westerly, onshore and offshore from winter through early spring.
6. Current speed decreased with depth.
7. The net annual movement of water in the study area was westerly.

CONCLUSIONS—CURRENTS

The currents of the near-shore continental shelf of south central Louisiana were variable throughout the course of this study. The currents varied with the seasons because regional winds produced an easterly flow during the summer months and a westerly flow from winter through early spring. Local winds caused some temporary changes in the direction of the currents. Onshore and offshore currents resulting from diurnal tidal action varied throughout the lunar cycle; the greatest effects were felt during the maximum declination of the moon. Current speed and direction varied with depth. The net annual movement of water in the study area was westerly for two reasons: The current flowed in that direction for nearly half the year, and westerly surface currents moved twice as fast as all other currents.

METHODS AND MATERIALS—HYDROGRAPHY

Seasonal and regional hydrographic measurements of salinity, temperature, transmissivity, dissolved oxygen, and current were made from surface to bottom along onshore-offshore traverses, along traverses trending roughly parallel with the shoreline, and along traverses including one or more offshore platforms. Water data were collected on four cruises in 1972, five in 1973, and one in January 1974, at 90 different sampling stations, many of which were reoccupied.

The water column and ancillary measurements were made with the following instruments:

Temperature and salinity. A Beckman Model RS5-3 direct reading portable salinometer was used to determine the *in situ* temperature, conductivity, and salinity. This instrument uses a thermistor probe in a balanced bridged circuit to measure temperature, and the conductivity is derived from the inductive coupling between the two coils in the probe. The salinity is automatically calculated by the readout unit from the temperature and conductivity measurements. The instrument is capable of the following accuracies:

Salinity ± 0.3 ppt (0-40 ppt)
Conductivity ± 0.5 millimhos/cm
Temperature ± 0.5° C

These accuracies were maintained by daily field calibration and verified by returning random samples from each sampling series for measurement using a Hytech, Model 6200, laboratory salinometer. The temperature measurements were calibrated by periodically checking the readings with a Kahlsico, Model 297WA105, porcelain scale thermometer.

Transmissivity. the light transmission properties were measured with a Hydro Products, Model 410BR, transmissometer with a 10 cm Model

411 head. This instrument provides a direct reading of the percentage of transmittance in an *in situ* water sample. Accuracy was maintained by "air calibration" at each station and by periodically calibrating the transmissometer to 100% in distilled water.

Dissolved oxygen. The dissolved oxygen in the water column was measured *in situ* with a Martek, Model DOA, dissolved oxygen analyzer. This instrument uses a gold/silver polarographic probe with a special Teflon membrane. Measuring accuracy is ± 1% of full scale on each of three selectable ranges: 0-2, 0-10, and 0-20 ppm. Calibration was maintained daily by checking the oxygen content of an air-saturated distilled water sample. Spot verification of *in situ* measurements was made using a standard Winkler titration method.

Water depth. The water depth at each sampling station was measured by an Edo Western, Model 353, piezoelectric transducer (34 kHz) and marked on a Hydro Products, Model 4000T, Gifft recorder. In addition, a metering wheel on the winch cable was used to establish water column sampling positions.

Wind velocity. Wind speed and direction were measured at each station from the same elevated vessel position with a Kahlsico, Model 03AM120, hand-held anemometer. The instrument accuracy is ± 1 knot in the range of 0-60 knots for speed and ± 10° for direction. The hand-held instrument was checked before and after each cruise with the laboratory wind speed monitoring system.

Air temperature. The thermistor probe of the Beckman, Model RS5-3, *in situ* salinometer was used to measure the air temperature. The readings were taken after the measurements in the water column to minimize the influence of solar heating of the probe head.

The R/V *Southwest Researcher I* was the principal sampling platform for the study. A total of 37 days of water column data was acquired during the 10 cruises. Measurements were made in August, September, October, and December 1972, in January, February, April, July, and October 1973, and January 1974.

The sampling program was designed to collect representative hydrographic data near an active production platform, at a sampling station outside the influence of petroleum activities, and at various other locations deemed necessary to provide the data required by other investigators. Whenever the vessel was on station for the purpose of sampling, observing, or measuring any parameter, water column information was acquired. The salinometer, transmissometer, oxygen probe, and current meter were fastened together to form the water quality package, which was attached to a winch-driven hydrowire. The extent to which the various water quality data were measured at each station was dependent on the duration of the primary sampling activity, and for current measurements, on whether or not the vessel was fixed by anchor or platform.

RESULTS AND DISCUSSION—HYDROGRAPHY

Seasonal trends

For each study period, all surface and bottom data for salinity, temperature, transmissivity, and dissolved oxygen were individually averaged from stations located in water 15 meters or deeper. Near-shore data were not included in the seasonal averages because of the added influence of local runoff, turbulence, etc.

Salinity. In general, throughout the year the surface salinity was lower than the bottom salinity. This was expected because of the surface effects of fresh water from local runoff and the Mississippi River. From August to December 1972, the surface salinity was only slightly lower than the bottom salinity, and some of the highest surface salinities were higher than some of the lowest bottom salinities (figure 6). During this period, no steep vertical salinity gradients were observed. The water column had little salinity stratification because the water was well-mixed throughout the fall and early winter.

By January 1973, the bottom salinity increased to near normal sea water salinity of 35-36 ppt. This appeared to be caused by a combination of the dominant onshore component of the bottom currents and the lessening of the fresh water runoff. Data from Houma, Louisiana, U.S. Weather Bureau, USDA Observer, show average monthly rainfalls of 6.1 inches (15.5 cm) for August-December 1972, and 3.3 inches (8.4 cm) for January-February 1973, hence a decrease in the source of fresh water during mid-winter. Throughout the period from January to September, the envelope of surface salinities and the envelope of bottom salinities did not intersect. January and February 1973 vertical salinity profiles show a pronounced salinity gradient beginning at about 10 meters, with a relatively well-mixed surface layer. Data gathered on January 13, 1974, were unusual in that the water column was temporarily mixed, the mixing probably caused by strong offshore winds. Because of these data and the lack of cruises in November and December, the graphic display for this period is dissimilar to that of the previous year.

On April 2, offshore surface salinities were as low as 13 ppt with a very steep gradient to 28 ppt at about 6 meters, and there was a gradual

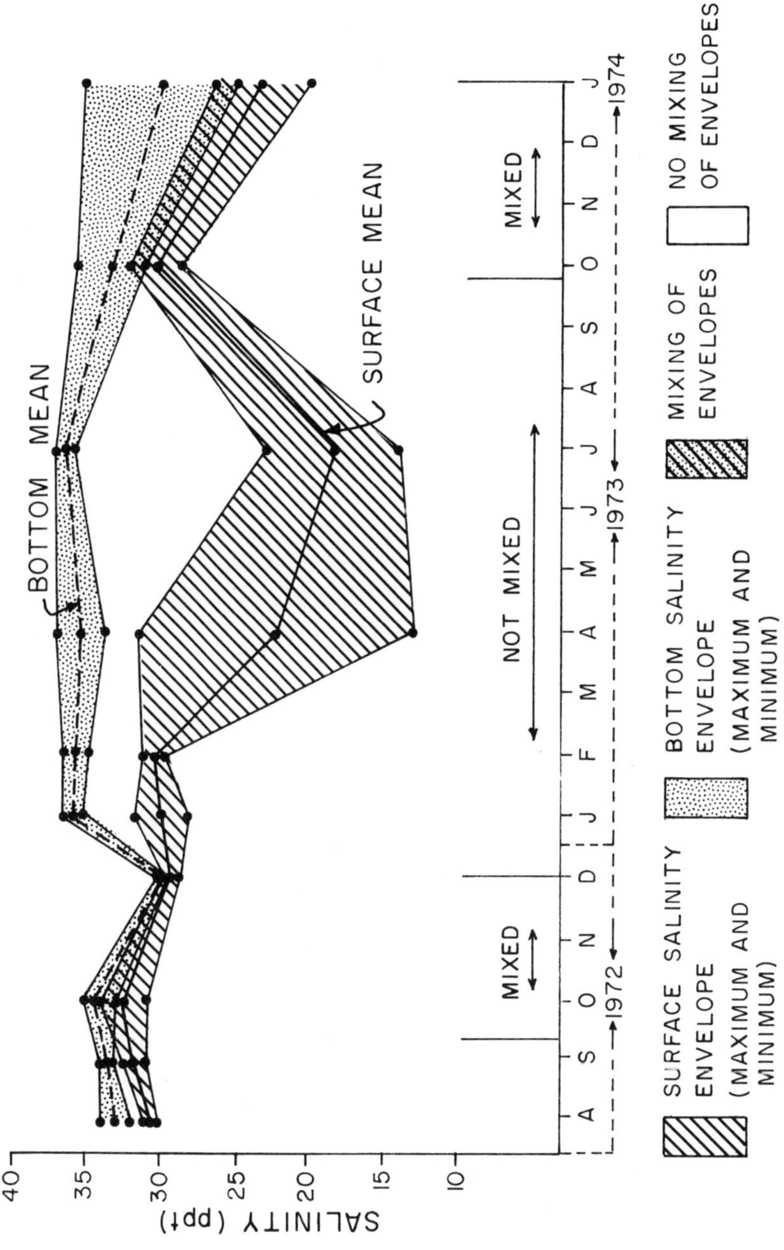

FIG. 6. MONTHLY MEANS OF SALINITY FROM SURFACE AND BOTTOM DEPTHS.

increase to near normal sea water salinities of 35 ppt at the bottom. The extremely low surface salinity in early April can be attributed to local runoff as well as to the spring flooding of the Mississippi River just 50 miles (93 km) to the east. March rainfall at Houma, Louisiana, was 9.5 inches (24.2 cm). During the late winter and spring, the surface currents on the nearshore Louisiana Shelf were predominantly westerly, with the result that the spring floods of the Mississippi River considerably diluted the surface waters west of its mouth. In July, the surface salinities were still low, averaging about 19 ppt with a steep gradient at 3-8 meters.

When offshore salinity data, taken during 1968-69, are compared with records of Mississippi River stage and discharge, they show a strong inverse correlation. Salinity was sharply reduced during the high spring stage and discharge of the Mississippi River in the area west of Southwest Pass (figure 7). The study found maximum offshore salinities of 29-30 ppt during November and December and minimum offshore salinities of 16-17 ppt, with the spring floods of the river from March through June (Barrett 1971). Our data indicate that offshore waters were mixed during the fall months, with surface and bottom salinities ranging from 20 to 35 ppt. Beginning with spring flooding, principally from the Mississippi River, surface salinity decreased sharply, and a pronounced salinity gradient developed at a depth of 3-12 meters. Bottom salinities increased to the level of normal sea water, indicating little exchange between surface and bottom waters.

Temperature. The offshore surface and bottom temperatures show seasonal trends that correlate with the seasonal changes in salinity. During the fall and early winter, when the water column was well mixed, bottom temperatures differed only slightly from surface temperatures (figure 8). Sometime during December or January 1972, the bottom temperature ceased to drop. This occurred with the increase in salinity of the bottom waters from about 30 to 36 ppt. The surface water temperature continued to decrease to a low of about 16° C, when the lowest bottom temperature observed was about 19° C.

Thus, during the cooler months (December, January, February, March, and perhaps part of April), the surface temperatures were lower than the bottom temperatures. By early spring, surface waters became warmer than bottom waters again. By July 1973, the surface water temperature reached an average of 30° C with bottom temperatures of 24° C. The large difference between surface and bottom water temperatures is in agreement with the salinity data, indicating little mixing of surface and bottom waters.

Dissolved oxygen. The seasonal trends of the surface and bottom

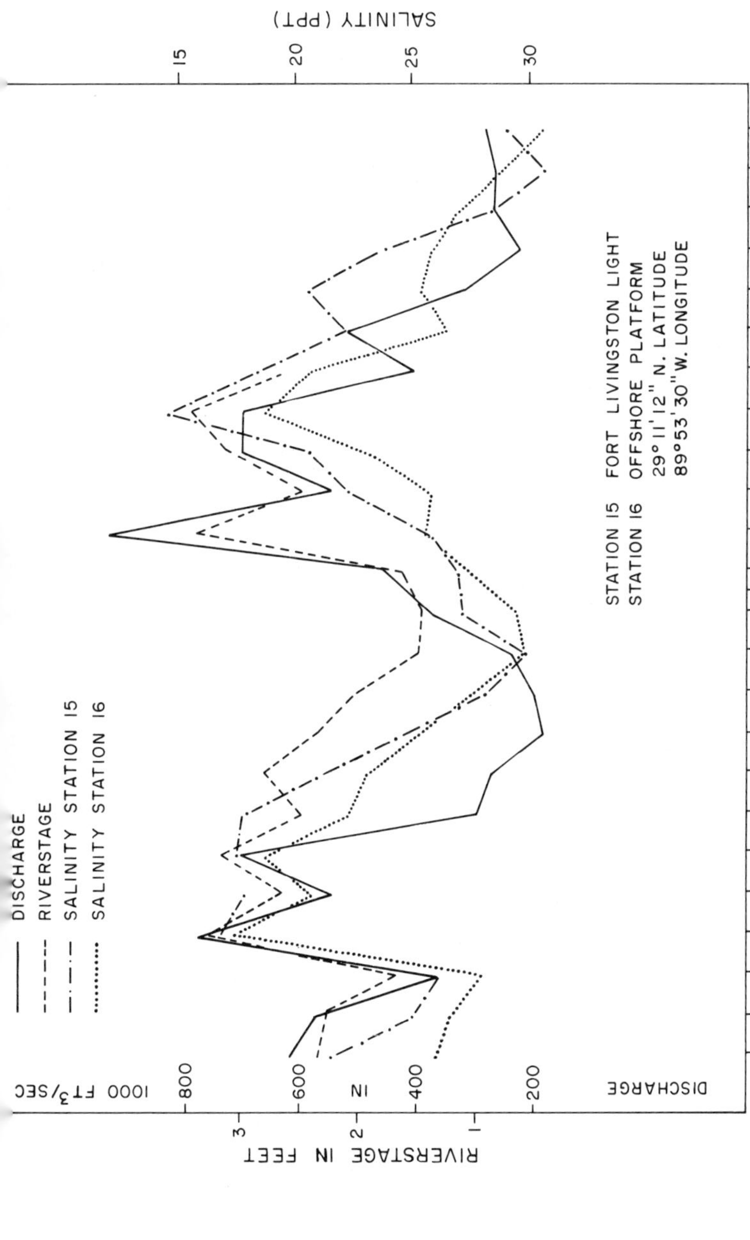

FIG. 7. RELATIONSHIP BETWEEN MONTHLY AVERAGE SALINITIES at stations 15 (Grand Terre slip) and 16 (Grand Isle offshore platform) and monthly average stage of the Mississippi River at head of passes and discharge at Tarbert Landing.

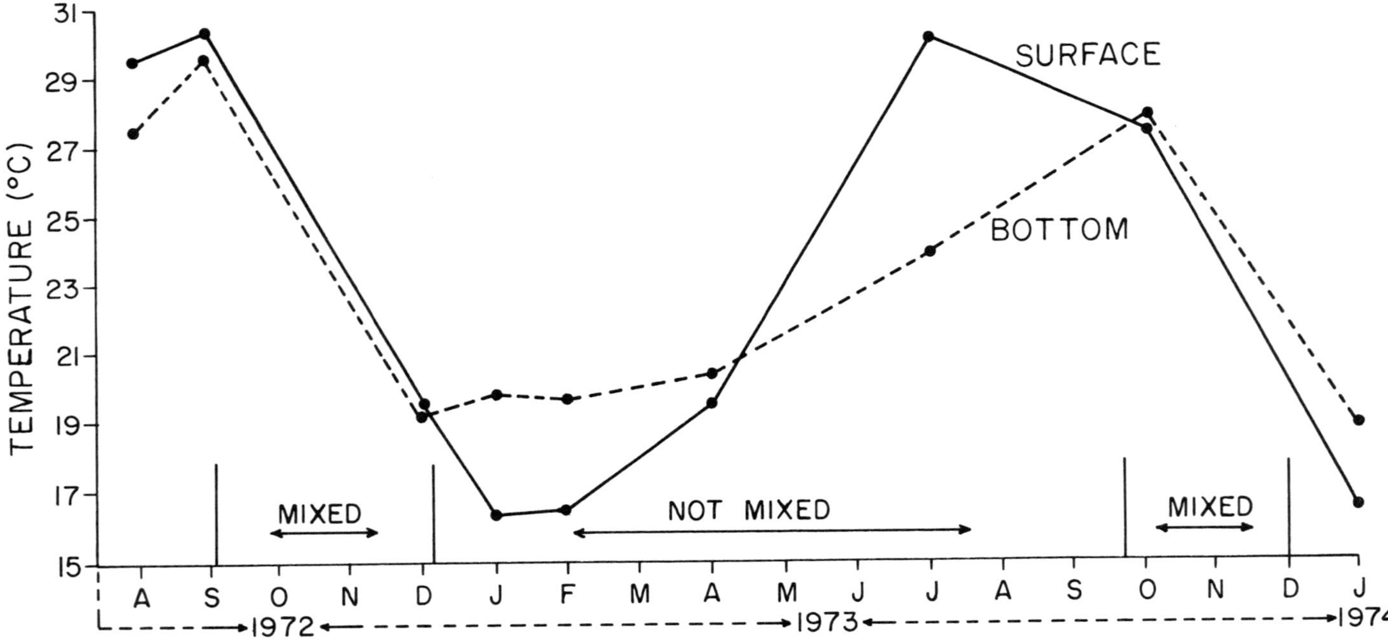

FIG. 8. MONTHLY MEANS OF TEMPERATURE FROM SURFACE AND BOTTOM DEPTHS.

dissolved oxygen in the offshore area reflect seasonal trends in temperature, mixing, salinity, respiration, and photosynthesis. High dissolved oxygen contents are promoted by low temperature, high wave turbulence, high rate of photosynthesis, and low salinity. Vertical mixing is a primary control on the vertical oxygen distribution.

Surface dissolved oxygen ranged from 6.0 ppm in the summer to 10.0 ppm in the winter. The highest dissolved oxygen figures corresponded to the period of coldest surface water (figure 9). Winter storms and the resulting wave action also increase the probability that surface waters will become saturated or supersaturated with oxygen. Figure 9 includes seasonal oxygen saturation index curves determined for the surface and bottom waters from monthly average salinities and temperatures, excluding mixing and biological effects. The rate of uptake by organisms probably has little effect on the surface dissolved oxygen because it can be easily replenished at the sea/air interface.

The seasonal variability in bottom dissolved oxygen was appreciable; dissolved oxygen reached a high of about 7.5 ppm during January and a low of less than 2.0 ppm during July 1973. Bottom dissolved oxygen shows almost a perfect inverse correlation with surface temperature and almost as good an inverse correlation with bottom temperature (figure 9). The major source of oxygen for the bottom waters was mixing with the surface waters above. Since organisms are ordinarily consuming at least part of the oxygen in the water column, the bottom waters are usually lower in dissolved oxygen than the surface waters, unless the column is well mixed or the bottom waters are from a completely different water mass.

The extremely low dissolved oxygen observed in the bottom waters during summer was probably a function of several factors. The near perfect inverse correlation with surface temperature implies that the bottom dissolved oxygen was controlled by temperature or by something that is temperature controlled. However, the less perfect correlation of bottom dissolved oxygen with bottom temperature and the lower variability of surface dissolved oxygen than of bottom dissolved oxygen suggest that the major factor controlling the variation of bottom dissolved oxygen from April to July was not solubility as controlled by temperature, but was respiration and decomposition.

Transmissivity. The transmissivity of the surface water, a measure of the water clarity, showed no recognizable seasonal trends. Throughout most of the year, the offshore surface transmissivity ranged from 80% to 90%. In water near the bottom, seasonal effects were detected; average transmissivity was highest during the summer (90-100%) and lowest in April (40%).

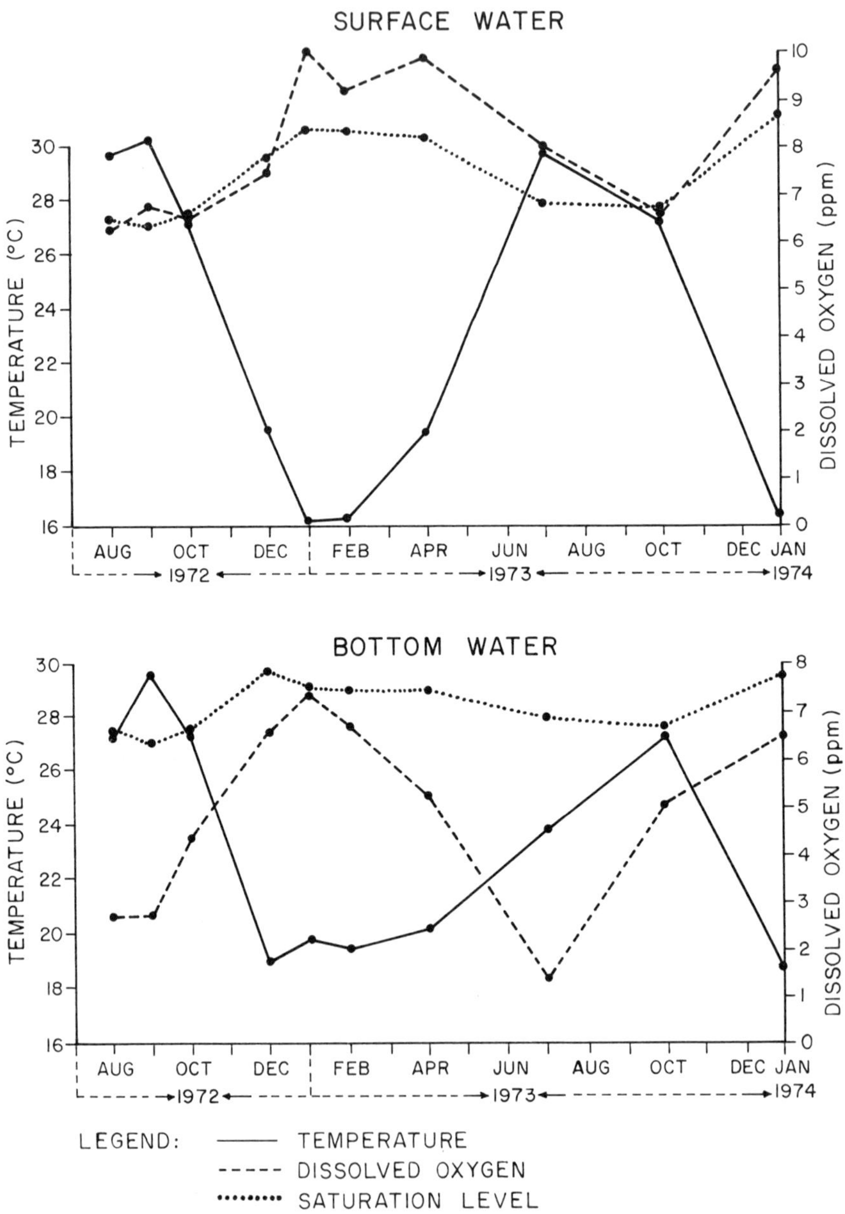

FIG. 9. MONTHLY MEANS OF TEMPERATURE, OBSERVED DISSOLVED OXYGEN, AND OXYGEN SATURATION SOLUBILITY.

The low transmissivity values in April showed a steep gradient near the bottom with clear water above. It appears that this may have represented downslope movement along the bottom of dense, sediment-laden waters. The coincidence in April of very low bottom transmissivities with spring flooding suggests that land-derived material suspended in runoff was the source of the turbid bottom layer. Also, strong flood-tide currents along the bottom may have been capable of some shoreward movement of a dense layer of suspended material. The maximum bottom current during the April cruise was moving northward at 0.34 knots (17.5 cm/sec.).

Although the surface waters were generally clearer than the bottom waters, the reverse was true in July 1973. Surface water runoff from the rivers and bays and high phytoplankton concentrations lowered the clarity of the surface waters. The strong density gradient in the upper waters may have been a deterrent to particle settling. The unusual clarity and the high salinity of the bottom half of the water column suggest that this water originated offshore.

Onshore-offshore trends

Salinity. Both surface and bottom salinity generally increased in the offshore direction. Near-shore salinities were usually lower because of the mixing of near-shore waters with local runoff coming through the tidal outlets. The salinity contours often dipped shoreward as a result of fresher, less dense water flowing seaward on the surface.

When the Mississippi River was in flood stage and the near-shore surface currents were flowing to the west, however, the offshore surface waters were significantly diluted by Mississippi River waters, and the surface salinity increased in the shoreward direction. A salinity profile taken on April 4, 1973, shows both of the trends noted above.

An anomalous condition occurred in the near-shore area in mid-January 1974. Strong north winds, which began on January 12, pushed cold brackish water out of the bays, and by January 13 this water dominated the water column on the shoreward side of platform HOR-ST-54A. In the area of the platform, the water remained stratified. The well-mixed water column was a temporary condition—stratification of the near-shore waters was re-established within a few days.

Temperature. Although surface temperature sometimes increased slightly in the offshore direction, it generally was constant regardless of onshore or offshore position on a given date. During the winter, bottom temperature increased in the offshore direction. By summer (July 1973), the bottom temperature trend was reversed; bottom temperatures decreased in the offshore direction.

Dissolved oxygen. The concentration of dissolved oxygen showed no well-defined trend in the onshore-offshore direction. Surface dissolved oxygen concentrations increased offshore, decreased offshore, or remained at the same level. Bottom dissolved oxygen decreased with depth and seemed to reflect local sources and sinks.

Transmissivity. Transmissivity increased in the offshore direction because there was more time for the land-derived particulate matter to settle. Also, the deeper offshore bottom was less frequently disturbed by wave action; consequently, mud was not locally introduced into the water column. In contrast, strong wave action in shallow water and muddy ebb tides and runoff leaving the bays caused greater sediment suspension in waters nearer shore.

Divers usually noticed a turbid layer near the bottom. This extensive low-transmissivity layer was well developed in the onshore-offshore profiles taken during April. The low-transmissivity zone was commonly more concentrated and thinner in the nearshore area and probably originated from dense turbid water leaving the bays or from wave turbulence on the near-shore bottom. The turbid water apparently flowed seaward along the bottom as a high density underflow. The salinity profile taken on April 4 shows evidence of seaward flow of low salinity water beneath more saline water. Apparently the density of this sediment-charged layer was sufficient to cause downslope movement.

Currents. The current direction diagrams accompanying most of the hydrographic profiles indicate one to five zones of water movement with depth. A correlation exists between the degree of mixing and the number of current direction zones in the water column. From August to December 1972, when there was little density stratification, three or fewer current zones were detected, but three or more were delineated when the waters were stratified. It appears that tidal currents and prevailing currents exist concurrently and independently of one another by occupying different density zones—ebb currents and east drift currents alternating from surface to bottom.

THE INFLUENCE OF OFFSHORE STRUCTURES ON HYDROGRAPHIC PARAMETERS

Many of the hydrographic profile transects include measurements taken at or near one or more offshore production platforms. Usually there was no detectable change caused by the platform, but on several occasions there were marked differences, which may have been caused by the production activities, in the water quality values.

A series of hydrographic changes occurred near platform HOR-ST-66D during the August 1972 cruise. On August 1, salinity, temperature, and dissolved oxygen appeared normal from surface to bottom: 30-32 ppt, 29-28.5° C, and 6.5-5.5 ppm, respectively. On August 2, there was a sharp drop in dissolved oxygen from 6.0 ppm at the surface to 1.8 ppm at the bottom. At a station 1500 feet (450 m) to the northwest, the bottom dissolved oxygen was 5.4 ppm, similar to that of the previous day. Neither salinity nor temperature was unusual. The two following days also showed low dissolved oxygen in the bottom waters throughout a 3-mile (5.5 km) distance from the platform. On August 3, however, the profile showed an increase in the salinity and a decrease in the temperature near the platform. In the middle water column, the maximum temperature decrease over 600 feet (180 m) was only about 0.4° C while the maximum salinity increase was about 1.2 ppt. The unusually low dissolved oxygen detected on August 2 was found by subsequent measurements to have had broad areal extent and does not appear to be associated with platform activities. It was probably a natural phenomenon, such as an intrusion of offshore waters, which happened to occur in the platform area at the time of the measurements.

On April 6, 1973, a profile transect passing near platform SH-ST-26C and extending about 6 miles (11 km) to either side of it showed unexpected large-scale changes in salinity, temperature, and dissolved oxygen. Transmissivity near the bottom increased in the vicinity of the platform and may have resulted from platform operations. Surface salinity increased from 21.6 ppt at 3 miles (5.5 km) from the structure to 32.7 ppt near the structure. Over the same distance the surface temperature increased 1.5° C. Unfortunately, the entire water column was not measured at closely-spaced intervals; consequently, it was impossible to determine the full area of influences of the higher salinity and temperature. Apparently associated with activities of the structure were low readings of dissolved oxygen, 5.0 ppm, relative to waters 5 miles (9.5 km) away, which had 10.3 ppm. Near the structure the dissolved oxygen profile is inverted, with lower concentrations of dissolved oxygen at the surface than at the bottom, indicating that the phenomenon was not affecting the bottom dissolved oxygen at that time.

With the exception of the data gathered near platform SH-ST-26C, no similar large-scale changes in water quality parameters attributable to the presence of offshore production platforms were detected.

SUMMARY—HYDROGRAPHY

1. The surface salinity was generally only slightly lower than bottom salinity during the fall months.

2. The water column was stratified from winter through summer; beginning in April, unusually steep vertical salinity gradients occurred in the upper column, reflecting the introduction of flood waters from the Mississippi River and elsewhere.

3. Salinity generally increased in the offshore direction but lenses of brackish water with higher salinity both landward and seaward were measured when westward drift and spring/summer flooding occurred.

4. Surface temperature was only slightly higher than bottom temperature from August to December, considerably higher from May to September, and significantly lower from January to March.

5. Surface dissolved oxygen normally ranged from 6.0 ppm in the summer to 10.0 ppm in the winter; at the same time, bottom dissolved oxygen ranged from 2.0 ppm to 7.0 ppm. Low temperature, high wave turbulence, and plankton blooms resulted in high dissolved oxygen concentrations. A stratified water column, high temperature, and biological activity were the primary causes of low dissolved oxygen levels in the bottom waters.

6. A seasonal transmissivity trend was observed in the bottom waters; it was highest in the summer and lowest in the spring. Clearest waters were generally found at mid-depth.

7. Transmissivities increased offshore, suggesting a near-shore origin of the suspended particulate material. Ebb tides were the probable downslope carrier of suspended muds, the likely source of the turbid layer. Strong flood tide currents along the bottom also may have been of sufficient strength to maintain particle suspension.

8. Only once throughout the course of the study was a significant change that appeared to be caused by activities of an oil production platform observed in the hydrographic parameters.

CONCLUSIONS—HYDROGRAPHY

The most impressive aspect of the hydrography of the study area was the large variation of salinity, temperature, dissolved oxygen, transmissivity, and currents with time, depth, and geographic location. Interactions of tides, the Mississippi River, runoff leaving the bays, and meteorological conditions were the primary agents of change. The study area was neither typically marine nor estuarine, but rather it was transitional. Consequently, organisms living there must be tolerant to large-scale variations in the environment.

Although some localized changes in hydrography linked to oil structures were measured, they were minor in comparison to vertical and seasonal changes caused by natural agents, especially the Mississippi River.

ACKNOWLEDGMENTS

This project was funded under Gulf Universities Research Consortium, Contract GU853-8; Southwest Research Institute, Corpus Christi, Texas.

For the section of the study on currents, we are indebted to N. P. Smith for his generous assistance and to T. Bilhorn for criticism of the manuscript.

For both sections of the study, we thank R. Case for criticism of the manuscript, R. Weed, E. Baker, and D. Krueger for their dedicated data acquisition, and L. Lakatos and M. Lynch for their aid in the preparation of the illustrations.

REFERENCES CITED

Barrett, B. B. 1971. Hydrology, coastal study area III. *In* Cooperative Gulf of Mexico Estuarine Inventory and Study, Louisiana. Publications of the Louisiana Wildlife and Fisheries Commission, pp. 42-67.

Curray, J. R. 1960. Sediments and history of Holocene transgression, continental shelf, northwest Gulf of Mexico. *In* Sediments, Northwest Gulf of Mexico. Tulsa, Oklahoma: American Association of Petroleum Geologists, pp. 221-266.

Kimsey, J. B. 1963. Currents on the continental shelf of the northwestern Gulf of Mexico. Annual Report Fiscal Year 1963 of the Fishery Research Biological Laboratory, Galveston, Texas, pp. 25-27.

Kimsey, J. B. and R. F. Temple. 1962. Currents on the continental shelf of the northwestern Gulf of Mexico. Annual Report Fiscal Year 1962 of the Fishery Research Biological Laboratory, Galveston, Texas, pp. 23-27.

MacMillan, D. H. 1966. Tides. New York: American Elsevier.

National Ocean Survey. 1972. Tide Tables, High and Low Water Predictions.

______. 1973. Tide Tables, High and Low Water Predictions.

U.S. Naval Oceanographic Office. 1965. Oceanographic Atlas of the North Atlantic Ocean. Publication No. 700, Section 1, Tides and Currents.

Watson, R. L. and E. W. Behrens. 1970. Nearshore surface currents, southeastern Texas Gulf Coast. Contributions in Marine Science, The University of Texas **15**:133-143.

ONSHORE HYDROGRAPHY OF TIMBALIER BAY, LOUISIANA

by Keith C. Price

ABSTRACT

Synoptic measurements of salinity, temperature, dissolved oxygen, and transparency were made in Timbalier Bay, Louisiana, over a two-year period (1972-1974) to assess the effects of petroleum-related activities upon these parameters. A total of 214 stations was sampled. No measurable effects attributable to such activities were found. Seasonal effects during the study period accounted for the greatest variations observed.

INTRODUCTION

Gulf South Research Institute (GSRI), under my direction, provided the inshore sampling platform for most of the OEI investigations and collected the inshore hydrography data. Investigators occupied a series of common stations, and in addition many investigators occupied extra stations chosen specifically to complement their research.

Measurements of water depth were made at each station. Conductivity (millimhos/cm), salinity (ppt), temperature (° C), dissolved oxygen (ppm), and light transmissivity (%) were measured near the surface and near the bottom at each station.

METHODS AND MATERIALS

Water depth at each station was measured with a fathometer to the nearest half-foot. Conductivity was measured with an electrodeless

At the time of this investigation, Keith Price was with Gulf South Research Institute, New Orleans, LA.

salinometer having a minimum level of detection of 0.5 millimhos/cm and an accuracy of ± 0.5 millimhos/cm. Salinity was measured with the same electrodeless salinometer to an accuracy of ± 0.3 ppt and a minimum level of detection of zero ppt. Temperature was also measured with the same electrodeless salinometer to an accuracy of ± 0.2° C. Dissolved oxygen levels were determined using a YSI model 51 oxygen meter with temperature and salinity corrections. The percentage of transmittance of light was also measured at each station.

The operating procedure at each station was to measure the hydrographic parameters as the various investigators on board collected samples for their individual research projects. Temperature, salinity, and conductivity were measured near the surface and near the bottom. Temperature and salinity corrections were then set on the oxygen meter and dissolved oxygen levels were measured near the surface and near the bottom. In those cases where the oxygen meter malfunctioned, water samples were collected with a Van Dorn sampler and the dissolved oxygen concentration was determined using the sodium azide modification of the Winkler method.

The bay area was sampled in July 1972, September-October 1972, December 1972, January-February 1973, April 1973, July 1973, October 1973, and January 1974.

RESULTS

The results obtained in this study agree generally with those reported by Barrett (1971) in the Cooperative Gulf of Mexico Estuarine Inventory and Study, Louisiana. In general, the Timbalier Bay area is less influenced by riverine activities than are other bays along the Central Gulf Coast. Salinities are higher and more stable, and there are fewer fluctuations in most of the other hydrographic parameters.

Salinity

The seasonal variability in salinity is shown in figure 1. This figure shows the data from all stations and all depths averaged together. The minimum figure for salinity was 10.1 ppt, while the maximum was 35.9 ppt. The effects of the 1973 flood can be seen clearly beginning in April 1973; salinities were depressed and far more variable than had been observed in previous surveys.

The spatial variability of salinity and the other parameters was analyzed by grouping the stations into five geographical areas (figure 2). Sector 1 includes all inner bay stations. Sector 2 is a mid-bay area in western Timbalier Bay and includes the platform area around Philo Brice Islands. Sector 3 is a mid-bay zone in eastern Timbalier Bay and includes

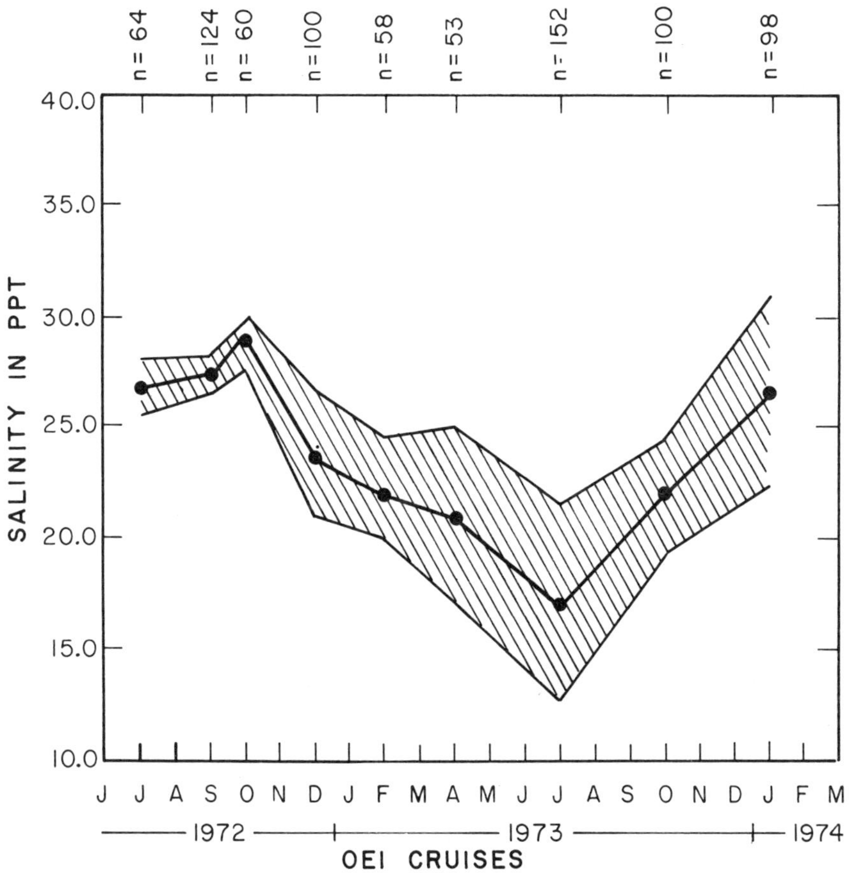

FIG. 1. TIMBALIER BAY SALINITY.

a control site. Sector 4 is the lower mid-bay region on the west and includes another control site. Sector 5 is the bay mouth area and includes the workover rig location where many samples were taken.

Sectors 2, 3, and 4 were very similar in salinity (figure 3). The mean value for all three sectors was near 24 ppt, and the minimum and maximum values were also similar, though they increased slightly from sector 2 to sector 4. Sector 1, the inner bay, had significantly lower salinities, with a mean of approximately 16 ppt and lower minimum and maximum figures than the other zones. Sector 5, the bay mouth, had slightly higher mean and maximum salinity values than the other sectors, but recorded a minimum figure lower than that recorded in sector 4. Examination of the data from individual cruises showed a similar pat-

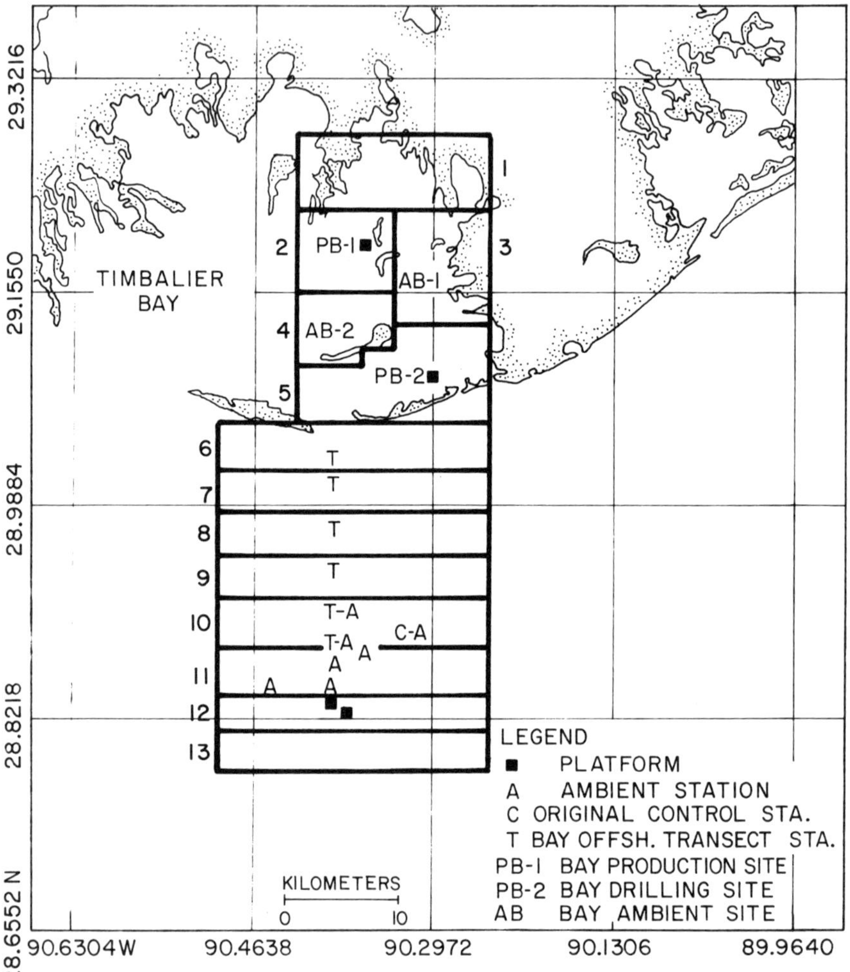

FIG. 2. GEOGRAPHICAL AREA OF THE OEI SHOWING DIVISION INTO SECTORS.

tern, though the mean values for sector 4 sometimes equaled or exceeded the mean values in sector 5.

Water temperature

Water temperatures varied in a manner typical for that area of coastal waters, with maximums in late summer and minimums in late winter (figure 4). The lowest temperature recorded was 8.0° C in September 1972. Spatial variability in water temperatures was minimal,

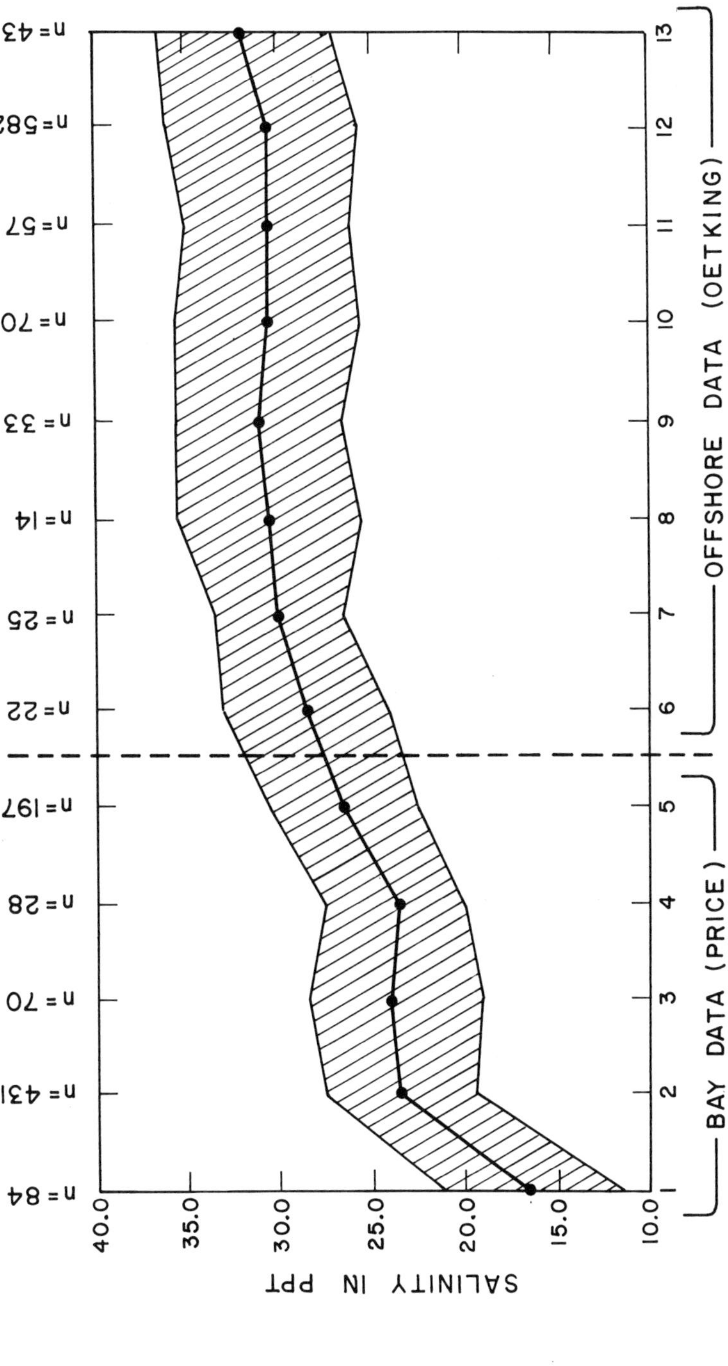

FIG. 3. BAY-OFFSHORE TRANSECT SALINITY.

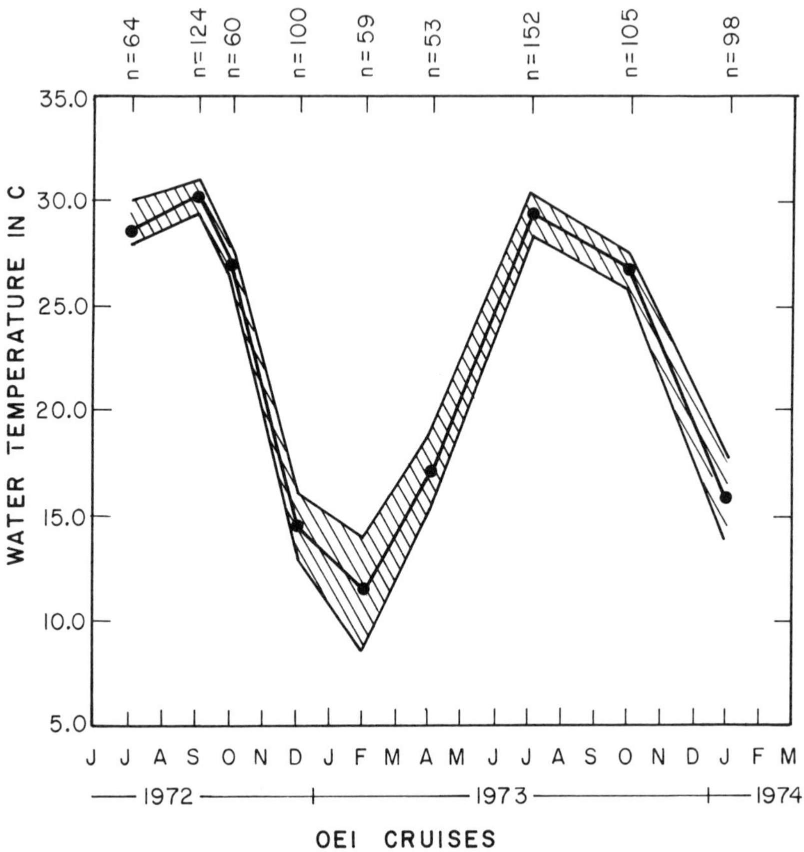

FIG. 4. TIMBALIER BAY WATER TEMPERATURE.

though the shallower, more inshore stations were more variable and more affected by the prevailing air temperatures (figure 5).

Dissolved oxygen

Dissolved oxygen concentrations showed more variability than would normally be expected. The highest readings, 13+ ppm, occurred in the winter months, and the lower reading, 1.3 ppm, occurred in July 1973 (figure 6). Mean values ranged from slightly less than 6.0 ppm to almost 11.0 ppm, except for the July 1973 cruise when the mean value was near 3.5 ppm. These low oxygen levels coincide with the flood of 1973.

The mean dissolved oxygen concentrations exhibited little spatial variability (figure 7). Sector 1, with a mean dissolved oxygen concentration of approximately 5.0 ppm, was the lowest, and sector 2, with

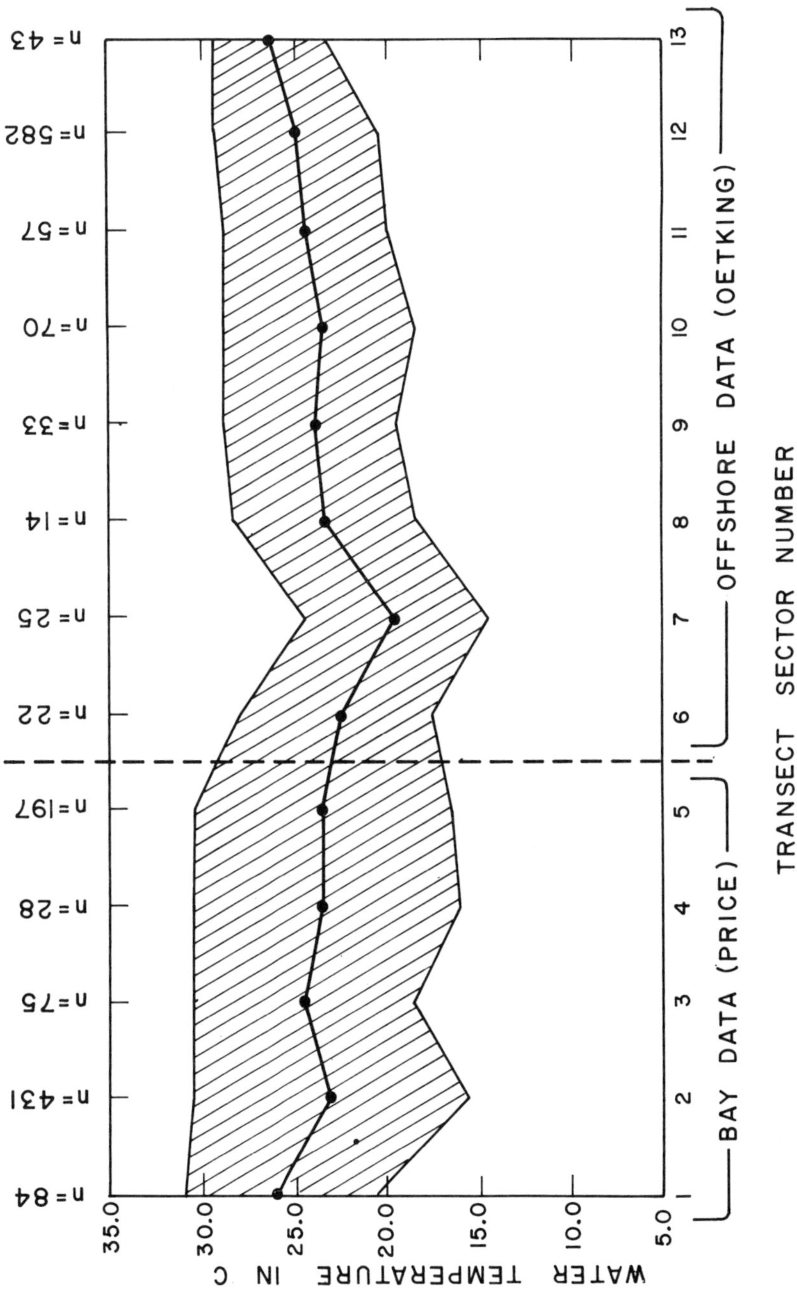

FIG. 5. BAY-OFFSHORE TRANSECT WATER TEMPERATURE.

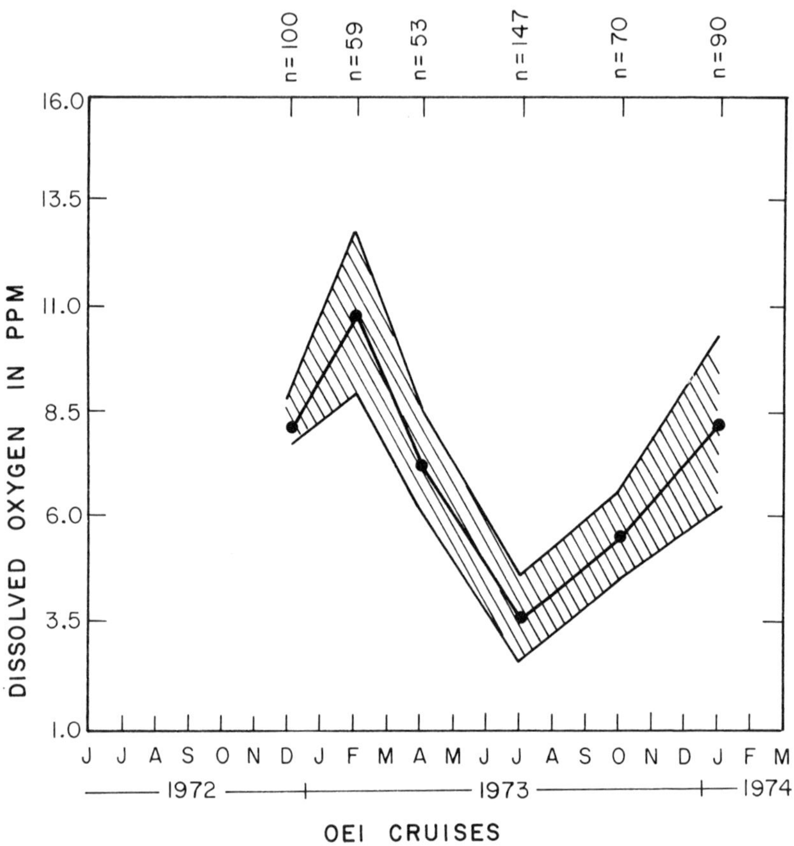

FIG. 6. TIMBALIER BAY DISSOLVED OXYGEN.

a mean of approximately 7.0 ppm, was the highest. Sectors 3, 4, and 5 all had mean dissolved oxygen values near 6.0 ppm. Examination of the mean dissolved oxygen values for the January 1973 and 1974 cruises and the July 1973 cruise also shows limited spatial variability (figure 8). The two winter cruises showed the mean dissolved oxygen values to be below the average for all stations in January 1973 and January 1974, though the mean for sector 1 in January 1973 was not the lowest mean of any of the sectors. The July 1973 data showed minimal variation in the mean value of any sector, the average value being below 4.0 ppm in all five sectors.

Transmissivity

Transmissivity showed a different seasonal pattern from that

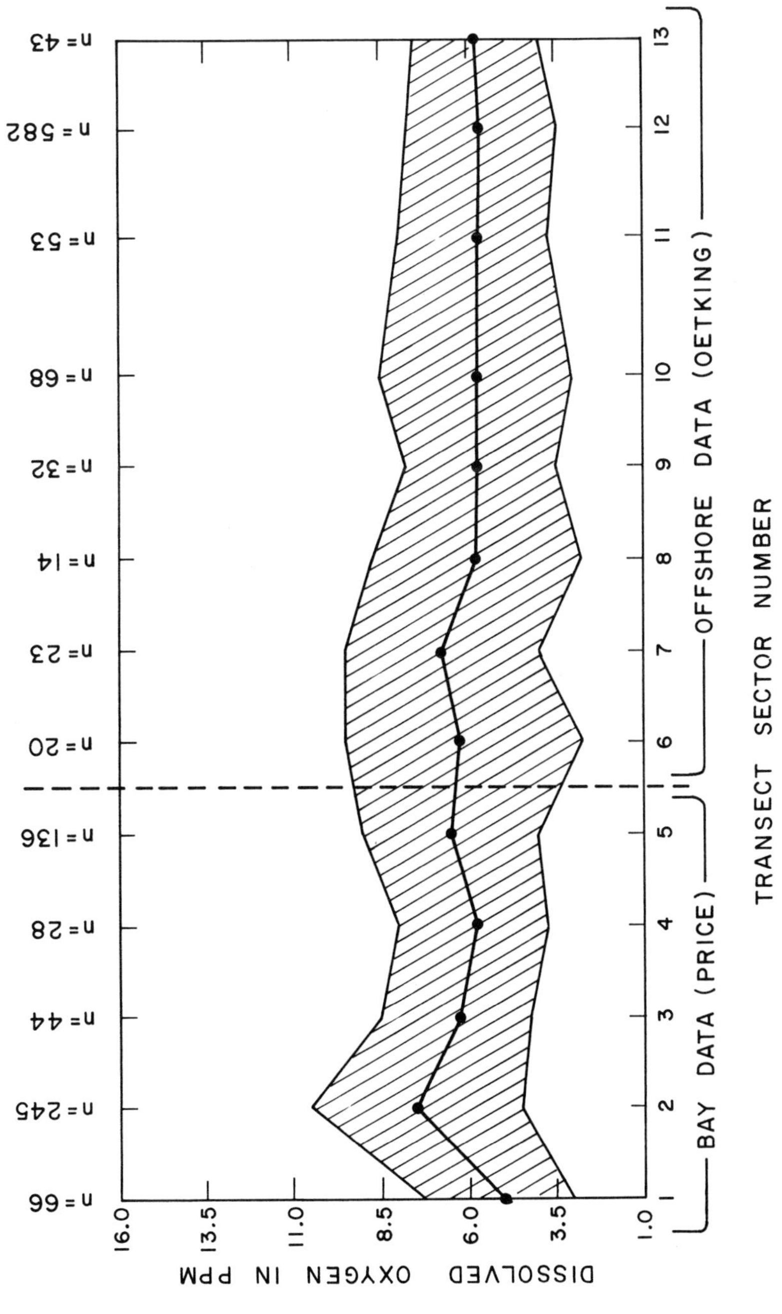

FIG. 7. BAY-OFFSHORE TRANSECT DISSOLVED OXYGEN.

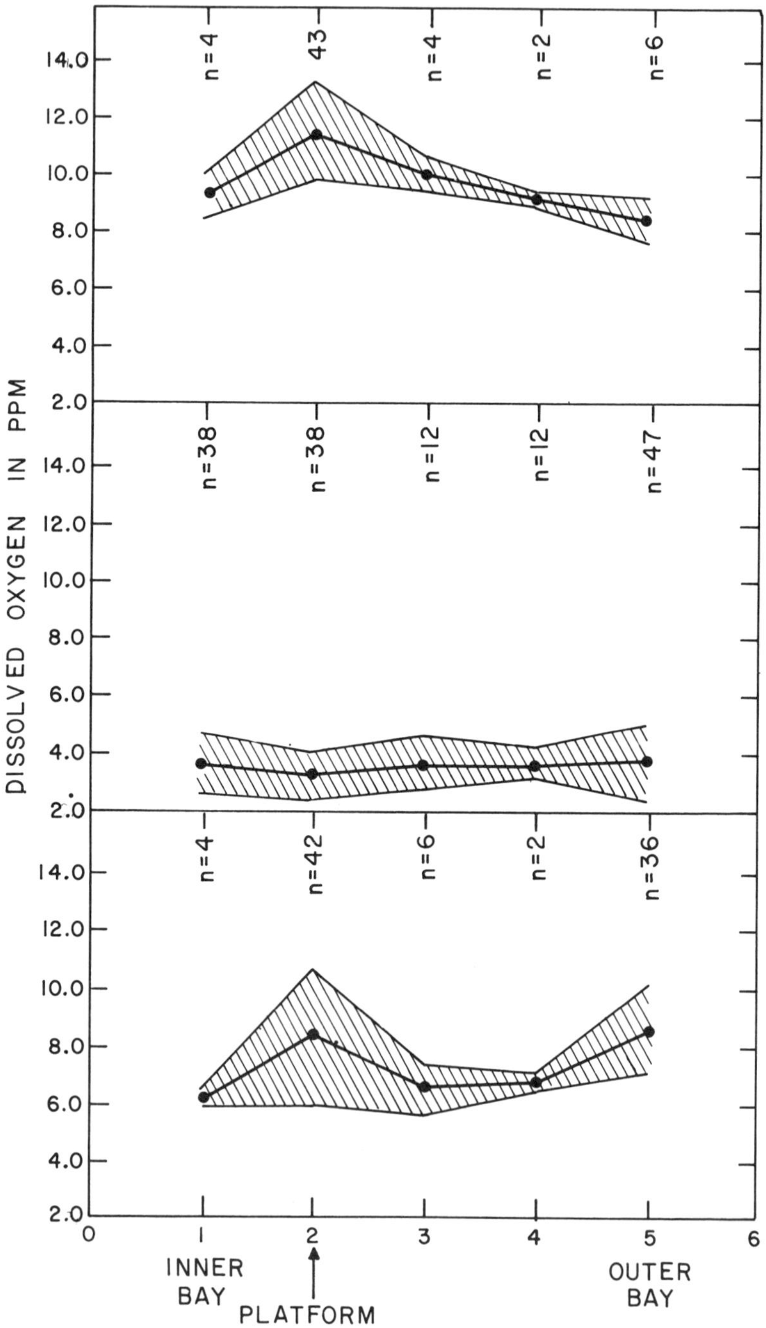

FIG. 8. INNER BAY-OUTER BAY TRANSECT DISSOLVED OXYGEN.

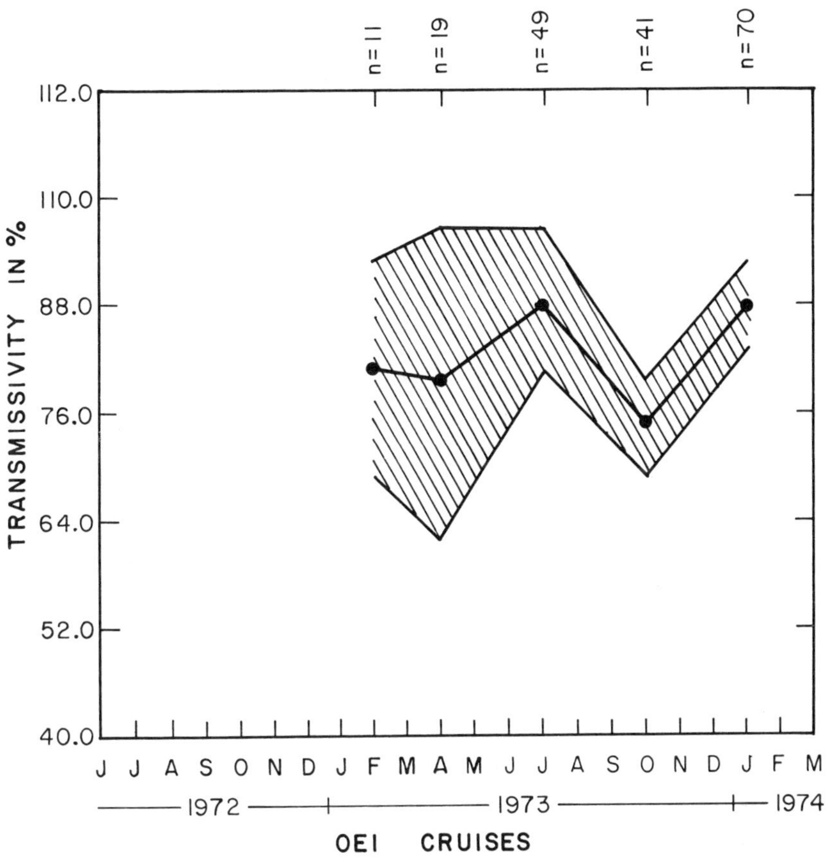

FIG. 9. TIMBALIER BAY TRANSMISSIVITY.

exhibited by the other parameters (figure 9). The maximum (99%) and highest mean (88%) were observed during the July 1973 cruise, which occurred during the 1973 flood. These same high transmittance values were nearly duplicated during the January 1974 cruise. Mean values during other seasons varied from approximately 76% to 80%. The lowest transmissivity, a reading of 42%, occurred during the April 1973 cruise.

Transmissivity had the lowest mean values at the two most inshore sectors (sectors 1 and 2) with mean values near 80% (figure 10). Transmissivity increased through sectors 3, 4, and 5, reaching a mean value of 88% in sector 5. High transmittance values were, however, recorded in all sectors, with values of 94% or greater being recorded from each sector.

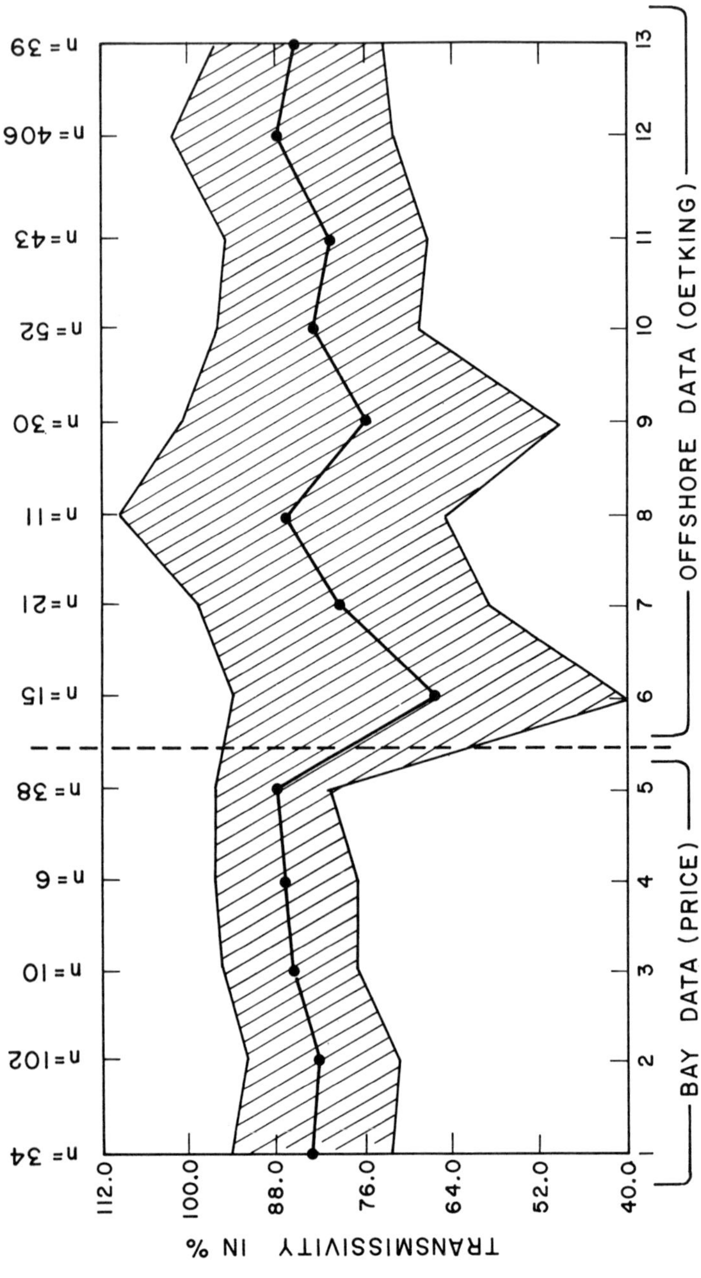

FIG. 10. BAY-OFFSHORE TRANSECT TRANSMISSIVITY.

DISCUSSION AND CONCLUSIONS

The hydrographic data from Timbalier Bay show the bay to be similar to many other bays along the Central Gulf Coast. It differs in that it is normally less affected by large rivers than are most bays and is thus more saline. Timbalier Bay shows less seasonal and spatial variability than is typical. The greater homogeneity resulting from this condition increases its value for the OEI experiments.

The major variation from normal conditions observed during this study was the major flood during the spring and summer of 1973. This flood affected Timbalier Bay despite the paucity of streams entering the bay and despite its relative isolation from the Mississippi River and the Atchafalaya River. As a result, some hydrographic conditions observed during this period were not typical of an average yearly cycle.

ACKNOWLEDGMENTS

I would like to acknowledge the assistance of many staff members at Gulf Research Institute, particularly Paul J. Conzelmann, who directed the field sampling program.

REFERENCE CITED

Barrett, B. 1971. Cooperative Gulf of Mexico Estuarine Inventory and Study, Louisiana—Phase II, Hydrology and Phase III, Sedimentology. Louisiana Wildlife and Fisheries Commission.

EVALUATION OF THE EFFECTS OF OIL PRODUCTION PLATFORMS ON THE TURBIDITY OF LOUISIANA SHELF WATERS

by George M. Griffin

ABSTRACT

As part of the Offshore Ecology Investigation (OEI) of the Gulf Universities Research Consortium (GURC), water clarity and particulate matter in the South Timbalier OCS (Outer Continental Shelf) area offshore of central Louisiana were studied from 1972 to 1974, to determine whether or not long-term oil production and associated platforms have altered the natural turbidity of continental shelf waters.

Turbidity of surface waters varied seasonally from a mean of 0.8 mg/l in autumn 1972 to 5.3 mg/l following the Mississippi River flood of 1973. Mid-depth waters were the clearest, typically containing 0.15 mg/l. A turbid bottom-water layer persisted throughout the whole period; turbidity in this layer varied from 1.5 to > 10 mg/l. No differences in turbidity existed between the area of numerous production platforms and an undeveloped control area.

Clay minerals filtered from surface, mid-depth, and bottom-water layers are quantitatively similar to Mississippi River suspended sediment, and dissimilar to the cohesive surface of the shelf sediments. Therefore, the bottom turbid layer is not generated by erosion of the cohesive bottom muds, but by the settled residuum of Mississippi River suspended sediment.

The turbid layer of bottom-water is actively by-passing the shelf at present. A process model is outlined, indicating that the

---

At the time of this investigation, George Griffin was Professor of Geology at the University of Florida.

suspended river sediment settles from the turbid surface plume through the mid-depth water, into the turbid bottom layer. There it migrates laterally toward shore, in a series of resuspensive pulses, merges with a surf-generated littoral turbid layer, and ultimately is trapped by vegetation in the prograding Chenier Plain of southwestern Louisiana. Offshore production platforms have no detectable effect on this process or on the turbidity of the central Louisiana shelf waters.

INTRODUCTION

For more than 20 years, the central Louisiana shelf has been the site of intensive petroleum exploration and production, resulting in one of the highest densities of offshore structures in the world. The primary purpose of this particular study was to determine whether or not long-term offshore operations of that magnitude have affected the turbidity of the surrounding waters significantly enough to be considered an environmental impact. It was one of 23 separate projects, mostly biological in nature, run concurrently as part of the Offshore Ecological Investigation (OEI) of the Gulf Universities Research Consortium (GURC). Sampling was conducted simultaneously for all the projects, and a computerized data base was erected for storage, retrieval, and correlation of diverse geological, biological, and physical-oceanographic data. Wherever practical, platform areas were compared with an undeveloped control area that was deemed ecologically similar. Results from all studies are available on microfilm from GURC. As the OEI progressed, explanation of a turbid bottom-water layer discovered on the August 1972 cruise assumed particular importance. Much of this study was devoted to identifying the source of that layer and determining whether or not it is part of the natural ecosystem of the area.

Details of the sampling and analytical methods, as well as tabulations of all data, are in Griffin and Ripy (1974), available from GURC, and are not repeated here. The principal *in situ* tool for measuring turbidity was a Hydro Products Model 612 Transmissometer. Light transmittance readings (% transmittance/meter) from that instrument have been converted into suspended-sediment concentrations in mg/l using samples of Louisiana shelf mud to construct the conversion curve (figure 1). The conversion curve for the Louisiana shelf correlates well with a previously published curve from the California shelf (figure 2), and allows quantitative measurements of suspended particulate matter to less than the 1 part per million range.

Data were collected on eight cruises between August 1972 and January 1974, the times selected to observe seasonal changes. Seven of the cruises were in the South Timbalier area, and the eighth, for comparative purposes,

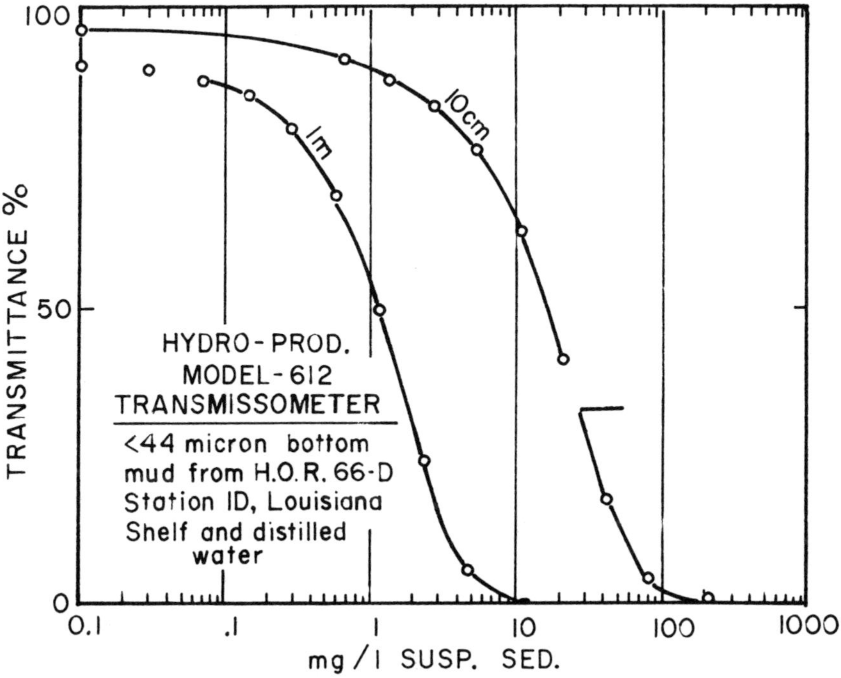

FIG. 1. TRANSMITTANCE VS. SUSPENDED SEDIMENT CONCENTRATION AS MEASURED IN A TEST TANK.

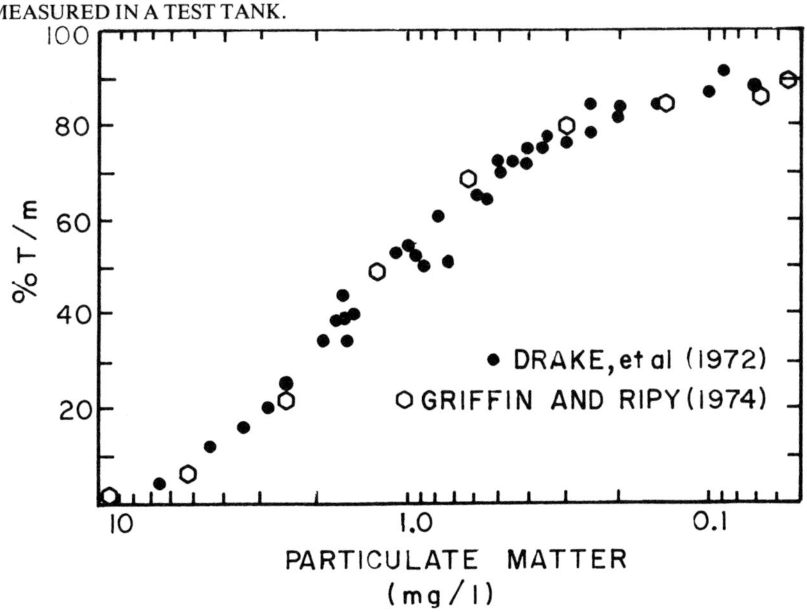

FIG. 2. COMPARISON OF TRANSMITTANCE VS. SUSPENDED SEDIMENT CONCENTRATION (• = California shelf, after Drake et al. 1972. ○ = Louisiana shelf, this study.).

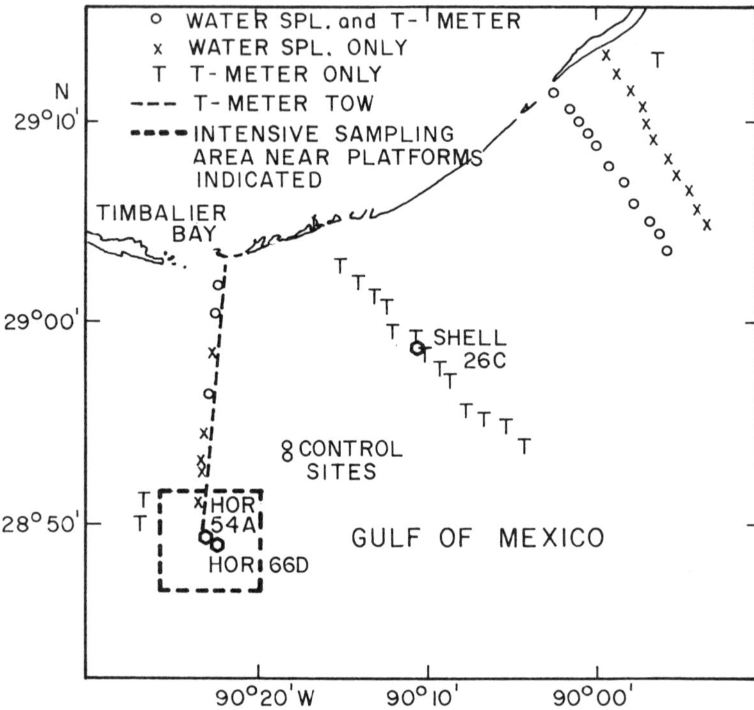

FIG. 3. STATION LOCATION MAP, SOUTH TIMBALIER OCS AREA, LOUISIANA.

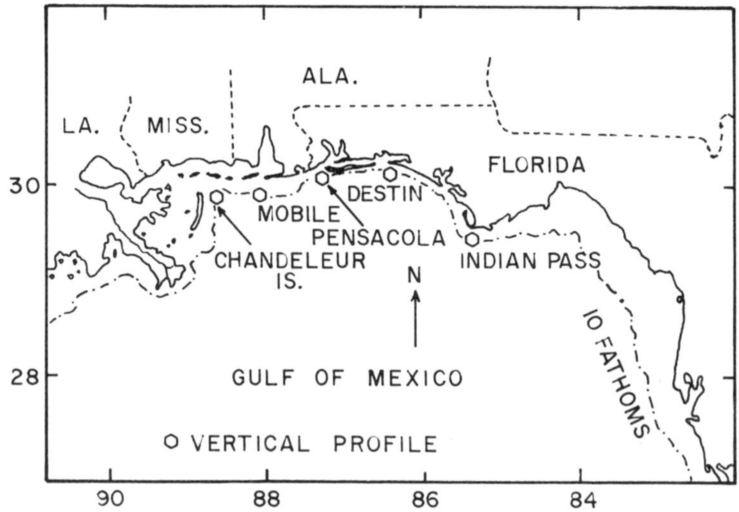

FIG. 4. LOCATIONS OF VERTICAL TRANSMISSIVITY PROFILES IN THE EASTERN GULF OF MEXICO (October 1973).

was east of the Mississippi Delta between Apalachicola, Florida, and the Chandeleur Islands. Stations are indicated on figures 3 and 4.

RESULTS

Seasonal variations in surface turbidity

The South Timbalier area lies 50-60 miles to the west, and in the lee, of the vast Mississippi Delta. This position produces considerable seasonal variations in surface-water turbidity that are correlatable with Mississippi River stages. High surface turbidities were noted especially during the summer of 1973, following a record Mississippi River flood (table 1, data for 7/73). At that time, a turbid plume spread through the area, and suspended sediment concentrations in the surface water reached an average of 5.3 mg/l in the immediate vicinity of Humble Oil and Refining Company (HOR) Platforms 54A and 66D, and 5.0 mg/l in the control area.

During lower river stages, surface concentrations declined to much lower values. In the summer and fall of 1972, only 0.8 to 1.2 mg/l were detected in the platform area, and 1.1 to 1.5 mg/l in the control area.

It is notable that the platform and control areas exhibited similar turbidity fluctuations; differences between the two areas were always less than 1 mg/l. On two cruises, the platform area was slightly more turbid, but on two other cruises, the control area was more turbid. None of these differences is considered significant.

On the whole, the surface water layer under non-flood conditions was clearer than might be expected off such a large river. The usual 1 to 2 mg/l range is, however, at least double the typical concentration level of water of the Florida Straits and other typical oceanic areas.

Turbidity of surface water adjacent to production platforms

Horizontal turbidity in the surface water within 6 km of HOR Platforms 54A and 66D was measured numerous times by towing the transmissometer along traverses radial to the platforms. Points extracted from the continuous chart record of September 18, 1972, are plotted on figure 5, which is representative of results from all tows. Turbidity on that day varied within a narrow range of 0.9 to 1.5 mg/l, with no effect from the platform at the center of the array.

A long traverse, from Platform 54A to near Grand Isle, was run on September 19, 1972 (figure 6). At the platform, the suspended-sediment concentration was only 0.8 mg/l, which is clear ocean water. The clear water continued shoreward for 13 km, after which the particulate content increased to 1.5 mg/l. Further shoreward, turbid plumes from Bayou Lafourche, Timbalier Bay, and Barataria Bay became evident, with a

TABLE 1. Suspended-Sediment Concentrations in Surface Water (<2m depth) Near H.O.R. Platforms 54A or 66D, and Comparisons with the Control Area. August 1972-January 1974. All Measurements with Hydro-Products Model 612 or 612A 1-m Transmissometer; Precision $\pm$ Percentage Points.

Optically-Measured Suspended Sediment Concentration (mg/l)												
	Near Platforms						Control Area					
Cruise Dates	8/72	9/72	10/72	7/73	1/74	All Cruises	8/72	9/72	10/72	7/73	1/74	All Cruises
η	8	14	6	19	3	50	0	2	1	1	1	5
$\bar{x}$	1.2	1.0	0.8	5.3	2.8	2.7	-	1.1	1.5	5.0	2.3	2.2
$s \pm$	0.4	0.03	0.2	0.3	0.6	2.1	-	-	-	-	-	1.6
1 s range	0.8-1.6	1.0-1.1	0.6-1.0	5.0-5.6	2.2-3.4	0.6-4.8	-	-	-	-	-	0.6-3.8
max.	2.1	1.2	1.2	6.0	3.1	6.0	-	-	-	-	-	5.0
min.	0.9	1.0	0.7	5.0	2.1	0.7	-	-	-	-	-	1.0

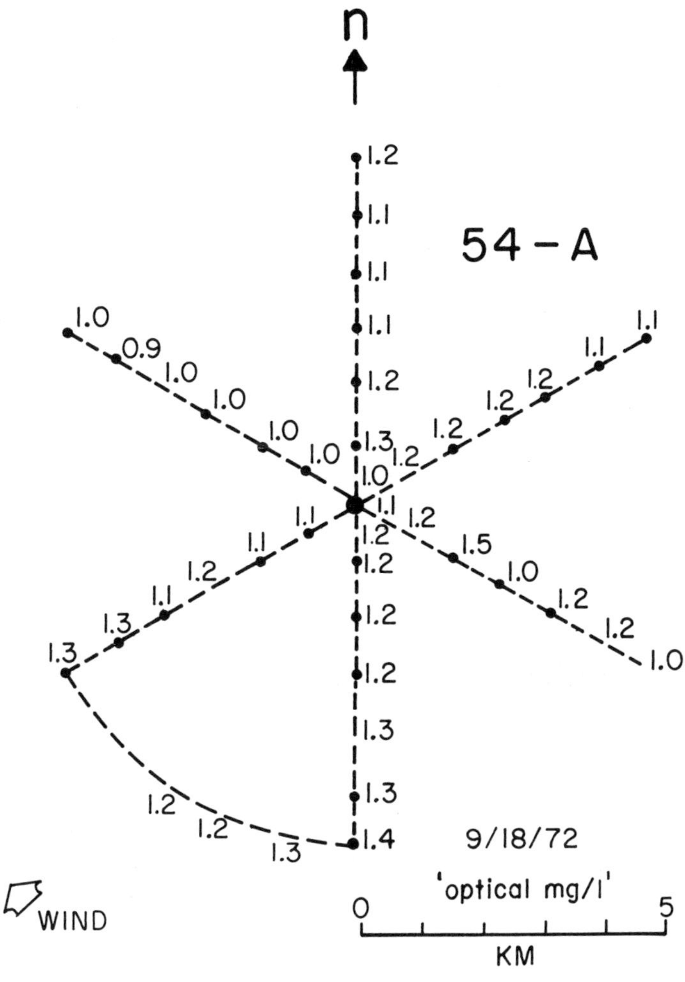

FIG. 5. SUSPENDED SEDIMENT CONCENTRATIONS (OPTICAL MG/L) IN SURFACE WATER (<0.3 M DEPTH). Derived from continuous recordings with an optical transmissometer, September 18, 1972. HOR Platform 54A was at the center of the pattern.

maximum of 17 mg/l reached 4.7 km off Barataria Pass in 6.1 m of water. The traverse terminated at that point.

On none of the traverses was there a detectable effect from the platforms. Such variations as were noted within several kilometers of the platforms were directly relatable to visible patches of plankton, the bubbles in tug boat wakes, and occasional surfacing of the transmissometer sensor.

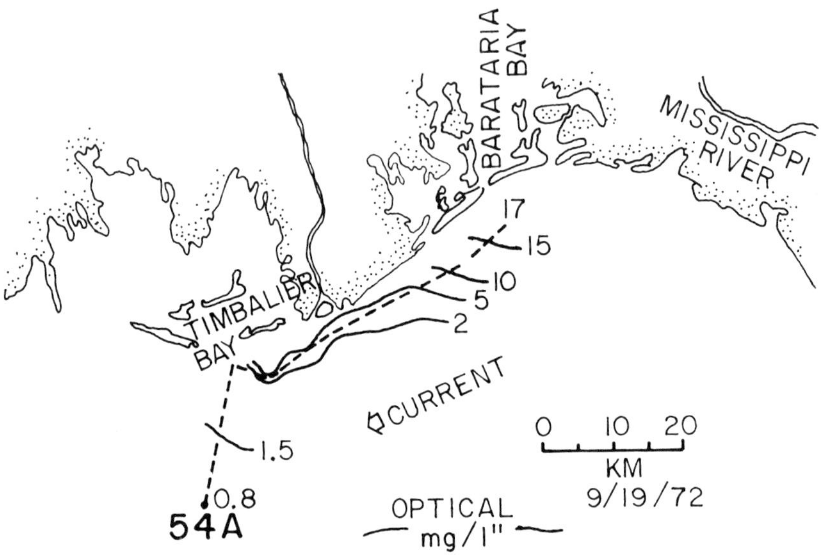

FIG. 6. SUSPENDED SEDIMENT CONCENTRATION (OPTICAL MG/L) IN SURFACE WATER (<0.3 M DEPTH) between HOR Platform 54A and the entrance to Barataria Bay. September 19, 1972.

Daily variations of approximately 1 mg/l were related to changes of current direction in the surface water. The greatest variations, on the order of 5 mg/l, were seasonal and attributable to the unusually large influx of Mississippi River floodwaters in the spring and summer of 1973.

Vertical Profiles of Turbidity

The central Louisiana shelf is characterized by a two- or three-layer stratified water column with a persistent turbid bottom-water layer. In detail, the profiles vary both daily and seasonally, but the turbid bottom-water layer was observed at 96% of the Louisiana stations regardless of season.

Diurnal variability

Figures 7 and 8 are profiles run on consecutive days at Platform 54A. Turbidity is plotted on the right-hand part of the figures as percentage of light transmission per meter of optical path length. In the discussion below, these values have been converted to mg/l by applying the working curve of

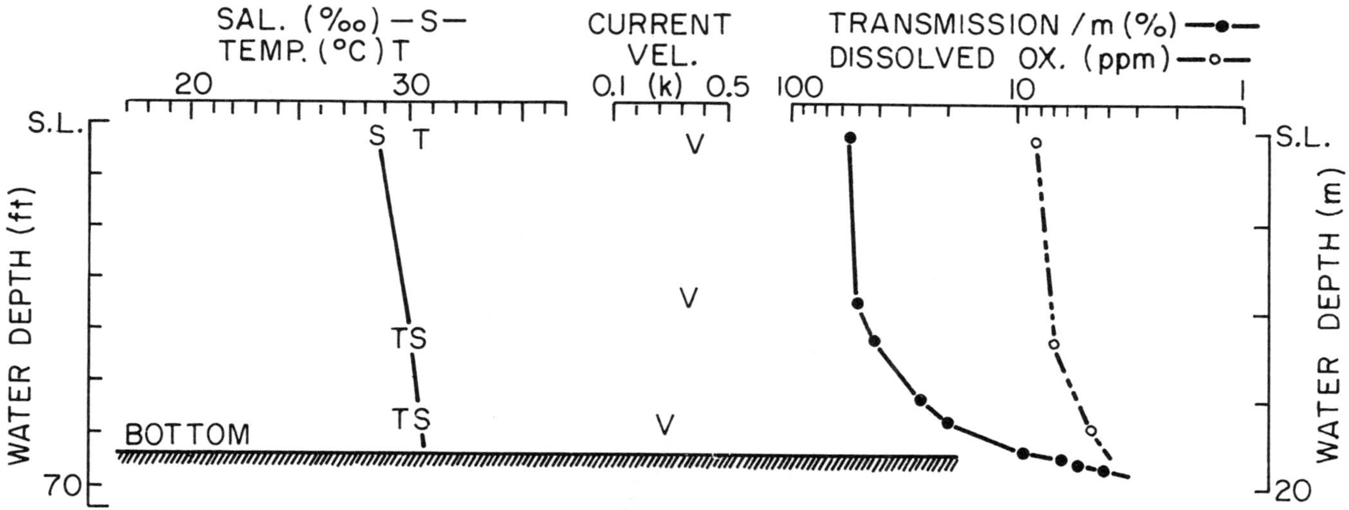

FIG. 7. TYPE 2 VERTICAL PROFILE OF TRANSMISSIVITY (% T/m) AND OTHER WATER MASS PARAMETERS. HOR Platform 54A. September 20, 1972.

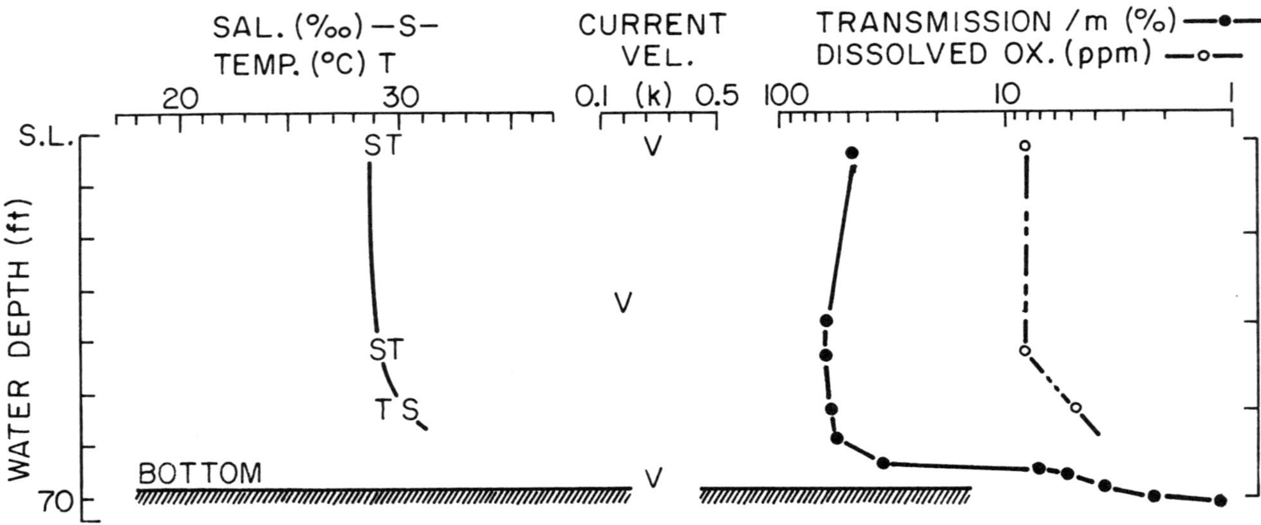

FIG. 8. TYPE 3 VERTICAL PROFILE OF TRANSMISSIVITY (% T/m) AND OTHER WATER MASS PARAMETERS. HOR Platform 54A. September 21, 1972.

figure 1. Approximately 60% of the Louisiana shelf profiles are of the types shown on figures 7 and 8.

On the September 20, 1972, profile (figure 7), turbidity increased exponentially with depth as particulate concentration increased from approximately 1 mg/l in the surface water to 6.4 mg/l in the turbid bottom water. The September 21, 1972, profile (figure 8) differs slightly in that a clearer mid-depth layer was present. On this day, the surface suspended-sediment concentration was near 1.2 mg/l, the mid-depth layer was near 0.7 mg/l, and the turbid layer contained as much as 10 to 15 mg/l.

The slightly differing profiles of September 20 and 21 correlated with a shift in current direction. On September 20, the weak (<0.5 k) current generally moved toward the north-northeast at all depths; on September 21, it moved toward the west-northwest at the surface and bottom, but toward the north in the mid-depth layer. Wind was less than 3 knots on both days.

Seasonal variability

Longer term, seasonal effects are illustrated by figure 9, taken at Platform 54A on July 9, 1973. A low-salinity, turbid surface layer is present, with a particulate concentration of 5 to 7 mg/l. This represents the turbid plume from the record Mississippi River flood of that year. It was approximately 20 feet thick at Platform 54A. Beneath the surface turbid layer, the mid-depth water was remarkably clear, with approximately 0.3 mg/l suspended matter, and a full marine salinity of 36 parts per thousand dissolved solids. Near the bottom, the turbid layer contained only 1.8 mg/l suspended sediment at this station. Surface water was moving toward the south at 0.55 knots, whereas the mid-depth water moved toward the east-northeast at 0.25 to 0.48 knots, and the

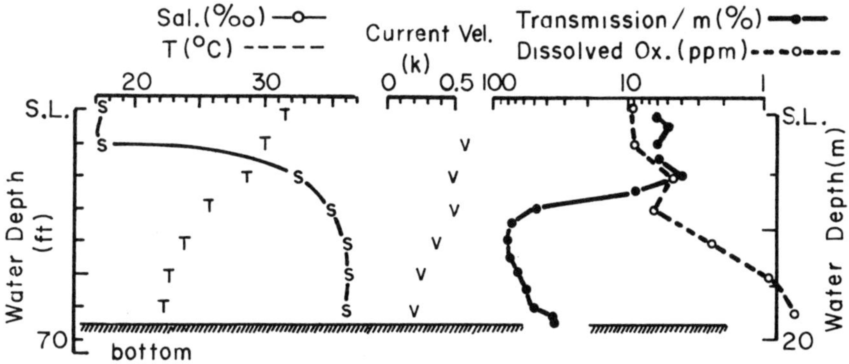

FIG. 9. TYPE 4 VERTICAL PROFILE OF TRANSMISSIVITY (% T/m) AND OTHER WATER MASS PARAMETERS. HOR Platform 54A. July 9, 1973.

bottom layer also moved east-northeast at 0.2 knots. Profiles similar to figure 9 were observed at 30% of the stations on the Louisiana shelf, all of them following the 1973 Mississippi River flood.

THE TURBID BOTTOM LAYER

Ecological significance

Upon discovery of the turbid bottom-water layer on the August, 1972, cruise, its potential ecological significance was quickly realized by several of the OEI project leaders. Biological stresses due to reduced light penetration through the bottom water were reported by Humm (1973) to be related to the lack of bottom-dwelling algae except on platform legs above the layer. George and Thomas (1973) found the depth of fouling organisms, such as barnacles, on platform legs to be limited by the average thickness of the turbid layer. Other biotic stresses and limitations were related to the very unstable nature of the sediment/turbid water-layer interface and to the sub-normal dissolved oxygen content of the layer. Although not all of these biological effects could be considered necessarily harmful to the environment, it was apparent that the turbid layer is an important governing factor in the ecology of the Louisiana shelf and that if it were caused by the platforms, it might be considered as an adverse environmental effect. Therefore, considerable effort was made to define various attributes of the layer including its thickness, distribution, concentration, dissolved oxygen content, and seasonality. Many of these features are summarized in table 2; other features are listed in Griffin and Ripy (1974).

Source of clastic particles in the turbid layer

Because the turbid layer is composed largely of suspended minerals, it was possible to collect filtrates from it and to use quantitive clay mineralogy as a natural tracer. In table 3, analysis of suspensates from the surface, mid-depth, and bottom turbid layers near Platforms 54A and 66D are compared with cohesive bottom sediments from the same area. Suspended sediment samples from the lower Mississippi River are included for comparison.

It is evident from the table that particulate matter in the turbid bottom layer, as well as in the surface and mid-depth waters near the platforms, is similar to the Mississippi River suspended sediment. On the contrary, cohesive bottom muds near the platforms differ mineralogically from the suspended sediment. This finding indicates that the inorganic particles in the turbid layer originate from the modern Mississippi River and are not generated by reworking of the surface of the Lafourche sub-delta, which forms the cohesive shelf bottom in this area.

TABLE 2. Summary of Characteristics of the Bottom Turbid Layer within 5.5 km of H.O.R. Platforms 54A and 66D and within 2-m of the Seafloor.

	n	Mean	Std.Dev.	1-s Range	Max.	Min.
Transmissivity (% T/m)	38	12.7	±14.0	0-26.7	53	0
Attenuation Coefficient (a/m)	38	>2.69	>±1.27	<1.4->3.96	>4.61	0.63
Concentration ("optical mg/l")	38	>5.0	>2.9	<2.1->7.9	>9.8	0.8
Concentration (filtered, mg/l)	10	6.0	3.9	2.1- 9.9	11.8	2.4
Dissolved Oxygen (% saturation)	95	61.4	±32.9	28.5-94.3	130	<1
Thickness (m)	44	6.2	±1.9	4.3- 8.1	10.1	1.5

TABLE 3. Clay mineralogy of suspended sediment samples from near Platforms 54A and 66D, cohesive bottom sediments from the same areas, and Mississippi River suspended sediments.

	Percent of the Clay Mineral Suite (±5%)				Number of
	montmorillonite	illite	kaolinite	chlorite	Samples
Group A - Near Platforms 54A and 66D.					
Surface water	33	42	11	14	7
Mid-depth water	29	47	7	20	4
Turbid bottom water	33	42	10	16	22
Group B - Control area.					
Turbid bottom water	31	43	9	16	1
Group C - Mississippi R. at Avondale					
Suspended sediment	33	34	13	21	4
Group D - Cohesive bottom sediments from near Platforms 54A & 66D	58	21	8	11	12

TABLE 4. Ratios of Bottom/Surface Suspended Sediment Concentrations, Based on Best Approximations, including Optical and Filtering Data of Various Types

Type of Water Body	Bottom/Surface Concentrations
A. Open Shelf	1-3X
B. Shelf, Near Po Delta	10-15X
B. Shelf, Near Mississippi Delta	10-54X
C. Delta Distributaries	5-30X
D. Estuaries	2-89X
E. Fjords	3X
F. Tidal Straits	3X
G. Non-Estuarine Bays	2-4X
H. Tidal Flat	2X
I. Hypersaline Lagoon	33-400X

Worldwide distribution of turbid layers

Particular attention was also given to a thorough search of the literature to establish whether such layers were common to similar geographic areas elsewhere in the world, especially to provide a comparison between areas with and without petroleum-production facilities. The complete compilation, including data on suspended sediment concentrations and proposed mechanisms of formation, can be found in Griffin and Ripy (1974). Table 4 of this report is a short summary indicating the range of environments represented, and comparing surface and bottom-water concentrations of suspended sediment.

It was apparent from this review that turbid bottom-water layers are far from rare. Some of the areas listed are off urbanized coasts, others are off largely uninhabited areas; however, the California shelf described by Moore (1972) is the only other one in which petroleum production is at all intensive. All of the more than 35 writers attributed the turbid bottom water to natural mechanisms, especially resuspension of fine particles by tidal currents or waves; biologic reworking was cited occasionally as a method for resuspending cohesive muds.

Comparison with the Eastern Gulf of Mexico Shelf

During October 1973, vertical profiles were run and suspended sediments collected from the bottom-water layer at four sites on the N.W. Florida, Alabama, and Mississippi shelf (figure 4). At all of these sites, the water was very clear and was well mixed throughout the water column. No evidence of a turbid layer or any other form of stratification was found. Particulate contents in 60-80 feet of water off Mobile Bay, Alabama, and Destin, Florida, were uniformly much less than 0.1 mg/l, and approached the limit of measurement with the one-meter transmissometer. Filtrates from eight samples of the near-bottom water average 20% montmorillonite, 53% illite, 9% kaolinite, and 16% chlorite. Thus, they differ from the turbid-layer samples of the Louisiana shelf and from the Mississippi River.

TURBID LAYER TRANSPORT ON THE LOUISIANA COAST

As a result of the results and reasoning given above, the bottom turbid layer on the central Louisiana shelf is attributed primarily to influx from the modern Mississippi River, rather than to erosion of the partly consolidated, cohesive bottom sediments of the study area. Artificial additions from the offshore petroleum industry were not detected; they are either absent, quantitatively insignificant, or of the same composition as modern Mississippi River mud.

The natural process sequence proposed for development of the turbid layer is as follows:

1. Turbid fresh-brackish water issues from the passes of the Mississippi Delta as a distinct surface plume, carrying approximately 40 to 500 mg/l suspended sediment (Coleman, Wright, and Suhayda 1972). The plume is most marked during floods, when it spreads west at least as far as the shelf south of Timbalier Bay; however, it is present to some extent during most river stages. The plume conforms generally to the hypopycnal plane-jet theory of Bates (1953), as modified by Coleman, Wright, and Suhayda (1972). Its remnants eventually result in the common salinity variations off western Louisiana noted by Geyer (1950), and the turbidity fluctuations noted by Scruton and Moore (1953) east of the delta, and by us to the west.
2. Because of the significant density differences between the brackish surface plume and the clearer mid-depth open-Gulf water, eddy diffusivity is limited, and the surface plume persists as a distinct water mass for tens of miles away from the delta. Slow erosion of the base of the turbid plume, by internal waves at the density interface, allows only small quantities of the suspended sediment to descend into the mid-depth layer at any point. The plume sediment is therefore dissipated slowly into the mid-depth water and diluted as the turbid surface plume drifts away from the delta with the prevailing current. Its initial movement is toward the west. This is a result of several factors: (1) the tide propagates from east to west and during the flood tide the plume is deflected strongly to the west (Coleman, Wright,

and Suhayda 1972); (2) the coreolis force tends to deflect the plume westward (Bates 1953); and (3) the net effective winds tend to produce a dominant westward flow of the surface layer along the Louisiana coast (Lipsey 1919; Scruton 1956).

The quantity of suspended material that settles through the mid-depth water at any one place is therefore small, and the relative clarity and low suspended-sediment concentration (0.1-1.0 mg/l) of the mid-depth layer is maintained. As I pointed out earlier, however, the clay mineralogy of the mid-depth water is quantitatively identical to the surface plume from which it evidently was derived.

3. The particles that pass downward into the mid-depth layer continue to settle slowly, and as that water mass migrates, they are distributed over a wider area. Eventually the sea floor is reached, and in the absence of any bottom motion, the particles would be sedimented. Bottom motion *does* normally occur, however, either through tidal action, stirring by long-period storm waves, upwelling of slope water, or the stirring action of fish, shrimp, and other near-bottom dwellers or benthos. Thus, the fine particles are not plated onto the sea floor permanently, but accumulate temporarily in a turbid near-bottom water layer that is subject to lateral migration and further distribution of the river-derived clay. Part of the suspended load of the turbid layer is undoubtedly temporarily sedimented onto the bottom during quiet times. Even after deposition, however, the initial high water-content of the freshly sedimented material inhibits columbic attraction between clay particles, which allows it to be easily resuspended by weak agitation. Thus, the tendency is for the sediment to form a bottom turbid-water layer that is similar in composition to its parent turbid-water layer from the Mississippi River, and which varies seasonally in sediment concentration, thickness, salinity, and temperature.[1]

   At times, the turbid bottom layer migrates laterally into areas in which the bottom waters are temporarily clear. If at those times the clear water is more dense, the turbid layer can override it, and the remnants of the original water mass then form a thin clear layer underlying the turbid layer. Such events are doubtless transitory but have been seen by divers (E. A. Shinn, personal communications 1973-74) and have been measured by me on three occasions.

4. From the current observations of Oetking (1973), we can predict some features of turbid layer movement across the Louisiana shelf, and the nature of the multiple resuspension process. These

data indicate a predominant migration of the bottom water toward the coast, often unrelated and sometimes contrary to surface or mid-depth movements. The average speed of the bottom current is only 0.21 knots (11 cm/sec), versus 0.7 knots (38 cm/sec) in the surface water. Bottom-water movement is affected by tidal motion, so that parcels of water describe elongated elliptical gyres, but the net result is shoreward movement.

Oetking (1973:8) states that in the summer regime, "Flooding occurs dominantly on the bottom, with ebb at mid-depth. . . ." The winter regime is less definite, with bottom water showing flood, ebb, and westerly influence (1973:9). Addition of flood tide movement to the net non-tidal flow tends to produce a pulsating bottom-water motion with a (tidal) period of 24 hours; this motion would tend to be most pronounced during the summer regime, when the floodtide reinforcement of the non-tidal flow is most pronounced. Thus, multiple resuspension of the freshly-sedimented, very hydrous, non-cohesive sediment particles from the turbid layer would tend to occur during the summer regime.

The dominant landward migration of the bottom turbid layer is important to note. The general case for landward migration of fine sediment has been presented for the Atlantic Coast by Meade (1972), and the data presented here indicate that it is also applicable to the Louisiana coast.

5. The stations occupied on the April 6, 1973, synoptic cruise allowed turbid-layer movement in the nearshore zone to be further defined (figure 10). Those stations suggest that the fine sediment transported landward in the turbid layer ultimately reached the inner shelf where it encountered a second "littoral turbid layer" of a different type. The littoral turbid layer occupied the entire 6-m water column at the most inshore station (2.8 km offshore). It protruded seaward under clearer surface waters as a relatively cool, low-salinity, turbid, *oxygen-saturated* wedge recognizable to about 14.8 km offshore. The clearer water at the 16.7 km juncture of the two turbid layers is not easily explained (figure 10), but seaward of 18.5 km, the shelf-generated "neritic turbid layer" is recognized by its warmer temperature, higher salinity, and lower oxygen content (<74% saturated). Presumably, the near-shore turbid materials in both the neritic and littoral layers tend to migrate westward as part of the littoral drift, and it is predicted that their distinctive characteristics are gradually blurred by diffusion near their juncture.

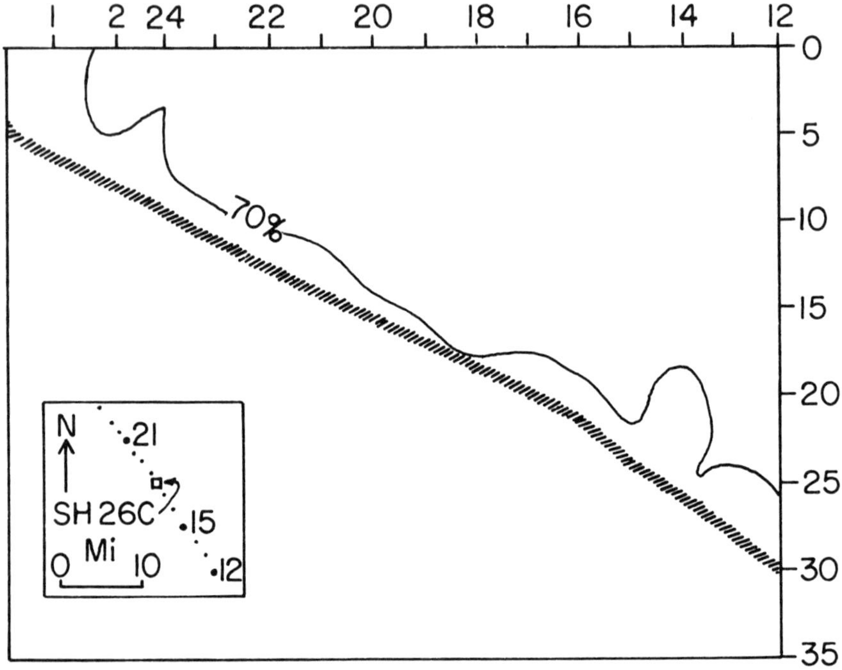

FIG. 10. CROSS-SECTION OF VERTICAL TRANSMISSIVITY (% T/10 cm) profiles along a transect passing close to Shell Platform 26C. April 6, 1973. A turbid layer is indicated between the seafloor and the 70% T/10 cm contour line. The seaward part of the turbid layer is believed to originate on the shelf, and the landward part is generated in the littoral zone. Note that a 10-cm path length was used on this cruise, as the 1-m instrument was being repaired.

6. The ultimate result of shoreward, then westward, littoral migration of the bottom turbid zone probably is an addition of fines to the western Louisiana coast (e.g., at Cameron, LA). There, much of the suspended material is ultimately trapped in the coastal marsh system, where it contributes to the progradation of the Chenier Plain mud flats, which is known to be occurring (Morgan, Nichols, and Wright 1958; Gould and McFarlan 1959; Gould and Morgan 1962). Besides documenting the significant trapping of clay during abnormally high tides, and the resulting formation of clay drapes overlying other coastal deposits, Morgan has subsequently found similar clay deposits

overlying Cheniers as old as 2000 years B.P. (personal communication 1974).

SUMMARY AND CONCLUSIONS

Louisiana shelf waters south of Timbalier Bay are characterized by a two- or three-layer water column with a persistent near-bottom turbid layer. The turbid bottom layer was detected at 96% of the vertical stations and at all seasons, although with large variations in thickness and concentration. The bottom turbid layer was absent from four stations on the Florida-Alabama shelf where the rate of clastic influx is much lower than on the Louisiana shelf. However, similar turbid layers are common to clastic depositional environments throughout the world, and the Louisiana shelf is quite normal in that respect.

The clay-mineral composition of the suspended sediment was identical in surface, mid-depth, and turbid bottom-water layers. It was also indistinguishable from Mississippi River suspended sediment. However, the suspended sediment differed significantly from the clay mineralogy of the cohesive, older bottom sediments of the study area. From this, I concluded that the suspended material in the turbid bottom layer and the water column as a whole is derived from the modern Mississippi River, without detectable additions from erosion of the older recent bottom sediments or from offshore petroleum operations.

A process model for turbid-layer transport of Mississippi River suspended sediment follows. The material originates in the surface turbid plume, settles slowly through the clear mid-depth water, and is distributed in accordance with net-westerly water movements that affect those upper two layers. Ultimately, through further settling, the particles enter a near-bottom zone where they accumulate as a nepheloid cloud of fine particles, here designated the *turbid layer*. The turbid water mass slowly migrates toward the coast via an often-interrupted multiple-resuspension process involving tidal gyres plus non-tidal movements. The shelf-bottom is by-passed by the fine sediment particles, which are neither permanently sedimented nor measurably augmented by erosion of the older, cohesive shelf sediments. Within 8 kilometers of the coast, the shelf turbid layer is joined by turbid water generated in the littoral zone. There, littoral drift carries the combined turbid water mass westward, where it can be trapped in the coastal marshes and mudflats of the prograding Chenier Plain of southwestern Louisiana.

Offshore production platforms of the type examined were found to have no detectable effect on the turbidity of the Louisiana shelf, or on the suspended sediment transport process described above. Turbidity parameters were identical in all respects adjacent to the platforms and in an ecologically similar, but undeveloped, control site.

NOTES

Reprinted with permission from Transactions—Gulf Coast Association of Geological Societies **28** (1978):163-171.

1. In water deeper than approximately 100 m off the active Mississippi Delta, the massive, clayey nature of the accumulating bottom sediment suggests that resuspensive forces are too weak to winnow the fines effectively, and the massive pro-delta clay of Fisk et al. (1954) accumulates. However, at the 20-30 m depth of the OEI area, resuspension is evidently active, so that winnowing and further transport is occurring.

REFERENCES CITED

Bates, C. C. 1953. Rational theory of delta formation. American Association of Petroleum Geologists Bulletin **37**:2119-2162.

Coleman. J. M., L. D. Wright, and J. N. Suhayda. 1972. Density gradients at river mouths. Naval Research Reviews **25**:9-20.

Drake, D. E., R. L. Kolpack, and P. J. Fischer. 1972. Sediment transport on the Santa Barbara-Oxnard Shelf, Santa Barbara Channel, California. *In* D. J. P. Swift, D. B. Duane, and O. H. Pilkey, eds., Shelf Sediment Transport, Process and Pattern. Stroudsburg, Pa.: Dowden, Hutchinson, and Ross, Inc., pp. 307-331.

Fisk, H. N., C. R. Kolb, and L. J. Wilbert, Jr. 1954. Sedimentary framework of the modern Mississippi Delta. Journal of Sedimentary Petrology **24**:76-99.

George, R. Y. and P. Thomas. 1973. Digest of preliminary data analysis by the Project Planning Council, January 4-5, 1974. *In* J. M. Sharp and J. Tyson, eds., The Offshore Ecology Investigation: First Project Council Report. Gulf Universities Research Consortium. Report No. 133, p. 15.

Geyer, R. A. 1950. The occurrence of pronounced salinity variations in Louisiana coastal waters. Journal of Marine Research **9**:100-110.

Gould, H. R. and E. McFarlan, Jr. 1959. Geologic history of the Chenier Plain, southwestern Louisiana. Transactions—Gulf Coast Association of Geological Societies **9**:261-270.

______ and J. P. Morgan, 1962. Coastal Louisiana swamps and marshlands. *In* Geology of the Gulf Coast and Central Texas and Guidebook of Excursions. Houston Geological Society, pp. 287-341.

Griffin, G. M. and B. J. Ripy. 1974. Turbidity, suspended sediments, and the origin of the turbid near-bottom water layer—Louisiana shelf south of Timbalier Bay—August 1972-January 1974—With comments on a process model for turbid layer transport. Final report of Project OG51H. Gulf Universities Research Consortium.

Humm, Harold. 1973. Benthic algae on the offshore platforms. Unpublished GURC-OEI Progress Reports. Gulf Universities Research Consortium.

Lipsey, T. E. L. 1919. Currents at or near the mouth, Southwest Pass, Mississippi River. U.S. Army Corps of Engineers, Professional Memoirs **11**:65-122.

Meade, R. H. 1972. Sources and sinks of suspended matter on continental shelves. *In* D. J. P. Swift, D. B. Duane, and O. H. Pilkey, eds., Shelf Sediment Transport: Process and Pattern. Stroudsburg, Pa.: Dowden, Hutchinson, and Ross, Inc., pp. 249-262.

Moore, D. G. 1972. Reflection profiling studies of the California continental borderland, structure and Quaternary turbidite basins. Geological Society of America Special Paper 107.

Morgan, J. P., L. G. Nichols, and Martin Wright. 1958. Morphological effects of Hurricane Audrey on the Louisiana coast. Louisiana State University, Coastal Studies Institute Technical Report no. 10, Part A.

Oetking, Philip. 1973. Currents on the nearshore continental shelf of south central Louisiana. Unpublished GURC-OEI Progress Reports. Gulf Universities Research Consortium.

Scruton, P. C. 1956. Oceanography of Mississippi Delta sedimentary environments. American Association of Petroleum Geologists Bulletin **40**:2864-2952.

——— and D. G. Moore. 1953. Distribution of surface turbidity off Mississippi Delta. American Association of Petroleum Geologists Bulletin **37**:1067-1074.

SURFACE AND SHALLOW SUBSURFACE SEDIMENTS OF THE NEAR-SHORE CONTINENTAL SHELF OF SOUTH CENTRAL LOUISIANA

by P. Oetking, R. Back, R. Watson, and C. Merks

ABSTRACT

High resolution acoustic profiling and sediment sampling were conducted in the study area over an 18-month period in 1972-1974, to map the bottom and sub-bottom and to examine the sediments near an offshore petroleum platform. Profiling records, verified by bottom samples, delineated a variety of sediment types and topographic bottom and sub-bottom features. More than 1000 kilometers of survey traverses were used in the preparation of a revised bathymetric map and the construction of a sediment distribution map. An analysis of information developed from the detailed profiling and sediment sampling conducted in the vicinity of an offshore petroleum platform showed no recognizable changes in the sediments attributable to drilling or production activities.

INTRODUCTION

The study area lies immediately offshore from the Holocene Lafourche lobe of the Mississippi Delta Complex. The outer limits are landward of the head of the Mississippi submarine trough and cross the axis of the buried Mississippi Trench, which was incised by the Mississippi River during Pleistocene low stands of sea level. With rising

---

At the time of this investigation, P. Oetking was Director and R. Back and C. Merks were Research Scientists at Southwest Research Institutes Ocean Science and Engineering Laboratory, Corpus Christi, TX. R. Watson was at the Marine Science Institute, The University of Texas, Port Aransas, TX.

sea level after Wisconsin glaciation, the site of active deposition retreated from the area of the present Continental Slope landward to the Continental Shelf. As a result, the study area is underlain by a thick sequence of delta plain facies including distributary channel sands, natural levee silts and sands, interdistributary marsh deposits, and near-shore marine sediments. In some places these relict sediments are covered by a thin veneer of modern prodelta clays (figure 1).

The modern Continental Shelf sediments are primarily the result of marine processes acting on the relict Late Pleistocene and Early Holocene delta sediments on the shelf. Near the shore, in water depths of less than about 30 feet (9 m), waves and currents actively modify the bottom by winnowing out the muds, leaving a sandy bottom with the mud content increasing with depth of water. Farther offshore, where turbulence on the bottom is significantly reduced, the sand-size sediments usually are not transported except during storms.

Water depth and bottom sediment data within the study area were obtained from about 600 miles (1100 km) of high resolution acoustic profiling traverses. In addition to the regional traverses, profile records and bottom samples were acquired near an offshore production platform in order to discern any evidence of sediment changes.

Mapping of subsurface and surface sediments, grain size analylses, and quantitative evaluation of wave, tide, and wind-induced currents provided criteria to place limits on the location of the maximum seaward advance of the Lafourche delta lobe. Analysis of the sediment transport capability of bottom current demonstrated that the medium and fine sands of the middle shelf are relict from lower stands of sea level.

METHODS AND MATERIALS

Bottom and sub-bottom acoustic records were made with three different frequencies: 34, 12, and 3.5 kHz. Each of the frequencies furnished records for different purposes. The two higher frequency systems are hull-mounted Edo Western, Model 353, transducers with Gifft, Model 4000T, recorders and are capable of high resolution profiles at vessel speeds of 18 knots. The highest frequency was used to define the water/sediment interface and to examine the upper few feet of the bottom sediments. The 12 kHz system normally showed detailed density variations in the upper 20 feet (6 m) of the sedimentary sections and occasionally penetrated to 40 feet (12 m) in the soft muds. When penetration requirements were greater than those attainable with these units, the towed 3.5 kHz profiling system (Hydro Products, Model 4600, transducer) was used in addition. Boat speed was considerably reduced

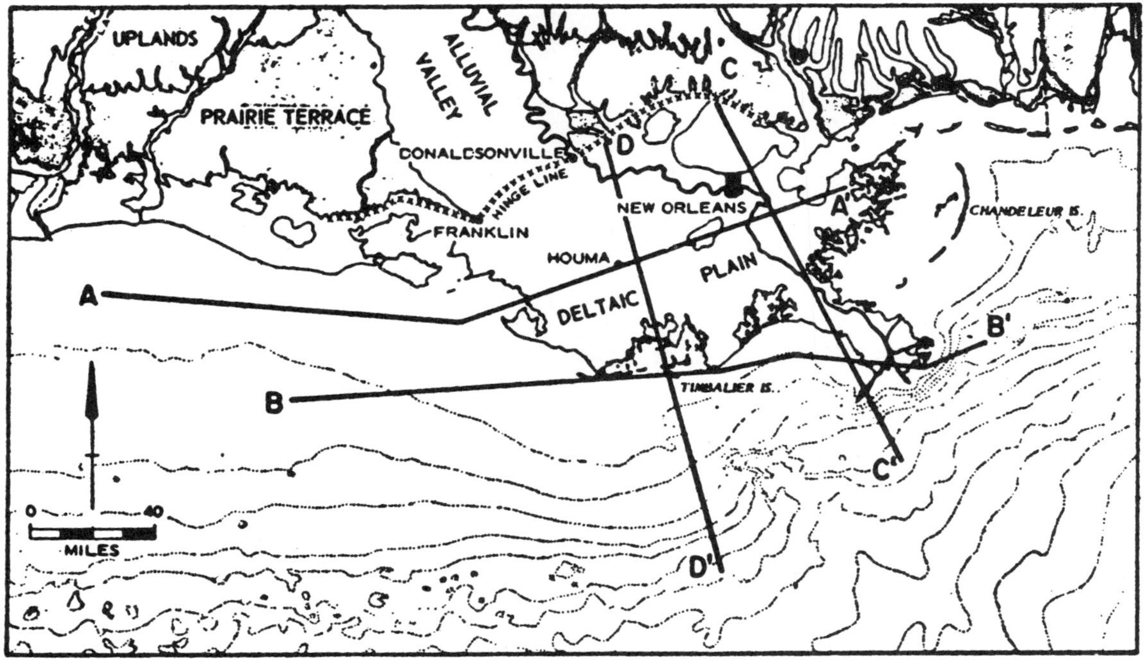

FIG. 1. CROSS-SECTIONAL RELATIONSHIPS AND FACIES OF THE LATE QUATERNARY DELTAIC MASS (continued overleaf). Reprinted by permission of the Geological Society of America from Fisk and McFarlan (1955).

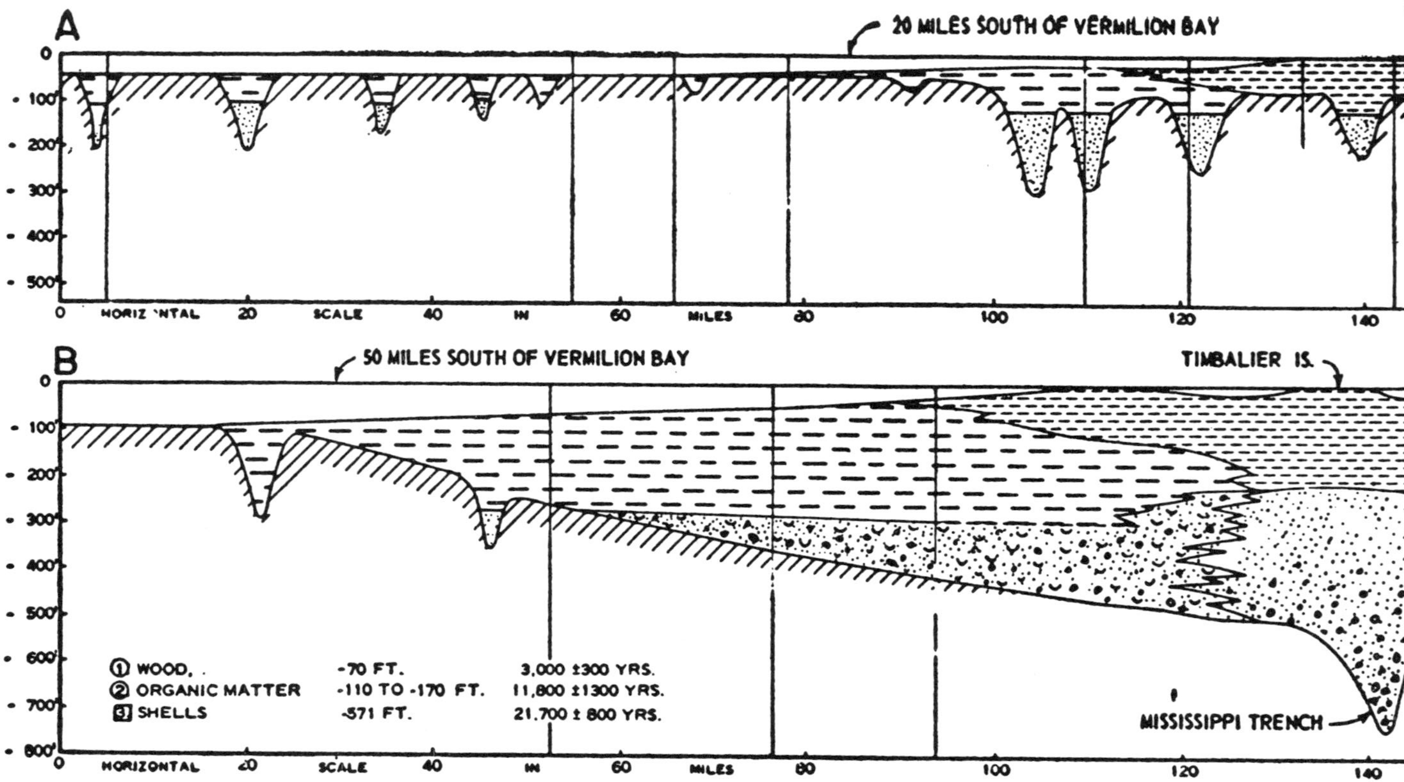

A
20 MILES SOUTH OF VERMILION BAY
0
100'
200'
300'
400'
500'
0
HORIZONTAL
20
SCALE
40
IN
60
MILES
80
100
120
140
B
50 MILES SOUTH OF VERMILION BAY
TIMBALIER IS.
0
100'
200'
300'
400'
500'
600'
700'
800'
① WOOD, -70 FT. 3,000 ±300 YRS.
② ORGANIC MATTER -110 TO -170 FT. 11,800 ±1300 YRS.
③ SHELLS -571 FT. 21,700 ± 800 YRS.
MISSISSIPPI TRENCH
0
HORIZONTAL
20
SCALE
40
IN
60
MILES
80
100
120
140

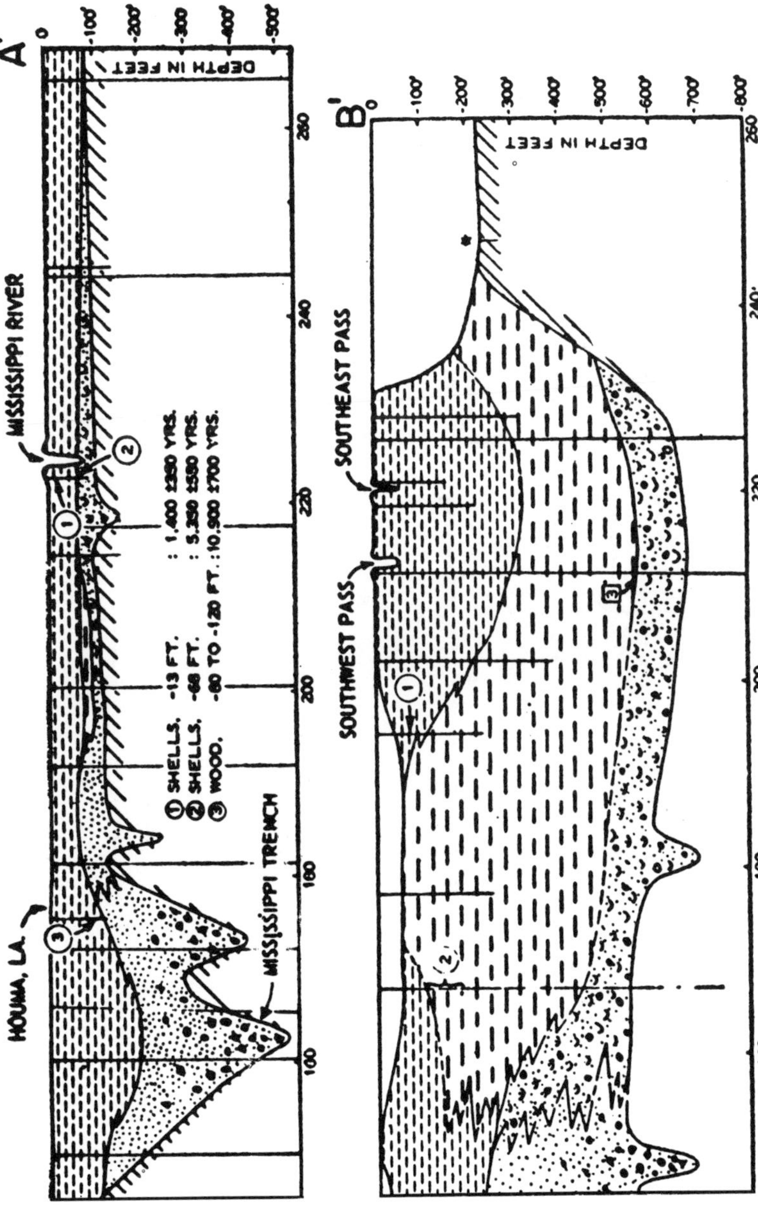

FIG. 1 continued.

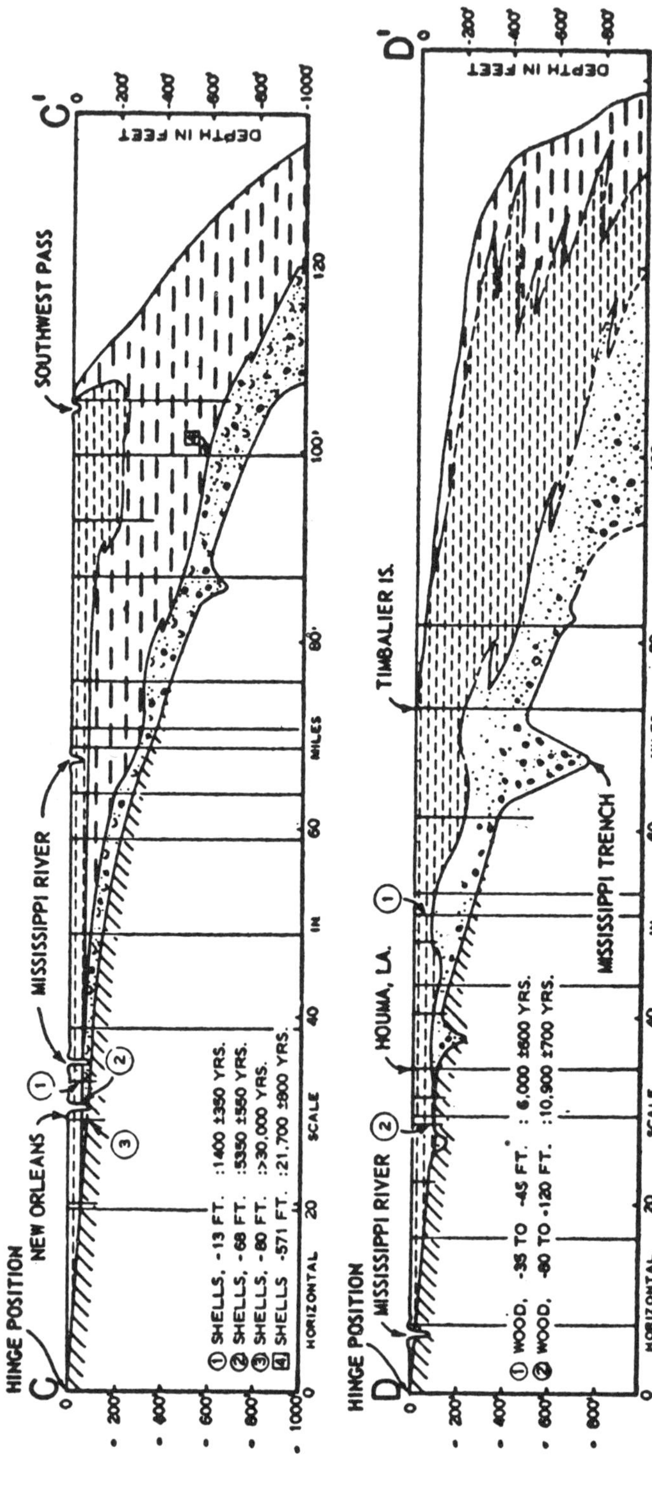

FIG. 1 continued.

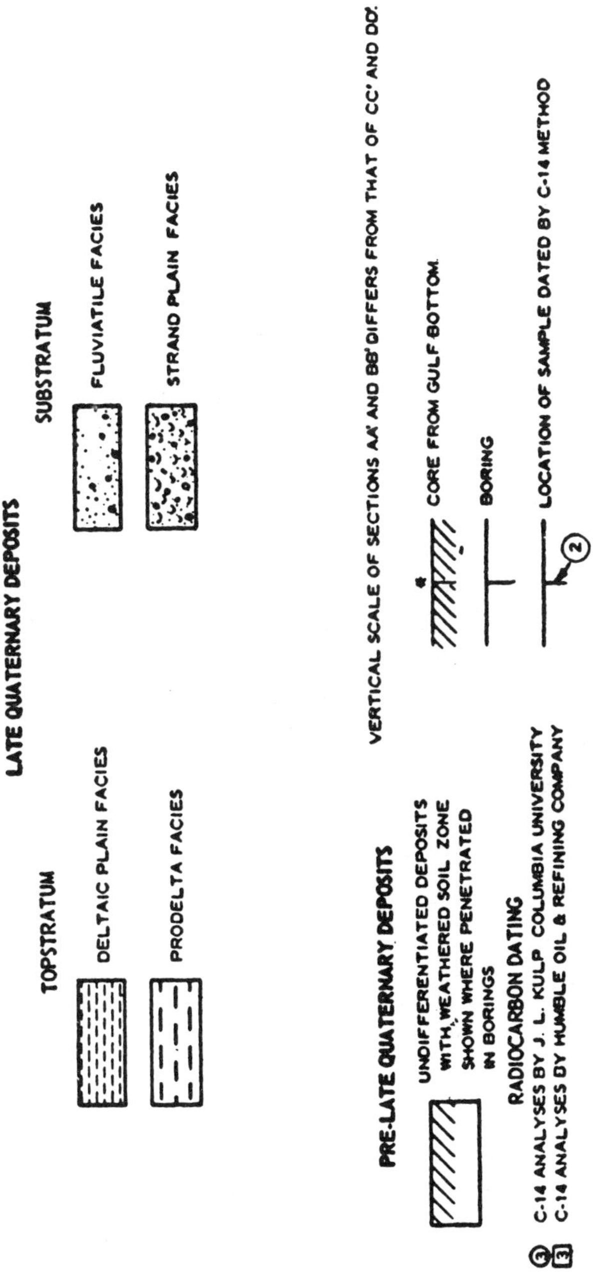
LATE QUATERNARY DEPOSITS
TOPSTRATUM
SUBSTRATUM
DELTAIC PLAIN FACIES
FLUVIATILE FACIES
PRODELTA FACIES
STRAND PLAIN FACIES
PRE-LATE QUATERNARY DEPOSITS
UNDIFFERENTIATED DEPOSITS WITH WEATHERED SOIL ZONE SHOWN WHERE PENETRATED IN BORINGS
RADIOCARBON DATING
3 C-14 ANALYSES BY J. L. KULP COLUMBIA UNIVERSITY
3 C-14 ANALYSES BY HUMBLE OIL & REFINING COMPANY
VERTICAL SCALE OF SECTIONS AA' AND BB' DIFFERS FROM THAT OF CC' AND DD'
CORE FROM GULF BOTTOM
BORING
2
LOCATION OF SAMPLE DATED BY C-14 METHOD

FIG. 1 legend.

with the operation of this instrument, but underway "scoopfish" sediment samples were obtainable during these traverses. The principal sediment samples were acquired with a Shipek sampler, Model 860 (1/25 square meter) and a Phleger gravity coring device.

The R/V/*Southwest Researcher I* was the principal vessel used in this study. The sampling plan for this project was designed to make acoustic profiles of the bottom and sub-bottom along the traverses depicted in figure 2. Most of the profiles were made while the vessel was en route to OEI sampling stations during six cruises from 1972 through 1974. A majority of the sediment grab and core samples were collected in July 1973. To determine if there was discernible evidence of sediment changes resulting from drilling or other platform activities, over 50 miles (90 km) of profiles were made and 54 grab and core samples were taken on concentric square traverses around platform HOR-ST-54A.

RESULTS

Bathymetry

Navigational chart data and 34 kHz profile records acquired during the study traverses show that the near-shore region, within a few miles of the present shoreline of the Lafourche subdelta, slopes seaward at 15 feet/mile (2.5 m/km). Most of the remainder of the study area slopes seaward at 4 to 6 feet/mile (0.7 to 1.0 m/km).

Platforms HOR-ST-66D and HOR-ST-54A are located on a flat region of the shelf that probably resulted from wave action on a submerged delta lobe during a lower stand of sea level. The sediments underlying this area will be discussed in conjunction with the examination of sub-bottom profiles.

Sub-bottom sediments

The recorded response characteristics of the various sediments to the profiling frequency provide the means of identifying different types of sediment. The five sub-bottom sediment types distinguished were:

Hard material (sand, stiff clay, or shell)
Sandy silts
Soft silt and clay
Multilayered soft silt and clay
Deformed multilayered soft silt and clay.

The limits of distribution of sediment types are delineated on the sub-bottom map shown in figure 3.

The near-shore area is underlain dominantly by sand and muddy sand, which probably represent the Lafourche delta front sheet sands

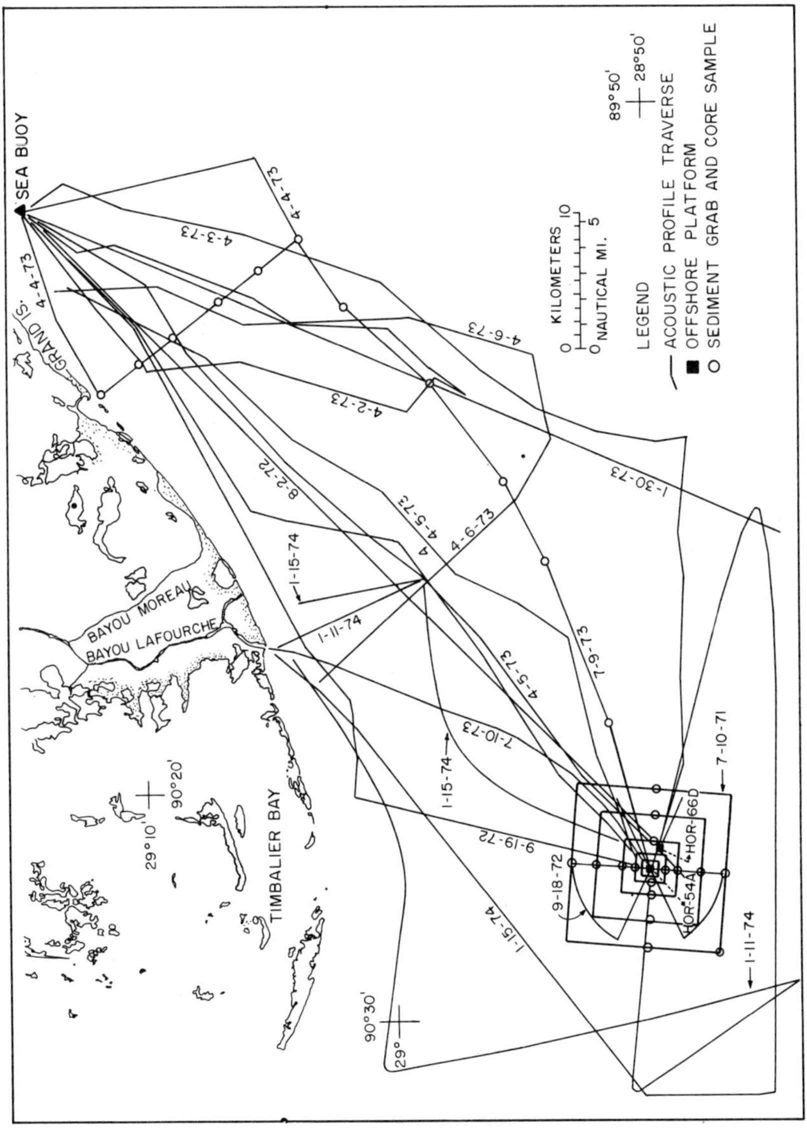

FIG. 2. INDEX MAP OF ACOUSTIC PROFILE TRAVERSES AND SEDIMENT SAMPLING SITES on the near-shore continental shelf of South Central Louisiana.

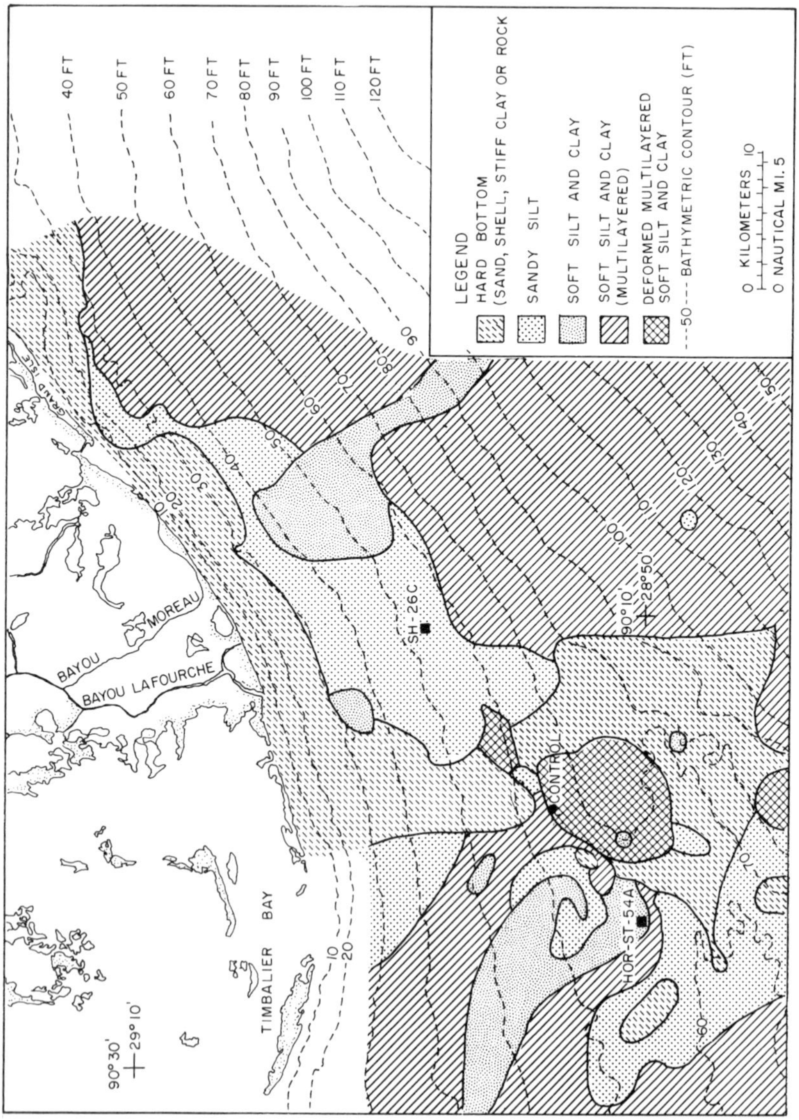

FIG. 3. SUB-BOTTOM SEDIMENT MAP.

deposited where the distributaries discharged into deeper water. Since the abandonment of the delta lobe, these sands have been further distributed into a narrow zone alongshore by waves and currents. In water depths greater than 30 feet (9 m), the sand content of the sediment progressively lessens and the silt and clay content increases. Farther offshore, the clay content increases and the sub-bottom shows pronounced layering. The multilayered silts and clays may have been deposited as part of the delta plain facies during lower stands of sea level. In the greater depths of the study area, there may be some deposition resulting from clays transported in suspension by high density underflows moving along the bottom. It should be noted that currents and low transmissivity measurements at numerous hydrographic stations provide evidence of near-bottom suspended particle movement.

Although the general trend in the offshore direction is decreasing grain size and firmness of the sub-bottom sediment, it is obvious from the distribution map that other influences disturb this pattern. Offshore from Bayou Lafourche, a zone of soft nonlayered clays and silts exists that appears to be composed of numerous small filled channels that tend to be perpendicular to the shoreline and directly downslope. The proximity of the near-shore end of this zone to the mouth of Bayou Lafourche and Bayou Moreau and the fact that the zone trends nearly downslope suggest that it may represent the site of shallow submarine channels.

The channels off Bayou Lafourche and Bayou Moreau vary from narrow steepsided channels to broad channels with gently sloping sides. Often a layer of soft multilayered silt and clay overlies the nonlayered channel sediments. Morgan (personal communication 1974) views the channels as "slump channels" and suggests the lack of layering is attributed to repetitious slumping coupled with sediment deposition. The profile records of this study do not show recognizable evidence of slumping associated with the channels. Platform HOR-ST-54A, a key station in this study, is located in one of the mapped channels filled with soft silt and clay (see figure 3). The lithology interpretation from the profile record is supported by a cored section in South Timbalier block 54 area where Fisk (1956) describes a 25 foot high water content/low bearing strength gray clay overlying a 50 foot sand body.

Gently deformed beds of soft material occur several miles (4-9 km) to the east and northeast of platform HOR-ST-54A and suggest slumping possibly related to sediment adjustments along the Mississippi Submarine Canyon (see figure 3). Surrounding the deeply penetrated deformed beds are firm sediments that exhibit only a few feet (1 m) of penetration with the high frequency signal.

Surface sediment distribution near Platform HOR-ST-54A

Twenty-four surface grab samples and numerous short cores were taken on a north-south and an east-west traverse 3 miles (5.5 km) to each side of platform HOR-ST-54A. These were acquired to determine the surface sediment distribution as well as to detect any characteristics in surface sediments that could be attributed to platform operations. Sand to silt-clay ratios were determined for each of the grab samples. The degree of penetration of the surface sediments by the profiler was used to indicate the density or, in this case, sand content, of the surficial sediments.

Sand percentages of the surface sediments are shown in figure 4. The general trend is increasing grain size in the offshore direction. The sediments range from 10% to 98% fine to medium sand. Most of the area south of the platform contains more than 67% sand, while the area to the north has less than 33% sand, with a transitional zone trending east-west in the vicinity of the platform. This sand body with modal sizes in the 0 phi to 2.9 phi range (1 to 0.12 mm) was mapped by Curray (1960).

DISCUSSION

Shown in figure 4 is a lobe of sandy silt and clay, which extends southward into the area immediately north and east of the platform, and which may have originated from fine sediments released during the oil well drilling. The dominant bottom currents in this area are to the north and northwest, and thus tend to be in the same direction as the finer sediments. Most of the sands in the vicinity of the platform are coarser than the very fine sands supplied by the modern Mississippi River, and are probably coarser than any supplied since the post-Pleistocene still-stand in sea level. Therefore, one could assume that the source of the fine and medium sand fractions could be attributed to drilling operations in the area.

On the other hand, consider the conditions that must prevail if the medium to fine sands found south of the platform are to be attributed only to drilling operations. In figure 4, the mapped area with more than 67% sand is approximately 20 square nautical miles (70 sq km) or 753×10^6 square feet (70×10^6 sq m). Short cores indicate surface sand deposits from 1 to 7 inches (2.5 to 18 cm) thick. Assuming an average sand thickness of only 0.1 foot (3 cm), the sand volume is 75.3×10^6 cubic feet (21.2×10^5 cu m). Using the mean sand content in the area as 83% and a medium to fine fraction of 87%, the total volume of medium and fine sand in the area would have to be at least 54×10^6 cubic feet (15×10^5 cu m). In contrast, the total volume of sediments produced by

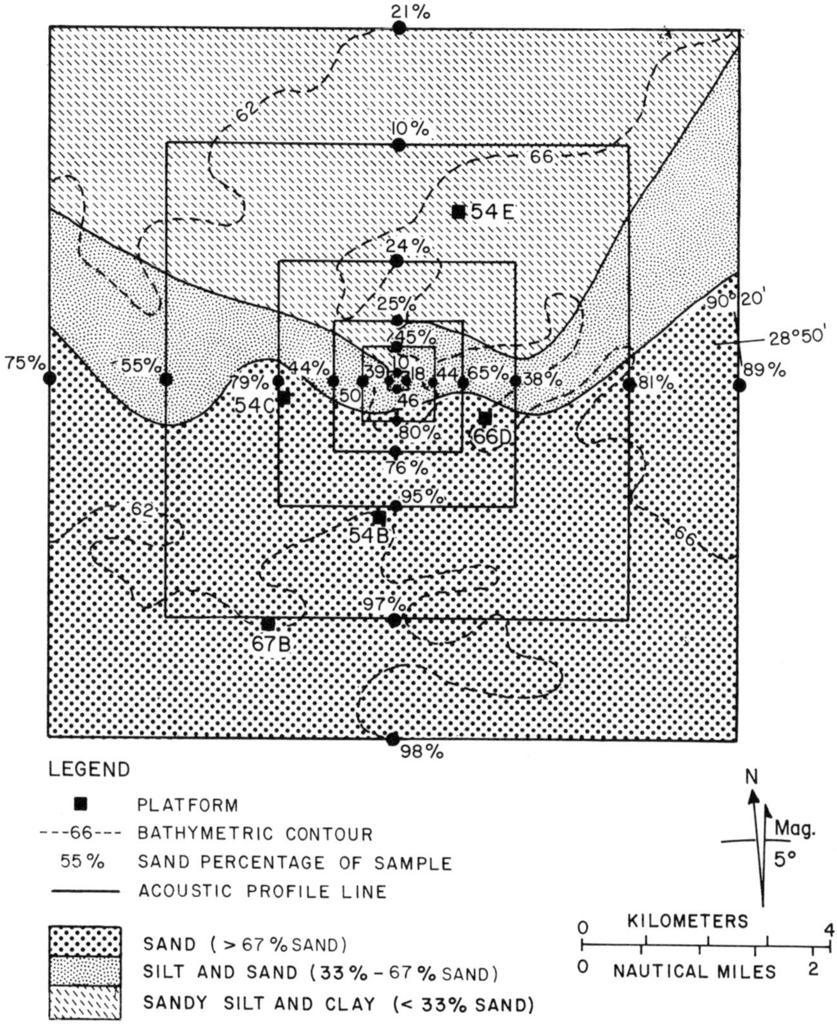

FIG. 4. SAND PERCENTAGE MAP OF THE SURFACE SEDIMENTS around Platform HOR-ST-54A on the near-shore continental shelf of South Central Louisiana.

a 10,000 foot (3050 m), 10-inch (25.4 cm) diameter boring is only 5440 cubic feet (154 cu m). If it were assumed that all of the sediment removed by boring was medium to fine sand, nearly 10,000 separate drillings would be required to yield this amount of medium to fine sand in an area of 20 square nautical miles (70 sq km).

It is also necessary to examine the dispersal mechanisms available to transport sediment away from the platforms. A current speed of about 1 ft/sec (30 cm/sec) is required to erode sand (Hjulstrom 1935). An examination of the continuous current records for the period from August 1972 through July 1973 shows that bottom currents greater than 1 ft/sec (30 cm/sec) at platform HOR-ST-66D existed for only 14 hours out of 66 days of measurements, amounting to only 1% of the time. Furthermore, these currents were in an offshore direction for only 5 out of 14 hours. The possibility also exists that the sand can be suspended by wave action and then transported by currents less than 1 ft/sec (30 cm/sec). The maximum horizontal orbital velocity at the bottom (Umax) produced by a sinusoidal wave is given by the following relationship:

$$\text{Umax} = H/\left(T\sinh\,(kh)\right)$$

where H is the significant wave height, T is the wave period, k is the wave number ($2\pi/L$), L is the wave length and equal to 5.12 T^2, and h is the water depth. For any given depth of water, the maximum horizontal bottom velocity can be determined as a function of a given wave height and period. If this computation is made based on the 60-foot (18 m) water depth at platform HOR-ST-54A (figure 5), the results indicate that either very high waves, waves with very long periods, or a moderate combination of both are required to disturb sand at that depth.

Wind velocity, duration, fetch, and decay distances were determined from 12-hourly synoptic weather charts for the Gulf of Mexico for the years 1950, 1952, and 1954, to calculate wave statistics for the region off of Burrwood, Louisiana (Bretschneider and Gaul 1956). These hindcast wave statistics, together with figure 5, were used to compute the following monthly and annual durations of wave-induced bottom currents greater than 1 ft/sec (30 cm/sec) at a depth of 60 feet (18 m):

Month	Hours
Jan.	16
Feb.	40
Mar.	40
Apr.	none
May	4
June	none
July	8
Aug.	10
Sept.	28
Oct.	20
Nov.	56
Dec.	60
Yearly Total	282 hours or 3.4% of the time

The figures given above are for waves moving in all directions. Waves moving offshore and thus capable of transport of sediment offshore occurred less than 54 hours per year (0.6% of the time). It seems unlikely that sand at a point source could have been transported very far under these conditions. The effect of severe tropical storms is unknown; Curray (1960), however, suggests that hurricane waves rework relict shelf sands with little or no net transport.

Size analyses of the sand fraction were performed on several sediment samples collected near platform HOR-ST-54A in an attempt to determine sand origins (figure 6). These analyses were conducted on a Woods Hole-

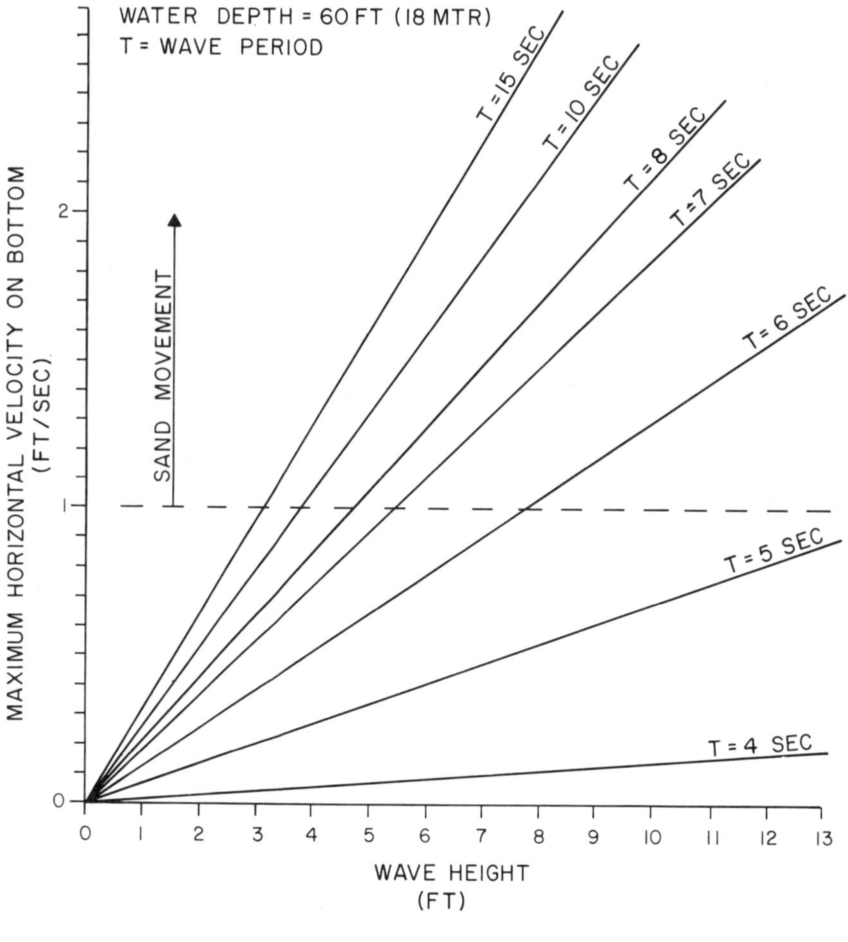

FIG. 5. MAXIMUM HORIZONTAL ORBITAL VELOCITY AT BOTTOM.

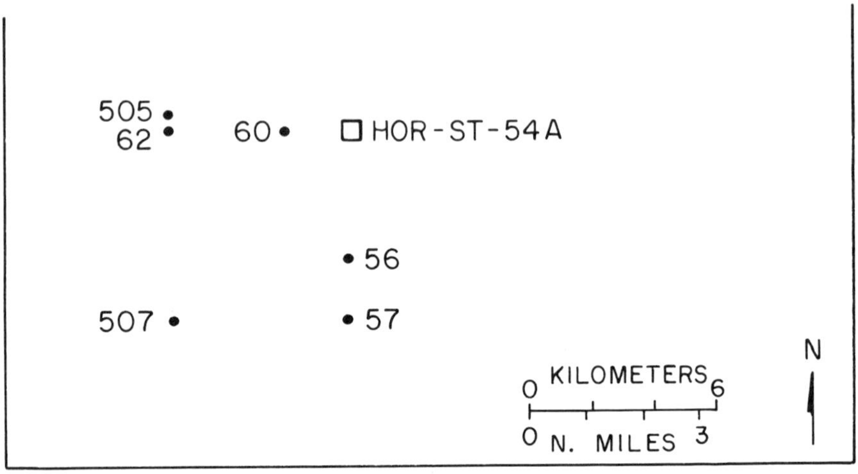

FIG. 6. SAND FRACTION SAMPLING STATIONS near Platform HOR-ST-54A, near-shore continental shelf of South Central Louisiana.

type rapid sediment analyzer (Schlee 1966). The grain-size distribution was determined by relating sediment settling velocity to grain-size (phi units) on the basis of a modified Rubey Equation for the settling velocity of spheres in a viscous fluid (Watson 1970). The mean size and standard deviation, computed by the graphic methods of Folk (1965), are presented in table 1.

The mean sizes of the sand fraction were in the fine sand range (2.0 to 3.0 phi). The samples contained an average of 25% medium sand (1.0 to 2.0 phi). No coarse sand (0.0 to 1.0 phi) was detected by the rapid sediment analyzer, although a very small amount was observed in some of the samples by microscopic examination. The phi standard deviations (ranging from 0.49 to 0.71 phi) indicate that most of the sands were moderately sorted. If medium sands brought to the surface by drilling operations had been mixed with very fine sands on the shelf, the standard deviations would have been higher because of poor sorting. The moderate sorting and the infrequent occurrence of bottom currents capable of transporting sand indicate that the sands are relict from a lower stand of sea level.

The Mississippi River deltaic plain, fashioned during the past approximately 5000 years since sea level reached its present stand, contains sands of only the very fine to fine size, such as those of the present river mouth bar and the destructional islands. Coarser sands would get into the delta only when the river gradients were steeper, i.e., during a lower stand of sea level. It can be concluded that the medium sands found on the continental shelf in 60 feet of water around platform HOR-ST-54A are most likely relict sands left from a period when sea level was

TABLE 1
SAND FRACTION CHARACTERIZATION

Sample Number	Grain Size Parameters		% sand in each size class			
	Size Mean (ϕ)	Standard Deviation	Coarse 0-1 ϕ	Medium 1-2 ϕ	Fine 2-3 ϕ	Very Fine 3-4 ϕ
Platform HOR-ST-54A Louisiana						
56	2.47	0.71		25	62	13
57	2.28	0.52		34	57	9
60	2.40	0.52		34	63	13
62	2.48	0.62		17	64	19
505	2.02	0.54		43	53	4
507	2.78	0.49			71	29
Platform GU-MU-889-1 Mustang Island, Texas						
516	3.51	0.40			10	90
517	3.46	0.36			12	88
518	3.33	0.36			21	79
519	3.39	0.32			15	85
520	3.14	0.52			36	64
521	3.28	0.59		5	24	71
522	2.02	1.39	43	19	11	27
523	3.43	0.47			19	81
524	3.42	0.37			15	85
525	3.51	0.40			11	89
526	3.50	0.33			13	87
527	3.44	0.40			17	83
528	3.49	0.34			11	89
529	3.37	0.33			15	85
530	3.36	0.39			19	81

significantly lower. Furthermore, it can be concluded that the Holocene Lafourche delta did not prograde into the area of HOR-ST-54A. Occurrence of the relict medium-grained sand fraction near the platform probably delimits the maximum seaward extent of the Holocene Mississippi delta (Lafourche). The "hard bottom" sand distribution of figure 3 suggests that the Holocene delta extended only a few miles offshore from the present destruction phase barrier islands.

In order to evaluate further the effects that an offshore petroleum platform has on the sediments surrounding it, a suite of samples was acquired at a platform off the Texas coast (Platform GU-MU-889-1, located offshore Mustang Island near Port Aransas, in 48 feet of water [15 m]).

The mean sizes of most of the sand fractions of the Texas samples were in the range of very fine sand (3.0 to 4.0 phi). For the most part, the sand fractions were well-sorted, as indicated by the low standard deviations in table 1. Several exceptions were found. Sample 522, nearest the platform on the northeast side, has a mean size of 2.02 phi, poor sorting, and a bi-modal sand fraction composed of coarse and medium as well as very fine sand. On the near southwest side of the platform, sample 521 has a mean size of 3.28 phi, a standard deviation of 0.59 phi, and about 5% medium sand. The rapid sediment analyses of the samples did not provide evidence of coarse sands at stations beyond the 40 foot (12 m) distance from the platform.

However, microscopic examinations of the sand fractions did reveal a trace amount of medium and coarse sand to 40 feet (12 m) southeast and northwest, 760 feet (230 m) northeast, and 1500 feet (460 m) southwest from the platform. Sample 522, with the greatest amount of coarse sands, has a high percentage of very angular grains of quartz and feldspar and some irregular shiny black grains with conchoidal fractures. The black fragments and most of the coarse and medium sand surrounding the platform may be blasting material used in the maintenance of the structure.

There is almost no detectable evidence of the drilling operations other than very near the Mustang Island platform, even though it is located in shallower water and in a more severe wave climate than the Louisiana platform HOR-ST-54A. Medium and fine sand, so abundant and extensive south of platform HOR-ST-54A, are minor constituents off the Texas platform and probably not a product of drilling operations.

SUMMARY AND CONCLUSIONS

1. The study area is underlain by Early Holocene deltaic deposits, by prodelta sediments of the Late Holocene Lafourche lobe, and by modern prodelta clays.

2. Sub-bottom sediments grade from sands near shore to multilayered silty clays offshore. This general pattern is altered by buried channels, areas of deformed sediments, and areas of thinly-covered firm sediments.

3. Sandy muds on the north side of platform HOR-ST-54A change to muddy medium sands on the seaward side. The amount of medium sands around the platform is too large to be accounted for by well cuttings.

4. Bottom current measurements and wave calculations indicate that

the transport of many sand particles seaward of the platform would be unlikely.

5. Sands in the vicinity of the platform are coarser than the fine and very fine sands supplied by the modern Mississippi River. Grain-size analysis of the sand fractions of the sediment samples near the platform showed an average of 25% medium sand. We conclude that the medium sands were deposited during a lower stand of sea level, about 5000 years ago, when the Mississippi River had a steeper gradient.

6. Sediments surrounding an offshore Texas platform that were similarly examined for effects of drilling and operational activities showed a dominance of very fine sand and only samples near the platform had particles larger than fine sand. The medium and coarse sands are thought to be blast sands.

7. The Holocene Lafourche delta lobe must have been located landward of the zone of medium sands near platform HOR-ST-54A. Its boundary may be represented by the seaward limit of the near-shore fine sands reworked from Lafourche delta front and distributary channel sands.

8. Drill cuttings and operational activities of an offshore platform are not significant factors in altering the bottom sediments except near the structure.

ACKNOWLEDGMENTS

This Project was funded under Gulf Universities Research Consortium Contract GU853-8; Southwest Research Institute, Corpus Christi, Texas.

We are indebted to D. E. Feray and the late E. Heuer for their generous assistance, to L. Rastrelli, R. Case, and J. P. Morgan for criticism of the manuscript, to R. Weed, E. Baker, and D. Krueger for their dedicated data acquisition, and to L. Lakatos and M. Lynch for their aid in the preparation of the illustrations.

REFERENCES CITED

Bretschneider, C. L. and R. D. Gaul. 1956. Wave Statistics for the Gulf of Mexico off Burrwood, Louisiana. Department of the Army, Corps of Engineers Beach Erosion Board, Technical Memo. 87.

Curray, J. R. 1960. Sediments and history of Holocene transgression, continental shelf, northwest Gulf of Mexico. *In* F. P. Shepard, F. B. Phleger, and T. H. van Andel, eds., Recent Sediments, Northwest Gulf of Mexico. Publications of the American Association of Petroleum Geologists, pp. 221-266.

Fisk, H. N. 1956. Nearsurface sediments of the continental shelf off Louisiana.

In Proceedings of the Eighth Texas Conference on Soil Mechanics and Foundation Engineering.

Fisk, H. N. and E. McFarlan, Jr. 1955. Late Quaternary deltaic deposits of the Mississippi River. Geological Society of America Special Paper **62**:279-302.

Folk, R. L. 1965. Petrology of Sedimentary Rocks. Austin, Texas: Hemphills.

Hjulstrom, F. 1935. Studies of the morphological activity of rivers as illustrated by the River Fyris. Bulletin Geologiska Institut Universitet Upsala **25**:1-298.

Morgan, J. P. 1974 (Personal communication). Louisiana State University, School of Geoscience, Baton Rouge, La.

Schlee, J. A. 1966. A modified Woods Hole rapid sediment analyzer. Journal of Sedimentary Petrology **36**:403-413.

Watson, R. L. 1970. Modified Rubey's Law accurately predicts sediment settling velocities. Water Resources Research **5**:1147-1150.

THE DISTRIBUTION AND ORIGIN OF BOTTOM SEDIMENTS IN TIMBALIER BAY, LOUISIANA, AND THE ADJACENT OFFSHORE AREA

by James I. Jones and Sam E. Williams

ABSTRACT

Sediment samples from Timbalier Bay, Louisiana, and the adjacent offshore area were analyzed for grain size distribution, percentage of organic carbon, percentage of carbonate, and selected elements. The observed sediment distribution in the study area was characterized by a naturally occurring transition between sand and mud. These sediments originate from the Mississippi River, from old submerged Mississippi River deltaic lobes, and from drill cuttings. The observable and documentable effects of drilling were generally limited to surface sediments within three miles of the drilling platform. Coarse, medium, and fine sand fractions were greatest adjacent to the platform because of the addition of drill cuttings to the naturally occurring sediment. In Timbalier Bay the organic and carbonate values increased shoreward from the barrier islands. Offshore, the carbon and organic carbon values were generally low. The carbonate values correlate well with the distribution pattern found for the granule size fraction. The most common clay minerals found, in decreasing abundance, were montmorillonite, illite, and kaolinite. With few exceptions, the measured elements demonstrated lower values in the sandy sediments and higher values in finer sediments and at the bottom of the cores. Of the measured elements only barium demonstrated values significantly greater than those measured from naturally occurring sediments reported in the literature. The high barium values may be attributed to the introduction of barite, a major component of

At the time of this investigation James Jones and Sam Williams were with the Department of Oceanography, Florida State University, Tallahassee, FL.

drilling mud, into the sediment. The results of this study indicate that no significant sedimentary alterations have occurred in Timbalier Bay or in the adjacent offshore area as a result of drilling activities. With the single exception of increased barium concentrations, no significant elemental concentrations of those elements measured have occurred as a result of drilling operations in this region.

INTRODUCTION

Timbalier Bay is a large, shallow, semi-enclosed bay located approximately 90 kilometers to the west of the present Mississippi River delta near Grand Isle, Louisiana. The bay has historically been influenced by at least two deltaic lobes of the Mississippi River (Coleman 1966). Since 1938, when the first producing offshore oil well was discovered here, the bay has been continuously drilled. Today it contains about 400 wells (Morgan 1973). The offshore area adjacent to Timbalier Bay in south central Louisiana was also studied. This area extends from the shoreline offshore to about the 30 meter depth contour, and covers an area of approximately 1000 square kilometers. The sediments in this area are influenced by the old Mississippi River deltaic lobes, the sediment presently transported by the Mississippi River, and the drill cuttings and drilling mud produced during the drilling of the offshore oil and gas wells.

The deep subsurface geology of the area is well known as a result of this drilling, but very little has been published about the surface and shallow subsurface geology. Baseline data for the recent sediments of this region prior to drilling are, therefore, not available. A previous study of the surface sediments of Timbalier Bay was conducted by the Louisiana Wildlife and Fisheries Commission in 1968 (Louisiana Wildlife and Fisheries Commission 1971).

The present study was undertaken to describe the sediments of Timbalier Bay and its adjacent offshore area, to determine the distribution and origin of these sediments, and to measure the effects, if any, of oil production and drilling on these sediments.

METHODS

Sediment samples were obtained by coring, grab sampling, and underway bottom sampling. Gravity cores were collected using a 60 cm core barrel with a 3.8 cm outside diameter core liner. Diver-collected hand cores consisted of a section of 6.4 cm OD core liner, which was pushed into the substrate. The advantage of hand coring over gravity coring is that the substrate maintains its structural integrity with little sample compaction, a significant problem with gravity cores. Each core was marked with an identifying number and an arrow to indicate the horizontal orientation of the core. Cores were refrigerated or frozen to retard bacterial action and were handled and stored vertically to minimize structural disturbance in the sample.

Grab samples were collected with a 0.25 square meter Van Veen bottom grab. Two subsamples of each Van Veen grab were taken for later analysis.

The scoopfish-underway bottom sampler was designed to obtain a sample without stopping the vessel. Its capacity is 164 cubic centimeters of bottom sediment. This sampling method was not used at speeds over 15 knots.

Each sample was labeled with an identifying number, position, date of collection, type of sample, water depth, and supplemental remarks. Sampling locations are shown in figure 1 for the offshore area and in figure 2 for Timbalier Bay.

ANALYTICAL PROCEDURES

Size analysis

Two methods of grain size analysis have been used in this study: standard sieve analysis, in which a nest of sieves and a rotap are used, and settling tube analysis. The method to be used was determined by the size of the collected sample. If the total sample size was less than 25 grams, the settling tube method was used.

Sieve Method. Samples to be used for grain size analysis were washed three times with tap water to remove salts. Twenty-four hours after the final wash a subsample was taken and placed into a pre-weighed beaker and dried in an oven at 75° C. While drying, the sample was checked for salt deposits on the surface of the sediment. If a deposit appeared, the sample was rewashed. When the sample was dry, it was placed into a desiccator for 8 hours to cool to room temperature. The sample was then removed from the desiccator and weighed to 3 decimal places. A dispersant was then added to the sample and it was allowed to sit for 24 hours. The sample was stirred occasionally or placed on an ultrasonic cleaner during this period to aid the dispersant process. The sample was then wet sieved through a 62.5 micron screen. That portion of the sample which remained on the screen after wet sieving was dried in an oven. After drying, the sample was placed in a nest of sieves, which ranged from −1 to 4 phi, at 0.5 phi intervals, and rotapped for 20

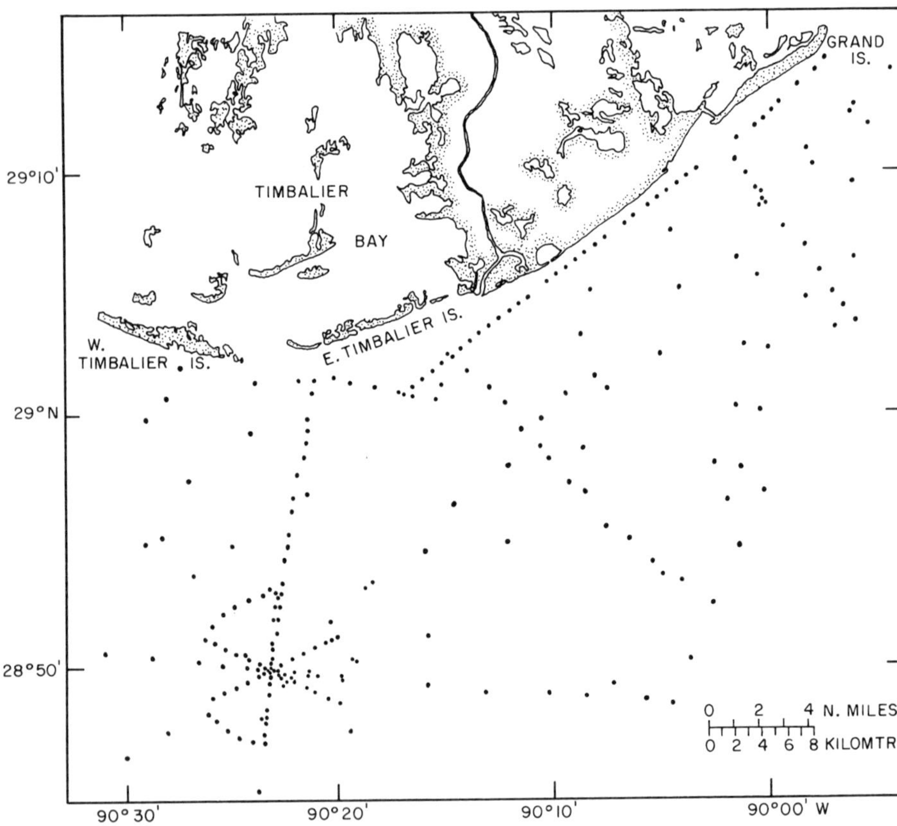

FIG. 1. SAMPLE LOCATIONS—OFFSHORE LOUISIANA.

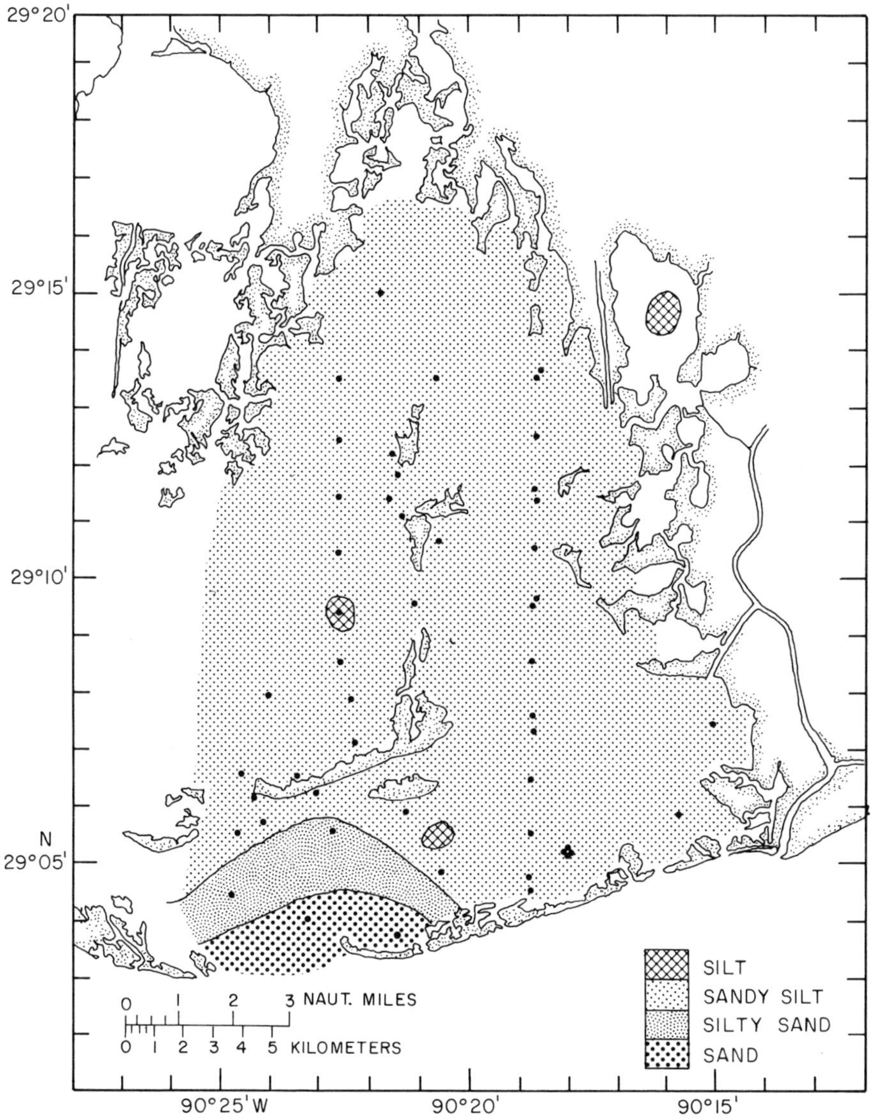

FIG. 2. SAMPLE LOCATIONS AND GENERAL SEDIMENT TYPES—TIMBALIER BAY.

minutes. Each fraction removed from the sieves was placed into a pre-weighed beaker and weighed.

Settling Tube Method. To determine the grain-size distribution for a sediment sample by use of the settling tube, the dispersed sediment sample was first wet-sieved through a 37 micron mesh screen, using tap water and a soft rubber spreader (policeman). That portion of the sample which remained on the screen was placed into a pre-weighed beaker, put into an oven at 75° C until it was dry, transferred to a desiccator for 8 hours to cool to room temperature, and then weighed to 3 decimal places. The grain-size analysis of the sedimentary material was done utilizing a UCLA settling tube (Ziegler et al. 1960; Schlee 1966; Felix 1969). This tube was one meter in length and equipped with a Cahn automatic electro balance and a chart recorder. Approximately 800 milligrams of sample were spread in a thin layer over the lower side of the lid of the settling tube and the lid placed over the settling tube just above the water level in the tube. When the unit was started, the lid was dropped the short distance to the water, where the sample washed off the lid and began to settle in the water column, at which time the chart recorder was started. The sample settled on a plane, and the weights were continuously recorded on the chart. The recorder continued to run until all the sample had settled out of the water column. The information recorded on the chart determined the percentage of the sample weight over the time of the run for the sample. From the information on the chart and the relationship between sphere size and settling velocity (Gibbs et al. 1971), the weight percentages at 0.25 phi size intervals were determined. These values were then used to calculate the statistical sediment parameters from the moments of frequency distribution of clastic sediments. The statistical sediment parameters included mean grain size, variance, standard deviation, degree of skew, and kurtosis. The formulas used in these calculations were those of Friedman (1961) and modified to fit the original moment method of Krumbein (1939).

Carbonate analysis

The percentage of carbonate in each sample was determined by weight loss after acidification (Carver 1971). A portion of the washed sample was dried and powdered and placed into a desiccator. Approximately 2 grams of powdered sample were placed into a pre-weighed beaker and weighed to 3 decimal places. Dilute HCl was added to the powdered sample. After the initial reaction had ceased, additional acid was added and the sample allowed to sit for 8 hours. Then, if no further reaction was observed and pH paper indicated an acidic condition, the excess acid was decanted off the sample. The sample was then washed with distilled water to remove $CaCl_2$ until pH paper indicated a neutral condition. The sample was then dried in an oven, placed in a desiccator for 8 hours and weighed, after which the necessary calculations were performed.

Organic analysis

The percentage of organic matter in each sample was determined by weight loss after ignition (Carver 1971). Approximately 1 gram of dried powdered sediment was placed in a pre-weighed crucible. The crucible was then placed in a muffle furnace and the sample heated to 550° C for 1 hour. Upon its removal from the furnace, the sample was partially cooled in air, and then placed in a desiccator to cool to room temperature. The sample was reweighed and the necessary calculations completed.

Microscopic analysis

The portion of any sample that was larger than 62.5 microns was examined through a binocular microscope to determine whether drill cuttings were present and to identify the minerals that were present.

X-ray analysis

The identification and relative abundance of clay minerals were determined by X-ray diffraction techniques. X-ray mounts were prepared by coating glass slides with several thin layers of suspended material, using those fine particles which had passed through the 37-micron screen during the wet sieving process. This produced a translucent coating on the slide. The slides were then X-rayed, using a K-alpha radiation and a G.E. XRD-5 diffractometer. Supplementary glycol and heat treatments were used as necessary to characterize the clay minerals. The patterns obtained were interpreted according to procedures previously published (Griffin 1971).

Elemental analysis

Elemental concentrations were measured by emission spectrography. The sample, as a dry powder, was mixed in a 1:1.3 ratio with an internal standard containing Lu, Pd, and Sr in a carbonate matrix, with powdered graphite. The prepared sample (30 μg) was loaded into a graphite electrode and ashed at 500° C for 15 minutes, and then burned in the emission spectrograph arc in a Stallwood jet, in an atmosphere of argon (90%) and oxygen (10%).

The detection limits were as follows: 3 μg/g for Sc and Y; 5 μg/g for B, Co, Cr, Cu, Ni, and V; 10 μg/g for La, Ti, and Zr; and 30 μg/g for Ba and Mo. The accuracy for elements present on the level of 3-4 times the detection limit is about ± 10%, and it is ± 20-25% for elements present in lower concentrations.

Al, Fe, Mn, and Si were determined by atomic adsorption with accuracies of 2% for these elements.

RESULTS

Timbalier Bay area

The dominant sediment type in Timbalier Bay was sandy silt, i.e., sand with >50% silt. Very fine sand (62.5-125.0 microns) was the dominant sand size in Timbalier Bay. The highest values for this sediment type occurred in the western part of the bay and near the pass (figure 2). Fine sand (125-250 microns) values were high near the pass, comprising 40-80% of the total. The remaining sand fractions had markedly lower values, usually less than 1% of total, with the higher values generally farthest from the pass. This size fraction was almost totally shell hash material. For comparison, figure 3 shows the sediment distribution reported by the Louisiana Fish and Wildlife Commission.

Drill cuttings, i.e., particles >125μ, were found in the ranges of fine to very coarse sand. The fine sand fraction was most common.

Maximum carbonate values in the bay generally occurred farthest from the pass in the barrier islands. The organic values follow this same pattern, with the highest values occurring farthest from the pass.

The most abundant element measured in the bay was silica (SiO_2), which ranged from 42% to 85% of total, with an average value of 57%. Al_2O_3 ranged from 7% to 16.5%, with an average of 10%. The average value for iron showed a tendency for lower values in the sandy areas. The values for iron at the bottom of the cores were slightly higher than those from the top of the cores.

TiO_2 ranged from 0.27% to 0.70%, with an average value of 0.6%. The lower values occurred in the sandy area. P_2O_5 values ranged from 0.08% to 0.17%, with an average of 0.12%. MnO ranged from 0.02% to 0.16% and had an average value of 0.08%. Both P_2O_5 and MnO exhibited values that were reasonably uniform throughout the Timbalier Bay area.

Barium values were generally high throughout the bay. These values ranged from 370 to 1400 μg/g with an average value of 861 μg/g. The higher values occurred in those sediments with a higher concentration of silt size materials. Although still high, the values at the bottom of the cores were significantly lower than those near the top. The average value at the bottom of the cores was 600 μg/g.

Chromium exhibited a range of 20-85 μg/g, with an average value of 55 μg/g. Boron ranged from 45 to 130 μg/g with an average value of 80

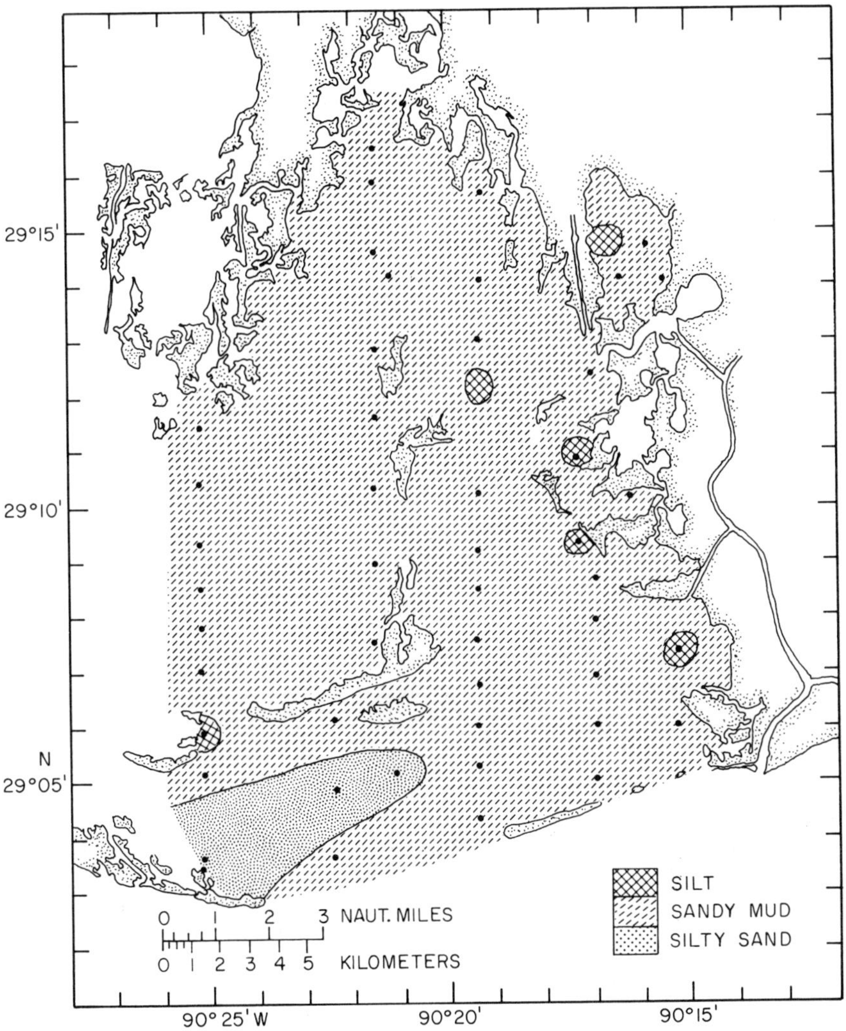

FIG. 3. GENERAL SEDIMENT TYPES—TIMBALIER BAY, 1968-69 (Louisiana Wildlife and Fisheries Commission 1971).

μg/g. Cobalt ranged from 5 to 25 μg/g and averaged 11 μg/g. These three elements all exhibited lower concentrations in the sandy areas of the bay. Nickel had relatively uniform values throughout the bay and in the cores, exhibited a range of 26-56 μg/g, and averaged 32 μg/g.

Copper had a range of 3-35 μg/g with an average of 13 μg/g. Vanadium ranged from 30 to 170 μg/g with an average value of 87 μg/g. Zirconium had a range from 90 to 410 μg/g, averaging 260 μg/g. The lowest values for each of these trace elements occurred in the sandy sediments near the pass. Unlike copper and vanadium, which showed higher values at the bottom of the core, zirconium exhibited highest values near the top of the cores.

Lanthanum values ranged from 12 to 45 μg/g and averaged 31 μg/g. They remained fairly uniform throughout the study area and did not exhibit any discernible distribution pattern. Both scandium and yttrium, which exhibited ranges of 3-15 μg/g and 6-30 μg/g, and average values of 7 μg/g and 21 μg/g respectively, exhibited a distribution pattern with the lower values in the sandy areas of the bay.

Area offshore Louisiana adjacent to Timbalier Bay

The dominant sediment types found in this area were silt, sandy silt, silty sand, and sand, as defined on the Wentworth Scale (Folk 1968). In the eastern portion, that area to the south of Grand Isle, Louisiana, west to a line running south from the mouth of Bayou Lafourche, the sediment types are generally sandy silt and silt. The only variation, a silty sand, occurred next to an oil well. The sandy portion of these sediments was generally a very fine sand (62.5-125 microns). In the western portion of the area sediment types ranged from silt to sand. The area with the greatest variety of sediment types was that around Exxon Platform 54A (figure 4). The entire range of sediments, from silt to sand, was found within a 1.6 kilometer radius of the platform. North of the platform the sediments were mostly sandy silt. Sand and silty sand were major sediment types to the southeast, south, and southwest of the platform.

The most common clay minerals in the area were, in relative abundance, montmorillonite, illite, and kaolinite. No general distribution pattern could be established.

The granule size fraction (2-4 mm) consisted of shell hash, drill cuttings, pieces of wood, and other organic matter. Shell hash was by far the dominant member of this size fraction, generally over 95% of the total. This fraction exhibited maximum values at the near-shore stations.

The coarse sand fraction (0.5-1.0 mm) was generally found within 4 kilometers of the platform and highest values occurred next to the platform (figure 5).

The only material found larger than a coarse sand was shell hash, although very coarse sand and some pieces of shale were found using a more selective sampling method. Some of the shale particles (drill cuttings) were larger than 4 mm, and found within a 6.4 kilometer radius of the platform, but never in large quantities. Just to the south of the eastern end of Timbalier Island a piece of material (larger than 4 mm) consisting of sand grains in an asphalt-like matrix was found. This was the only petroleum-base conglomerate material found in the whole study.

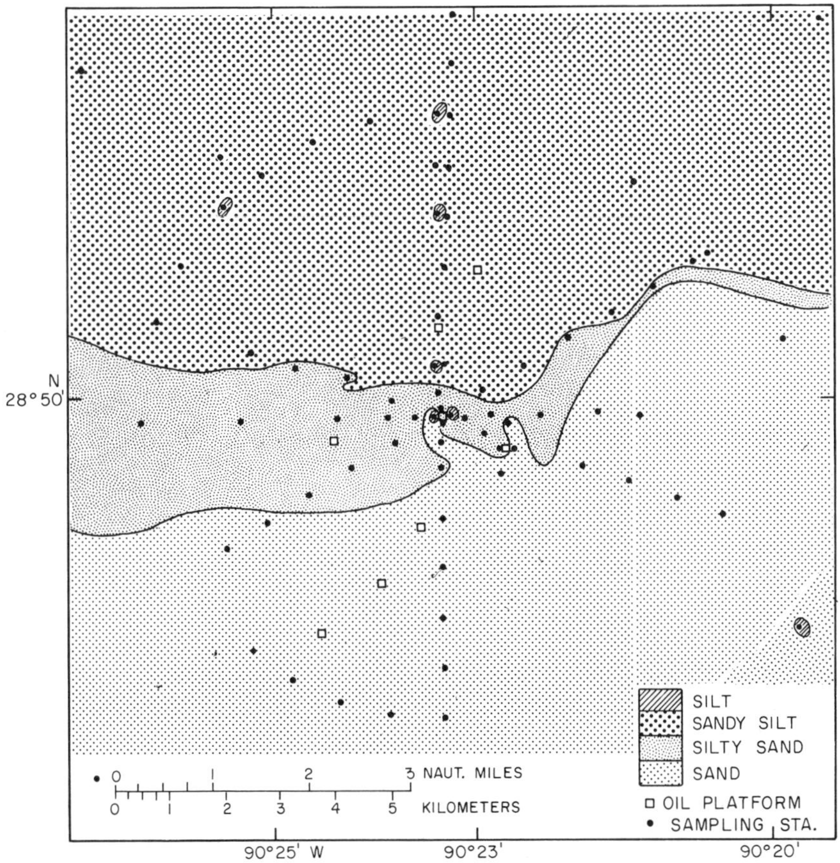

FIG. 4. GENERAL SEDIMENT TYPES AROUND EXXON PLATFORM 54A.

Cores collected from around Exxon Platform 54A, out to a distance of 4.8 kilometers, show a sediment distribution pattern at the surface in which there is a narrow band of silty sand running generally east-west across the platform location. To the north of this band, about 0.8 kilometers north of 54A, the sediment types are sandy silt and silt. To the south of this band, about 1.6 kilometers south of 54A, the sediments are sandy (see figure 4). Around the platform there is a complete range of sediment types from silt to sand, with the majority being silty sand or sand. At the 10 cm and 20 cm depth levels the sediment types are sandy silt and silt. The sediment distribution pattern found at the surface was not observed at depth, although the silty sediments generally occur north of 54A or next to a platform.

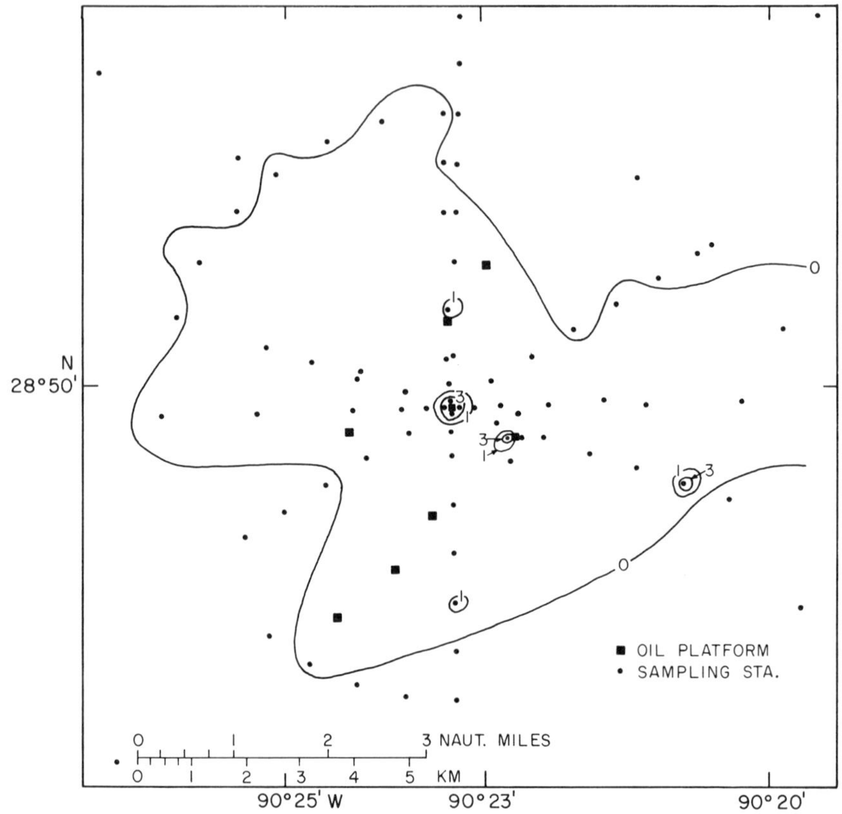

FIG. 5. PERCENTAGE OF COARSE SAND (0.5-1.0 mm).

The percentage of total sands (figure 6) shows distribution patterns similar to those shown by the general sediment types (figure 4). Eight-tenths of a kilometer north of 54A the values for percentage of total sands were generally less than 40%, 0.8 kilometers to the south the values are generally greater than 80%.

The very coarse sand fraction and the coarse sand fraction showed very low values, generally less than 1%, throughout the cores, and with the exception of those samples taken next to the platform consisted of shell hash and organic matter. Those samples taken next to 54A showed higher values (1-8%) for fractions of these sizes.

The values for the medium sand fraction at the surface were generally less than 8% in the area northeast, north, and northwest of Platform 54A. Within 2.4 kilometers to the east and west, and to the southwest, south, and southeast, the values were generally greater than 10%. At the

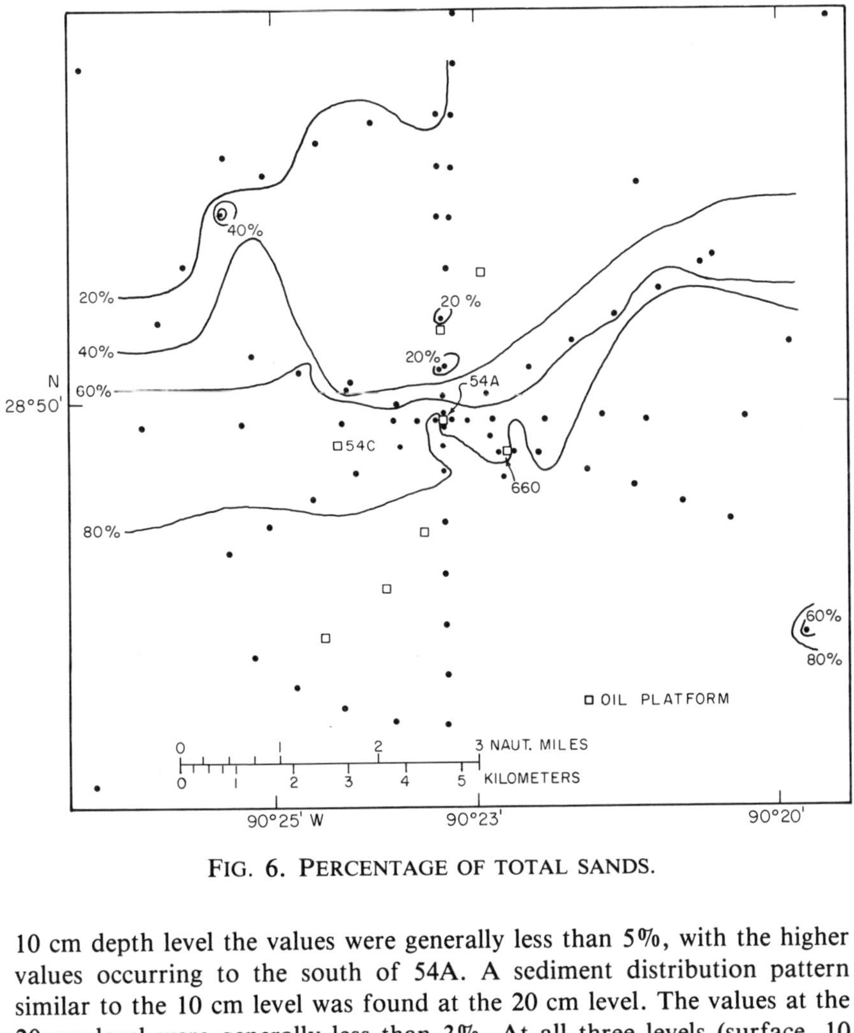

FIG. 6. PERCENTAGE OF TOTAL SANDS.

10 cm depth level the values were generally less than 5%, with the higher values occurring to the south of 54A. A sediment distribution pattern similar to the 10 cm level was found at the 20 cm level. The values at the 20 cm level were generally less than 3%. At all three levels (surface, 10 cm, and 20 cm) the higher values occurred next to the platform.

The values for the fine sand fraction generally increased with increased distance offshore. This was true for all depths. High values were found next to the platform. Very fine sand was found throughout the cores in small amounts, generally less than 10%.

Generally, the percentages of very coarse sand, coarse sand, medium sand, and fine sand all decreased with depth.

The carbonate values were generally less than 10%, with the higher values usually occurring near shore. The carbon values and the granule size fraction values correlate very well. The total organic carbon values were usually low, 4-12%. The sandy area south of 54A had values that were usually less than 3%.

Only one sample had a characteristic petroleum odor. It was collected next to a producing platform.

Silica (SiO_2), the most abundant element measured, ranged from 42% to 85% of total, with an average value of 61%. No surface distribution pattern could be established, but values were generally lower at the bottom of the cores. Al_2O_3 ranged from 3.5% to 18%, with an average of 10.8%. Slightly higher values occurred in those sediments with a higher percentage of mud. The values at the bottom of the cores were generally slightly higher. Iron (Fe) is found in formation water brines and drilling muds, and it accumulates in heavy oils and asphalts. It ranged from 0.9% to 7.5%, with an average value of 2.8%. Lower values occurred in sandy areas and slightly higher values occurred at the bottom of the cores.

TiO_2 ranged from 0.20% to 0.90%, with an average value of 0.60%. Lower values were found in sandy areas and slightly higher values occurred at the bottom of the cores. P_2O_5 values ranged from 0.06% to 0.20% and averaged 0.10%. The values obtained were fairly uniform throughout the study area. MnO values from 0.02% to 0.15% were recorded, and averaged 0.08%.

Barium (Ba) values were generally high throughout the offshore area, with the lower values occurring in the sandy areas. Barium values ranged from 520 to 2100 μg/g, and averaged 741 μg/g. Values at the bottom of the cores were generally lower than those at the top.

Oils are usually poor in chromium and molybdenum. In the study area chromium exhibited a range of 21-155 μg/g, and averaged 60 μg/g. Molybdenum was always less than 30 μg/g. Both chromium and molybdenum had their lowest values in the sandy sediments.

Boron (B) is commonly concentrated in oil field brines. In this study it ranged from 33 to 120 μg/g and averaged 70 μg/g. Cobalt (Co) value ranged from less than 5 to 30 μg/g and averaged 10 μg/g. Both boron and cobalt had their lowest values in sandy sediments, and at the top of the cores. Nickel (Ni) ranged from less than 5 to 75 μg/g, with an average of 38 μg/g.

Copper (Cu) had a range of 3 to 60 μg/g and averaged 20.2 μg/g. Zirconium (Zr) had a range of 110-1200 μg/g and an average of 250 μg/g. Although these three elements varied in their ranges, they all exhibited their lowest values in the sandy sediments. Where copper and vanadium showed higher values at the bottom of the cores, zirconium showed a reverse pattern with the higher values at the top of the cores.

Lanthanum, scandium, and yttrium all had their lower values in the sandy sediments, and at the top of the cores. Their measured values were: lanthanum, 10-80 μg/g, average 31 μg/g; scandium, less than 3 to 19 μg/g, average 8 μg/g; and yttrium, less than 5 to 60 μg/g, average 22 μg/g.

DISCUSSION

Timbalier Bay

A comparison between a 1968-1969 study by the Louisiana Wildlife and Fisheries Commission and the study reported here shows that the sediment distribution pattern of Timbalier Bay and its surrounding area has varied little over this time period. Both studies show the main sediment to be sandy silt, with silty sand and sand in the pass between the barrier islands of Timbalier and East Timbalier. The major difference between the two studies was the change in the shape and size of East Timbalier Island (figures 2 and 3). The present study shows the island to be much larger than in the former. In addition, some smaller islands in the bay no longer exist. Since this type of environment is highly dynamic, the differences are presumed to be due to natural causes, such as the reworking of bottom sediments and erosion by wave and current action. The values found for the silt, sand, and granule size fractions vary little between the two studies. Those small differences which are observed are probably due to redistribution of sediments by wave action, currents, boat propeller wash, shrimping activities, dredging, and drilling and well rework activities.

It should be noted that neither this study nor that completed by the Louisiana Wildlife and Fisheries Commission provides a true natural baseline for the parameters measured in the Timbalier Bay region, because oil and gas wells have been drilled in the bay since 1938, with the heaviest drilling activity having taken place in the mid-1950s and early 1960s (Morgan 1973).

Drill cuttings are those materials brought to the surface during the drilling of a well. In this report, cuttings are defined as those sand and shale particles coarser than a very fine sand (125 microns). This size limitation is based on a 1938 study by the U.S. Army Corps of Engineers (1939), which stated that no material coarser than a very fine sand was being transported as suspended sediment by the Mississippi River, and on the opinion of Dr. J. P. Morgan (personal communication), who believes that this has been the case for the past 5,000 years. Therefore, it has been assumed that anything coarser than a very fine sand has derived from drilling operations that have brought the deeper, coarser sediments to the surface.

The only additional source of sands in the Timbalier Bay region is the old deltaic lobes of the Mississippi River. When the relative amount of sands available from these old deltaic lobes and from drill cuttings are compared, and the composition of the sediment types coarser than a very fine sand examined, it is evident that the effect of oil drilling operations on the size distribution of the sediments in Timbalier Bay is insignificant.

The reworking of the bottom sediments by wave and tidal action near the pass between the barrier islands has resulted in the fine size particles and light materials being winnowed out. This explains, to some degree, the observed distribution patterns of the carbonate and organic carbons. It is assumed that the marshes bordering the bay have provided essentially all of the organic carbon found in the bay.

The sands are predominantly quartzose. This explains the observation that silica was the most abundant element of those measured.

The values for chromium and iron were expected to be higher than those values reported from non-drilling areas because they are two of the principal components of ferrochrome lignosulfonate, a component of drilling muds. The values for these two elements, as well as those for most of the other elements measured, however, fall within or close to the limits and averages reported in the literature for these elements in non-drilling areas. Any enrichment of these elements by drilling operations appears almost nonexistent, and certainly is insignificant.

The sources of barium in the bay are naturally occurring barium and barium contained in barite, one of the major constituents of drilling mud. Although baseline data for the naturally occurring barium in Timbalier Bay could not be determined, the average value of barium in the study area was found to be many times greater than the average value for it in sediments from outside the study area. Those sediments in Timbalier Bay had an average value of 861 μg/g, while those from outside the study area to the east of the Mississippi River delta had an average value of 57 μg/g, some 15 times less. The sediments analyzed from these areas outside the Timbalier Bay region were sandy, and sandy sediments in general have lower trace element values than finer sediments. One must not overlook the fact, however, that even when the sandy sediments from outside Timbalier Bay are compared to sandy sediments within the bay, there is still a very large difference, with the concentrations within the bay some ten times greater. When those sediments in Timbalier Bay were compared to silty sediments collected from the Mississippi River delta, the average barium values were still significantly higher. Sediments from the delta area, collected at the mouth of the Mississippi River, had an average value of 762 ppm. Although the difference in these values is not so large as differences for the other values from outside the study area, it must be realized that the possibility of enrichment to the Mississippi River sediments from the large number of nearby platforms and from accidental spillage of drilling mud from barges and work boats on the river is certainly significant, and that no definitive natural (pre-drilling) baseline could be obtained. These high concentrations of barium are therefore attributed to the effect of the barite component of the drilling mud.

With few exceptions, the measured elements demonstrated lower values in the sandy sediments. This is because clay minerals tend to adsorb many elements, and in the sands the finer particles, such as silts and organic matter, have been removed by winnowing wave action. The exceptions to this are SiO_2, P_2O_5, Ni, and La. These elements demonstrated relatively uniform values throughout their range of distribution.

Offshore area adjacent to Timbalier Bay

The great variety of sediment types found around Exxon Platform 54A may be due to the fact that this area is a natural transition zone between a silty area and a sandy one, caused by the addition of drill cuttings to the sediments, or the result of a combination of these conditions.

The types and relative abundance of the clay minerals in the offshore study were expected because the rivers that empty into the Gulf of Mexico in this region carry predominantly montmorillonite clay assemblages, with subordinate illite and little kaolinite.

The silt size fraction (<62.5 - $>4\mu$) in the study area may have originated both from the Mississippi River and from the drilling muds discharged from the platform during drilling. Scruton (1956) reported that in the Mississippi River no suspended sediment in transport contained materials coarser than a very fine sand (62.5-115μ). Moreover, only a very small percentage (2%) of this material actually fell within the very fine sand range. Barite ($BaSO_4$) can be classified as a sandy silt, with about 75% of the barite falling into the silt size fraction, and the sandy fraction being almost entirely a very fine sand. Therefore, it is impossible to separate the barite mud from the naturally occurring sediments by grain size analysis.

The very fine sands suspended in the Mississippi River drilling mud and old submerged deltaic lobes of the Mississippi River are most probably the sources of sand in the study area. Much of the very fine sand and fine sand probably comes from the old Lafourche deltaic lobes of the Mississippi River. This feature, which covered much of the adjacent region, has been winnowed out. Morgan (1955) reported that Tiger, Trinity, and Ship Shoals are all residual deltaic remnants that have subsided below sea level. When sediments from Ship Shoal, which is located just to the west of the study area, were compared to those sediments in the study area south and southwest of 54A, both exhibited similar fine to medium sand ratios. This suggests that the sediments in the area south and southwest of 54A were deltaic in origin. The increase in the mean phi size between the study sediments south of 54A and those sediments next to 54A is most likely due to the addition of larger particles from drilling operations.

Drill cuttings may cover the entire sand range, the size being dependent on the composition of the substrata through which the well is drilled. In the present study area, only drill cuttings can account for sands coarser than a fine sand.

The greatest number of large cuttings (>1 mm) was found at the base of the platform, where they were relatively common. At any distance from the platform the number of large cuttings was quite small. Since the average current velocity in the area is not strong enough to carry these cuttings any distance from the platform, it must be assumed that they were moved by stronger currents that developed during a severe storm or hurricane.

Sediment distribution patterns show both the effects of drill cuttings and a natural transition zone. Introduction of drill cuttings made the bottom sediments coarser. This can be seen by the increase in the mean phi size next to the platform and by the distribution patterns exhibited by the large cuttings and sands, which all had relatively high values next to the platform. The natural transition of sediment types is demonstrated by the distribution patterns of fine and medium sand to the south and west.

Generally, most of the elements measured had higher values near the bottom of the cores. This may be because sediments containing a higher percentage of silt-size particles (<62.5 - $>4\mu$) were present in greater concentrations in the lower sections of the cores. The exceptions to this trend were silica and barium. The higher values for silica were due to the greater amount of quartz sand in the surface sediments. We presumed that barite, a major component of drilling mud, was responsible for the high barium values; however, this hypothesis could not be confirmed because the natural background level for barium in the study area could not be determined because of the previous history of the region. This statement is based on the number and location of oil and gas wells drilled in the region and on the settling time of barite.

SUMMARY

Timbalier Bay Area:

1. The types of sediment found in Timbalier Bay were silt, sandy silt, silty sand, and sand.
2. The sediment type found most commonly throughout the bay was

sandy silt, but near the pass between the barrier islands, sand was dominant.

3. Silt composed the largest percentage of the total weight of the samples.
4. Very fine sand was the predominant sand size found in the study area. It has its origin in the old deltaic lobes of the Mississippi River and from drill cuttings.
5. Barium was the only element measured that showed values significantly higher than those commonly reported in the literature for sediments in non-drilling areas.
6. In most cases the measured elemental values were lower in the sandy areas.
7. Carbonate and organic carbon values tended to increase as distance from the pass between the barrier islands increased, i.e., these values tended to increase shoreward from the barrier islands.

Area Offshore Louisiana Adjacent to Timbalier Bay:

1. The dominant sediment types found in the eastern portion of this area are sandy silt and silt. In the western portion of the study area the sediment types range from silt to sand. In the naturally occurring transition zone the sediments tend to become coarser with an increase in distance from shore.
2. The silt and the sands found in the study area originate from the Mississippi River and from drilling operations. In addition, some of the sands originate from an old submerged deltaic lobe of the Mississippi River.
3. With the exception of those sandy areas to the south and southwest of Exxon Platform 54A, the shallow sub-surface sediment types are sandy silt and silt throughout the study area.
4. Very fine sand is generally the most dominant and widespread of the sands.
5. Very coarse, coarse, and medium sand exhibited highest relative values next to the platform.
6. The most common clay minerals found in the study area are, in decreasing relative abundance, montmorillonite, illite, and kaolinite.
7. Although some large drill cuttings were found 6.4 km from the nearest platform, most of these cuttings were within 4.8 km of the platform. The highest concentrations occurred adjacent to the platform.
8. The major identified influence of oil and gas drilling on the texture of the sediments in the study area has been to make the sediments within 4.8 km of the platform slightly coarser, with the greatest

effect taking place immediately adjacent to the platform. Only one petroleum-base conglomerate and only one sample that smelled strongly of petroleum were found.

9. The carbonate and organic carbon values were generally low. The carbonate values exhibited a high positive correlation with the distribution of the granule size fraction.
10. With the exception of silica, all the measured elements had lower values in the sandy sediments.
11. The higher values for most of the measured elements occurred at the bottom of the cores.

CONCLUSIONS

Although the drilling of oil and gas wells in the two areas studied has added millions of cubic feet of cuttings and mud to the sediments in this region, the net change in the composition of the sediments caused by this addition is insignificant, except immediately adjacent to the drilling sites, where enrichment by drill cuttings has produced coarser than normal sediments. The amount and type of sediments from the Mississippi River, and from old Mississippi River deltaic lobes, are similar to the sedimentary materials added by drilling, although many times greater in volume and significance.

With the exception of barium and silica, the measured elemental values found for the sediments are within or near those naturally occurring ranges reported in the literature for those elements. It is, therefore, assumed that the high concentration of barium in the sediment was due to artificial enrichment by drilling mud. The high silica values are explained by the high concentrations of quartz in the sands of the area, a natural phenomenon.

The results of this study indicate that no significant sedimentary alterations have occurred in Timbalier Bay or in the adjacent offshore area as a result of drilling activities. The results also show that, with the single exception of increased barium concentrations, no significant elemental concentrations of those elements measured have occurred as a result of drilling operations in this region.

REFERENCES CITED

Carver, R. E. 1971. Procedures in Sedimentary Petrology. New York: John Wiley.

Coleman, J. M. 1966. Recent Coast Sedimentation: Central Louisiana Coast. Louisiana State University Coastal Studies Institute, Technical Report No. 29.

Felix, D. W. 1969. An inexpensive recording settling tube for analysis of sands. Journal of Sedimentary Petrology **39**:777-780.

Folk, R. 1968. Petrology of Sedimentary Rocks. Geological Publications of The University of Texas, Austin, Texas.

Friedman, G. M. 1961. Distinction between dune, beach, and river sands from their textured characteristics. Journal of Sedimentary Petrology **31**:514-529.

Gibbs, R. J., M. D. Matthews, and D. A. Link. 1971. The relationship between sphere size and settling velocity. Journal of Sedimentary Petrology **41**:7-18.

Griffin, G. M. 1971. Interpretation of x-ray diffraction data. *In* R. E. Carver, ed., Procedures in Sedimentary Petrology, a Symposium. New York: John Wiley, pp. 541-569.

Krumbein, W. C. 1939. Tidal lagoon sediments on the Mississippi delta. *In* P. Trask, ed., Recent Marine Sediments. Tulsa, Oklahoma: American Association of Petroleum Geologists, pp. 178-194.

Louisiana Wildlife and Fisheries Commission. 1971. Cooperative Gulf of Mexico Estuarine Inventory and Study, Louisiana. Phase II, Hydrology and Phase III, Sedimentology. New Orleans, La.

Morgan, J. P. 1955. A geographical and geological study of the Louisiana coast with emphasis upon establishment of the historic shoreline. Coastal Studies Institute, Contribution No. 55-3.

———. 1973. Memorandum—cutting deposition in Timbalier Bay, Louisiana. Prepared for Offshore Ecology Investigation of the Gulf Universities Research Consortium.

Schlee, J. 1966. A modified Woods Hole rapid sediment analyzer. Journal of Sedimentary Petrology **36**:403-413.

Scruton, P. C. 1956. Oceanography of Mississippi delta sedimentary environments. American Association of Petroleum Geologists Bulletin **40**(12):2864-2952.

U.S. Army Corps of Engineers. 1939. Study of materials in transport passes of the Mississippi River. Technical Memo. 158-1. U.S. Corps of Engineers, New Orleans and Waterways Experiment Station, Vicksburg, Miss.

Ziegler, J. M., G. G. Whitney, and C. R. Hayes. 1960. Woods Hole rapid sediment analyzer. Journal of Sedimentary Petrology **30**:490-495.

NUTRIENT CONCENTRATIONS IN TIMBALIER BAY AND THE LOUISIANA OIL PATCH

by H. P. Burchfield, R. J. Wheeler, and W. Subra

ABSTRACT

The purpose of this study was to determine if petroleum drilling and/or production activities in Timbalier Bay and in the Louisiana oil patch contributed certain nutrients to the environment.

Samples were collected from August 1972 through January 1974. Water samples were analyzed for total alkalinity, dissolved silica, orthophosphate, ammonia nitrogen, and nitrate and nitrite nitrogen. Sediments were analyzed for total alkalinity and dissolved silica of the interstitial water and for total Kjeldahl nitrogen and total phosphate in bulk sediments.

Sampling locations included areas near oil platforms, at various distances and directions from the platforms, and at locations remote from the oil platforms.

Seasonal variations were shown for all of the nutrients studied.

Statistical comparisons between the platform sites and the ambient areas in the bay were possible only on three occasions. During these periods, no significant differences were shown among study areas.

Variations in most parameters were occasionally observed in close proximity to the platforms. Both elevations and reductions from ambient occurred; it is believed that they resulted from the structures' affecting mixing processes within the water column and from introductions by platform activities.

When this investigation was conducted the authors were at the Gulf South Research Institute, New Iberia, LA.

INTRODUCTION

This segment of the OEI study was designed to assess the effects of oil drilling and production activities on the nutrient concentrations of waters and sediments in Timbalier Bay and the Louisiana oil patch.

In the bay, sampling stations were established around a non-drilling production platform and an actively drilling workover rig. In addition, stations were established throughout the bay to assess natural seasonal variations and to serve as control sites.

In the offshore region, samples were collected near a production platform, at various distances from the platform, and along a transect from Timbalier Bay to the offshore platform.

Samples were collected from August 1972 through January 1974. Water samples were analyzed for total alkalinity, dissolved silica, orthophosphate, ammonia nitrogen, and nitrate and nitrite nitrogen. Sediments were analyzed for total alkalinity and dissolved silica of the interstitial water and for total Kjeldahl nitrogen and total phosphate in bulk sediments.

METHODS

Laboratory

Ammonia nitrogen was determined on a 50 ml sample of water, which was steam distilled in the presence of magnesium oxide for 3 minutes (Bremner 1965). The distillate was collected on 0.1 N HC1 and the resulting solution was tested for ammonia content after either Nesslerization or diazotization (Strickland and Parsons 1965).

Nitrite and nitrate nitrogen were also determined as ammonia, as outlined above, after steam distillation in the presence of magnesium oxide and Devardo's Alloy for a period of 5 minutes.

Orthophosphate was determined on a 100 ml sample treated with sulfuric acid and ammonium molybdate. The resulting mixture was extracted with ethyl acetate and the aqueous layer discarded; the remaining fraction was then treated with ascorbic acid and reacted with a tartrate compound to produce a blue color (Ho et al. 1970).

Silica was determined on a 25 ml sample, using the method described by Strickland and Parsons (1965).

Quantification was accomplished using a Beckman DU spectrophotometer for colorimetric measurements on all of the nutrients listed above.

Total alkalinity was determined by titration of a 100 ml sample to a pH of 4.5 and 4.2 with 0.05 N sulfuric acid.

In sediments, organic nitrogen was determined on a 0.5 g dried sample after digestion with sulfuric acid in the presence of K_2SO_4, $CuSO_4$, and Se. Sodium hydroxide was added and the mixture was steam distilled for 5 minutes. The distillate was then collected and quantified as described above for ammonia nitrogen.

Total phosphorus was determined as orthophosphate after digestion of a 1.0 g sample of dried sediment with sulfuric acid on a steam bath for 2 hours.

Field

The general locations of the stations sampled in Timbalier Bay and along a transect into the offshore region are shown in figure 1. More samples were taken in the immediate vicinity of the platforms, i.e., within 1,000 meters, than at the control sites.

RESULTS

Bay

Seasonal variations in the levels of nutrients in the bay are shown in figures 2 and 3 (pp. 227f). Nitrogen levels peaked during the winter months (November 1972 and January 1973), decreased during the spring (April 1973), increased in July in response to the heavy summer rainfall, decreased slightly in October, and then increased during the winter (January 1974). Phosphate levels (figure 3) followed a somewhat different pattern. The highest levels were observed during the fall (October 1972). They then decreased and remained relatively constant throughout the remainder of the study. Silica concentrations were highest during October of both 1972 and 1973 and minimal during the winter months (November 1972 and January 1973).

Levels of total phosphate, Kjeldahl nitrogen and interstitial water concentrations of dissolved silica are shown in figure 4. Phosphate levels peaked during the fall of both years and were remarkably similar at the control and production platform sites. Nitrogen levels increased during the winter of 1972, a period when the phosphate levels were decreasing, but followed a pattern similar to phosphate in the summer and fall of 1973. Enrichment of nitrogen in the sediments around the platforms was indicated during the winter of 1972 and the spring of 1973. Silica levels increased at both study sites throughout the period.

Offshore

Seasonal variations in the levels of nitrogen, phosphorus, and silica were observed along a transect from Timbalier Bay to the offshore platforms (figures 5, 6, and 7).

(text continues on p. 231)

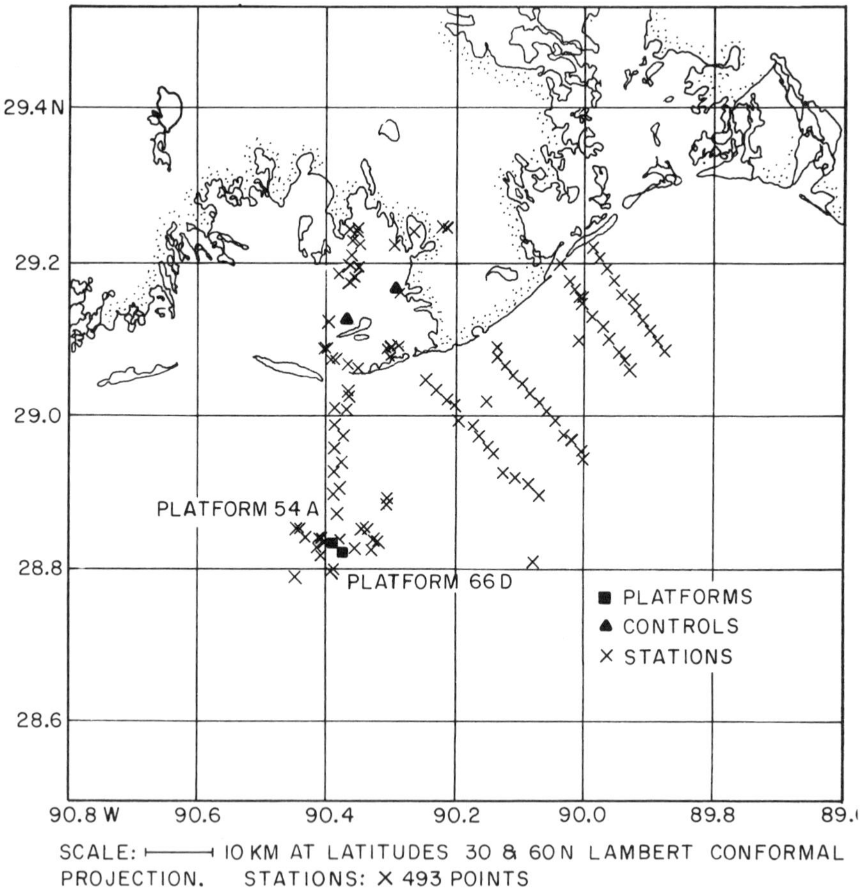

FIG. 1. NUTRIENT SAMPLING STATIONS.

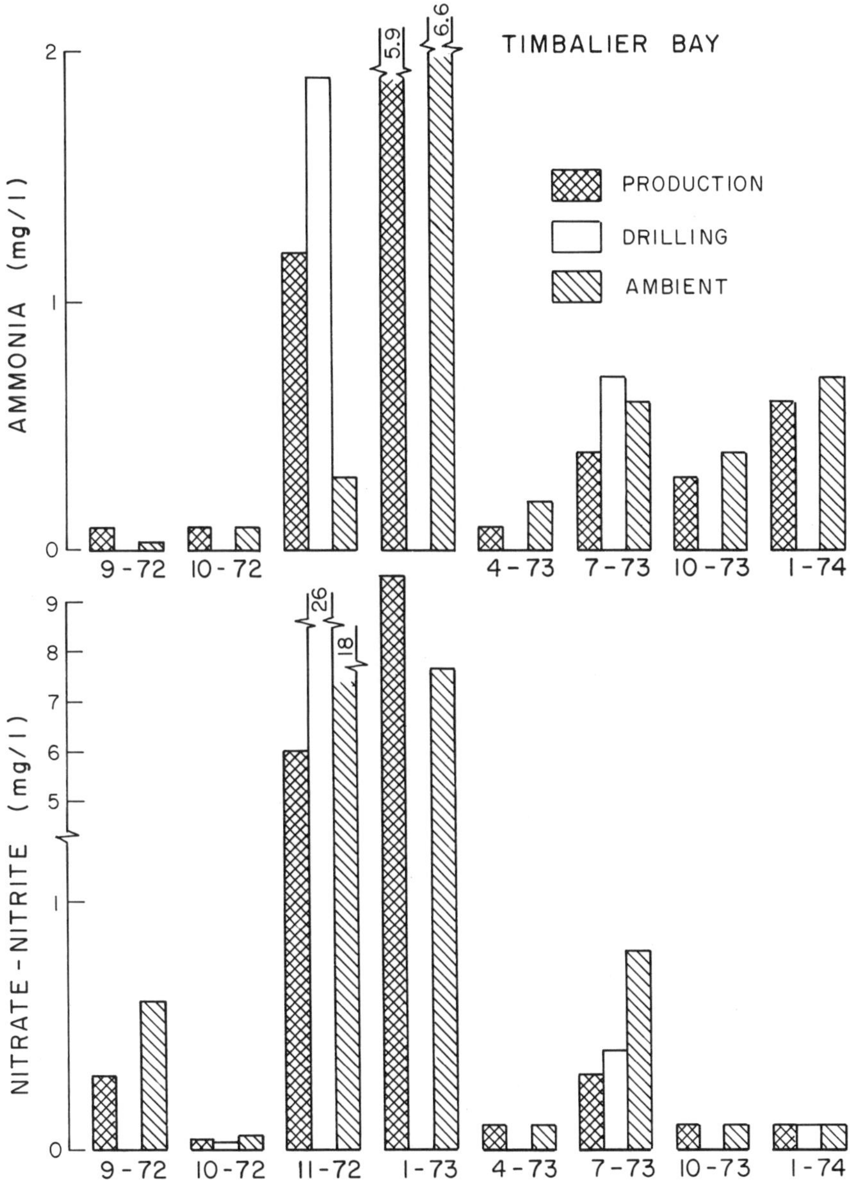

FIG. 2. SEASONAL VARIATION IN WATER COLUMN LEVELS OF AMMONIA AND NITRATE-NITRITE NITROGEN.

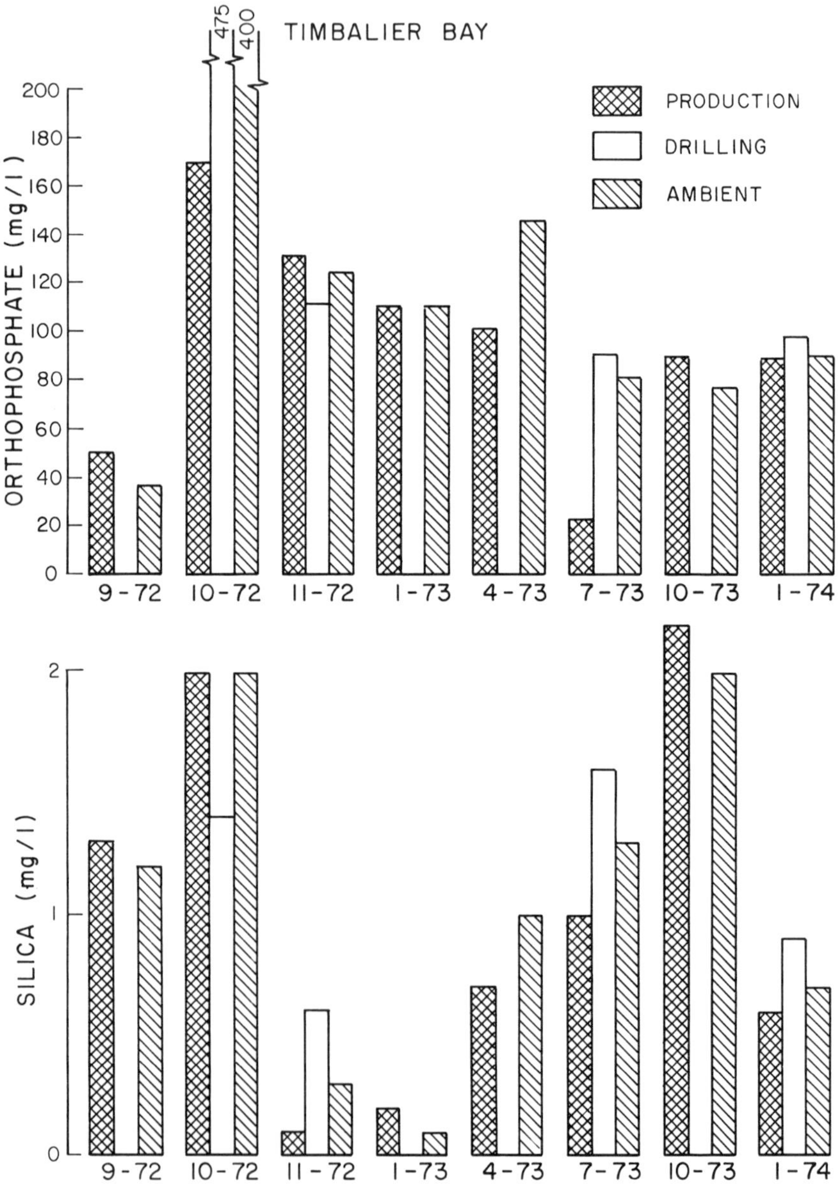

FIG. 3. SEASONAL VARIATION IN PHOSPHATE AND SILICA LEVELS.

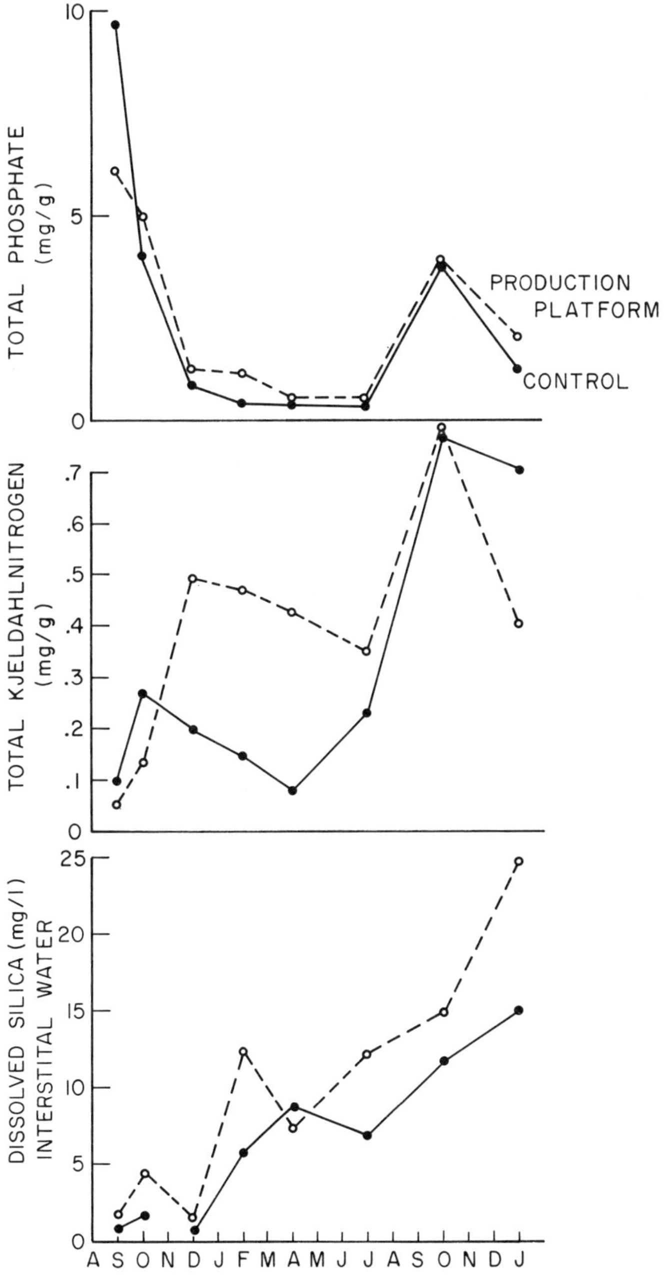

FIG. 4. SEASONAL VARIATION IN SEDIMENT AND INTERSTITIAL NUTRIENT LEVELS.

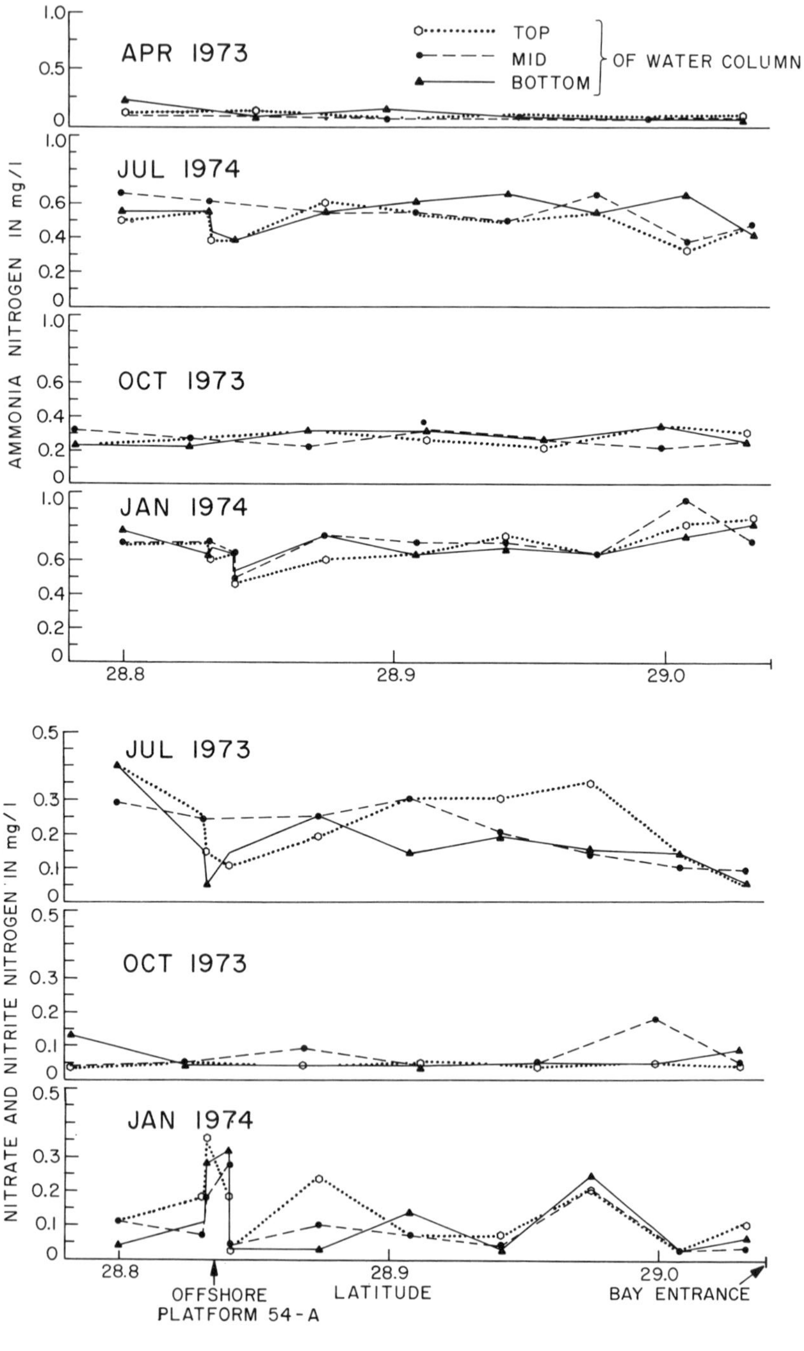

FIG. 5. AMMONIA NITROGEN ALONG BAY TO OFFSHORE TRANSECT (upper diagrams). NITRATE AND NITRITE NITROGEN ALONG BAY TO OFFSHORE TRANSECT (lower diagrams).

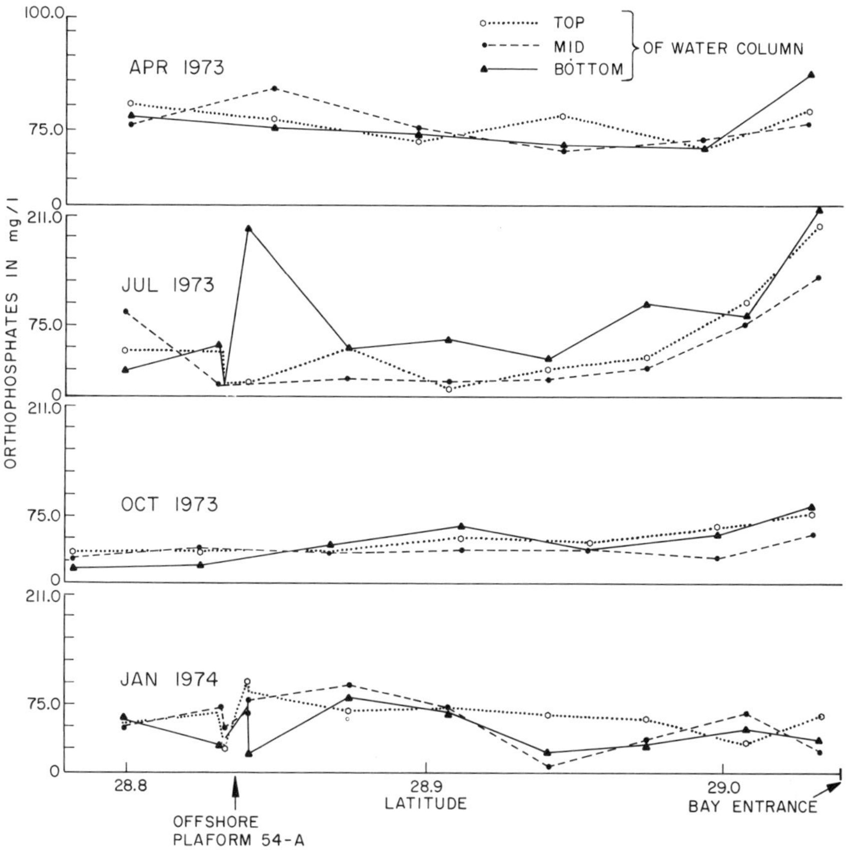

FIG. 6. ORTHOPHOSPHATE ALONG BAY TO OFFSHORE TRANSECT.

Ammonia and nitrate-nitrite nitrogen levels were low in April, increased in July apparently in response to the Mississippi River flood, decreased to near April levels during October, and then increased again in January.

Silica levels followed, in general, the trends observed for nitrogen, with concentrations low in April, increasing during the July flood period, decreasing in October, and then increasing in January. A relatively large increase in silica occurred in the bottom waters near the platform in July, and a similar increase was noted for phosphorus.

DISCUSSION

The seasonal variations in nutrient concentrations observed in this

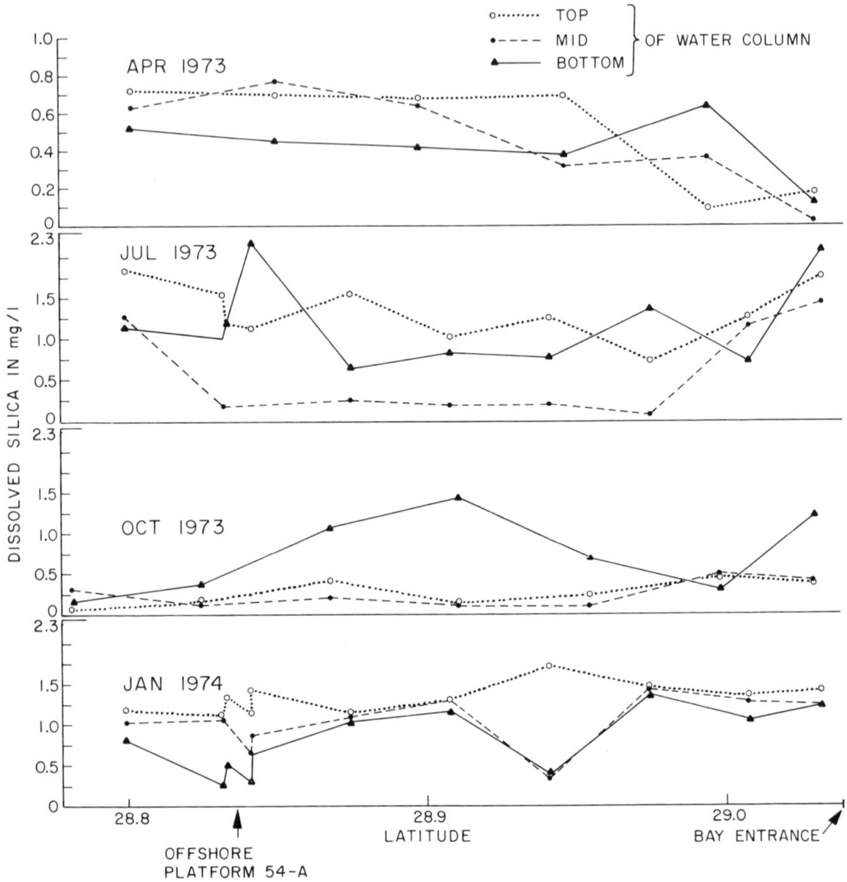

FIG. 7. DISSOLVED SILICA ALONG BAY TO OFFSHORE TRANSECT.

study of Timbalier Bay were generally comparable to the data collected by Barrett (1971) at two stations in the bay from January 1968 though June 1969. In this study, inorganic phosphate levels in south Timbalier Bay peaked in March 1968. Inorganic nitrogen levels were elevated in January 1968, then decreased progressively through February 1969.

Ho (1971) and Ho et al. (1970) reported seasonal variations of water and sediment nutrients in 1969 for three locations in Barataria Bay. Inorganic nitrogen concentrations in the bay in 1969 were high from late fall through early spring, decreased in April to a low that continued throughout the summer, and corresponded to the trends observed in Timbalier Bay from August 1972 through January 1974.

Phosphate in Barataria Bay ranged from a low of 2 μg/l in April 1970 to a high of 120 μg/l in July 1970. Phosphate was higher during the summer months of 1970 and lower during the colder months. In contrast, the orthophosphate values of Timbalier Bay were generally lowest during the summer months and higher during the colder months.

REFERENCES CITED

Barrett, B. B. 1971. Cooperative Gulf of Mexico estuarine inventory and study, Louisiana, Phase II, hydrology and Phase III, sedimentology. Louisiana Wildlife and Fisheries Commission.

Bremner, J. M. 1965. Inorganic forms of nitrogen, total -N and semi micro-Kjeldahl. *In* C. A. Black, ed., Methods of Soil Analysis. American Society of Agronomy **2**:1169-1178.

Ho, C. L. 1971. Seasonal changes in sediment and water chemistry in Barataria Bay. Louisiana State University Coastal Studies Bulletin **6**:67-84.

———, E. H. Schweinsberg, and L. Reeves. 1970. Chemistry of water and sediments in Barataria Bay. Louisiana State University Coastal Studies Bulletin **5**:41-56.

Strickland, J. P. H. and T. R. Parsons. 1965. A manual of sea water analysis. Fisheries Research Board of Canada Bulletin No. 125.

CONCENTRATIONS OF Hg, Pb, Zn, Cd, AND As IN TIMBALIER BAY AND THE LOUISIANA OIL PATCH

by J. G. Montalvo, Jr., and D. V. Brady

ABSTRACT

The purpose of this two-year study was to determine if petroleum drilling and/or production activities in Timbalier Bay and in the Louisiana oil patch contributed certain toxic metals to the marine environment—specifically, the extent of the contribution, if any, to the water column.

Samples were analyzed for total mercury, lead, cadmium, zinc, and arsenic. Sampling locations included areas near oil platforms, at various distances and directions from the platforms, and at locations remote from the oil platforms.

Over the sampling period, in bay water column samples, mercury ranged from <0.3 to 6.9 μg/l with an average of 0.6 μg/l. Cadmium ranged from <0.1 to 22.5 μg/l with a mean of 0.6 μg/l; lead from <0.5 to 50 μg/l with an average of 2.8 μg/l. Zinc varied between <0.2 and 66 μg/l with a mean of 8.5 μg/l. All of the arsenic values were below 10 μg/l, the limit of detection.

Over the study period in the offshore area, mercury ranged from <0.3 to 7.5 μg/l with an average of 0.4 μg/l. Cadmium varied between <0.1 and 20 μg/l with a mean of 0.3 μg/l. Lead ranged from <0.5 to 50 μg/l with a mean of 1.5 μg/l; zinc from <0.2 to 66 μg/l with an average of 6.8 μg/l. All of the arsenic values were below the 10 μg/l limit of detection.

When this investigation was conducted, J. G. Montalvo, Jr., was manager of the Analytical Chemistry Department and D. V. Brady was Chief Chemist at the Gulf South Research Institute, New Orleans, LA.

On several occasions the levels of zinc were found to decrease with increasing distance from the immediate vicinity of the oil platforms.

INTRODUCTION

The purpose of this study was to determine if petroleum drilling and/or production activities in Timbalier Bay and in the Louisiana oil patch contributed certain toxic metals to the environment, and the extent of the contribution, if any, to the water column.

This toxic metal program was limited to the determination of the levels of mercury, arsenic, zinc, cadmium, and lead in the water column. Samples were collected near oil platforms, at various distances and directions from the platforms, and at locations remote from the oil platforms.

Another purpose of the program was to establish scientifically sound baselines, but since oil production has been going on in the general area off the Louisiana Coast for many years, it is questionable if the values obtained, even in areas remote from oil platforms, can be considered truly representative of conditions that existed before oil exploration began. The data can, of course, be usefully compared to data obtained in the future.

Toxic metals could conceivably be contributed to the environment from accidental spillage of petroleum products, drilling fluids and cuttings, lubricating oils and fuels, brines, and the corrosion of platform metals.

The fate and effects of trace metals in marine and estuarine waters have until recently been studied mainly in laboratory investigations. It is known that certain metals concentrate in sediments and bioconcentrate in marine life (Horne 1969; Bender et al. 1972).

El-Sayed et al. (1972) presented available data on various metals in the Gulf and noted that the data were exceedingly sparse. For example, they presented data on only 15 samples for iron, nickel, and lead, and 63 samples for zinc. Essentially all of these samples were from the Texas Coast.

METHODS

Laboratory

Arsenic was determined by the silver diethyldithiocarbamate method, in which inorganic arsenic is reduced to arsine, which in turn reacts with silver salt forming a soluble red complex that is suitable for photometric measurement (American Public Health Association 1971). Quantitation was performed photometrically and the detection limit for the method was 10 μg/l.

Cadmium, lead, and zinc were determined by atomic adsorption spectrophotometry after complexation with ammonium pyrolidine-dithiocarbamate (APDC) at pH 4 and extraction with methyl isobutyl ketone (MIBK), as described in Standard Methods (American Public Health Association 1971).

Mercury was determined by cold vapor atomic adsorption spectrophotometry. Samples were oxidized with sulfuric acid, nitric acid, potassium permanganate, and potassium persulfate. Reduction to elemental mercury was achieved with sodium chloride-hydroxylamine sulfate and stannous sulfate.

With these methods, the limits of detection were 0.3 μg/l for Hg, 0.1 μg/l for Cd, 0.2 μg/l for Zn, and 0.5 μg/l for Pb.

Metal recoveries as determined from spiked water samples were as follows: Cd - 95%; Pb - 77%; Zn - 89%; and Hg - 116%. Relative standard deviations expressed as percentage of the mean were 22% for Cd, 16% for Pb, 14% for Zn, and 11% for Hg.

Field

The general locations of the stations sampled in Timbalier Bay and along a transect into the offshore region are shown in figure 1. More samples were taken in the immediate vicinity of the platforms, i.e., within 1,000 meters, than at the control sites.

RESULTS

Bay

Over the sampling period in the bay water column samples, mercury ranged from <0.3 to 6.9 μg/l with an average of 0.6 μg/l. Cadmium ranged from <0.1 to 22.5 μg/l with a mean of 0.6 μg/l; lead ranged from <0.5 to 50 μg/l with an average of 2.8 μg/l. Zinc varied between <0.2 and 66 μg/l with a mean of 8.5 μg/l. All of the arsenic values were below 10 μg/l, the detection limit.

Shown in table 1 are comparisons for the bay ambient stations (AB-1 and 2) and the platform sites during two periods for which at least 5 samples were available from each location. An analysis of variance performed on these data indicated no significant differences between locations at the 0.05% level.

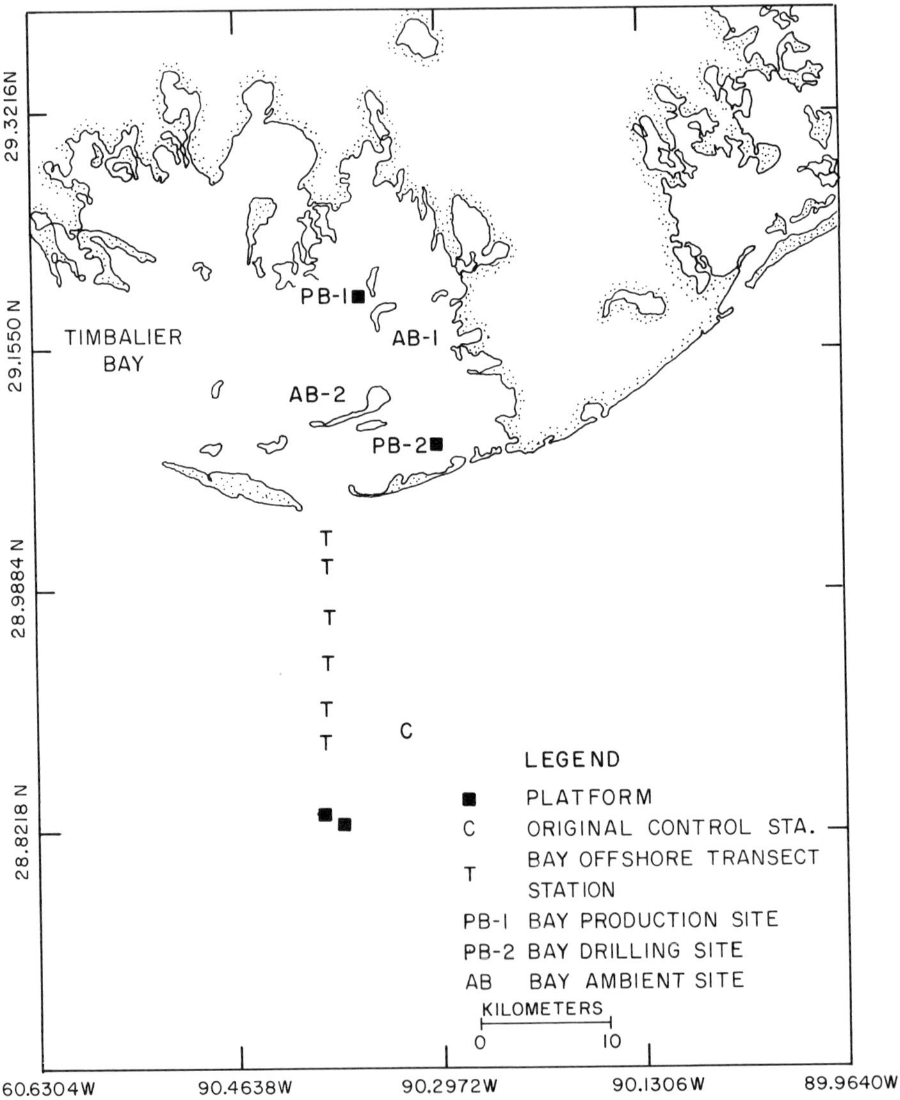

FIG. 1. THE OEI STUDY AREA SHOWING THE STATIONS SAMPLED.

TABLE 1
TIMBALIER BAY—METAL LEVELS ($\bar{x}$ in $\mu g/l$)

	September 1972			October 1972		
	Production	Drilling	Ambient	Production	Drilling	Ambient
Zn	4.3	4.5	4.3	10.8	10.3	11.1
Cd	0.2	0.3	0.2	0.2	0.5	0.8
Pd	1.1	1.7	1.1	4.5	6.9	6.6
Hg	1.6	0.7	1.0	0.5	0.5	0.4

On occasion in the offshore region, zinc levels were shown to be elevated in the immediate vicinity of the platforms. As shown in figure 2, these elevations were of limited extent and probably represent metals from the sacrificial zinc anodes.

Variations in metal levels from the bay to the offshore platforms are shown in figures 3 and 4. These data represent the metal concentrations determined within the sectors shown in figure 5 over the study period. The platforms were located in sectors 2, 5, and 12.

Levels of cadmium and zinc (figure 3) were generally higher in the bay, decreasing along the transect into the Gulf. No clear platform effects are indicated from these data. As shown in figure 4, however, lead levels, although extremely variable, did appear elevated in the sector encompassing the workover platform in the bay. Mercury levels as shown in figure 4 were generally higher in the Gulf than in Timbalier Bay.

DISCUSSION

Data collected in close proximity to the platforms did indicate increased levels of Zn in the offshore area and Pb near the workover rig in the bay. Even these localized elevations were below those estimated as chronically toxic to marine life by the National Academy of Sciences (1972). For example, they estimated safe levels for zinc to be 20 $\mu g/l$ and 10 $\mu g/l$ for lead. As shown in figures 3 and 4, these levels were not exceeded, on the average, at the platform sites. In addition, it must be remembered that the Academy criteria are based on dissolved metals, and the values reported here are totals. Cadmium levels are estimated to be hazardous to marine life if they exceed 10 $\mu g/l$. The highest mean value for cadmium reported in this study was 0.6 $\mu g/l$ at the bay production platform.

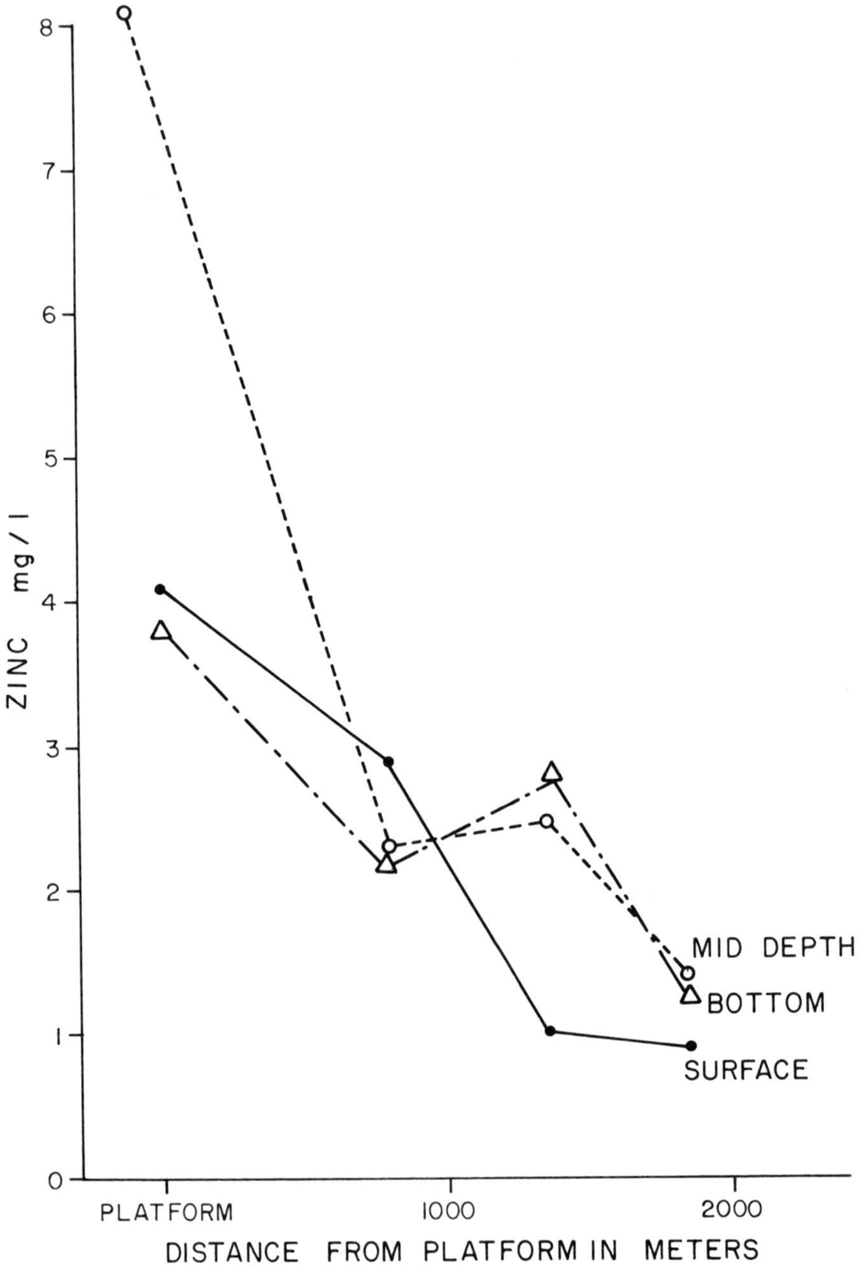

FIG. 2. WATER COLUMN ZINC LEVELS AT PLATFORM HOR-54A, SEPTEMBER 1972.

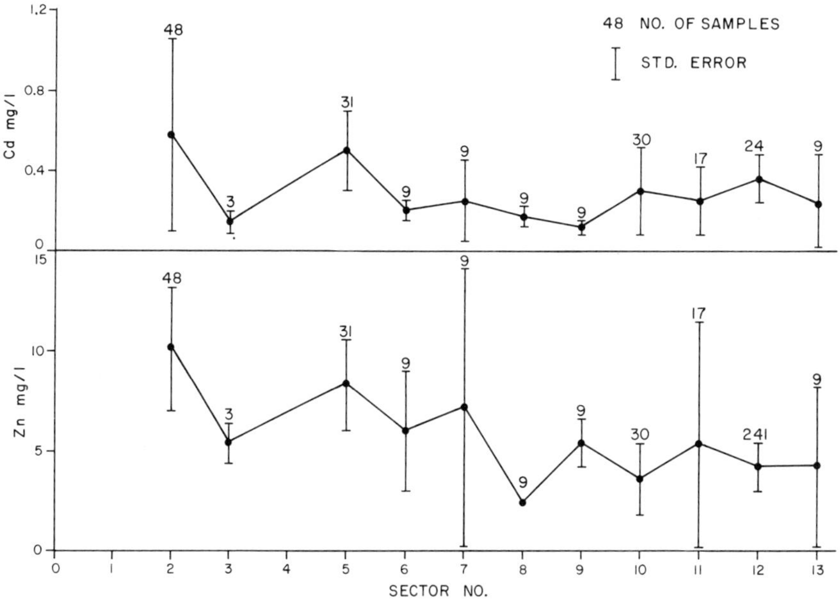

FIG. 3. OEI—BAY TO OFFSHORE.

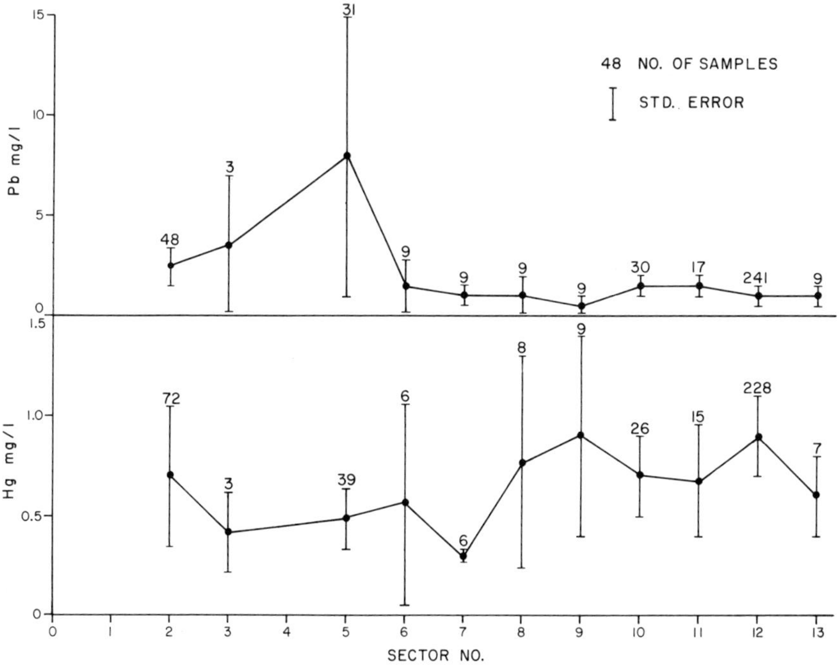

FIG. 4. OEI—BAY TO OFFSHORE.

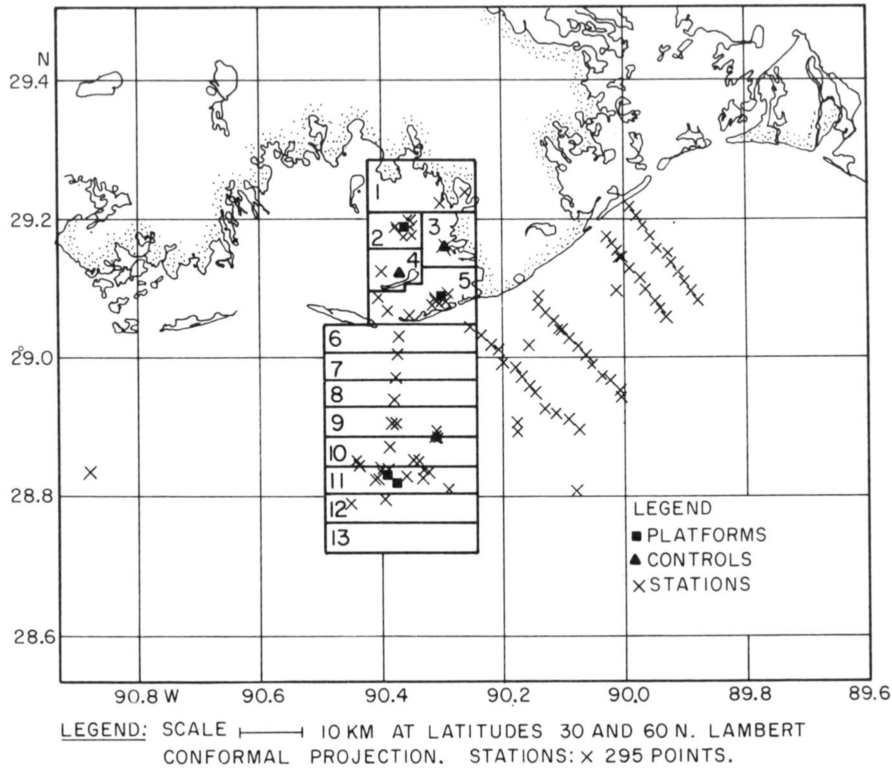

FIG. 5. METAL CONCENTRATIONS.

CONCLUSIONS

Utilizing the criteria set forth by the Academy and the data reported here, we do not predict a risk of chronic biological effects due to the metals zinc, lead, or cadmium originating from oil production platforms.

REFERENCES CITED

American Public Health Association, Standard Methods, 13th Edition. 1971. Washington, D.C.

Bender, M. E., R. J. Huggett, and H. D. Slone. 1972. Heavy metals—an inventory of existing conditions. The Fate of Chesapeake Bay. Journal of the Washington Academy of Science **62**:144-153.

El-Sayed, S. Z. and K. W. Fucik, 1972. Chemistry, primary productivity and benthic algae of the Gulf of Mexico. American Geographical Society, Folio 22.

Horne, R. A. 1969. Marine Chemistry: The Structure of Water and the Chemistry of the Hydrosphere. New York: Wiley—Interscience.

National Academy of Science and National Academy of Engineering. 1972. Water Quality Criteria. Washington, D.C.: U.S. Government Printing Office.

ORGANIC CARBON, INORGANIC CARBON, AND RELATED VARIABLES IN OFFSHORE OIL PRODUCTION AREAS OF THE NORTHERN GULF OF MEXICO

by Charles R. Brent, Howard P. Williams, W. A. Bergin, John L. Tyvoll, and Tommy E. Myers

ABSTRACT

Analyses of over 700 surface and water column samples taken from Timbalier Bay and the offshore area from the South Timbalier Field eastward toward Grand Isle, Louisiana, revealed an ecosystem rich in organic carbon (OC). Total organic carbon (TOC) averaging 5 mg/l is experienced on the near continental shelf of this region, in sharp contrast with the waters of the open Gulf, which average 1.5 to 2 mg/l.

Another large increase in TOC is experienced as one moves inshore into Timbalier Bay where mean values exceed 9 mg/l, occasionally ranging from 15 to 20 mg/l. During certain seasons or on certain sampling trips, the presence of a production platform or workover rig can be noted by increased OC values or biochemical oxygen demand (BOD), particularly in Timbalier Bay where surface sampling is possible.

Several local factors peculiar to this area of the Gulf contribute to (1) a heavy organic load, (2) effects on dissolved oxygen (DO), and (3) occasional ecological instabilities on the continental shelf off Grand Isle. The Mississippi River Plume, the extensive tidal marshes of South Louisiana, the large shallow water bays, the oil industry, and a clockwise eddy current off Grand Isle all contribute to conditions under which oxygen depletion can occur on the continental shelf. In the summer of 1973, following heavy spring flooding, "fresh" water from the

The authors are with the Institute of Environmental Science, University of Southern Mississippi, Hattiesburg, MS.

Mississippi plume overlay the Gulf waters below it to a depth of 7 m, isolating it and allowing the oxygen content below 7 m to sag to less than 2 mg/l over most of the study area. This value is below the mean minimum DO in the deep Gulf, indicating that it is a local phenomenon, not caused by up-welling from the deep Gulf.

Vigilance should be maintained to minimize effects from oil platform effluents and operational spills because of the inherent instability of this ecosystem during spring and summer months. Plans for diverting even more water from the Mississippi River through Southwest Pass and into this area should be considered carefully before implementation.

INTRODUCTION

The dissolved organic carbon (DOC) and particulate organic carbon (POC) levels in the Atlantic Ocean and the Gulf of Mexico are difficult to reconcile on the basis of known input mechanisms of the carbon cycle. Estimates by Fredericks and Sackett (1970), based on their own results of 1961 and the data of Menzel and Ryther (1964, 1968) and of Menzel (1967), indicate DOC levels in the upper 90 m of water some 16 times the annual productivity of the Gulf of Mexico reported by El Sayed et al. (1972).

The highest values of DOC plus POC reported by Fredericks and Sackett were near 4 mg/l. The observed gradients in DOC in the upper 90 m of the Gulf could best be explained if OC inputs from rivers and estuaries were in the neighborhood of 10 to 15 mg/l. This level of OC exceeds that reported in rivers and estuaries up to the beginning of this study in 1972. Our study revealed that TOC levels emanating from terrestrial sources, estuaries, rivers, and the continental shelf, when combined with all other OC sources, are of the right order of magnitude to explain DOC gradients observed near the mouth of the Mississippi River and along the Louisiana shores.

The offshore and inshore operations areas off the Louisiana coast were chosen as areas for intensive ecological study by an interdisciplinary team of scientists belonging to member institutions of the Gulf Universities Research Consortium (GURC) early in 1972. The purpose of the Offshore Ecology Investigation (OEI) was to determine the effects, if any, of platform oil operations on the ecology of Timbalier Bay and the continental shelf immediately to the south in what is commonly referred to as the South Timbalier Field. The primary thrust of our particular project was to determine OC and related variables in surface samples in both Timbalier Bay and the offshore oil production fields. A large number of water column samples and some bottom samples were analyzed as well, for the sake of completeness. The TOC measurement was chosen because of the expected high productivity of the area and the fact that we could not predict whether oily materials would show up in the DOC or the POC.

METHODS AND PROCEDURES

Water column samples from the offshore area were taken in 2 liter Van Dorn or 5 liter Niskin water samplers. Two types of surface samplers designed and constructed at the Institute of Environmental Science, University of Southern Mississippi, worked with little or no success in the offshore area. The second sampler worked with nominal success in Timbalier Bay; however, most surface samples in Timbalier Bay were taken by lowering a bottle or cubitainer so that its neck was 0.5 cm below the surface and allowing it to fill.

Samples were quick frozen in 1-liter polyethylene bottles, 1-quart cubitainers, or polypropylene culture tubes with a capacity of 17 ml. These samples were transported still frozen to the Institute's Laboratory, where they were stored frozen, then processed within 1 or 2 days.

Processing included quick thawing (under warm water with stirring), adjusting the temperature to 25° C, blending for 30 seconds in a Waring blender, saturating with DO, and splitting the samples for BOD, OC, inorganic carbon (IC), and chemical oxygen demand (COD) determinations. One BOD bottle was used for an initial DO determination in the BOD test, and one was seeded and incubated at 20° C for 5 days for the final BOD determination. One 30 to 50 cc sample was acidified and frozen for a later COD determination, and one small sample in a 17 ml culture tube was frozen for OC and IC determinations. These tubes were used only after careful studies showed that they had no effect on TOC values of a variety of water samples, including distilled deionized CO_2-free water.

The OC and IC determinations were made on a Beckman Model 915 TOC analyzer. Twenty μl samples of OC and IC standards containing 100 mg/l carbon were injected successively into the high temperature (950° C) channel and the low temperature (150° C, acidified with H_3PO_4) channel of the TOC analyzer until the adjusted instrument provided consistent readings of 100 ± 1 ppm. Duplicate or triplicate injections of 20 μl seawater samples were then made on each channel of the TOC analyzer. After each five sample determinations, standards were injected again to check for drift or deterioration of column packings. A series of

100, 50, 30, 20, and 5 mg/l standards was injected during each continuous period of measurements to determine calibration curves or correction factors where they were significant.

The standard deviation of duplicate or triplicate total carbon determinations was ± 0.6 mg/l, with IC values agreeing to within ± 0.3 mg/l. Since OC is obtained by the difference of these results, the standard deviation is ± 0.7 mg/l carbon.

BOD determinations were made using procedures in *Standard Methods* (Menzies 1973), except that a YSI Model 54 dissolved oxygen meter was used to follow the oxygen sag during BOD tests on the last two sets of samples.

COD was determined (with little success as expected) according to procedures in *Standard Methods* on selected samples for which a high OC or BOD value was measured. Although blanks for artificial seawater were low, 25 to 30 mg/l, no consistent COD values were obtained on seawater samples.

RESULTS

Some 719 water samples taken by personnel involved in the OEI study were analyzed for organic and inorganic carbon content. Inshore samples were taken from Timbalier Bay at standard stations located around a stationary platform near Philo Brice Island, around a roving workover rig, and within a "control" area in the bay. Offshore samples were concentrated around two or three platforms in the South Timbalier field, a control area located 6 miles from the nearest rig, and two synoptic transects located between the South Timbalier platforms and Grand Isle (figure 1). Results of these analyses, which are voluminous, are stored in the GURC computerized data bank.

DISCUSSION

General observations on the whole body of data reveal that the offshore and inshore areas are two distinct but interacting systems. OC values in Timbalier Bay are nearly twice as high as values measured in the offshore region (table 1). Seasonal fluctuations in OC values and local variations around production and drilling platforms in Timbalier Bay tend to obscure the differences between control areas and areas of intensive oil operations.

Mean values of OC in Timbalier Bay are significantly higher than those measured by Brent et al. (1973) in the Bay of St. Louis, a coastal bay in Mississippi considered to be a "clean" ecosystem. That Timbalier

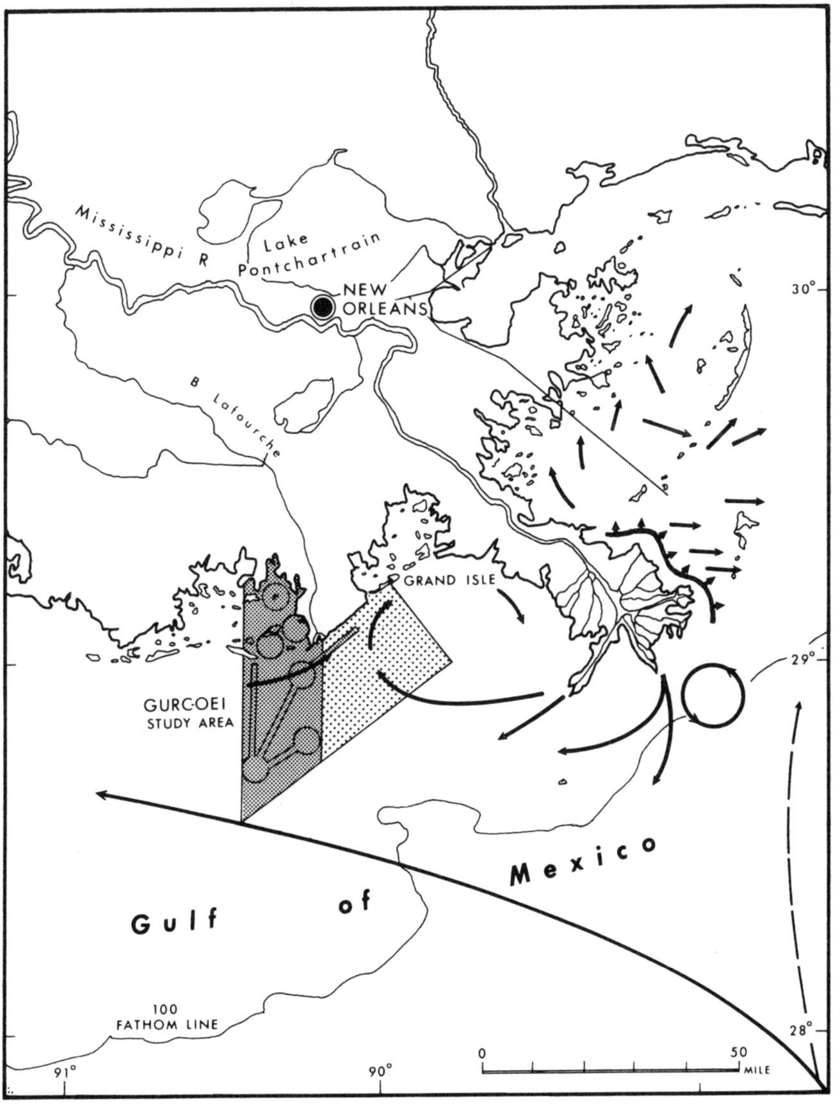

FIG. 1. GURC OEI STUDY AREA, TIMBALIER BAY, LOUISIANA. (after Harris [1974])

Bay is richer in OC can be attributed to two major factors: the extensive marshes, which feed detritus into the bay, and the presence of a major oil industry operating in the bay. That both are significant contributors to the high OC and biological productivity of the bay is shown by ancillary studies in this consortium investigation by Marum (1974), Farrell

TABLE 1. Organic Carbon and Related Variables in Selected Areas of the GURC/OEI Study

Location	Organic Carbon (mg/l)		Inorganic Carbon (mg/l)		BOD (mg/l)		COD (mg/l)	
	Mean	σ	Mean	σ	Mean	σ	Mean	σ
<u>Timbalier Bay</u>								
Stations Surrounding Philo Brice Platform A089 Lat 29° 11' 30" Long 90° 21' 40"	13.1	4.0	27.5	7.4	2.6	1.4	82	31
Control Area E003 Lat 29° 09' 42" Long 90° 17' 32"	9.3	1.6	27.8	3.3	1.8	0.8	102	26
Roving Work Over E042-43 Rigs South Timbalier Bay	7.8	4.2	22.6	5.7	2.0	1.0	93	29
<u>Offshore</u>								
Platform 54A Lat 28° 49' 57.7" Long 90° 23' 24.6"	4.5	2.0	25.2	7.6	1.1	0.8	105	39
Platform 66D Lat 28° 49' 32.8" Long 90° 22' 24.7"	5.2	3.0	22.0	5.9	1.3	1.0	68	17
Platform 26S Lat 28° 58' 46" Long 90° 10' 33"	6.1	1.5	23.2	4.7	1.4	1.1	160	45
Control Area H 062 Lat 38° 53' 30" Long 90° 18' 30"	4.9	2.2	24.6	4.4	1.2	1.1	120	34

(1974), Humm (1974), Fish and Lewis (1974), and Kritzler (1974). Although Laseter and Ledet's (1974) gas chromatography/mass spectroscopy data provided a measure of hydrocarbons present, indicating that the hydrocarbons represent only a small fraction of the measured OC, it is difficult to assess the contribution of the oil industry because of the abundance of oil-degrading bacteria populating the waters of the study sites (Oppenheimer 1974). The petroleum hydrocarbon input is certain to lose its identity as it enters the food chain. The "control" area within Timbalier Bay appears not to be isolated enough from oil industry operations to avoid all of the wind- and wave-borne effects. Clearer estimates of effects come from contours of data taken around platforms on a specific day, as will be discussed later.

Mean organic values in the offshore area of the OEI are higher by nearly 60% than those reported by Fredericks and Sackett. Since they sampled at the mouth of the Mississippi River, during only one season, the oil industry contribution to the increased OC load cannot be ascertained from comparison of their data with ours with any reasonable degree of confidence. An occurrence on the near continental shelf during the spring and summer of 1973 dramatically illustrated the importance of the interaction of OC load and spring flooding on the Mississippi. During April 1973 a synoptic study in the offshore area revealed a heavy overlay of "fresh" water in the study area (Oetking et al. 1974), with high nutrient concentrations (Burchfield et al. 1974) and evidence of an algal "bloom" in the upper layer (El-Sayed 1974). In July 1973, the "fresh" (23 g/l) water still overlay the study area to a depth of 5 to 7 m, isolating the high salinity (34 to 35 g/l) Gulf waters below that depth. Apparently, wave action was not sufficient to cause mixing below 7 m during this period. The isolation of the deep waters (7 to 20 m) and the high oxygen demand from OC loading in this area combined to produce a very low DO content, ranging from 2 mg/l immediately below the halocline to 0.5 mg/l near the bottom. Long residence times of water in this area due to a clockwise current, which was observed by Harris (1974), prevented the replenishing of DO from open Gulf waters (see figure 1). Upwelling from the deep Gulf can be excluded as a source of oxygen-depleted water because DO in the open Gulf does not fall much below 2.5 mg/l at any depth between 0 and 2500 m (El Sayed et al. 1972). Evidence of oxygen near-depletion (0.1 mg/l) also occurred in the summer of 1972, near the base of a study platform. The geological coincidence of the most highly productive offshore oil industry in the world existing in an eddy current downstream from the plume of one of the world's largest rivers leads to an ecosystem strained to the breaking point to maintain a viable DO content near the bottom during summer months following heavy flooding of the Mississippi River.

Even though the 1973 discharge from the Mississippi River was unusually high, the adjacent near continental shelf is ripe for summer seasonal fish kills, such as the one reported at the entrance of Timbalier Bay during the summer of 1973 by sampling personnel of this study. Efforts should be made to minimize OC loading and oxygen-depleted discharges near the continental shelf from Southwest Pass beyond the Timbalier Bay region, especially after heavy spring flooding. Alternatives to plans for diverting more of the outflow of the Mississippi River through Southwest Pass should be carefully considered by the best ecological expertise available before implementation of a project that might further strain this ecosystem's ability to assimilate an increased OC load. Runoff predictions would indicate that an occurrence similar to that observed in July 1973 may also have occurred in the summer of 1974.

To answer the question "does the offshore and inshore (Timbalier Bay) oil industry affect the ecosystems in which it operates?" one has to look carefully at individual sets of data taken on a given day or within the same week and at the spatial distribution of results around production and drilling platform structures. The comparison of near-platform data with control data fails to answer this question because of seasonal fluctuations and because the "control" areas are in basically the same ecosystem as the platforms. The control areas are probably exposed to spills and leakage from platforms, storage, and piping facilities through wind-driven wave transport, along with tidal and prevailing current transport.

Surface sampling, which was the only type of water column sampling done in Timbalier Bay for this study, is subject to a high degree of variability affected mostly by wave action and sampling technique. MacIntyre (1974) points out the high variability of surface samples taken by different sampling techniques as well as the wide variations in reports on the nature of the organic matter in the upper millimeter of ocean surface.

Our results show that sampling of the upper 0.5 cm of the surface can be done with reasonable consistency during calm days on Timbalier Bay. The data give smooth contours pinpointing the sources of organic pollution (figures 2, 3, 4, and 5).

The effect of the Philo Brice Platform and tank battery, Lat. 29° 11.50′, Long. 90° 21.68′, was noted on all days for which extensive OC data were taken around that site. Figure 2 shows the contours of OC data taken in September 1972 around that platform. Contours are drawn by interpolating values between measured points at 0.5 or 1.0 mg/l intervals and connecting points having the same value with smooth continuous curves. General contour shapes were checked by computer

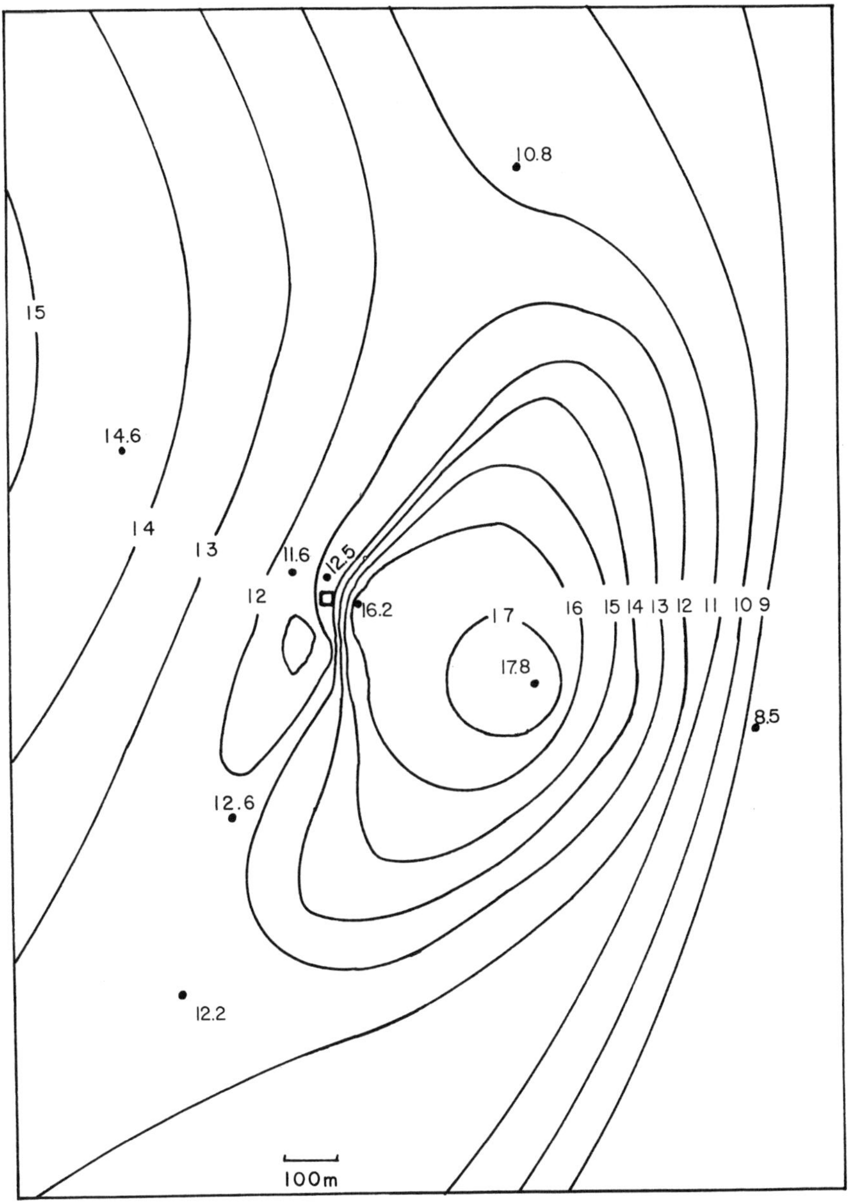

FIG. 2. SEPTEMBER 1972 ORGANIC CARBON ISOPLETHS AROUND PHILO BRICE PLATFORM, an operating platform in Timbalier Bay, Louisiana. Isopleths show that the platform and a source 400 m to the southeast are significant contributors of organic carbon.

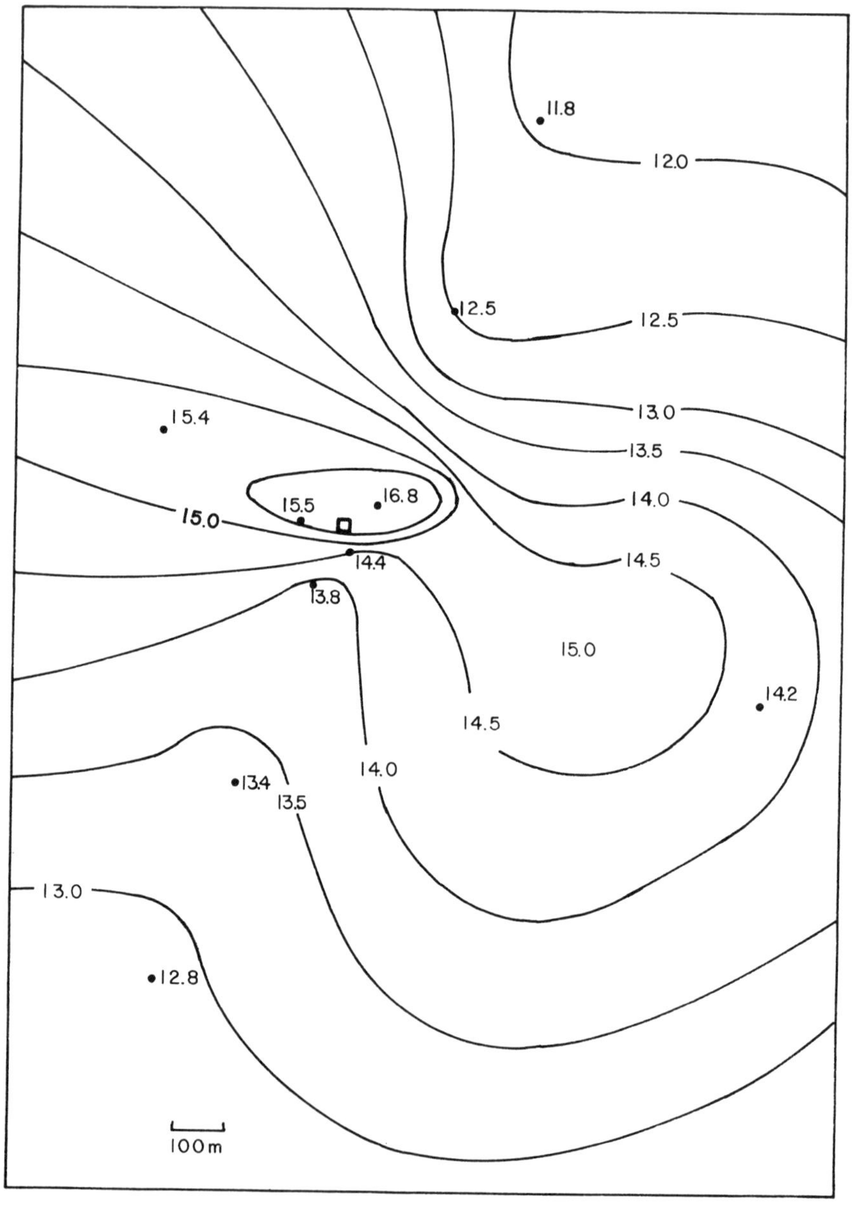

FIG. 3. DECEMBER 1972 ORGANIC CARBON (mg/l) ISOPLETHS AROUND PHILO BRICE PLATFORM indicate that the platform and a source 400 m to the southeast contribute organic carbon to the ecosystem.

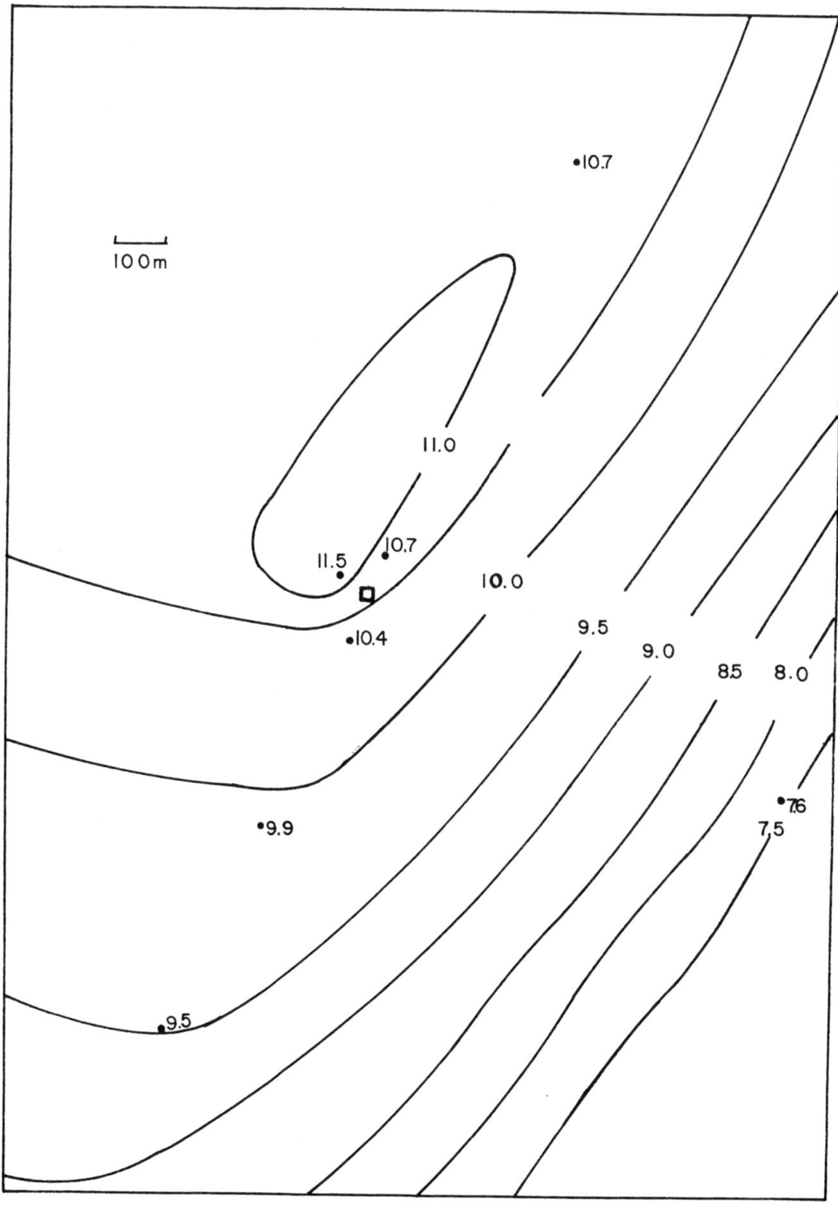

FIG. 4. JANUARY 1974 ORGANIC CARBON (mg/l) ISOPLETHS AROUND PHILO BRICE PLATFORM indicate that the platform is a source of organic material.

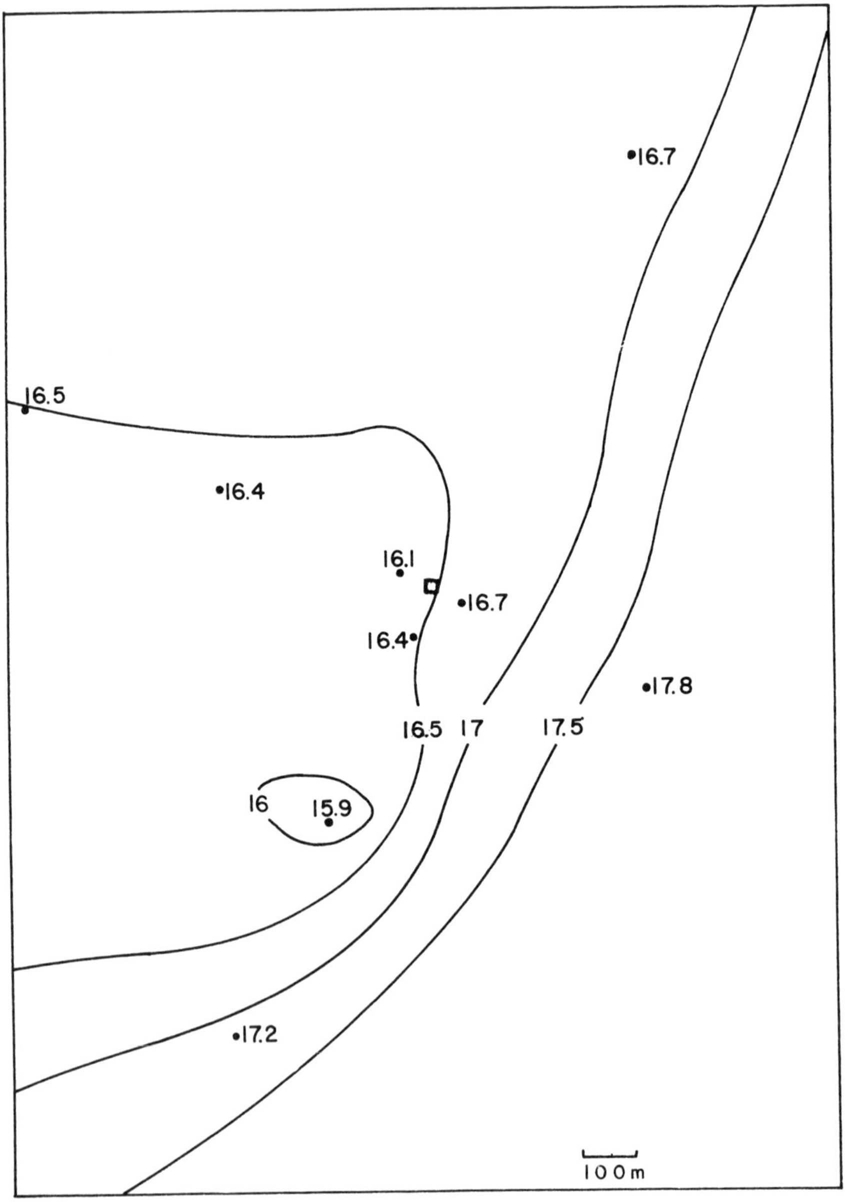

FIG. 5. JANUARY 1973 ORGANIC CARBON (mg/l) ISOPLETHS AROUND PHILO BRICE PLATFORM. Rather uniform high values indicate that the platform is not a major source of organic carbon.

contouring in five intervals of the range of results. The two contour maps were similar. As one proceeds from the east-southeast to the west-northwest along a sampling route, one encounters a rise in OC of 9.3 mg/l in the southeast quadrant to a peak value of 17.8 mg/l, a minimum value of 11.0 mg/l just southwest of the platform and a general increase to 15.5 mg/l in the quadrant northwest of the platform. The effect of oil operations around this platform extends 800 m to the southeast and an indeterminate distance to the northwest, but at least 800 m.

A similar distribution of OC values in this area occurred in December 1972 (figure 3); in this case, however, the maximum increase within the data points was only 5 mg/l to a maximum value of 16.8 mg/l in the vicinity of the platform. Figure 4, indicating the distribution of values obtained in January 1974, more clearly shows the effects of the platform alone. Though the increase noted is only 3.9 mg/l, it is obvious that the platform is the source of this rise in OC levels found primarily in the quadrant northwest of the source.

One set of OC data (figure 5), measured in the Philo Brice Platform area in January 1973, was so uniform that contours did not reveal any significant variations in the OC values. These data represent the highest set of OC values measured in an area this large during the OEI study, 16.6 ± 0.4 mg/l. Wind and wave action was heavy on the bay when these measurements were made.

The OC data collected in Timbalier Bay indicate that an active platform could either increase the surface OC to levels as high as 20 mg/l as in the case of Philo Brice Platform, or decrease the OC far below the mean values measured at the more distant stations, as noted around one of the workover rigs in South Timbalier Bay in January 1974 (figure 6). The latter effect could possibly be explained by dilution processes from well operations or the herding of the surface microlayer neuston away from the platform as an oil slick spreads from its source. The effect often extends 800 m from the rig location. Timbalier Bay production and workover rigs could be described as perturbations in the bay ecosystem causing data collected near structures to have higher standard deviations than data from distant stations showed.

In the offshore domain, the OC data taken alone revealed the presence of platforms on only a few occasions. However, comparing salinity, DO (Oetking et al. 1974), and OC data for a given excursion confirmed that platform effects were being observed. For example, during August 1972, the highest value for OC occurred at the base of Platform 66D (Lat. 28° 49′ 32.8″, Long. 90° 22′ 24.7″). Oetking's dissolved oxygen data also showed that the minimum dissolved oxygen level of 1.0 mg/l occurred near the platform on that date. The general trend of high OC values (greater than 5 mg/l) being associated with low DO measurements

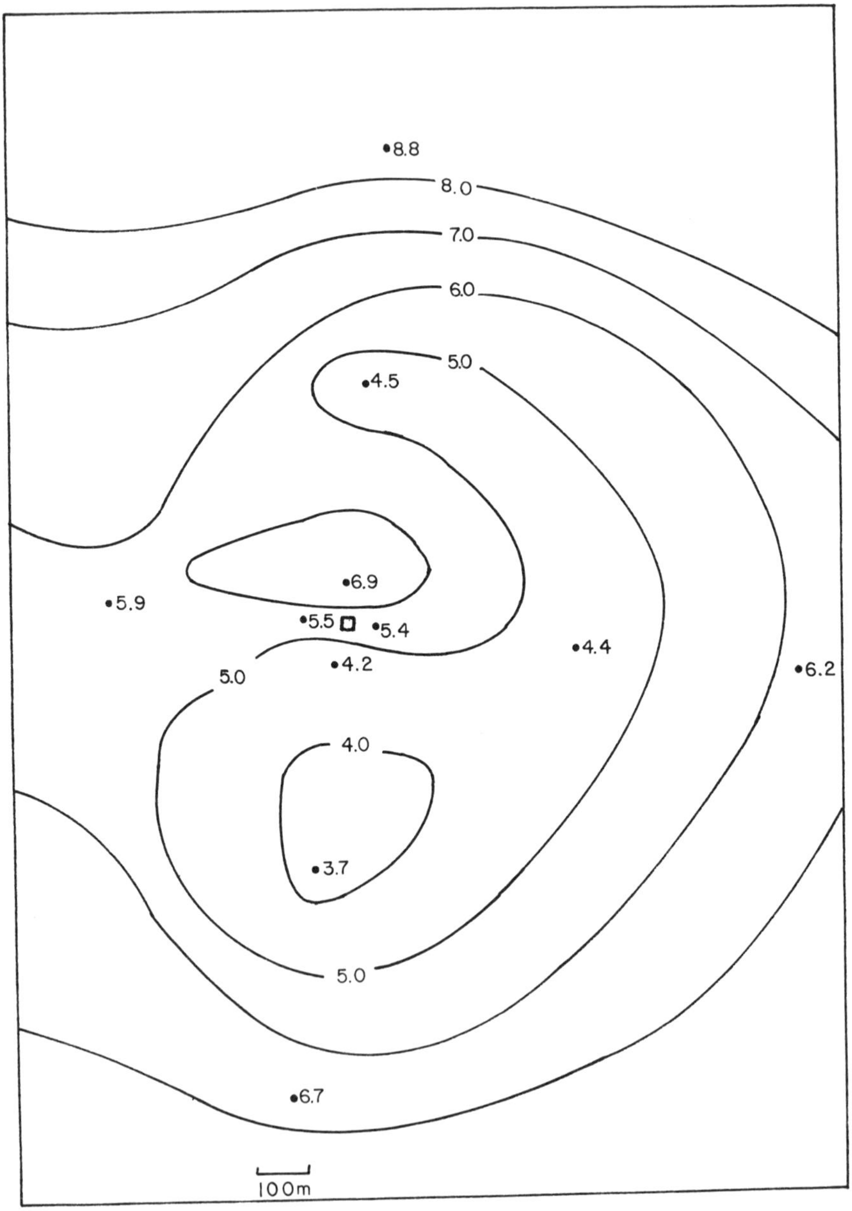

FIG. 6. JANUARY 1974 ORGANIC CARBON (mg/l) ISOPLETHS AROUND A WORKOVER RIG IN SOUTH TIMBALIER BAY. The rig represents a perturbation in the bay ecosystem.

was observed offshore, particularly in the summer months when oxygen depletion rates are high. Data collected around Platform 54A (Lat. 28° 49′ 57.7″, Long. 90° 23′ 24.6″) on July 11, 1974, illustrate this to a degree (figure 7). One of the higher OC (9.8 mg/l) levels occurred at the surface just south of Platform 54A, and some of the lowest DO values occurred beneath that area at a depth of 5 to 7 meters. A high OC value of 9.3 occurred near the bottom at station 30 (Lat. 28° 56.28′, Long. 90° 22.46′) and low dissolved oxygen values accompanied it. A high OC value of 8.8 mg/l occurred near the bottom at station 22 and again low values of DO occurred in this area. Although fewer OC determinations (15 total) were made compared to DO measurements, some of the stratification is indicated by horizontal contours at the 5.0 mg/l OC level at the same depth (5 to 7 m) as an oxygen sagged wedge indicated by Oetking's data.

Platform effects were observable in April 1973 around Platform 26S

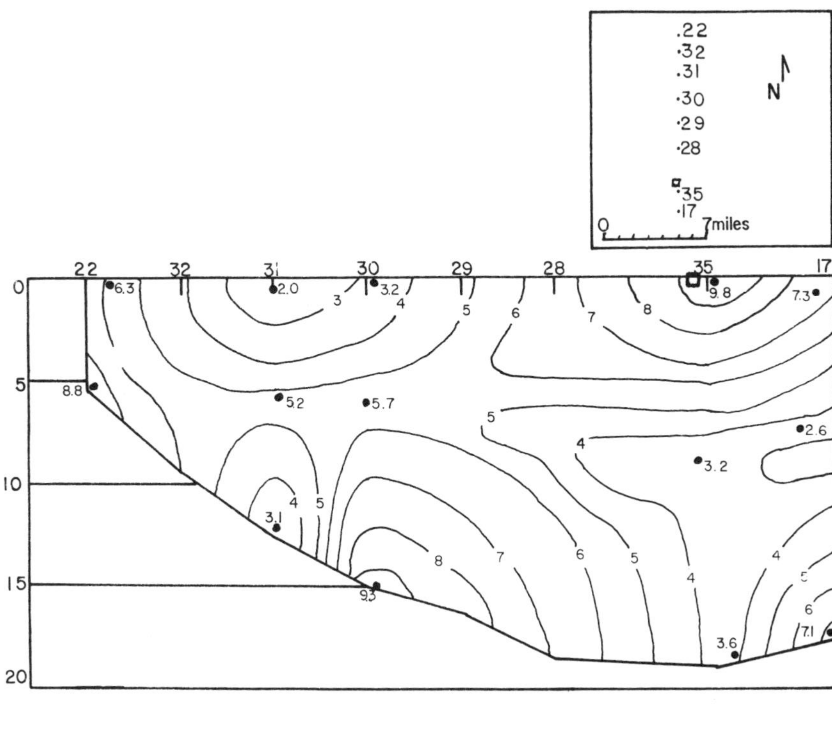

FIG. 7. ORGANIC CARBON VALUES (mg/l) ALONG A LINEAR TRANSECT on the continental shelf passing through Platform 54A sampled on July 11, 1973. Highest surface value measured near platform.

(Lat. 28° 58′ 46″, Long. 90° 10′ 33″) in a synoptic study of two transects that pass through the highly developed oil industry off Grand Isle (figure 8). Comparison of OC contours with salinity, temperature, and DO data (Oetking et al. 1974) revealed general stratification along

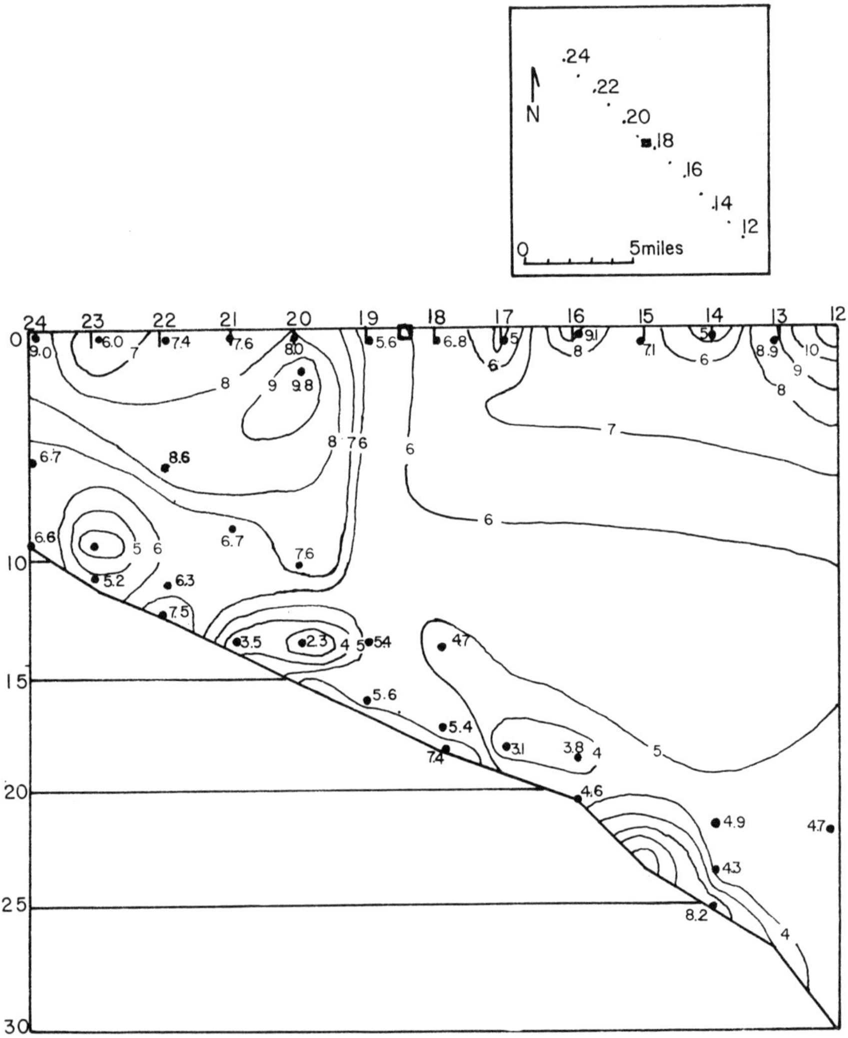

FIG. 8. ORGANIC CARBON VALUES (mg/l) MEASURED ON A TRANSECT BY PLATFORM SH26C, April 6, 1973. Temperature, salinity, and DO values show even more vividly a uniform column of water between stations 18 and 19.

the SE-NW transect except in the vicinity of platform SH26C. There, high salinities extended from the bottom to the surface and all other variables, OC, T, and DO, were nearly uniform from the surface to near bottom. There are at least three plausible reasons for this condition: (1) thermal convection of high salinity waters from the bottom, (2) mixing caused by currents flowing through a platform or platforms, or (3) discharge of brines from well pumping operations. The fact that the high salinity column is so extensive, 2 miles in width, would probably exclude consideration (2) or (3), leaving only thermal convection as a mechanism for the formation of this large vertical column of uniform high salinity water. Whether heat from platforms, heat from a salt dome, or some other mechanism is affecting salinity would require further extensive study. The fact that the phenomenon is associated with a platform suggests that it is caused by the platform.

Except for upper water column measurements where oxygen replenishing occurs, high OC values are accompanied by low DO levels. Platform effects on OC values were usually limited to a maximum distance of 800 to 1600 m.

Unusually high values of OC up to 39.3 mg/l were measured at scattered points within the offshore region during the synoptic study of April 1973. Values as high as 15.4 mg/l were measured for OC in the vicinity of Platform 66D in July and October 1973.

Five-day BOD measurements were made on a large number of selected samples. Oxygen uptake was apparently very slow, resulting in rather low BOD values (0.5 to 5 mg/l). Poor correlation between BOD and OC was observed.

Inorganic carbon values were dependent mostly on pH and salinity in the open Gulf, and upon salinity and stirring in Timbalier Bay.

COD determinations were attempted using mercuric chloride to tie up chlorides and using blanks and standards of equal chlorinity with the set of samples being run. Further, only samples with high OC values were chosen during months when the work load was heavy. COD data did not correlate well with the other variables measured. The poor precision of this measurement obscured real fluctuations in both ecosystems, which were more apparent in the BOD and OC data. This was expected and it confirms the statement in the Environmental Protection Agency's Methods for Chemical Analysis of Water and Waste (1971) that COD's less than 250 are unreliable in saline waters.

SUMMARY AND CONCLUSIONS

OC measurements using a Beckman Model 915 TOC analyzer

provide a useful, convenient measure of the total organic load on a marine ecosystem. While DOC and POC determinations (Jeffrey and Hood 1958; Menzel and Vaccaro 1964) may be necessary to detect subtle, low-level changes in OC in the open Gulf, TOC measurements provide a useful assessment of organic enrichment of ecosystems with high organic productivity or loading of a magnitude of 4 mg/l or greater. TOC measurements are very useful when one is dealing with surface samples taken from the top half-centimeter of water surface, where OC values range as high as 20 to 40 mg/l. The TOC analysis in saltwater has become much more convenient and reliable since the incorporation of a factory modification using a ceramic combustion tube in the high temperature channel of the Beckman Model 915 TOC analyzer.

Systematic surface sampling in Timbalier Bay provided data contours whose range from 2 or 3 mg/l to 20 mg/l clearly outlined stationary sources of OC in the vicinity of an operating platform near Philo Brice Island. Other contour maps around workover rigs indicate perturbations in the background OC levels, usually a lowering of OC in the vicinity of the rig. Effects of bay platforms and rigs extended for at least 800 m and beyond.

Offshore investigations revealed OC loading of the marine system from operating platforms and wells ranging from 2 to 4 mg/l above background levels. The effects usually extend some 800 to 1600 m from the platform. The highest value of OC measured in the offshore region was 39 mg/l. Comparing the OEI values with those measured in other marine or aquatic systems (Riley 1937; Duursma 1961; Livingstone 1963; Wilson 1963; Garrett 1967; Williams 1967; Skopints 1968; Brooks et al. 1971; El-Sayed et al. 1972; and Brent et al. 1973) indicates that the offshore area near Grand Isle, Louisiana, is one of the most organic-rich ecosystems in the world of its type, quite atypical for an "open" continental shelf region.

The April 1973 and July 1973 synoptic studies revealed an offshore ecosystem approaching the conditions often found in a freshwater lake or a closed marine system like the Black Sea. Stratification and isolation of deeper waters had occurred along with an algal bloom in April that led to widespread oxygen depletion in waters deeper than 7 m in July. This condition is somewhat surprising since the continental shelf would appear to be an open Gulf ecosystem.

REFERENCES CITED

Brent, C. R., H. P. Williams, W. A. Bergin, and J. L. Tyvoll. 1973. Chemical Definition of an Ecosystem—St. Louis Bay, Mississippi—A Sea Grant Report. Universities Marine Center.

Brooks, J., W. Sackett, and A. Fredericks. 1971. Organic carbon as a pollution parameter in Brazos River estuary and Gulf of Mexico. Abstracts of Papers, American Chemical Society **56**:71.

Burchfield, H. P., R. J. Wheeler, and W. Subra. 1974. Nutrient analysis of water and sediment samples from the Gulf of Mexico south of Timbalier Bay, Louisiana. Gulf Universities Research Consortium. Appendix to Report No. 138.

Duursma, E. K. 1961. Dissolved organic carbon, nitrogen and phosphorus in the sea. Netherlands Journal of Sea Research **1**(1-2):1-47.

El-Sayed, S. Z. 1974. Primary productivity and standing crop of phytoplankton in the Gulf of Mexico. Gulf Universities Research Consortium. Appendix to Report No. 138.

———, W. M. Sackett, L. M. Jeffrey, A. D. Fredericks, R. P. Saunders, P. S. Conger, G. A. Fryxell, K. A. Steldinger, and S. A. Earle. 1972. Serial Atlas of the Marine Environment-Chemistry, Primary Productivity, and Benthic Algae of the Gulf of Mexico. American Geographical Society, folio 22, p. 6.

Environmental Protection Agency. 1971. Methods for Chemical Analysis of Water and Wastes. Cincinnati, Ohio: EPA Water Quality Office, p. 24.

Farrell, Douglas. 1974. Benthic communities in the vicinity of producing oil wells in Timbalier Bay, Louisiana. Gulf Universities Research Consortium. Appendix to Report No. 138.

Fish, A. G. and P. L. Lewis. 1974. A study of the effects of environmental factors upon the distribution of the Foraminifera, nematode and polychaetous annelid meiofauna of Timbalier Bay, Louisiana. Gulf Universities Research Consortium. Appendix to Report No. 138.

Fredericks, A. D. and W. M. Sackett. 1970. Organic carbon in the Gulf of Mexico. Journal of Geographic Research **75**:2199.

Garrett, W. D. 1967. The organic chemical composition of the ocean surface. Deep-Sea Research **14**:221-227.

Harris, A. 1974. Marine environment conditions that could affect Louisiana shrimp production. *In* Proceedings of the 15th Annual Convention, Louisiana Shrimp Association, New Orleans, Louisiana, March 15.

Humm, H. J. 1974. The effect of the offshore oil and gas wells on the benthic marine plants of Louisiana. I. Timbalier Bay, II. Offshore. Gulf Universities Research Consortium. Appendix to Report No. 138.

Jeffrey, L. M. and D. W. Hood. 1958. Organic matter in seawater: An evaluation of various methods of isolation. Journal of Marine Research **17**:247-271.

Kritzler, Henry. 1974. Oil production and polychaetous annelids in a Louisiana estuary. Gulf Universities Research Consortium. Appendix to Report No. 138.

Laseter, J. L. and E. J. Ledet. 1974. Hydrocarbons and free fatty acids associated with the air/water interface, sediments, and beaches of the Timbalier Bay and offshore Louisiana area. Gulf Universities Research Consortium. Appendix to Report No. 138.

Livingstone, D. A. 1963. Chemical Composition of Rivers and Lakes. Data of Geochemistry. Professional Paper 440-G, U.S. Geological Survey: G1-G64.

Marum, J. P. 1974. Effects of long-term oil drilling and production on the zooplankton of Louisiana coastal waters. Gulf Universities Research Consortium. Appendix to Report No. 138.

MacIntyre, F. 1974. The top millimeter of the ocean. Scientific American **230**:62-77.

Menzel, D. W. 1967. Particulate organic carbon in the deep sea. Deep-Sea Research **14**:229-238.

——— and J. H. Ryther. 1964. The composition of particulate organic matter in the western North Atlantic. Limnology and Oceanography **9**(2):179.

——— and J. H. Ryther. 1968. Organic carbon and the oxygen minimum in the South Atlantic Ocean. Deep-Sea Research **15**:327-337.

——— and R. F. Vaccaro. 1964. The measurement of dissolved organic and particulate carbon in sea water. Limnology and Oceanography **9**:138-142.

Menzies, R. J. 1973. A Handbook of Procedures and Methods Employed in the Offshore Ecology Investigation (OEI). Gulf Universities Research Consortium. Report No. 121.

Oetking, Philip, R. Back, R. Watson, and C. Merks. 1974. Currents, hydrography, surface and shallow subsurface sediments on the nearshore continental shelf of South Central Louisiana. Gulf Universities Research Consortium. Appendix to Report No. 138.

Oppenheimer, Carl. 1974. Hydrocarbons in seawater and organisms and microbiological investigations. Gulf Universities Research Consortium. Appendix to Report No. 138.

Riley, G. A. 1937. The significance of the Mississippi River drainage for biological conditions. Journal of Marine Research **1**:60-74.

Skopints, B. A. 1968. Organic carbon in waters of the Norwegian Sea and of the Northeast Atlantic. Oceanolog, USSR **8**:178.

Williams, P. M. 1967. Sea surface chemistry: Organic carbon and inorganic nitrogen and phosphorus in surface films and subsurface waters. Deep-Sea Research **14**:791-800.

Wilson, R. F. 1963. Organic carbon levels in some aquatic ecosystems. Publications of the Institute of Marine Science, The University of Texas **9**:64-76.

HYDROCARBONS AND FREE FATTY ACIDS ASSOCIATED WITH THE AIR/WATER INTERFACE, SEDIMENTS, AND BEACHES OF THE TIMBALIER BAY AND OFFSHORE LOUISIANA AREA

by J. L. Laseter and E. J. Ledet

ABSTRACT

Some 5,800 compounds were isolated and analyzed from 224 sediment and air/water interface samples taken during the Offshore Ecology Investigation. The objective was to characterize the nature and distribution of hydrocarbons and related organic substances and to distinguish them from pollution-derived hydrocarbons resulting from petroleum activity in the area. The results of these investigations, including the sampling methods developed and the results of the air/sea interface studies, were published in three papers. These papers are reprinted herein, with minor changes, as they originally appeared in *Analytical Chemistry, Analytical Letters,* and *Science*. The authors and the Gulf Universities Research Consortium are grateful to the original publishers for permission to reprint the papers.

I. NEW SAMPLING DEVICE FOR THE RECOVERY OF PETROLEUM HYDROCARBONS AND FATTY ACIDS FROM AQUEOUS FILMS

by Russell Miget, Howard Kator, Carl Oppenheimer, John L. Laseter, and Enoch J. Ledet

---

At the time of this investigation, John Laseter and E. J. Ledet were at the Department of Biological Sciences, University of New Orleans, New Orleans, LA.

Replacement of coal by petroleum as the major world source of energy has resulted in increasing amounts of petroleum products being released into the environment. The need to quantify and to identify petroleum hydrocarbons in natural waters has presented methodological difficulties—especially with hyrocarbon films at the air/water interface. Since petroleum spills generally result in surface slicks of non-uniform thickness that can cover relatively large areas, a surface film sampler was required that would permit rapid, consistent, and efficient retrieval of surface hydrocarbons.

We have developed an inexpensive sampler which fulfills these requirements, and which can be operated under moderately rough surface conditions. Field comparison of this sampler with the screen technique (Garrett 1965) and the sorbent-in-a-can sampler (Estes et al. 1973) showed it to be easier to use and to require less sampling time. While we have used this sampler for both inshore and offshore field work for the past year under a variety of sea surface conditions, the purpose of this report is to discuss laboratory studies performed to determine the selectivity of the sampler for hydrocarbons and related compounds as a function of molecular structure, and the efficiency of recovery for petroleum films of varied thicknesses.

EXPERIMENTAL

Sampling apparatus and procedure

The surface film sampler (figure 1) consists of a disc of 2-mm Teflon attached to a 4-mm marine aluminum backing by means of 24 aluminum countersunk bolts. The original prototype tested by one of us (Miget) utilized a "heat cured" epoxy coated wooden backing plate. It was later found, however, that the epoxy would contribute small quantities of aromatic compounds in the 220- to 524-molecular weight range to the samples collected. Substitution of marine aluminum eliminated the contamination problem and produced a stronger and lighter weight device. A unidirectional hinge is located in the center of the backside of the aluminum disc. For field use, a wooden pole is attached to the hinge at such an angle that the Teflon face and water surface are parallel.

In order to collect a sample, the device is first touched to the water surface as lightly as possible, and the disc is then rested vertically on a large glass funnel. Organics adhering to the disc are then washed through the funnel into a container by a gentle stream of CCl_4. The sampler is slowly

FIG. 1. PHOTOGRAPH OF THE SAMPLING DEVICE.

rotated and CCl_4 applied in a fine stream from the center of the disc downward to the washed surface. A 30- to 40-ml solvent rinse is generally sufficient, although unusually thick films require more solvent. When crude oil slicks are sampled, the dark oil contrasts against the white Teflon, aiding in an efficient removal. Other than for allowing the solvent to dry, no additional treatment of the Teflon surface is required before taking the next sample. Thus, several samples may be collected at the same site and washed into the same tube for a pooled sample. Further treatment of the samples for qualitative and quantitative analysis will be described in the following sections.

Gas chromatography and mass spectrometry

Gas chromatographic separation of both the paraffinic hydrocarbons from Louisiana crude oil and paraffinic standards was achieved using a stainless steel capillary (270-m × 0.05-cm i.d.) column: initial 100 m coated with 3% Apiezon L and the remainder with 3% OV-17. The column was programmed from 125 °C to 280 °C/min with an initial hold of 15 min. and a final hold for 45 min. The flow rate of 5-6 ml/min was measured at 125°C. A 7620A Hewlett-Packard gas chromatograph equipped with a flame detector was employed. The injector was at 250°C and the detector was at 350°C. No injection split was used. Separation of the aromatic compounds was carried out using a 3.25-m × 2-mm i.d. glass column employing 1% GE SE-30 on 800/100 mesh Chromosorb Q and temperature programmed from 50°C to 200°C at 4°/min. A 5750 Hewlett-Packard gas chromatograph equipped with a flame detector was employed. The injector was at 250°C and the detector at 350°C.

A 9.5-m × 0.318-cm i.d. stainless steel packed column employing ethylene glycol succinate (12%) on Chromosorb P at 180° isothermal was used to separate the fatty acids. A Hewlett-Packard 5750 gas chromatograph was employed with the injector at 250°C and the flame detector at 350°C. The flow was at 18 ml/min at 180°C. The identity of each recovered compound was determined with gas chromatography-mass spectrometry. The system consisted of a Hewlett-Packard model 5750 gas chromatograph connected by means of a jet-type separator to a model 21-491 duPont double focusing mass spectrometer as described previously (Laseter, Lawler, and Griffin 1973). The transfer line and separator were maintained at temperatures of 225° to 250°C. All spectra were taken at 70 eV and recorded by use of an oscillographic recorder and a Digital Equipment Corporation PDP-12 LDP computer. Identifications were made by comparison with authentic laboratory standards. Pristane was obtained from K & K Labs., Inc. (Plainview, N.Y.), whereas phenanthrene and diphenylmethane were obtained from Aldrich Chemical Company (Milwaukee, Wis.). All other standards were purchased from Applied Science Laboratories (College Park, Pa.). Spectrograde solvents were obtained from Matheson Coleman & Bell (Norwood, Ohio).

Qualitative recovery of petroleum paraffinics

The crude oil employed was obtained from the 5380-m depth production zone of a Louisiana offshore well operated by Exxon, located in South Timbalier Block 54 approximately 16 km southwest of Grand Isle, La., at latitude 28.5 N, longitude 90.3 W. The petroleum is typical of the lighter crudes produced in the Southern Louisiana area. Crude oil (0.1 ml) was carefully pipetted onto deionized water (27°C) contained in a stainless steel tray. The water was left undisturbed for 20 minutes prior to sampling, during which time the petroleum spread to form a thin film.

The Teflon surface of the sampler was then gently placed in horizontal contact with the water surface and immediately removed. The petroleum was washed from the disc using spectrograde CCl_4. The solvent was evaporated under reduced pressure at 40°C and the residual material taken up in 5 ml of *n*-heptane for fractionation on silica gel (activated at 250°C for 24 hours) in a 1- × 20-cm column. Prior to use, the column was washed with three column volumes of *n*-heptane. The paraffinic hydrocarbons were eluted with 40 ml of *n*-heptane. The resulting fraction was taken to dryness under a stream of purified nitrogen. A duplicate 0.1-ml volume of crude oil used as a control was taken up in 100 ml of CCl_4 and treated as above, with the exception that the sampler was not employed. In both cases, the residues were dissolved in 50 μl of *n*-pentane and injected directly on a capillary column for gas chromatographic analysis (figure 2).

Qualitative recovery of unsaturated and branched chain paraffins, aromatics, and free fatty acids

Pristane (2, 6, 10, 14-tetramethylpentadecane), 1-hexadecane, *n*-hexadecane, 3-methylhexadecane, and 2-methylhexadecane were dissolved in *n*-hexane. Aliquots containing from 50 to 100 μg of each hydrocarbon were carefully pipetted onto the water surface and recovered using the sampler. Following solvent removal as described above, the residue was injected onto a capillary column (figure 3). An aliquot of the original mixture was used as the non-sampled control. Recovery of aromatic hydrocarbons was likewise tested using diphenylmethane and phenanthrene with *n*-octadecane as an internal standard. After solvent removal, the organics in the recovered residue were dissolved in benzene and injected directly onto a packed column which employed 1% GE SE-30 (figure 4). An aliquot of the original mixture was chromatographed as the control. Approximately 250-750 μg each of free stearic, oleic, and linoleic acids along with *n*-eicosane were also recovered from the water surface using the sampler. The recovered residue after solvent removal was taken up in 3 ml of BF_3-methanol and methylated (Laseter, Lawler, Walkinshaw, and Weete 1973). The methyl esters and *n*-C_{20} standard recovered from the methylation procedure were taken up in 50 μl of benzene and samples directly injected onto a packed gas chromatographic column employing 12% ethylene glycol succinate (figure 5). A portion of the original fatty acid-alkane mixture was methylated directly and served as the control.

Quantitative recovery of a Louisiana crude oil

The crude oil was artificially "weathered" by evaporation under low heat (50 °C) for 2 hours. This process essentially removes all compounds with boiling points below *n*-C_{16}. A small volume of the "weathered" petroleum (0.1 ml) was mixed with 5 ml of benzene. The oil in 50 μl of this mixture formed an incomplete sheen on distilled water contained in an aluminum pan (30-cm diam. × 20-cm deep) after the benzene evaporated. A 100-μl aliquot left a slightly broken sheen, whereas the oil in 200- and 300-μl aliquots covered the water surface with a continuous film. A sampler with the same diameter as the pan was used to recover the oil so that the uneven distribution of the films would not affect the

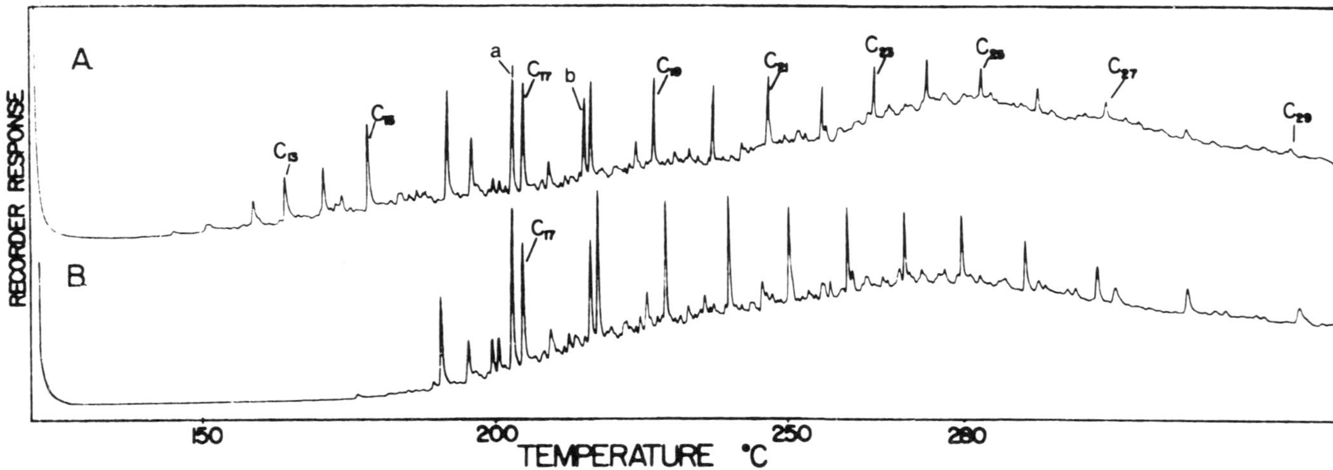

FIG. 2. GAS CHROMATOGRAPHIC SEPARATION OF THE PARAFFINIC HYDROCARBONS from Louisiana crude oil (Trace A) and from the same crude recovered from water surface by means of the sampling device (Trace B). Pristane and phytane are indicated by 'a' and 'b,' respectively.

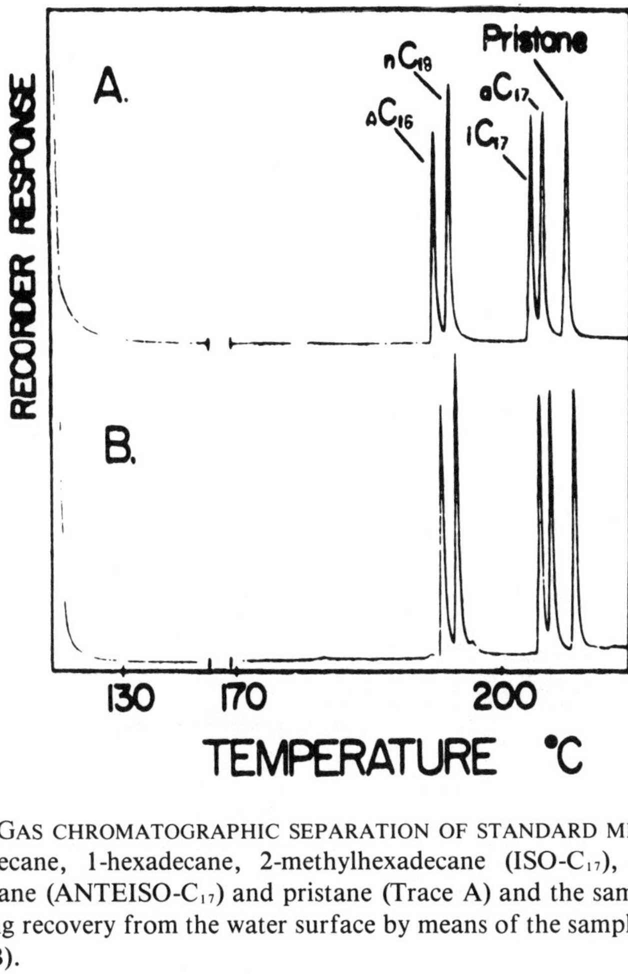

FIG. 3. GAS CHROMATOGRAPHIC SEPARATION OF STANDARD MIXTURE of *n*-hexadecane, 1-hexadecane, 2-methylhexadecane (ISO-C_{17}), 3-methylhexadecane (ANTEISO-C_{17}) and pristane (Trace A) and the same mixture following recovery from the water surface by means of the sampling device (Trace B).

recovery efficiencies. Sets of eight replicate samples for each of four selected film thicknesses were collected. The pan was washed with soap and repeatedly rinsed with cold water between samples. For controls, identical quantities of "weathered" oil from each set were carefully added directly to the Teflon disc using a 50-μl syringe, the solvent was allowed to evaporate (20 °C) and the residual crude washed off with CCl_4 using the recovery procedure described.

Each CCl_4 washing was collected in a glass test tube, the water drawn off the top with a glass pipet, and the solvent plus oil carefully transferred to a tared aluminum weighing dish. The solvent was slowly evaporated (from the dish) on a heating tray (40 °C) in a fume hood.

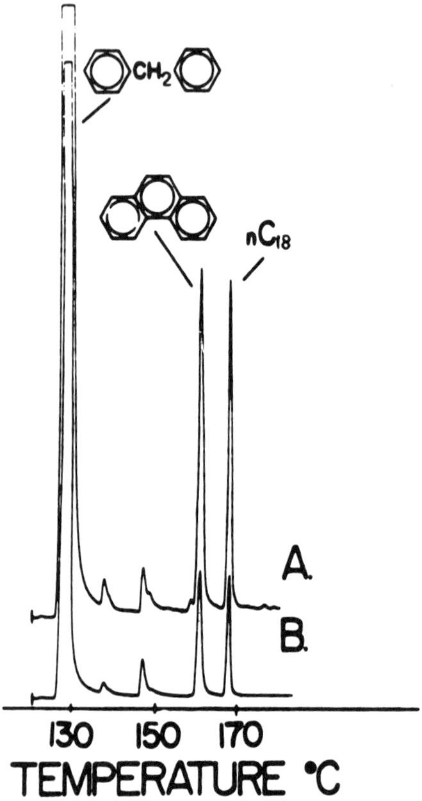

FIG. 4. SEPARATION OF DIPHENYLMETHANE, PHENANTHRENE, AND n-C_{18} STANDARD MIXTURE (Trace A) along with that same mixture following recovery from the water surface by means of the sampling device (Trace B).

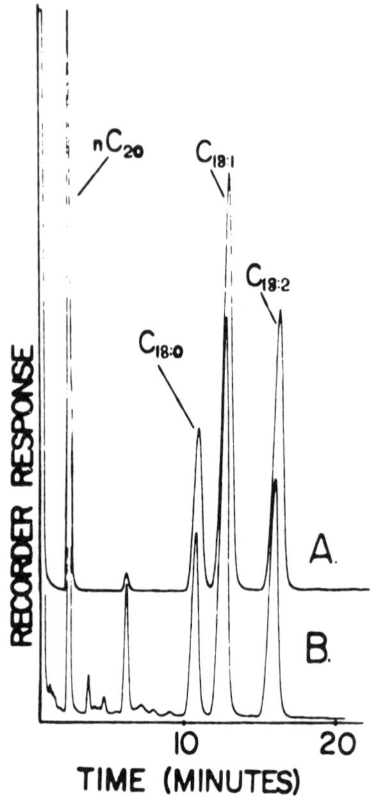

FIG. 5. SEPARATION OF A STANDARD MIXTURE of stearic ($C_{18:0}$), oleic ($C_{18:1}$), and linoleic ($C_{18:2}$) fatty acid methyl esters plus n-eicosane (Trace A) and the same mixture recovered from the water surface by means of the sampling device (Trace B).

After the solvent evaporated, the dishes were placed in a desiccator and weighed until constant weights were obtained. Blanks using CCl_4 were run by evaporating 50 ml of solvent (approximate amount used for each wash) in tared dishes. In a similar experiment, three sets of relatively thicker oil films were quantitatively collected using the sampler. After evaporating the solvent under reduced pressure at 40 °C, each residue was taken up in exactly 100 μl of n-heptane and analyzed using gas chromatography. Analysis was carried out using a Hewlett-Packard 7620A Gas

Chromatograph with flame ionization detectors and 1/8-in. × 6-ft. stainless steel columns packed with 10% UCCW-982 on Chromosorb W. N_2 carrier flow was 40 ml/min. Temperature was programmed from 100° to 300°C at 4°C/min. Peak heights of *n*-paraffins from *n*-C_{17} to *n*-C_{26} were measured, corrected for attenuation factor, and summed. Controls again consisted of oil added directly to the sampler. The efficiency of recovery was calculated using the following formula:

$$\% \text{ Efficiency} = \frac{\text{Av. } \Sigma \text{ peak heights of recovered oil}}{\text{Av. } \Sigma \text{ peak heights of control oil}} \times 100$$

RESULTS AND DISCUSSION

The ability of the sampler to recover crude oil compounds from aqueous surface films is illustrated in tables I and II and figure 2. Not included in table I is the information from control experiments where the weathered crude oil was added directly to the disc surface (100 μl) with an equal volume of oil added to a weighing pan containing approximately 50 ml of CCl_4. The oil was collected from the disc in the usual manner. Following evaporation of the solvents in both cases, the weight of residual oil in the weighing pan and that recovered from the disc were found to be essentially the same, considering the limits of error of the microliter syringe used. The chromatograms in figure 2 clearly demonstrate that low molecular weight alkanes below *n*-C_{16} are not recovered efficiently by the sampler. However, there appears to be no discrimination on either a qualitative or quantitative basis with respect to alkanes above *n*-C_{16}. All major compounds were identified by mass spectrometry.

Figures 3, 4, and 5 respectively illustrate the ability of the sampler to recover normal alkanes, aromatics, and free fatty acids. In each of these experiments, a normal alkane was included as an internal standard to determine if there was selective discrimination between it and the other organics being evaluated by the Teflon surface. The data suggest that the sampler does not discriminate with respect to branching, saturation, or aromatic nature of the hydrocarbons tested. Also, the presence of a carboxyl group on a hydrocarbon chain containing one, two, or no double bonds does not appear to influence the recovery of the compound. In all instances, the quantitative recovery of the various classes of organics tested in the experiments ranged from 87% to 96%. It should be noted that extensive secondary washing of the Teflon surface did not yield additional organics.

Table I. Retrieval Efficiency of the Surface Film Sampler Using a "Weathered" Louisiana Crude Oil—Gravimetric Analysis

Amount of oil added, mg, control	Calculated film thickness, nm[a]	Appearance	Amount of oil retrieved, mg	Std dev, mg	Efficiency, %
0.623	10.4	Partial sheen	0.551	±0.035	88
1.216	20.2	Continuous sheen	1.110	±0.045	91
2.532	42.2	Dull sheen	2.165	±0.114	85
4.181	69.6	Continuous film	3.596	±0.251	86

[a] Density = 0.85 g/ml.

Table II. Retrieval Efficiency of the Surface Film Sampler Using a "Weathered" Louisiana Crude Oil—Gas Chromatographic Analysis

Amount of oil added, control, mg	Calculated film thickness, nm[a]	Appearance	Peak height control	Peak height retrieved	Std dev	Efficiency, %
4	66.6	Continuous film	158	130	±7.77	82
13	216.5	Continuous film	309	277	±13.43	89
30	499.6	Continuous film	232	205	±9.19	88

[a] Density = 0.85 g/ml.

Table I shows the quantitative recovery of a "weathered" Louisiana crude oil using the sampler. The chromatograms in this experiment were similar to those in figure 2 and revealed no discrimination in the recovery of alkanes above n-C_{16}.

These laboratory studies indicate that the sampler should prove to be valuable for both quantitative and qualitative retrieval of surface film paraffins, aromatics, and fatty acids in the environment. The sampler has been employed in actual field exercises to collect surface organics from offshore Louisiana, offshore Florida, and Lake Pontchartrain. Analogously, the ability of the sampler to collect petroleum-type organics suggests that it will also be useful in collecting related surface film organics such as pesticides.

ACKNOWLEDGMENT

The authors wish to thank D. Carlisle, C. W. Schuler, and R. Evans for technical assistance.

REFERENCES CITED

Estes, J. E., P. G. Mikolaj, R. R. Thaman, and L. W. Senger. 1973. Volumetric determination of marine oil spills using coordinated airborne and surface sampling data. *In* Proceedings of the 1973 Conference on Prevention and Control of Oil Spills. Washington, D.C.: American Petroleum Institute.

Garrett, W. D. 1965. Collection of slick-forming materials from the sea surface. Limnology and Oceanography **10**(4):602-605.

Laseter, J. L., G. C. Lawler, and G. W. Griffin. 1973. Influence of methyl substitution on mass spectra of diphenylmethane: analytical applications. Analytical Letters **6**(8):735-744.

———, G. C. Lawler, C. H. Walkinshaw, and J. D. Weete. 1973. Fatty acids of *Pinus elliottii* tissues. Phytochemistry **12**(4):817-821.

II. A COMPARISON OF TWO SAMPLING DEVICES FOR THE RECOVERY OF ORGANICS FROM AQUEOUS SURFACE FILMS

by Enoch J. Ledet and John L. Laseter

INTRODUCTION

Observance of monomolecular films associated with the air-sea interface have been described by investigators for a number of years. It has been established that organics at an interface will exert an influence on many important exchange processes with the atmosphere by changing the surface potential, tension, and viscosity (Garrett 1972). The need to quantify and identify petroleum hydrocarbons in natural waters has presented methodological difficulties—especially with hydrocarbon films at the air-water interface. Both laboratory and offshore studies have been performed to determine the selectivity of a Teflon disc sampler for hydrocarbons and related organics and the efficiency of recovery for petroleum films of varied thicknesses (Miget et al. 1974; Ledet and Laseter 1974). Methods for removal of monolayers of oleic acid, stearic acid, and oleyl alcohol from the surface of both distilled water and synthetic sea water by using a screen sampling technique have also been reported (Garrett 1965). Reports describing the use of a wire screen device to recover pesticides, hydrocarbons, and fatty acids from the air-sea interface have recently been published (Duce et al. 1972; Bioleman and Olney 1974).

This paper will report on a preliminary comparison between the Teflon disc and wire screen techniques on the basis of consistency, efficiency, and general utility of each for retrieval of typical organics at the air-sea interface.

EXPERIMENTAL

Apparatus

Gas chromatographic analyses were carried out with a Hewlett-Packard 7620A chromatograph equipped with a flame ionization detector, connected to a Hewlett-Packard 3370A integrator. Separation of hydrocarbons collected from surface films was achieved on a 243m × .05 cm steel capillary coated with 3% Apiezon L (Applied Science, College Park, Pennsylvania). The column was programmed from 155° to 285°C at 2°/min. with a post-injection hold of 10 minutes and a final hold of 45 minutes. The flow rate measured at 150°C was 6 ml/min. The chart speed of the recorder was set at 0.63 cm/min. The injection port was maintained at 250°C while the flame detector was set at 325°C. Separation of aromatics and fatty acid methyl esters was accomplished on 2.3 m × 0.33 cm steel column coated with UCW-98 (Hewlett-Packard, Avondale, California) programmed from 85° to 230°C at 6°/min. with an initial flow rate of 40 ml/min. The injection port and detector were maintained at 250°C and 325°C respectively.

Sampling devices and procedures

Surface samples from Lake Pontchartrain (a large, brackish body of water opening to the Gulf of Mexico) were obtained concurrently at six different locations by means of a Teflon disc and a 16-mesh stainless steel screen, measuring 27.9 cm in diameter, incorporated in a circular aluminum frame. A 16-mesh size screen was selected because of the high recovery efficiencies reported for monomolecular layers of oleic acid on synthetic sea water (Garrett 1965). Both disc and screen devices were touched to the water's surface as lightly as possible and then rested vertically on a large polyethylene funnel. Organics adhering to the surface of both devices were washed through the funnel and collected in two separate screw top 500 ml narrow-neck glass bottles by a gentle stream of CCl_4 from a polyethylene squeeze bottle. Details of construction and the sampling procedure using the Teflon device are reported in the preceding reprint from *Analytical Chemistry*.

Laboratory recovery experiments consisted of pipetting, either independently or as a mixture, 300 to 500 μg of dodecane, hexadecane, octadecane, palmitic acid, stearic acid, oleic acid, napthalene, and phenanthrene onto deionized water (27 °C) contained in a stainless steel tray (53.3 cm × 47.7 cm × 3.0 cm deep). Benzene was used as the solvent. Each experiment was repeated in triplicate. The disc covered approximately 53% of the tray's surface area while the screen covered approximately 25%. The water was left undisturbed for 20 minutes prior to sampling, during which time the organics spread to form a thin film. Sampling and collection procedures for the laboratory were the same as those described above for the field work. After each collection had been accomplished, the solvent was evaporated under reduced pressure at 40 °C and the residual material taken up in 2 ml of *n*-heptane for fractionation on silica gel (activated at 250 °C for 24 hours and washed with 3 volumes of *n*-hexane prior to use) in a 1 × 20 cm column. Hydrocarbons, aromatics, and free fatty acids were eluted with 40 ml of *n*-hexane, benzene solvent, and methanol respectively. An aliquot of each of the experimental compounds dissolved in benzene was used as a control. The solvent was removed and the residue taken up in 150 ml of CCl_4 and treated as above with the exception that the sampling devices were not employed. Free fatty acids were methylated prior to injection into the gas chromatograph (Morrison and Smith 1964).

In both cases, the residues collected by each device were dissolved in 10 μl of *n*-heptane and injected directly onto a capillary or packed steel column for gas chromatographic analysis. Differences between the chromatograms of the control and collected samples would demonstrate the extent of selectivity of each device for the various organics employed. Percentage of recovery was based upon the contact surface area for each of the sampling

TABLE I

ORGANICS RECOVERED FROM THE SURFACE FILM FOR LAKE PONTCHARTRAIN (mg/m^2)[a]

Sampling Device	Hexane Eluate	Benzene Eluate	Methanol Eluate
Screen	0.50	0.14	0.22
Teflon Disc	0.45	0.28	0.56

[a]Averages of six concurrent sample collections for each device.

devices. The Teflon disc was approximately 2.12 times greater in diameter than the screen and therefore a compensation factor was taken into account when percentage of efficiency was calculated.

RESULTS AND DISCUSSION

Both the Teflon disc and wire screen were relatively easy to operate. Field and laboratory comparisons between the two collection devices, however, showed the disc to require substantially less sampling time. A longer time period was necessary with the screen in order to prevent sample loss and spillage.

The information in table I suggests that the screen is approximately as effective as the Teflon disc in the recovery of hydrocarbons from the air-sea interface. However, care must be taken in placing a strict interpretation on this observation due to the fact that elemental sulfur will also appear in the *n*-hexane fraction. It therefore becomes necessary to inspect the *n*-hexane fractions in chromatographic detail in order to determine if any qualitative differences exist in the hydrocarbons collected. A typical chromatographic separation is shown in figure 1. One will note that initially there is marked discrimination of the *n*-alkanes in the low molecular weight portion of the sample collected by the Teflon disc. This observation was reported previously (Garrett 1972). The screen does not appear to be as restricted with respect to *n*-alkanes below 16 carbons. The screen shows hexadecane as the major *n*-alkane, whereas it is octadecane for the Teflon disc. Table II supports these observations derived from the field samples. For example, the screen is 3 times more effective at retrieval of dodecane, whereas the Teflon disc is over twice as efficient as the screen at recovery of octadecane.

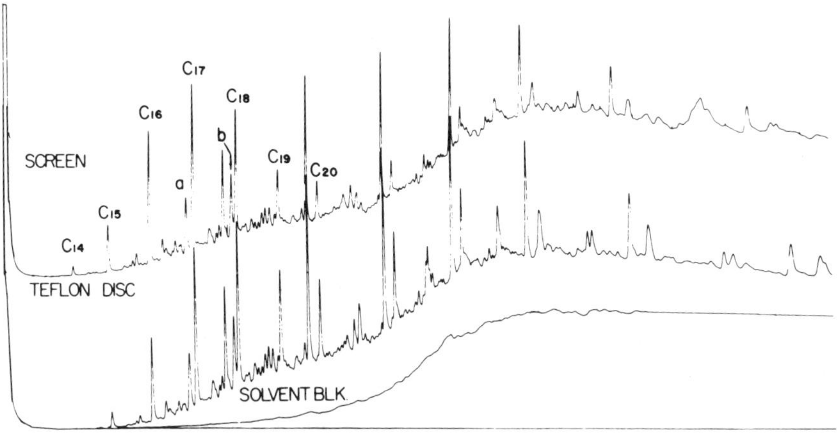

FIG. 1. GAS CHROMATOGRAPHIC SEPARATION OF HYDROCARBONS found in the *n*-hexane eluate from a typical surface sample collected by the screen (upper trace) and Teflon disc (middle trace). Details of chromatographic operation are described in the text. Pristane is indicated by 'a' and phytane by 'b.'

Also of interest is the retrieval capacity of normal alkanes with respect to the methyl branched alkanes. Pristane (2, 6, 10, 14-tetramethylpentadecane) and phytane (2, 6, 10, 14-tetramethylhexadecane) ratios appear to be about the same in each case, as is their respective

TABLE II

PERCENT EFFICIENCY FOR RECOVERY OF ALKANES, AROMATICS, AND FREE FATTY ACIDS FROM SURFACE FILMS ON DEIONIZED WATER

Compound Recovered	%Efficiency[a] Wire Screen	Teflon Disc
Alkanes	6.80	2.20
n-dodecane	7.98	8.90
n-hexadecane	25.05	60.89
n-octadecane		
Aromatics		
napthalene	4.50	12.80
phenanthrene	32.99	29.45
Fatty Acids		
palmitic	65.95	99.53
stearic	87.31	71.64
oleic	44.08	71.47

[a]Values represent an average of triplicate experiments.

relationship to hexadecane. Sampling devices used to study the environment must provide an accurate picture of such ratios. This is because significance is frequently attached to such ratios as an aid in differentiating between petroleum-derived and indigenous alkanes.

It is of interest to note in table I that the efficiency of the recovered organics from the surface of Lake Pontchartrain, in which the Teflon disc was employed, increased in relationship to increased polarity of the test compound. Table II shows that for the aromatics employed in this study, napthalene is recovered approximately 3 times more efficiently by the Teflon sampler. No significant difference was noted between the two devices for phenanthrene. In the case of the free fatty acids, however, the Teflon disc was more efficient at recovering the saturated acids. There appears to be a relationship between chain length and the efficiency of recovery with both sampling devices. One *cis* double bond markedly improves the ability of the wire screen to retrieve a free acid when compared to a saturated acid of equal carbon number. No significant difference in efficiency between oleic and stearic acids was observed when the disc was employed.

In summary, it appears that the wire screen sampling device is more effective in recovering low molecular weight ($<C_{16}$) *n*-alkanes from surface films in the concentrations employed in this study. Little discrimination is observed in higher molecular weight normal or methyl branched alkanes by either device. Both samplers are somewhat limited in their ability to recover polycyclic aromatics. The Teflon disc is superior at recovering saturated free fatty acids, whereas both devices were about equal with respect to their ability to recover oleic acid. The values observed for percentage of efficiency of oleic acid from synthetic sea water were similar to those reported previously for a 16-mesh screen (Garrett 1965).

Our findings suggest that care should be exercised in selecting any device for environmental sampling of organics associated with aqueous surface films. Because of the ease of operation and ruggedness of the Teflon disc, it appears to have several advantages over the wire screen used in this study. However, in any study of the organics at the air-water interface the particular device employed should be calibrated for the compound or compounds to be collected in order to ensure as much sampling accuracy as possible.

REFERENCES CITED

Bidleman, T. F. and C. E. Olney. 1974. Chlorinated hydrocarbons in the Sargasso Sea atmosphere and surface water. Science **183**(4124):516-518.

Duce, R. A., J. G. Quinn, C. E. Olney, S. R. Piotrowicz, B. J. Ray, and T. L. Wade. 1972. Enrichment of heavy metals and organic compounds in the surface microlayer of Narragansett Bay, Rhode Island. Science **176**(4031):161-163.

Garrett, W. D. 1965. Collection of slick-forming materials from the sea surface. Limnology and Oceanography **10**(4):602-605.

———. 1972. Impact of Petroleum Spills on the Chemical and Physical Properties of the Air-Sea Interface. NRL Report 7372. Washington, D.C.

Ledet, E. J. and J. L. Laseter. 1974. Alkanes at the air-sea interface from offshore Louisiana and Florida. Science **186**:261-263, reprinted in this volume, pp. 000-000.

Miget, R., H. Kator, C. Oppenheimer, J. L. Laseter, and E. J. Ledet. 1974. New sampling device for the recovery of petroleum hydrocarbons and fatty acids from aqueous surface films. Analytical Chemistry **46**:1154-1157, reprinted in this volume, pp. 000-000.

Morrison, W. R. and L. M. Smith. 1964. Preparation of fatty acid methyl esters and dimethylacetyls from lipids with boron fluoride-methanol. Journal of Lipid Research **5**(4):600-608.

III. ALKANES AT THE AIR-SEA INTERFACE FROM OFFSHORE LOUISIANA AND FLORIDA

by Enoch J. Ledet and John L. Laseter

In recent years there has developed considerable interest in the distribution and impact of hydrocarbon residues introduced into the marine environment by petroleum production and maritime activities. Studies of the C_1 to C_5 hydrocarbons dissolved in the surface waters of the Gulf of Mexico show that the highest concentrations are apparently associated with shipping and petroleum activities (Brooks et al. 1973). The presence of dissolved paraffins of higher molecular weights has also been reported in the Gulf of Mexico (Parker et al. 1972). Koons and Monaghan (1973) concluded that the hydrocarbon content of the water in the northern Gulf of Mexico is probably less than 7 μg/l. Concentrations of total dissolved hydrocarbons in the area of the east central Atlantic vary from 10 to 140 μg/l (Barbier et al. 1973), and 2 to 13 μg/l is reported for the Nova Scotia vicinity (Levy 1971). In the last case alkanes from C_{14} to C_{37} were present, and there was no preference for odd or even carbon numbers.

Garrett (1967), using both a screen technique to sample the air-sea interface and a bucket sample, found that a variety of organics was present at a number of locations. More recently, samples collected at the air-sea interface in the area of the Sargasso Sea by use of a stainless steel screen were found to contain a variety of chlorinated hydrocarbons, and the highest concentrations were associated with the surface microlayer (150 μm) (Bidleman and Olney 1974). Duce et al. (1972), using similar techniques, showed that hydrocarbons and fatty acids, as well as pesticides, are more concentrated in the surface microlayer of Narragansett Bay than in water as deep as 20 cm below the surface.

It has been established that organics affect the physical properties of the ocean surface and many important exchange processes between the ocean and the atmosphere (Garrett 1968). For example, they reduce the capillary wave spectrum and thereby contribute to the production of sea slicks. Also, during any type of petroleum-related accident or oil spill it is the surface layer that is initially disrupted. Because of the importance of this microlayer and the general lack of qualitative and quantitative data, it is important to characterize the indigenous chemical components found at the air-sea interface. This report deals with the nature and distribution of alkanes at the air-sea interface in selected areas of Timbalier Bay and offshore Louisiana and Florida. Samples were collected at five intervals between August 1972 and July 1973.

An aluminum-backed Teflon disc was employed for sample collection. Approximately 90% of weathered crude can be recovered by this technique, and there appears to be little or no discrimination in retrieving petroleum paraffins above C_{15} (Miget et al. 1974). Residues collected were fractionated by using a silica gel column (Laseter, Lawler, and Griffin 1973; Laseter, Lawler, Walkinshaw, and Weete 1973), and the alkanes were eluted with *n*-heptane. After removal of the solvent under a stream of purified nitrogen, the residue was weighed and further fractionated by gas chromatography. A portion of each sample eluted from the gas chromatographic column was passed through a DuPont 21-491 mass spectrometer attached to a PDP-12 computer. Alkanes from local production facilities and fuel and lubrication oils from the ships used to collect samples were also analyzed as possible sources of sample contaminants. The average (dry) weight of the *n*-heptane eluate was 0.70 mg per square meter of sea surface for residues collected from offshore Louisiana during five field experiments, whereas the 43 samples from the two exercises in Timbalier Bay averaged 0.36 mg/m^2.[1] Similar fractions averaged 0.18 mg/m^2 for Florida, while ten samples collected at the same time from Louisiana averaged 0.21 mg/m^2. No significant relation between proximity to a drilling or production platform and the quantity of alkanes at the surface was noted.

Figure 1 illustrates a typical distribution of compounds observed in a surface sample; more than 90% of the compounds are alkanes, and 118

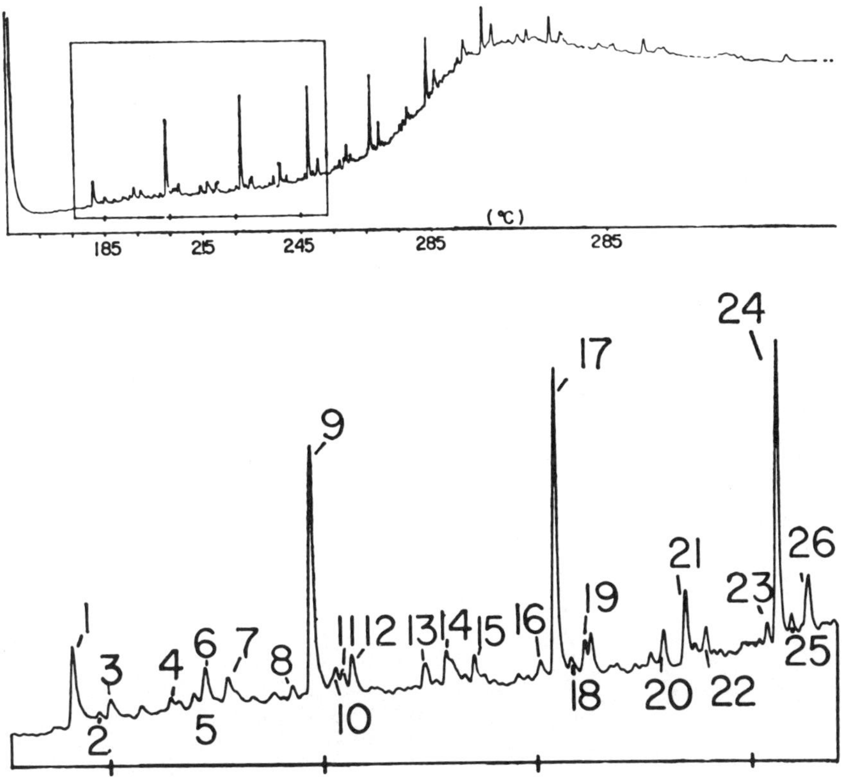

FIG. 1. GAS CHROMATOGRAPHIC SEPARATION OF A TYPICAL PARAFFIN FRACTION collected from the air-sea interface of offshore Louisiana in October 1972. Separation was on a stainless steel capillary column 250 m long, 0.5 mm in inner diameter, and coated with Apiezon L. A flame detector was used. The chromatograph was held at 155 °C for 10 minutes and programmed from 155° to 285 °C at 2 °C per minute. The injector was at 250 °C and the detector at 325 °C. Numbered components are identified in table 1.

such samples were analyzed. Samples obtained during a single collection trip showed only minor variations in the distribution of alkanes, qualitatively or quantitatively. Treatment with a molecular sieve to remove normal alkanes followed by hydrogenation to remove the unsaturated compounds failed to alter significantly the observed chromatographic pattern. The sample represented in figure 1 contained about 70% branched alkanes, of which 50% by weight are 3-methyl branched, 13% are cycloalkanes, and 3% are normal alkanes. The components num-

Table 1. Mass spectrometric identification of the components in Fig. 1.

Chromatogram peak number	Compound
1	3-Methylpentadecane
2	Nonadecylcyclohexane
3	*n*-Hexadecane
4	3-Methylhexadecane
5	Decylcyclohexane
6	Pristane
7	2,6-Dimethylpentadecane
8	Dimethyloctadecane
9	3-Methylheptadecane
10	Undecylcyclohexane
11	Phytane
12	*n*-Octadecane
13	3-Methyloctadecane
14	Dodecylcyclohexane
15	2,6-Dimethylheptadecane
16	Dimethyleicosane
17	3-Methylnonadecane
18	Tridecylcyclohexane
19	Dimethyleicosane
20	3-Methyleicosane
21	Tetradecylcyclohexane
22	2,6-Dimethylnonadecane
23	Dimethyldocosane
24	3-Methylheneicosane
25	Pentadecylcyclohexane
26	Dimethyldocosane

bered in figure 1 are identified in table 1. To date, some 50 compounds ranging from C_{16} to C_{36} in carbon number have been identified by mass spectrometry. When normal alkanes were present and in sufficient concentration to be detected, they usually ranged from C_{16} to C_{21} and exhibited no odd-even carbon preference. In only a few samples, collected during August 1972 from offshore Louisiana and during October 1972 from Timbalier Bay, did the quantity of normal alkanes exceed 20% by weight of the paraffinic fraction. Samples collected at approximately the same time and location from a 10-m depth were found to contain predominantly normal alkanes from C_{14} to C_{36} (Oppenheimer et al., unpublished). Tar ball samples from the Gulf of Mexico, even though they vary considerably in chemical composition, show substantial proportions of normal alkanes (Koons and Monaghan 1973).

Our data suggest that the branched and cyclic paraffins comprise the bulk of alkanes at the air-sea interface, whereas the normal paraffins accumulate in the water column. There are several possible explanations for this enrichment, such as selective removal by autooxidation or photooxidation, emulsification by wave action, and adsorption on particulate matter. However, one of the more important factors may be biological. Marine bacteria preferentially oxidize normal alkanes (Kator 1972; Jobson et al. 1972), and microbial attack is most effective against oil in thin films or adsorbed on solid particles (Miget et al. 1969). The pristane/phytane ratio in the alkane fraction at the air-sea interface (which varies from 1.5 to 2.3) and the presence of a large number of dimethylalkanes suggest that crude oil or petroleum products may serve as the hydrocarbon precursor pool. If so, however, it is difficult to explain the presence of such high concentrations of 3-methyl branched alkanes with even carbon numbers. Certain plants contain substantial quantities of 3-methyl branched alkanes with even carbon numbers ranging from C_{28} to C_{34} (Weete et al. 1971), and it has been suggested that such alkanes occur in marsh plants that are common to the Gulf Coast (Lytle et al. 1973). We suggest that the alkanes at the air-sea interface are probably entering the marine environment from a wide variety of biological sources as well as from petroleum released by natural seeps and man's activities, and that branched and cyclic alkanes accumulate at the interface as a result of a complex combination of physical and biological influences. The data that have been collected so far show that the types and amounts of hydrocarbons in the surface film along the northern Gulf Coast have remained fairly constant during the study period.

NOTE

1. Gas chromatographic methods revealed that the alkanes present in the heptane eluate averaged 35 $\mu g/m^2$. This apparent discrepancy with the total weight values reported for this fraction is probably due to a number of factors: (i) a substantial portion of the alkanes present are not chromatographically resolved and are expressed as a base-line "hump"; (ii) some of the components may have molecular weights above C_{40} and do not emerge below 280°C; and (iii) elemental sulfur (produced in large amounts in the vicinity) will concentrate in the heptane eluate.

REFERENCES CITED

Barbier, M., D. Joly, A. Saliot, and D. Tourres. 1973. Hydrocarbons from sea water. Deep-Sea Research **20**:305-314.

Bidleman, T. F. and C. E. Olney. 1974. Chlorinated hydrocarbons in the Sargasso Sea atmosphere and surface water. Science **183**(4124):516-518.

Brooks, J. M., A. D. Fredericks, W. M. Sackett, and J. W. Swinnerton. 1973. Baseline concentrations of light hydrocarbons in the Gulf of Mexico. Environmental Science and Technology **7**(7):639-642.

Duce, R. A., J. G. Quinn, C. E. Olney, S. R. Piotrowicz, B. J. Ray, and T. L. Wade. 1972. Enrichment of heavy metals and organic compounds in the surface microlayer of Narragansett Bay, Rhode Island. Science **176**(4031):161-163.

Garrett, W. D. 1965. Collection of slick-forming materials from the sea surface. Limnology and Oceanography **10**(4):602-605

———. 1968. The influence of monomolecular surface films on the production of condensation nuclei from bubbled sea water. Journal of Geophysical Research **73**:5145-5150.

———. 1972. Impact of Petroleum Spills on the Chemical and Physical Properties of the Air-Sea Interface. NRL Report 7372. Washington, D.C.

Jobson, A., F. D. Cook, and D. W. S. Westlake. 1972. Microbial utilization of crude oil. Applied & Environmental Microbiology **23**(6):1082-1089.

Kator, H. 1973. Utilization of crude oil hydrocarbons by mixed cultures of marine bacteria. *In* D. A. Ahearn and S. P. Myers, eds., Proceedings of Microbial Degradation of Oil Pollutants. Atlanta: Georgia State University Press, p. 47.

Koons, C. B. and P. H. Monaghan. 1973. Petroleum derived hydrocarbons in Gulf of Mexico waters. Transactions of the Gulf Coast Association of Geological Societies **16**(8):170-181.

Laseter, J. L., G. C. Lawler, and G. W. Griffin. 1973. Influence of methyl substitution on mass spectra of diphenylmethane: analytical applications. Analytical Letters **6**(8):735-744.

———, G. C. Lawler, C. H. Walkinshaw, and J. D. Weete. 1973. Fatty acids of *Pinus elliottii* tissues. Phytochemistry **12**(4):817-821.

Ledet, E. J., and J. L. Laseter. 1974. Alkanes at the air-sea interface from offshore Louisiana and Florida. Science **186**:261-263.

Levy, E. M. 1971. The presence of petroleum residues off the east coast of Nova Scotia, in the Gulf of St. Lawrence, and the St. Lawrence River. Water Research **5**(9):723-733.

Lytle, T. F., J. S. Lytle, and P. L. Parker. 1973. Geochemical study of a marsh environment. Gulf Research Report **4**(2):214.

Miget, R. J., H. Kator, C. Oppenheimer, J. L. Laseter, and E. J. Ledet. 1974. New sampling device for the recovery of petroleum hydrocarbons and fatty acids from aqueous surface films. Analytical Chemistry **46**:1154-1157, reprinted in this volume, pp. 265-275.

———, C. H. Oppenheimer, H. I. Kator, and P. A. LaRock. 1969. Microbial degradation of normal paraffin hydrocarbons in crude oil. *In* Proceedings of the Joint Conference on the Prevention and Control of Oil Spills. New York: American Petroleum Institute, pp. 327-331.

Oppenheimer, C., R. Miget, H. Kator. Unpublished results.

Parker, P. L., J. K. Winters, and J. Morgan. 1972. Baseline Studies of Pollutants in the Marine Environment and Research Recommendations: The I.D.O.E. Baseline Conference. Washington, D.C.: National Science Foundation.

Weete, J. D., S. Venketeswaran, and J. L. Laseter. 1971. Two populations of aliphatic hydrocarbons of teratoma and habituated tissue cultures of tobacco. Phytochemistry **10**(5):939-943.

ECOLOGICAL RELATIONSHIPS BETWEEN MARINE MICROORGANISMS AND HYDROCARBONS IN THE OEI STUDY AREA, LOUISIANA

by Carl H. Oppenheimer, Russell Miget, and Howard Kator

ABSTRACT

Seven cruise exercises were conducted as one component of the Offshore Ecology Investigation (OEI) to determine the relationships between microorganisms and hydrocarbons in surface waters. Techniques were developed to take surface samples and to determine bacterial numbers, hydrocarbon content, and the response of indigenous microorganisms to various added hydrocarbon molecules. The numbers of heterotrophic bacteria as determined by dilution techniques varied from 10 to 10,000 per ml. The hydrocarbon-oxidizing organisms varied from 10 to 100,000 per liter with maxima in the month of January. Hydrocarbon concentrations varied from 0.03 to 0.6 $\mu g/l$. There was no correlation between the numbers of oil degraders and hydrocarbon concentrations found. This may be the result of the low concentrations of hydrocarbons *in situ*, which would limit microbial response. The numbers of bacteria and concentrations of hydrocarbons in the OEI study area were similar to those found in control areas in the eastern and western parts of the Gulf. Biological oxygen demand (BOD) experiments conducted to measure the response of the microorganisms to various low and high molecular weight paraffinic and aromatic hydrocarbons

---

Carl Oppenheimer is with the Marine Science Institute, the University of Texas at Port Aransas, Port Aransas, TX. Russell Miget is with the Texas A&M University Research and Extension Center, Corpus Christi, TX. Howard Kator is with the Virginia Institute of Marine Science, Gloucester Point, VA.

indicated that all hydrocarbons were oxidized by some of the mixed indigenous microorganisms. These data suggest that an adequate inoculum was naturally present to respond to oil contamination in nature.

The dominant saturated hydrocarbons and isoprenoids ranged from C_{15} to C_{41}. Water from Timbalier Bay generally contained higher concentrations than offshore samples and exhibited a more pronounced preference of odd over even in the range C_{24} to C_{33}. Gas Chromatograph-Mass Spectrometer (GC-MS) analysis of the dominant paraffins revealed a characteristic bimodal distribution dominated by C_{17} and pristane and C_{25} to C_{35} with an unresolved envelope. No buildup of specific *n*-paraffin molecules was found, a finding supported by the BOD test results where hydrocarbons were added to indigenous microflora.

INTRODUCTION

The ecological significance of the occurrence and distribution of hydrocarbons in a marine environment cannot be considered without discussing the role of microorganisms. While such hydrocarbons in the form of crude oil have been commercially exploited from geological formations since the turn of the century, it was not until the advent of the gas chromatograph that hydrocarbons were found in recent sediments (Smith 1952). ZoBell (1945, 1946a, 1946b, 1950, 1952) and Oppenheimer (1965) have discussed the role of microorganisms in the formation and transformation of hydrocarbons in the marine environment and sediments. Stone and ZoBell (1952) first showed that up to 0.25% of the dry weight of petroleum was produced by several species of microorganisms. Hydrocarbons found in organisms such as fungi (Weete 1972), plants (Stransky et al. 1967), and marine phytoplankton (Blumer et al. 1971) clearly show that hydrocarbons are a common constituent of all living organisms at levels from 10 ppm to 1%.

Meinschein (1959, 1969) has reviewed the origin of hydrocarbons and the natural diversity of saturated, unsaturated, and aromatic species. A summary of world distribution of hydrocarbons in marine waters is given in Myers and Gunnerson (1976).

Up to the time of the OEI, little published work was available on the distribution of hydrocarbons and hydrocarbon-oxidizing microorganisms as related to any given ecological system.

ZoBell (1969) reported on the distribution of hydrocarbon bacteria in water and sediment in Barataria Bay, Louisiana, the Texas coast, and the California coast. The numbers of bacteria ranged from 0.1 to 10^8 per ml of water or mud. Other reports on oil-microbe distribution are presented in the Symposium on the Microbial Degradation of Oil Pollutants (Ahearn and Myers 1973) and table 9 of this paper.

The calculated yearly potential of hydrocarbon production by marine phytoplankton for the world's oceans, using a level concentration of 10 ppb, is approximately 8×10^7 bbls. Estimates of man's introduction by all sources to the oceans is 9×10^6 bbls per year. The report by Myers and Gunnerson cited above indicates a residual hydrocarbon concentration in the waters of the world's oceans to be about 10 ppb or 8.6×10^9 bbls. This is roughly 100 times the annual production from phytoplankton and 1,000 times the annual input from human sources.

It is generally accepted by microbiologists working in the field that the action of hydrocarbon-degrading microorganisms prohibits the accumulation of hydrocarbons in aquatic systems. Davies and Hughes (1968), Crow et al. (1974), Byrom and Beastall (1971), and others have demonstrated the mechanism and rate of degradation of crude oils by various microorganisms. Several reviews on the subject have been published, including those by Butler and Berkes (1972), Gholson et al. (1972), and McKenna and Kallio (1965).

Most of the experimental work on hydrocarbon-oxidizing microorganisms has been accomplished in the laboratory using cultures obtained from various areas historically exposed to oil. No known mechanisms have been shown for anaerobic decomposition; supporting the hypothesis that there is none is the persistence of large quantities of oil in ancient sediments.

This natural production and the accidental addition of hydrocarbons in the marine environment have been balanced by the oxidative activities of several species of microorganisms (Davis 1967). A review of the distribution of hydrocarbons in the waters of the world's oceans by Myers and Gunnerson (1976) suggests persistent levels at the threshold of microbial activity. The low levels of individual molecules, that is, 1 to 10 ppb, suggest that there has been no buildup through the ages of any persistent molecules. It is generally believed that microbial activities have alleviated accumulation of hydrocarbons at high levels except, of course, for the accumulation of crude oil and natural gas in ancient sediments.

The objective of the OEI investigation was to determine the relationship between the distribution of microorganisms and hydrocarbons (1) to detect and evaluate any differences between the offshore platforms and the control site and between the OEI area and areas remote from oil activities, and (2) to evaluate the responsiveness of the OEI area to biological degradation of crude oil contamination.

For most of the microbiological studies, samples were taken at depths of approximately 1.0 m. It was expected that the upper meter of the water column would be well mixed and would probably contain higher concentrations of hydrocarbons than would be found at greater depth in the water column. Yet it was considered desirable to avoid using samples from the air-water interface because it was assumed that, although greater concentrations of hydrocarbons would occur there, the higher spatial and temporal variabilities would reduce the value of comparisons. Also, more detailed investigations of hydrocarbons occurring at the interface were being made under another OEI component investigation in order to determine possible platform-control differences. These "surface" samples were used, then, to enhance the probability of observing differences in the populations of hydrocarbon degraders and other heterotrophs that might be the result of higher hydrocarbon concentrations than could exist near the platforms. Of particular interest were possible increases in the population of hydrocarbon degraders with increasing oil concentration. These studies were supplemented with BOD determinations using samples taken from comparatively large areas of the interface (depths of less than a few cm). In addition to making the platform-control comparisons, we compared the populations of hydrocarbon degraders and other heterotrophs and concentrations of hydrocarbons in the OEI area with those in areas remote from oil activities, using our measurements and those reported by others. Laboratory studies of microbiological activity using water/oil/microbe samples taken from the OEI study area were conducted further to evaluate responsiveness to microbial degradation.

For comparions of hydrocarbon concentrations *per se*, measurements were made at depths of 1.0 and 4.0 m. These were compared with measurements for remote areas in the Gulf and in the world's oceans. The remote areas were far removed from oil drilling and production activities. Again, it was considered preferable to avoid the high spatial and temporal variabilities that would occur if surface samples were used.

Hydrocarbons, being hydrophobic and insoluble, will remain at the air-water interface until they are physically absorbed by heavier particles, reduced to more dense asphaltic compounds, emulsified in the water, or ingested by larger organisms. Devices were developed to take surface samples for both hydrocarbons and microorganisms. Analyses for microorganisms were conducted aboard the research vessel. BOD experiments to determine the response of indigenous microorganisms to added hydrocarbons were also conducted on board. Hydrocarbon analyses and the testing of degradation rates and oil changes by mixed cultures obtained from the enumeration studies were conducted at the Port Aransas Laboratory.

The research area selected for the OEI is one of the most prolific offshore oil-producing areas in the world. Development began in the late

1930s, and in August 1972 there were nearly 6,000 wells working from 1,900 platforms. Through the end of 1973, approximately 3 billion barrels of oil had been produced in the area. In addition, the area has been historically exposed to continuous tanker traffic. During World War II, sinkings in the area from May through July 1942 released an estimated 700,000 barrels of oil. The historical picture of oil production and exposure in the 400 square mile study area provides an excellent setting in which to determine the effects of oil production on a coastal environment.

This paper provides data on (1) the distribution and ecology of hydrocarbon-oxidizing and other heterotrophic microorganisms in the Louisiana coastal environment, and (2) the content and the molecular forms of hydrocarbons in the water column.

METHODS AND MATERIALS

Water sampling methods and stations

The locations of stations where water samples were taken for microorganisms and hydrocarbon analyses are presented in tables 1 and 2 and figures 1 and 2 (pp. 294-295).

Considerable emphasis was placed on obtaining Gulf control stations (50 to several hundred miles from the study sites) for comparative purposes. The designated control area, being only 5 miles from the platform site, might not accurately reflect baseline or background differences in microbial populations or metabolic activities free from the influence of oil drilling and production.

Water samples for hydrocarbons were collected at depths from 1 to 4 m from the surface. Hydrocarbons in surface films were collected with an apparatus described in Miget et al. (1974). All possible care was exercised in using techniques designed to prevent contamination from the ship or from handling.

Hydrocarbons in the water were sampled in a clean 5-gallon glass bottle in a metal frame. The device, attached to hydrographic wire, was dropped into the water over the side of the vessel and allowed to fill. On return to the surface, 100 ml of benzene were added to preserve the sample and a stopper coated with Teflon was used to close the bottle. The sample with benzene was mixed and not opened until extraction took place.

Surface water for BOD studies and bacterial enumeration was collected using a sterilized sampler developed as part of this investigation. The sampler, depending on the sea surface condition, collected the topmost 0.5 to 1.5 cm of the water column. The sampler

TABLE 1 - OEI MICROBIOLOGICAL PROGRAM SAMPLING STATIONS

YEAR	MONTH	STA. #	LOCATION	AREA	GURC LOC.
1972	June	1	28.34 N 93.00 W	N.W. Gulf coast	None
		2	28.32 N 90.18 W	OEI study area	P072
		3	27.37 N 88.30 W	N. Central Gulf	None
		4	27.11 N 85.40 W	N.E. Gulf	None
		5	27.20 N 84.11 W	E. Central Gulf	None
		6	24.46 N 83.16 W	Dry Tortugas area	None
		7	23.53 N 84.08 W	S.E. Gulf	None
		8	24.49 N 88.00 W	Central Gulf	None
1972	Sept.	1	28.51 N 92.10 W	S. Marsh Is. area	None
		3	28.49.53 N 90.23.18 W	S. Tim. 54A	K007
		4	28.53.13 N 90.19.30 W	Desig. control	H062
		6	28.48.30 N 90.27.30 W	S.W. Plat. 54A	K023
		8	28.49.53 N 90.23.18 W	S. Tim. 54A	K007
		9	28.53.13 N 90.19.30 W	Desig. control	H062
1973	Jan.	1	28.49.53 N 90.23.18 W	S. Tim. 54A	K007
		4	28.53.13 N 90.19.30 W	Desig. control	H062
		5	29.12.20 N 90.21.30 W	Tim. Bay	A079
		6	28.54 N 90.45 W	Ship Shoal area	None
		7	29.39 N 93.11 W	East of Sabine R.	None
		8	29.30 N 94.01 W	West of Sabine R.	None
1973	April	1	29.06 N 93.59 W	West of Sabine R.	None
		2	28.50 N 91.20 W	Ship Shoal area	None
		3	28.49.53 N 90.23.18 W	S. Tim 54A (upstrm)	K007
		5	28.49.53 N 90.23.18 W	S. Tim 54A (dnstrm)	K007
		6	29.13 N 90.00 W	11 mi off Grand Is.	C070
		7	28.49.53 N 90.23.18 W	S.T. 54A (Dnstrm)	K007
		8	28.53.13 N 90.19.30 W	Desig. control	H062
		16	29.12.20 N 90.21.30 W	Tim. Bay	A079
1973	July	1	28.55 N 95.05 W	Off Galveston, Tx.	None
		2	29.13 N 93.42 W	West of Sabine R.	None
		4	28.30 N 91.10 W	Ship shoal area	None
		5	28.49.53 N 90.23.18 W	S. Tim 54A	K007
		7	28.53.13 N 90.19.30 W	Desig. control	H062
		8	29.12.20 N 90.21.30 W	Tim. Bay	A079
1973	Oct.	1	29.07 N 92.52 W	East Cameron area	None
		4	28.49.53 N 90.23.18 W	S. Tim. 54A	K007
		7	29.12.20 N 90.21.30 W.	Tim. Bay	A079
		6	28.53.13 N 90.19.30 W	Desig. control	H069
1974	Jan.	1	28.53 N 94.06 W	N.W. Gulf	None
		4	28.49.53 N 90.23.18 W	S. Tim. 54A	K007
		5	28.53.13 N 90.19.30 W	Desig. control	H062
		6	29.12.20 N 90.21.30 W	Tim. Bay	A079

TABLE 2 - n-PARAFFINS IN SEAWATER (Depths average from 1 to 4 meters, water unfiltered)

Date	Station	Location	Concentration Total, g/l	Unresolved Envelope	Major Paraffins and Isoprenoids*
9-72	K040 S.E. 54A	28°50' 90°30'	0.35	present	C-25 to C-33
9-72	H062 Control	28°53' 90°19'	0.21	n.a.	C-15,16,17+pristane, 22,26,27, 28,29
9-72	K040 S.E. 54A	28°50' 90°30'	0.49	present	C-24 to C-33
9-72	K007 54A	28°49' 90'23'	0.34	present	17+pristane,26,27,28,29,30, 31,32
1-73	Timbalier Bay	29°12' 90°21'	0.46	present	C-25,27,28,29,30,31,32,33
1-73	Timbalier Bay	29°12' 90°21'	0.38	present	C-25,26,27,28,29,30,31,32,33
1-73	H062	28°53' 90°19'	0.05	present	C-22,25,26,27,28,29,30,31,32,33
1-73	H062	28°53' 90°19'	0.20	present	C-17+pristane, 18+phytane 19,20,28
1-73	E.Sabine R.	29°40' 93°11'	0.12	present	C-17+pristane, 18+phytane 19,20,21,22
1-73	W.Sabine R.	29°30' 94°01'	0.12	present	C-24,26,27,28,29,30,31,32,33
1-73	Ship Shoal	28°54' 90°45'	0.19	present	C-20,21,22,29,30,31,32,33,34,35
1-73	W.Sabine R.	29°30' 94°01'	0.22	present	C-18+phytane,20,21,22,23,24
1-73	K014 S.W.Plat. 54A	28°50' 90°30'	0.60	present	C-28,29,30,31,32,33,34,35
1-73	K014	28°50' 90°30'	0.19	present	C-17+pristane, C-18+phytane, C-19,20,28,29,30,31,32
1-73	K007 54 Plat.	28°49' 90°23'	0.13	present	C-17+pristane, C-18+phytane, C-19,20,21,22,23,28
4-73	Timbalier Bay	29°12' 90°21'	0.10	present	C-17+pristane, 18+phytane 19,20,22
4-73	Timbalier Bay	29°12' 90°21'	0.37	present	C-17+pristane, 19,27,29,31,33
5-73	P.Aransas 5 mi off-shore	27°46' 92°52'	0.07	absent	C-18+phytane, 28,29,30,31,32, 33,34
7-73	H062	28°53' 90°19'	0.05	absent	15,17+pristane, 18+phytane
7-73	29·03 95 52	29°03' 95°52'	0.09	absent	17+pristane, 18+phytane
7-73	W.Sabine R.	29°30' 94°01'	0.03	absent	17+pristane, 29,31,33,34,35,37
7-73	1.8 mi from Tim.Is.	- -	0.05	absent	17+pristane, 31,34,35,38
7-73	Browns-ville	26°10' 96°40'	0.09	present	15,17+pristane, 18+phytane
10-73	Timbalier Bay	29°12' 90°21'	0.20	present	15,17+pristane, 18+phytane, 19,21,23
1-74	Timbalier Bay	29°12' 90°21'	0.08	present	17+pristane, 18+phytane, 23,25,29,31
1-74	K007 54A	28°49' 90°23'	0.08	absent	17+pristane, 18+phytane, 19,28,29,30,31,32
1-74	H062	28°53' 90°19'	0.09	present	17+pristane
4-74	P.Aransas 5 mi off-shore	27°46' 92°52'	0.07	present	C-17+pristane, 18+phytane, 31,32,33,34,35

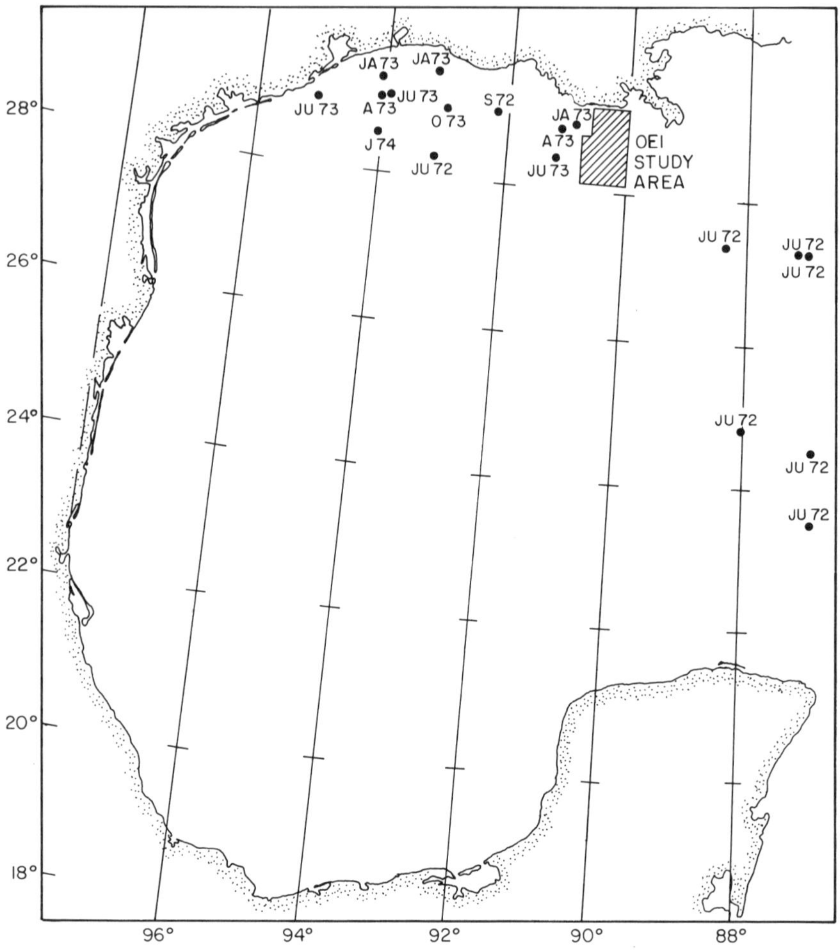

FIG. 1. MICROBIOLOGICAL GULF CONTROL STATIONS.

consists of a stainless steel funnel perforated around its upper perimeter with 0.2 mm diameter holes, which were sandwiched between two circular discs of epoxy-coated plywood that provided flotation at the surface. The apex of the funnel was connected to 50 feet of 1/4″ I.D. silicon tubing, which was connected to a vacuum flask aboard ship. A 1/16″ line connected to the sampler was attached to a boom amidships. It was then gently dropped onto the water and pulled to the end of the boom amidships. The ship was oriented so that the sampler was continuously pushed into clean water. When the floating sampler was at the desired position, a vacuum was pulled on the sterile sampling flask

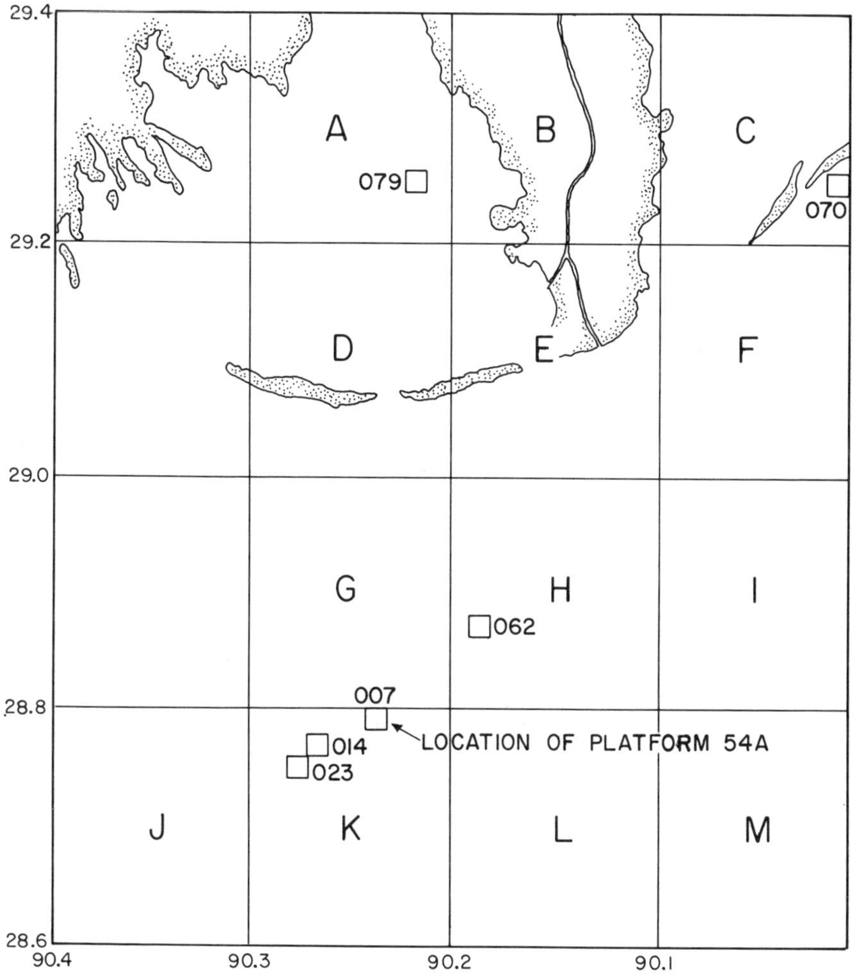

FIG. 2. MICROBIOLOGICAL STATIONS WITHIN OEI STUDY AREA. The numbered quadrants identify the areas where samples were collected.

for bacteria or the 20-liter bottle for BOD, causing the surface water to pass through the perforated top of the funnel between the plywood discs into the flask or bottle.

The same silicon tubing and vacuum flask arrangement was used for subsurface water sampling for bacteria. A sterile glass tube was inserted into the end of the tubing and a sterile weight (hydrocast messenger) was attached to the end of the glass tubing. A tension line was attached to the weight and tubing was dropped to the desired depth.

The tubing was flushed by discarding the first two liters of water, after which the desired volume of sample water was drawn into a sterile vacuum flask.

Hydrocarbon extraction techniques

Hydrocarbons were collected by a liquid-liquid extraction technique performed in the collection bottle. The mouth of the bottle was tapered for a Teflon collar. Heptane (300 ml) was added to the refluxing chamber and heat was applied. The vapors of the heptane passed up the column and down into the bottom of the mixed water. The heptane was released in small bubbles and mixed through the water. It slowly coalesced and floated to the top, whence it returned to the reflux chamber through a constant leveling device. Hydrocarbons were extracted during a 24-hr reflux period. Contamination from the air was prevented by a silica gel filter at the top of the reflux unit.

All solvents were double-distilled to remove contaminants. A 20 μl aliquot of a 0.1 μg/μl nC_{36} internal standard solution was added to the sample. Upon completion of the extraction, the heptane in the extraction flask was evaporated to near dryness on rotary evaporator (40-50° C). The residual solvent was transferred to a 100 ml pear-shaped flask and a small volume of heptane (300 μl) was added. The extract was passed through a pre-washed chromatographic column containing 1 ml of activated silica gel (100-200 mesh, high purity grade, heated for 12 hrs at 200° C, and deactivated with 5% water). Saturated paraffins were eluted with 1.5 ml of heptane (3 column volumes) and aromatic-naphthenic hydrocarbons with 1.5 ml of benzene. Ninety to 95% of *n*-paraffins in a standard mixture with chain lengths greater than nC_{18} were routinely recovered from seawater.

Hydrocarbon extracts were analyzed with a Finnigan 1015C GS-MS for compound identification. Using multiple channel ion monitoring, we selected ions for routine analysis: paraffin ions, 57 m/e and 71 m/e; isoprenoid ion, 113 m/e; olefin ion, 69 m/e; and tropylium ion, 91 m/e. This procedure is useful in the analysis of complex hydrocarbon extracts. Hydrocarbons are identified by a combination of their characteristic retention indices on columns of differing polarity with respect to authentic standards, multiple channel ion monitored mass chromatograms, and mass spectra when feasible.

Enumeration of oil-oxidizing marine bacteria and "total" heterotrophs in the field

Microbiological assays were performed aboard ship immediately after the water samples were collected. The "total" (aerobic-proteolytic) microbial population and the number of hydrocarbon-degrading bacteria were enumerated. "Total" heterotrophs were determined on spread plates and/or membrane filters using 2216E marine medium (ZoBell 1946a). Colonies were counted after one week incubation in the dark at 22-24° C. Populations of hydrocarbon-degrading bacteria were enumerated using the following modified dilution technique: (1) Sample volumes of 0.01 ml or smaller were diluted serially (1:10) in sterile seawater and the appropriate dilution tubes were then added to 50 ml of sterile nutrient salts enriched seawater (ESW) contained in 125 ml incubation flasks (Miget et al. 1969); (2) sample volumes of 0.1 to 10 ml were added directly to 50 ml of ESW in incubation flasks; and (3) sample volumes greater than 10 ml were filtered through sterile 0.45 micron membrane filters, then the filters were cut in small pieces with sterile scissors and added to the incubation flasks containing 50 ml ESW. All flasks, including uninoculated controls, then received 0.5 ml of sterilized crude oil (obtained from a wellhead in ST 54), and were incubated on a gyrotary shaker at the ambient temperature on the ship (22-24° C) for four days. Positive growth (scored minus 4 to plus 4) was indicated by visible oil emulsification, or by increase in turbidity of the water accompanied by a physical change in the oil sheen—all relative to the appearance of the oil and water in the control flasks. Although positive flasks were scored on the completeness of emulsification, the number of oil-degrading microorganisms for each sample was determined by a dilution technique based on replicate dilution flasks showing any change in appearance of the oil or water compared with control flasks.

Evaluation of bacterial hydrocarbon degradation using the biochemical oxygen demand (BOD) technique

BOD studies were carried out as follows: 19-liter sterile glass carboys were filled with sample water from the surface at a given station, shaken vigorously to insure uniform oxygen dispersion, then carefully siphoned into sterile 300 ml standard BOD bottles. Nutrient salts [$(NH_4)_2SO_4$ at 1 g/l and K_2HPO_4 at 0.01 g/l] were added to each bottle from a sterile concentrated stock solution. Hydrocarbon substrates (either pure hydrocarbons or crude oils) were added to the replicate bottles in one of the following two ways: (1) Hydrocarbons that were solid at 20° C were dissolved in benzene and added to approximately 0.1 g of clean, silicon coated, sterile micro glass beads (590-840 micron diameter) contained in small pieces of solvent-cleaned aluminum foil. The solvent was allowed to evaporate in a fume hood and the coated beads were placed in BOD bottles immediately prior to filling with seawater; (2) hydrocarbons that were liquid at 20° C were added directly to filled bottles using a microliter syringe. Hydrocarbon substrates were rapidly injected into the nutrient-enriched water near the bottom of the bottle; the syringe was withdrawn and the bottle was stoppered before the hydrocarbon floated to the surface. The bottles were incubated upside down. All ground glass stoppers were coated with a thin film of silicon grease to insure an airtight seal. Replicate BOD bottles of water from each carboy were immediately preserved and the oxygen concentration referred to as T+O values.

The oxygen concentration in each sample was determined after incubation in the dark by the Winkler method as described in the OEI handbook of procedures and methods (Menzies 1973).

RESULTS

Hydrocarbons in seawater

The concentrations and several salient features describing *n*-paraffins and selected isoprenoid hydrocarbons for the study period are listed in tables 2, 3, 4, 5, and 6 and in figures 3 and 4. Generally, *n*-paraffins were the predominant hydrocarbons found. The isoprenoid hydrocarbons, pristane and phytane, were detected in almost all samples. Despite differences in the distribution of *n*-paraffins and isoprenoids within any one sample, the total *n*-paraffin concentrations for offshore samples were within a relatively narrow range of 0.6 to 0.03 μg/l. Estimates of total saturated paraffins based on integration of the total signal area above the solvent baseline were around one order of magnitude larger than the total paraffins, or in the 5 μg/l range. The range of *n*-paraffins found in seawater was from nC_{15} to nC_{41}. These values are similar to those presented by Myers and Gunnerson (1976).

If any seasonal trend can be discerned, it is that the total *n*-paraffin concentrations decreased from the 9-72 collection period to a minimum in 7-73. This observation is derived from data both from samples in the study area and from relatively large distances away (see table 6 for Sabine River, Port Aransas, and Brownsville). The near-shore waters of the Gulf of Mexico appear to be rather homogeneous. Values for control and platform stations were very similar during the same sampling exercises. Hydrocarbon levels in Timbalier Bay (table 5) were generally higher than offshore. This is undoubtedly related to the high levels of suspended material (algae and organic detritus) in these waters. Carbon Preference Indices (nC_{21} to nC_{33}) reveal carbon preference of odd over even for hydrocarbons in the nC_{25} to nC_{33} region in water from Timbalier Bay (figure 3). A similar preference is observed in recently synthesized hydrocarbons derived from terrestrial marsh plants such as *Batis maritima* and *Salicornia bigelovii* (Lytle, Lytle, and Parker 1973). Crude

(text continues on p. 310)

TABLE 3 - TOTAL AMOUNT AND PERCENT COMPOSITION OF n-PARAFFINS ON DIFFERENT DATES OF COLLECTION PLATFORM 54A AREA

9-72 - K007 54A Platform	4M	1-73 - K007 54A	1M	1-74 - K007 54A	1M
Total n-Paraffins	0.34 μg/l	Total n-Paraffins	0.13 μg/l	Total n-Paraffins	0.08 μg/l
Minor unresolved envelope		Broad envelope present		No Envelope Present	
n-Paraffins	%	n-Paraffins	%	n-Paraffins	%
15	2.0	15	3.4	15	0.0
16	0.0	16	4.0	16	0.0
17+pristane	13.1	17+pristane	8.4	17+pristane	17.6
18+phytane	2.0	18+phytane	8.4	18+phytane	10.8
19	1.0	19	7.8	19	5.6
20	1.1	20	7.6	20	4.6
21	0.7	21	5.4	21	3.8
22	1.1	22	6.6	22	2.9
23	1.3	23	5.3	23	3.4
24	3.7	24	4.4	24	3.3
25	3.4	25	4.4	25	4.9
26	8.7	26	4.4	26	4.9
27	9.5	27	4.5	27	3.8
28	9.8	28	5.7	28	5.6
29	9.5	29	4.5	29	6.1
30	8.6	30	2.6	30	6.4
31	7.9	31	3.4	31	5.1
32	6.2	32	1.4	32	5.7
33	4.6	33	2.1	33	3.4
34	2.9	34	1.7	34	3.3
35	1.8	35	1.9	35	2.2
36 Internal Std.		36 Internal Std.		36 Internal Std.	
37	0.7	37	1.9		
38	0.5				

(continued on next page)

9-72 - K040 S.E. 54A 1M
Total n-Paraffins 0.49 μg/l
Peaked envelope from C-17 on

n-Paraffins	%
17+pristane	1.7
18+phytane	0.8
19	1.8
20	2.8
21	3.0
22	4.1
23	4.5
24	5.0
25	6.3
26	7.1
27	6.2
28	7.2
29	6.5
30	6.6
31	7.6
32	6.6
33	5.9
34	4.5
35	3.6
36 Internal Std.	
37	2.1
38	1.8
39	1.7
40	1.6
41	1.0

9-72 - K040 S.E. 54A 4M
Total n-Paraffins 0.35 μg/l
Broad envelope from C-17 to C-36

n-Paraffins	%
17+pristane	1.2
18+phytane	0.2
19	2.3
20	3.5
21	2.1
22	4.6
23	4.5
24	5.1
25	6.2
26	8.9
27	6.8
28	7.5
29	6.9
30	7.1
31	7.8
32	6.6
33	6.4
34	4.9
35	4.3
36 Internal Std.	
37	3.3

1-73 - K014 S.W. 54A 1M
Total n-Paraffins 0.60 μg/l
Envelope from C-16 to C-36

n-Paraffins	%
16	1.3
17+pristane	1.9
18+phytane	0.5
19	0.7
20	0.5
21	0.6
22	0.6
23	0.6
24	0.5
25	0.6
26	0.9
27	2.8
28	5.4
29	5.1
30	5.9
31	11.3
32	12.8
33	13.9
34	11.7
35	10.4
36 Internal Std.	
37	4.9
38	3.9
39	2.5
40	1.8

1-73 - K014 S.W. 54A 4M
Total n-Paraffins 0.19 μg/l
Envelope from C-17 to C-35

n-Paraffins	%
15	2.9
16	2.5
17+pristane	7.0
18+phytane	6.8
19	6.2
20	5.3
21	3.3
22	3.4
23	2.9
24	4.1
25	4.3
26	4.6
27	3.1
28	6.6
29	6.9
30	6.7
31	8.1
32	5.5
33	3.5
34	2.1
35	2.2
36 Internal Std.	
37	1.3
38	1.2

TABLE 4 - TOTAL AMOUNT AND PERCENT COMPOSITION OF n-PARAFFINS ON DIFFERENT DATES OF COLLECTION
CONTROL AREA H062

9-72	1M	1-73	1M	1-73	4M
Total n-Paraffins	0.21 μg/l	Total n-Paraffins	0.20 μg/l	Total n-Paraffins	0.05 μg/l
		Broad envelope present from C-17 to C-36		Broad envelope present from C-20 to C-35	
n-Paraffins	%	n-Paraffins	%	n-Paraffins	%
15	8.7	15	2.2	17+pristane	0.0
16	6.3	16	3.6	18	0.0
17+pristane	16.3	17+pristane	6.7	19	0.0
18+phytane	3.4	18+phytane	15.5	20	0.0
19	4.3	19	6.1	21	3.9
20	0.0	20	6.4	22	10.0
21	2.6	21	4.1	23	2.5
22	5.8	22	4.5	24	3.3
23	3.6	23	4.9	25	6.0
24	2.3	24	3.6	26	8.8
25	4.4	25	4.0	27	7.4
26	5.1	26	3.6	28	5.3
27	8.1	27	4.0	29	9.3
28	5.0	28	5.3	30	8.2
29	6.2	29	3.7	31	11.6
30	4.5	30	4.3	32	9.5
31	4.2	31	4.9	33	6.8
32	3.1	32	3.7	34	4.7
33	3.0	33	3.4	35	2.8
34	2.6	34	2.4	36 Internal Std.	
35	0.4	35	1.5		
36 Internal Std.		36 Internal Std.			
		37	1.3		
		38	0.7		

Table 4 continued.

7-73	1M
Total n-Paraffins	0.05 μg/l
No envelope	
n-Paraffins	**%**
15	21.4
16	0.0
17+pristane	35.0
18+phytane	12.7
19	4.2
20	2.6
21	2.2
22	2.6
23	1.6
24	1.2
25	1.1
26	1.0
27	1.8
28	1.6
29	1.4
30	1.8
31	2.0
32	1.6
33	1.6
34	1.8
35	1.2
36 Internal Std.	

1-74	1M
Total n-Paraffins	0.09 μg/l
Envelope nC-17 to C-26	
n-Paraffins	**%**
15	0.5
16	2.6
17+pristane	9.5
18+phytane	3.3
19	1.7
20	2.4
21	3.6
22	4.7
23	8.4
24	8.4
25	6.5
26	7.9
27	7.7
28	6.4
29	6.2
30	4.7
31	4.4
32	5.0
33	1.9
34	1.7
35	1.1
36 Internal Std.	
37	1.4

TABLE 5 - TOTAL AMOUNT AND PERCENT COMPOSITION OF n-PARAFFINS ON DIFFERENT DATES OF COLLECTION TIMBALIER BAY

1-73 Total n-Paraffins 0.46 μg/l Broad envelope C-17 to C-30		1-73 Total n-Paraffins 0.38 μg/l Envelope present		4-73 Total n-Paraffins 0.37 μg/l Broad envelope present	
n-Paraffins	%	n-Paraffins	%	n-Paraffins	%
17+pristane	1.4	15	2.2	15	1.7
18+phytane	2.6	16		16	3.2
19	0.3	17+pristane	1.7	17+pristane	6.7
20	0.8	18+phytane	2.8	18+phytane	3.4
21	0.9	19	3.7	19	5.2
22	3.0	20	3.7	20	3.6
23	2.0	21	4.2	21	4.0
24	4.3	22	4.7	22	3.2
25	7.3	23	2.4	23	2.1
26	4.1	24	4.9	24	2.2
27	12.7	25	5.1	25	3.6
28	7.6	26	6.0	26	3.4
29	13.3	27	5.7	27	5.7
30	6.2	28	6.1	28	3.4
31	13.7	29	7.2	29	9.4
32	6.8	30	7.8	30	2.0
33	7.2	31	8.8	31	9.4
34	3.6	32	7.9	32	2.5
35	2.3	33	5.7	33	6.1
36 Internal Std.		34	4.0	34	1.7
		35	3.0	35	1.3
		36 Internal Std.		36 Internal Std.	
		37	1.1		
		38	1.1		

Table 5 continued.

4-73
Total n-Paraffins 1.10 μg/l
Large envelope present
C-17 to C-38

n-Paraffins	%
15	2.7
16	5.1
17+pristane	12.3
18+phytane	7.6
19	6.7
20	6.0
21	4.0
22	5.2
23	4.6
24	4.4
25	4.6
26	4.0
27	4.3
28	3.0
29	4.4
30	2.5
31	4.4
32	2.4
33	3.4
34	1.8
35	2.6
36 Internal Std.	
37	2.6
38	1.3

10-73
Total n-Paraffins 0.20 μg/l
Envelope C-20 to C-28

n-Paraffins	%
15	18.8
16	1.3
17+pristane	29.0
18+phytane	7.2
19	5.6
20	4.3
21	5.1
22	4.7
23	5.1
24	3.0
25	1.8
26	1.7
27	1.8
28	0.0
29	1.9
30	1.5
31	1.8
32	1.1
33	1.1
34	0.6
35	1.2
36 Internal Std.	

1-74
Total n-Paraffins 0.08 μg/l
Broad envelope present

n-Paraffins	%
17+pristane	19.4
18+phytane	6.2
19	3.6
20	3.2
21	3.9
22	3.6
23	7.3
24	4.7
25	5.8
26	4.7
27	4.9
28	4.6
29	5.8
30	4.9
31	5.0
32	2.8
33	2.8
34	2.3
35	1.8
36 Internal Std.	
37	1.6
38	0.4

TABLE 6 - TOTAL AMOUNT AND PERCENT COMPOSITION OF n-PARAFFINS ON DIFFERENT DATES OF COLLECTION

SABINE RIVER, TEXAS

29 40; 93 11		24 30; 94 01	
1-73	1M	1-73	4M
Total n-Paraffins	0.12 µg/l	Total n-Paraffins	0.12 µg/l
Large envelope from C-15 to C-35		Broad envelope from C-17 to C-30	
n-Paraffins	%	n-Paraffins	%
15	1.1	17+pristane	3.2
16	4.5	18+phytane	2.5
17+pristane	11.4	19	4.2
18+phytane	11.1	20	3.7
19	12.4	21	3.2
20	10.9	22	4.6
21	7.0	23	4.1
22	6.9	24	5.3
23	4.8	25	1.0
24	3.7	26	6.0
25	4.5	27	6.6
26	2.9	28	9.0
27	2.3	29	7.8
28	3.9	30	6.5
29	2.0	31	9.0
30	1.3	32	5.3
31	2.7	33	6.0
32	1.4	34	3.7
33	1.6	35	3.3
34	2.1	36 Internal Std.	
35	1.6	37	2.3
36 Internal Std.		38	2.5

PORT ARANSAS, TEXAS

5-74	1M	5-73	1M
5 miles offshore from Marine Science Institute 27 46; 96 52		26 meter contour 5 miles offshore	
Total n-Paraffins	0.07 µg/l	Total n-Paraffins	0.07 µg/l
Minor envelope C-19 to C-28		No envelope	
n-Paraffins	%	n-Paraffins	%
15	2.9	17+pristane	1.7
16	4.3	18+phytane	8.2
17+pristane	22.2	19	1.7
18+phytane	9.5	20	2.1
19	3.8	21	2.1
20	2.0	22	3.0
21	2.2	23	3.5
22	1.1	24	3.0
23	0.9	25	3.6
24	0.9	26	2.7
25	0.9	27	3.2
26	0.6	28	6.6
27	1.1	29	6.5

(continued on next page)

Table 6 continued.

28	0.8	30	8.5
29	2.0	31	10.1
30	2.5	32	9.2
31	6.2	33	7.3
32	6.3	34	6.5
33	8.3	35	4.7
34	7.2	36 Internal Std.	
35	6.5	37	3.6
36 Internal Std.		38	3.9
37	3.2		
38	2.6		
39	0.9		
40	1.1		

BROWNSVILLE, TEXAS

7-73
Total n-Paraffins 0.09 μg/l
Small envelope C-20 to C-30

n-Paraffins	%
15	20.7
16	4.0
17+pristane	29.4
18+phytane	6.7
19	2.9
20	1.6
21	1.1
22	1.1
23	1.1
24	1.1
25	1.5
26	1.0
27	1.4
28	1.2
29	1.9
30	2.3
31	3.4
32	2.3
33	2.4
34	2.2
35	2.6
36 Internal Std.	
37	1.2
38	2.1
39	1.7
40	1.7
41	0.8
42	0.6

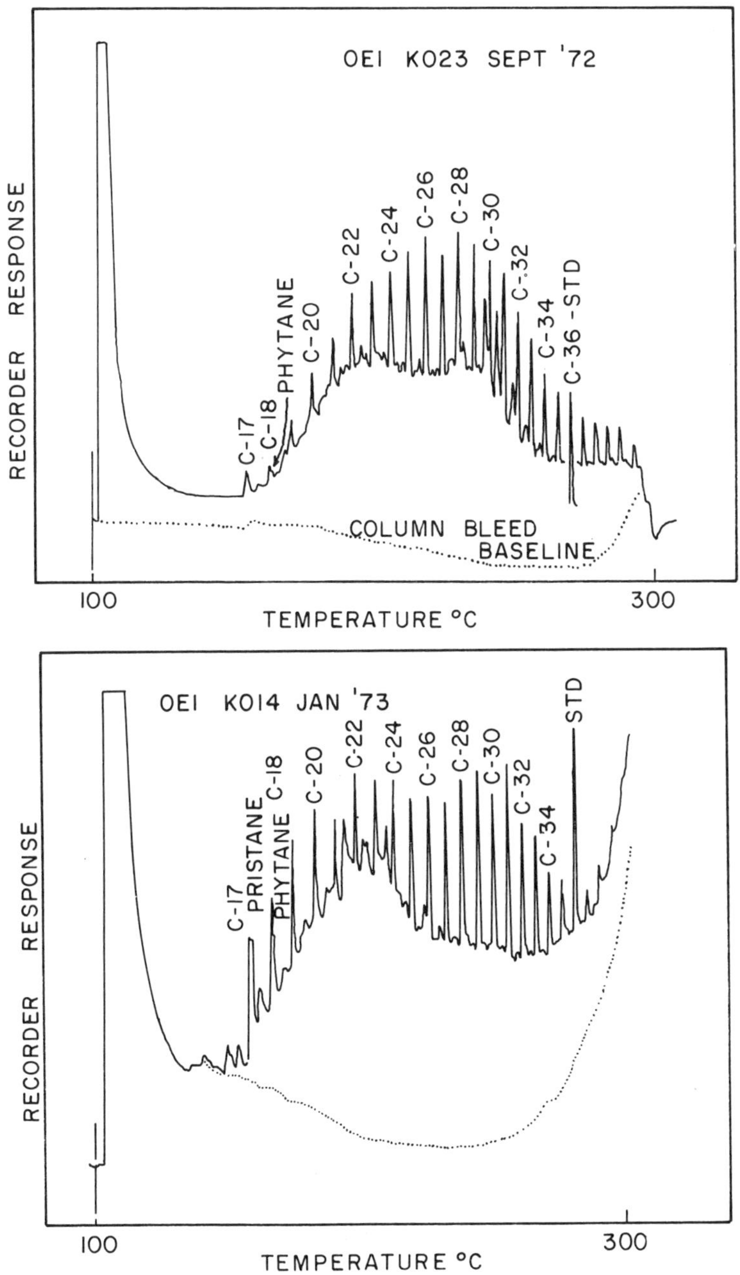

FIG. 3. CHROMATOGRAMS: HYDROCARBON COMPOUNDS IN TIMBALIER BAY (continued next page).

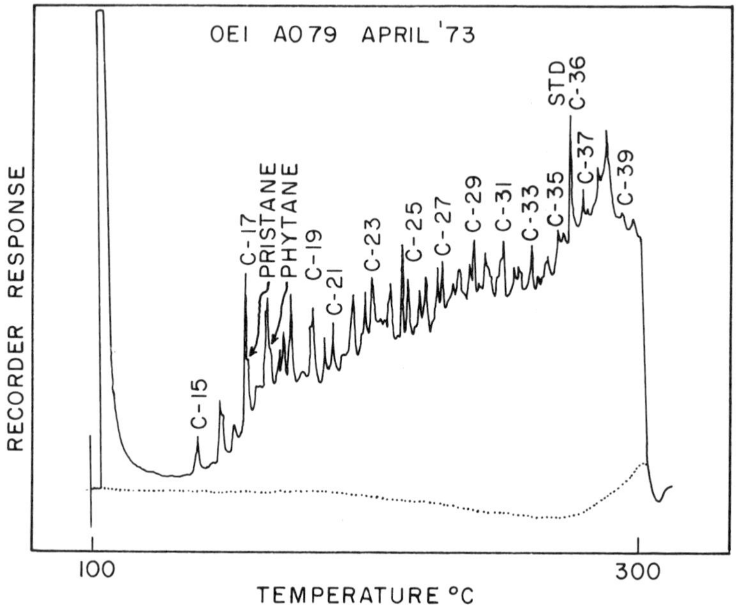

FIG. 3 continued.

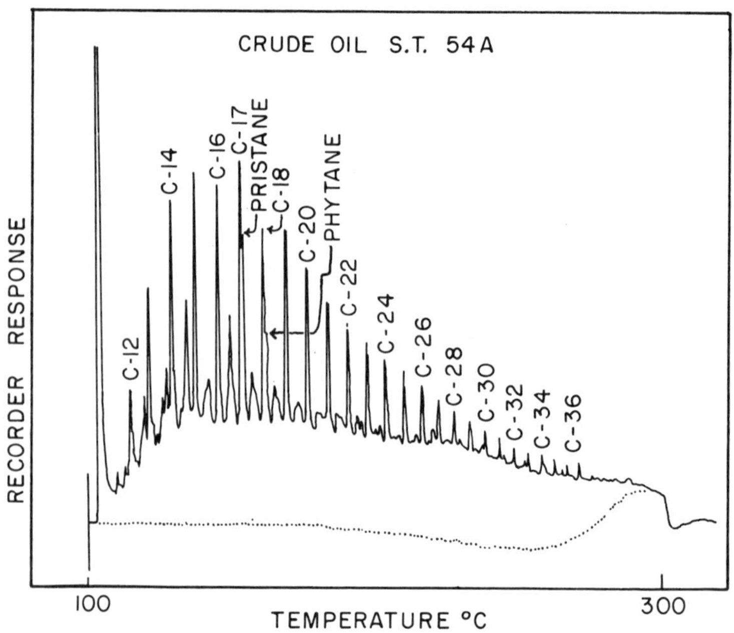

FIG. 4. CHROMATOGRAMS: MICROBIAL DEGRADATION OF CRUDE OIL FROM PLATFORM 54A (continued on next page).

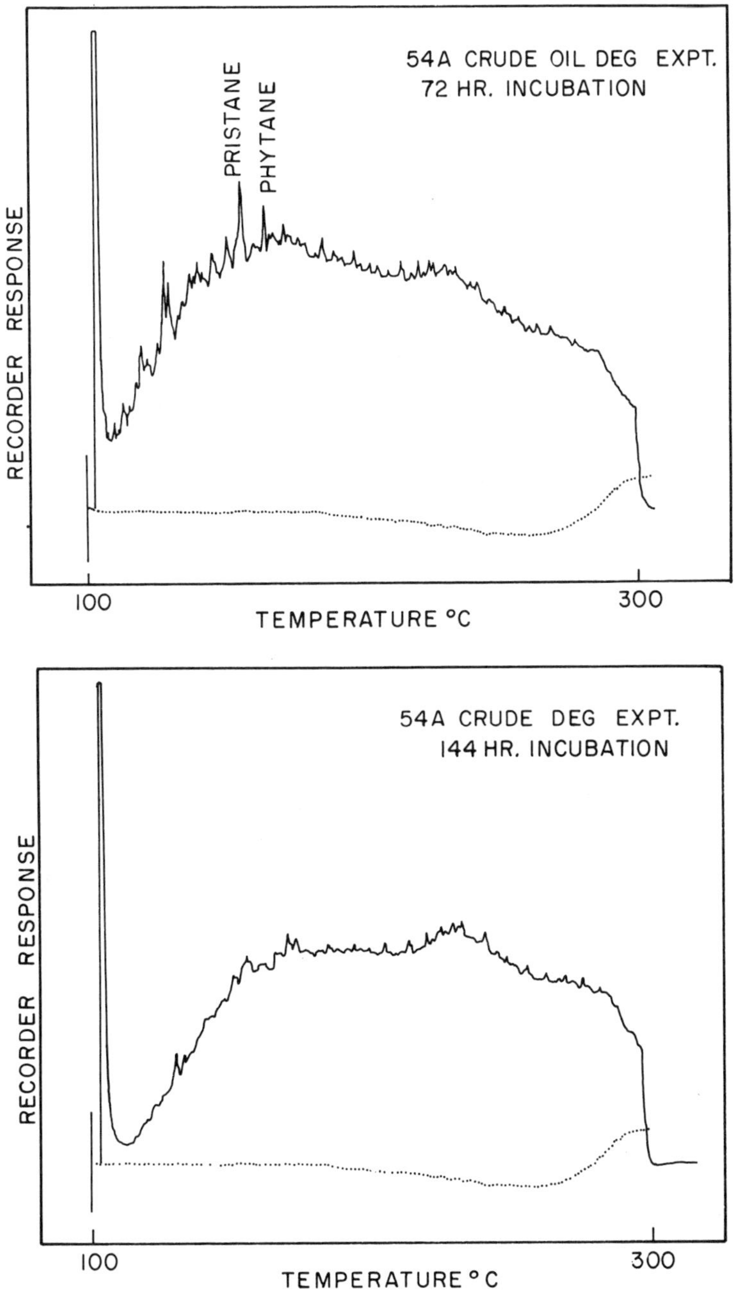

FIG. 4 continued.

oils and many of the other offshore water samples exhibit little or no odd-even carbon preference. Many of the samples possessed envelopes or unresolved baselines, indicating the presence of complex mixtures of hydrocarbons not resolved on packed columns. While this observation refers to almost all samples examined, the most complex chromatograms were generally encountered in Timbalier Bay samples. Figure 3 shows a chromatogram from a Timbalier Bay sample and reveals the presence of many overlapping peaks with retention indices intermediate of *n*-paraffins. Marsh plants are known to produce complex mixtures of branched as well as normal paraffins.

The previous chromatograms all possessed unresolved baseline envelopes, which indicate the presence of mixtures too complex to resolve with packed columns. Figure 4 shows a Louisiana crude oil (54A) which is typical of crudes produced in the area with its rather high *n*-paraffin content and recognizable *n*-paraffin fingerprint. Figure 4 illustrates the change in *n*-paraffins resulting from microbial degradation following 72 and 144 hours of incubation with mixed marine bacteria in nutrient-salts enriched seawater from the study area (Miget et al. 1969). All *n*-paraffins and major recognizable isoprenoids have been reduced to the level of an essentially undifferentiated baseline hump. Very little is known about degradation of these compounds, especially in the field where the nutrient regimes may not favor the rapid rates of degradation measured in the laboratory. It is therefore possible that the unresolved baseline encountered in hydrocarbon extracts from seawater in the Gulf of Mexico are derived from partially degraded crude oils or fuel oils and, if this is so, the widespread occurrence of this envelope in geographically diverse samples suggests that the near-shore Gulf is in some type of dynamic equilibrium with very complex mixtures of hydrocarbons such as those found in crude oils, petroleum-derived refined products, and tar balls.

The distributions of dominant *n*-paraffins and isoprenoid hydrocarbons in seawater extracts were generally similar. A bimodal distribution was typical with the nC_{17} (pristane), nC_{18} (phytane), nC_{19}, and nC_{20} paraffins composing the first dominant mode of the chromatogram and the paraffins nC_{25} through nC_{35} composing the second. Generally, a slight odd over even carbon chain length preference is exhibited in this latter region, suggesting the influence of terrestrially-derived hydrocarbons. Samples with a paraffin distribution dominated by nC_{17} + pristane, nC_{18}, nC_{19}, and nC_{20} (E. Sabine R. 1-73, K007 54A 1-73, Timbalier Bay 4-73, and K014 1-73) in approximately equal abundance suggest microbial degradation of a crude oil or a refined fuel. Finally, a large number of samples contain a C_{18} isoprenoid that is characteristically found in crude oils; certain workers consider the presence of phytane as indicative of contamination by fossil fuels (Blumer and Snyder 1965).

While there are relatively few data on hydrocarbons in seawater, those values reported for *n*-paraffin concentrations in Gulf of Mexico and Caribbean samples (Lytle, Lytle, and Parker personal communication) are in the same range as the values obtained in this study. Estimates of total saturates based on integration of the total saturated paraffin signal also fall within the ranges reported by others (Brown et al. 1973; Levy 1971).

Distribution of bacteria

Figure 5 shows the results of the OEI microbiological survey from June 1972 through January 1974. Hydrocarbon content of water taken at the same sample site and time is also shown. The data are presented by cruise. The June 1972 data from a trip throughout the Gulf of Mexico on the Texas A&M ship, *Alaminos*, can be considered "baseline" data with which to compare subsequent values from the OEI study areas. With the exception of the two near-shore stations (1 and 2), the remainder of the open Gulf samples were consistent in the "total" heterotroph populations and hydrocarbon degraders—resulting in a 0.1% value for hydrocarbon-degrading bacteria. Water samples for hydrocarbon extraction were not collected on this cruise.

Subsequent trips to the OEI study areas offshore Louisiana and in Timbalier Bay showed correspondingly low percentages of hydrocarbon-oxidizing microorganisms. The percentage of hydrocarbon degraders not only did not vary significantly or consistently with location in the June 1972 open Gulf samples, but additional control samples ("Gulf" controls) taken en route to the study area from Port Aransas, Texas, failed to indicate a significant difference in the percentage of hydrocarbon degraders in near-shore, non-oil production areas.

The assumption that minor fluctuations in the low percentages of hydrocarbon degraders did not constitute significant differences was based largely on a comparison of values for the designated control and platform areas. These locations differ by only 5 miles, which, when offshore samples are being analyzed, is not a large difference. It can be seen that for most of the sampling trips these locations fluctuated in the percentage of hydrocarbon degraders as much as additional "Gulf" control stations taken en route hundreds of miles from the OEI study sites.

Except for the July 1973 samples, there was essentially no difference in the "total" heterotroph or oil-degrading populations at the surface and at 6 m depth. In July 1973 a massive plankton bloom, concentrated in the top few meters of the water column throughout the offshore study

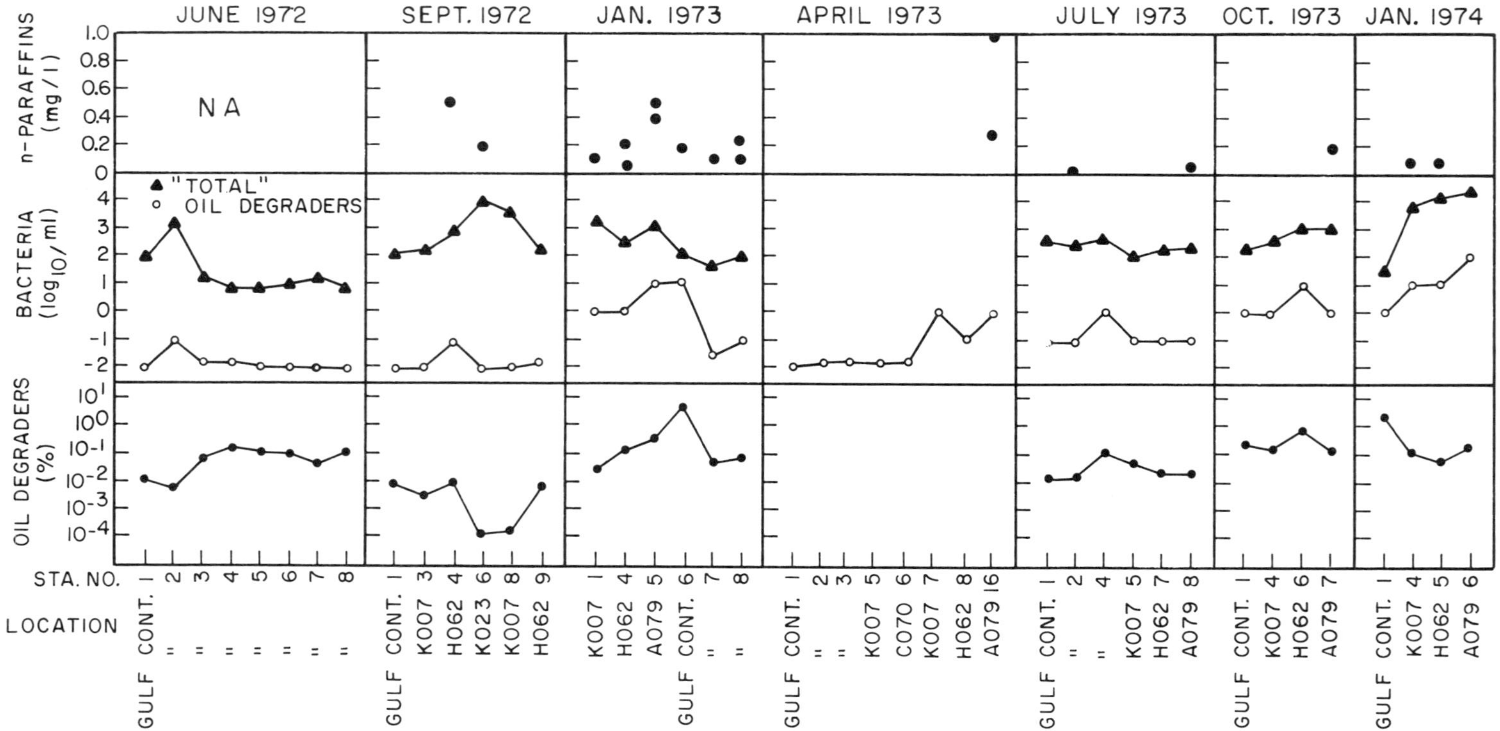

FIG. 5. BACTERIOLOGICAL SURVEY DATA.

area, was probably responsible for the order of magnitude difference observed in both "total" heterotrophs and hydrocarbon-degrading populations.

The microbial data, therefore, when considered along with the low values for hydrocarbons in the water (relative values of up to 0.8 g/l benzene extractable petroleum reported by Colwell et al. [1973]) would indicate that the OEI study area was not grossly contaminated with hydrocarbons and that the hydrocarbon-degrading populations of microorganisms were generally at a "Gulf baseline" level; i.e., they were comparable to those in non-oil production areas.

Two additional remarks concerning the microbial data are warranted. First, the numbers of "total" heterotrophs in the OEI study areas seemed anomalously low relative to organic carbon concentrations and plankton populations. For example, Jannasch and Oppenheimer (1962) found populations of 10^6 bacteria per ml ("total" heterotrophs) in near-shore waters off Port Aransas, Texas. Thus, the consistent enumeration of relatively low numbers of heterotrophs (generally 100 to 1000/ml) in the productive waters offshore Louisiana and in Timbalier Bay needs further study.

Second, as stated in the methods section, the numbers of hydrocarbon-degrading microorganisms shown in figure 5 were calculated based on changes in the appearance of either the oil or the water in replicate test flasks relative to control flasks. A re-examination of the dilution flask data showed that for certain sampling trips almost all of the flasks, including 300 ml filtered samples, showed only a "flaky" oil emulsification and/or cloudy water with no oil emulsification. During other times of the year nearly all of the flasks in a dilution series contained well emulsified oil, and only one or two "flaky" flasks as the sample volume was increased. Although a lack of replicate samples of both hydrocarbon-degrading microbial populations and hydrocarbon concentrations from the study areas prevented a statistical treatment of these two parameters, it should be noted that the "emulsifying" populations occurred in January 1973, April 1973, and January 1974, and to some extent in September 1972 and October 1973. All July 1973 samples exhibited the "flaky" type of oil degradation.

BOD studies

Table 7 shows BOD test data from samples collected in January 1974. These are typical of the BOD results and indicate the potential activities of microbial populations *in situ* in response to added hydrocarbons.

These data show that crude oil, paraffin oil, hexadecane, benzene,

TABLE 7 - JANUARY 1974 HYDROCARBON BOD DATA
(Platform 54A surface water)

SUBSTRATE[1]	INCUBATION (days)	O_2 REMAINING[2] (ml/l) X S.D.	X cont. -X test[3] (ml/l)
Control	10	3.83 + 0.21	--
Control	20	2.80 + 0.15	--
Control	43	2.55 + 0.11	--
Benzene	10	0.19 + 0.12	3.64
Toluene	10	0.14 + 0.08	3.69
Ethylbenzene	10	0.20 + 0.12	3.63
Propylbenzene	10	0.16 + 0.11	3.67
Hexadecane	10	0.48 + 0.39	3.35
Paraffin oil	20	1.30 + 0.39	1.50
54A crude oil	20	0.40 + 0.56	2.40
Isooctane	43	0.30 + 0.52	2.25
Cyclooctane	43	0.82 + 1.97	1.73
Anthracene	43	1.45 + 1.57	1.10
Triphenylene	43	0.80 + 0.96	1.75
Naphthalene	43	1.34 + 0.82	1.21
Fluoranthene	43	1.88 + 1.27	0.67
Phenanthrene	43	1.71 + 1.54	0.84
Benzanthrene	43	1.44 + 1.97	0.81

[1] Substrate concentrations were 10 µl/300 ml sample water for hydrocarbons liquid at room temperature and 100 mg/300 ml sample water for hydrocarbons solid at room temperature and dissolved in benzene as described in the Methods section.

[2] Determined chemically using the Winkler titration method as outlined in the Analysis of Seawater Handbook (Strickland and Parsons 1968).

[3] All values are reported even though the standard deviation sums exceeded the difference in oxygen concentration for some of the higher molecular weight aromatics. See text for justification.

and alkylbenzenes were metabolized by indigenous bacteria. On the basis of previous laboratory investigations using active hydrocarbon-degrading cultures, the rates of oxidation of the crudes were considered to be primarily a function of the composition of the crude oils and the solubilities of their more biologically labile components.

The heptane fraction of #8 crude oil (containing primarily *n*-paraffins) was oxidized at approximately the same rate as the whole crudes, indicating perhaps that the oxidation of *n*-paraffins was primarily responsible for oxygen utilization in these water samples. Biological oxidation of *n*-paraffins in crude oils has been previously reported from this laboratory (Miget et al. 1969). In order to verify this hypothesis, however, one of the most active mixed cultures isolated on a crude oil substrate from the study area offshore was grown on #8 crude oil in nutrient salts enriched seawater (ESW). The sequence of gas chromatograms of extracted hydrocarbons from replicate cultures after various periods of incubation shows the degradation of *n*-paraffins in whole crude oil (figure 4).

These data, however, do not necessarily imply a preferential utilization of *n*-paraffins in crude oils by naturally occurring microbial populations, as low boiling aromatic hydrocarbons (benzene and the alkylbenzenes) were also oxidized by indigenous microorganisms (table 7). It is obvious that these alkylbenzenes not only were non-toxic at this concentration (10 μl/300 ml water), but, in part because of their high solubilities relative to paraffins, were rapidly metabolized by the indigenous microorganisms.

The branched paraffin, isooctane, the cycloparaffin, cyclooctane, and all of the heavier molecular weight aromatic hydrocarbons tested showed greatly reduced biological oxidation.

Summarizing the BOD field data, we can say that microorganisms capable of metabolizing a variety of hydrocarbons (*n*-paraffins, crude oils, low molecular weight aromatics, and quite probably heavier polynuclear aromatics) were always present in the OEI study areas. All metabolic assays employed nutrient salts enriched seawater (ESW), so rates of metabolic activity of indigenous microorganisms reported are not comparable with *in situ* metabolism of the same hydrocarbons.

It should be re-emphasized that several of the low boiling aromatic hydrocarbons found in crude oil that have been shown to be highly toxic to marine organisms (Blumer 1969) were readily metabolized by the native microflora in the offshore Louisiana area. This is especially significant with regard to the "self-cleansing" potential of the areas, as GC-MS analyses of 30 offshore Louisiana crude oils obtained at the well-heads showed a composition averaging 20% benzene plus alkylbenzenes.

TABLE 8 - HYDROCARBONS IN SEAWATER LITERATURE REVIEW

INVESTIGATOR	SAMPLE LOCATION/NO.	VALUES OBTAINED
Brown et al. 1973	Atlantic (18)	Total - 1 to 12 μg/l Saturates - 5.7 μg/l Aromatics - 1.0 μg/l
GURC, 1973	Gulf of Mexico (16)	n-Paraffins - 0.30 μg/l Saturates - 3.7 μg/l
Levy, 1971	Nova Scotia area	Total - 2 to 13 μg/l
Parker et al. 1975	Gulf of Mexico (6)	n-Paraffins - 0.23 μg/l
Barbier et al. 1973	E. Central Atlantic	Total - 10 to 137 μg/l

DISCUSSION

A thorough literature review at the onset of this project suggested that total naturally occurring hydrocarbons in seawater are found at relatively low concentrations (table 8) and hydrocarbon-oxidizing bacteria are found in all habitats tested (table 9). We, therefore, developed an analytical technique that effectively extracted seawater for hydrocarbons at the nanogram/liter level and used for analyses a Finnigan GC-MS, an instrument of relatively high sensitivity.

As data accumulated, some tentative conclusions could be made. One conclusion was that one could reliably detect hydrocarbon components in any sample at the parts-per-trillion range, as shown by the similarity of replicate samples. Triplicate aliquots of the extracts from each 20-l sample were analyzed. The analytical technique, then, permitted a routine sampling and analysis of a sufficiently large number of samples to provide statistically meaningful ecological data. Sampling distribution and frequency, rather than laboratory analysis, were the limiting factors.

Ecologically, the hydrocarbon data are quite interesting. In the water column we found the average total saturate (and normal hydrocarbon) content to be 2 μg/l. In the water extracts no relatively high peaks (e.g., high concentrations of specific compounds) were found by chromatography, etc., that would indicate any specific concentration of hydrocarbons from oil spills or their degradation products or from primary production. This was true, also, of the test and control areas off Port Aransas, Texas, which were used for additional comparisons. As one reviews the possible mechanisms of hydrocarbon oxidation, these determinations would indicate that the biota has maintained the

TABLE 9 - INDIGENOUS HYDROCARBON DEGRADING BACTERIAL POPULATIONS

LOCATION	"TOTAL" HETEROTROPHS/ml	HYDROCARBON DEGRADERS/ml	HYDROCARBON DEGRADERS	HYDROCARBONS IN WATER	COMMENT	REFERENCE
Moscow River						
1. Upstream refinery	NA	10 to 9×10^4	< 5	NA	Oil visible on water near refinery	Vorosilova, 1950
2. Downstream	NA	$> 10^7$	up to 12,800			
North Sea						
1. Helgoland Mar.St.	NA	0.9 to 460	0.5 to 16.8	NA	Data are averaged over a 14 mo. sampling period. No sig. diff. in # of H.C. degrading bacteria in surface film and at 1 m.	Gunkel, 1973
2. 25 mi. offshore from the mouth of the Elbe R.	90	3	3.3	NA		
Cook Inlet, Alaska	NA	Av. = 1.0 (n=32)	Av. = 10%	NA	No sig. diff. in # of H.C. degrading bacteria in the surface film and at 1 m.	Miget, et al. 1974
Raritan Bay, NJ						
1. Arthur Kill	NA	3.4	NA	NA	Classified as a "polluted" area by authors.	Bartha, 1973
2. Seaward fringe of Raritan Bay	NA	0.06	NA	NA	A seasonal fluctuation in # of H.C. degrading bacteria was noted.	

Table 9 continued.

Cheasapeake Bay										
1. East Bay	10-72 / $1.5x10^3$	11-72 / $5.0x10^1$	10-72 / $5.0x10^1$	11-72 / $5.0x10^0$	10-72 / 3.3	11-72 / 12.5	10-72 / 0.002%w/v	11-72 / NA	Described as an unpolluted commercially productive area.	Colwell, et al., 1973
2. Colgate Creek	$1.5x10^4$	$5.0x10^5$	$6.0x10^2$	$5.0x10^4$	4.0	8.9	0.08%w/v	NA	Adjacent to Baltimore Harbor. Not commercially productive.	
Cornish Coast (Torrey Canyon)										
1. Sennen		$8.5x10^5$		$5.0x10^4$		6%		NA	No oil visible in area.	Gunkel, 1968
2. Trenow Cove		$2.5x10^8$		$7.0x10^8$		280%		NA	Oil was visible on water surface and in sediments.	
Gulf of Mexico										
1. Barataria Bay		NA		≥ 0.1		NA		NA	Over 300 stations sampled during a three-year period.	ZoBell, 1966
2. Texas coast Brownsville to Matagorda Bay		NA		≥ 0.1		NA		NA	Seven of 24 stations were negative for H.C. degrading bacteria in 10 ml samples.	

NA - Not Available

hydrocarbons in the water column at a very low level. In fact, the level appears to be too low for biological uptake; i.e., the combination of population, grazing range, and concentration are such that probability of uptake is very small. The probable, but still preliminary, conclusion is that the biota has decomposed the hydrocarbons down to an insignificant level in respect to their ecological effect.

Data for the distribution of hydrocarbon oxidation, rates of hydrocarbon degradation, and specific molecule uptake by BOD experiments support the low water concentrations and the fact that we were not able to detect any specific buildup of hydrocarbon molecules. It is pertinent to note that such compounds as benzene and toluene, considered by many to be toxic, are readily utilized by the indigenous microorganisms from the offshore Louisiana area. The hydrocarbons in Timbalier Bay organisms and water appear to reflect land influence, as our data show a general odd-even carbon ratio similar to hydrocarbons of land plants.

Data obtained for hydrocarbons $>C_{10}$ in the water column are also supported by other investigators (see table 8), who found comparable hydrocarbon levels in seawater in the Gulf of Mexico and in Atlantic oceanic water. It would appear that the values for open ocean waters would be lower than inshore waters, particularly in the vicinity of producing oil platforms, such as those in the OEI test area. The finding that levels in the OEI area are comparable to those in the open ocean appears to be consistent with the previous conclusion regarding biological oil grazing activities.

Information developed in this investigation also can be correlated with data obtained by other investigators in the OEI program. For example, Brent (1979) found total carbon in the water column to be in the range of 1 to 5 ppm. Total hydrocarbons reported in this investigation are approximately 0.1% to 1% of Brent's total organic matter. This value is also relevant to the concentrations of hydrocarbons in living systems that have been reported (from 0.001% to 1% by dry weight) in the literature.

Further, the data can be compared to the average daily rates of primary productivity reported by Fucik and El-Sayed (1979): 1.06 gC/m^2 at 54A and 1.03 gC/m^2 at the control station. If, of this carbon, 0.01% is hydrocarbon, then approximately 0.1 $mg/m^2/day$ are fixed (this percentage is based on laboratory experiments where algae grown in a hydrocarbon-free environment produce hydrocarbons). From the data it can be estimated that approximately one million barrels of oil-equivalent are produced annually by primary production in the Gulf of Mexico.

The total values and ratios of hydrocarbon-utilizing bacteria to total heterotrophs indicate a low hydrocarbon level. The rates of hydrocarbon uptake by indigenous microbial populations in the test area also substantiate the conclusion that the grazing effect results in the very low hydrocarbon levels in the test area.

Laseter and Ledet (1979) reported that the air-water interface offshore had an average of 0.7 mg/m^2 of paraffins. They did not report any correlation between surface hydrocarbons at Platform 54A and the control area. They did, however, show that the hydrocarbons in the surface slick of the test area were higher than for north Florida waters. This difference could be due in part to the higher primary production and the hydrocarbons resulting from the Mississippi River drainage in addition to local spills from oil operations in the oil producing field. We have determined the relationship of Laseter's findings of surface oil concentrations to water column concentrations. If one assumes that Laseter's sampling technique would sample the top tenth of a millimeter of the surface, the value of 0.7 mg/m^2 is equivalent to 70 μg/1.

By using various OEI data, the content of 100 m^2 of water 10 m deep in the test area offshore can be calculated as follows:

Author	*Basis*	*Amount*
Laseter	Surface content paraffins (based on 0.70 mg/m^2)	0.7 g
Kator	Content of paraffins (in water based on 1 μg/)	1.0 g
El-Sayed	Daily primary productivity (based on 1.0 g Carbon fixed/day/m^2	100.0 g
Brent	Organic carbon (based on 5 mg/l)	5000.0 g

As Laseter's, El-Sayed's, and our work indicates, the distributions and types of hydrocarbons in seawater will be influenced by the normal productivity of photosynthesis, the spillage of hydrocarbons through production activities, and microbial degradation. In addition to these seasonal trends, however, there may be daily fluctuations. The BOD results show a rapid uptake of hydrocarbon, indicating the transitional nature of residuals and oxidation products.

Some correlations show interesting relationships. For example, Laseter gave data that the surface oil in the test area was 0.70 mg/m^2 compared to 0.18 mg/m^2 in the area offshore from north Florida. El-Sayed studied the same areas and found primary productivity in the water column at the test site to be approximately 85 mgC/m^2/hr as compared to approximately 5 mgC/m^2/hr off Panama City, Florida. Such a difference in productivity could explain the differences in surface hydrocarbons between the two areas, as the data also indicate a higher primary productivity in the Fertile Crescent off Louisiana. Neither Laseter's nor our work definitely fingerprints the presence of crude oil in the carbon present in seawater. This suggests that it will be difficult to separate the input of hydrocarbons from the Mississippi River, from production operations, from primary productivity, and last, from the microbial degradation products of these materials.

CONCLUSIONS

The year-and-a-half study of the distribution of hydrocarbons and bacteria in surface and subsurface waters in the Louisiana area provided the basis for an environmental evaluation. The distribution pattern of low numbers of oil-degrading bacteria and low hydrocarbon concentrations, and the response of indigenous bacteria to added hydrocarbons, were indicative of a balanced ecosystem. Hydrocarbons released during oil production and ship operations, as well as from the Mississippi River, had not accumulated to concentrations greater than in other areas of the oceans. The low amount of hydrocarbons and numbers of bacteria suggest that in the previous months little contamination had been present. Otherwise, one would expect to find larger numbers of bacteria. The BOD data clearly indicate that the bacteria in the area are capable of oxidizing a wide variety of low and high molecular weight paraffinic and aromatic hydrocarbons. The rate of activity as related to oxygen uptake is indicative of a heterogeneous population, which has degraded contaminating oil to the very low level of a few parts per billion. There was also no evidence that specific molecules were more resistant to degradation, thus resulting in their concentration in water samples. Individual molecules occurred in the samples analyzed in the parts-per-trillion range.

Quantitative and qualitative data regarding hydrocarbons extracted from seawater and organisms collected both in the OEI study areas and in non-production Gulf control stations suggest that a large portion, if not all, of the Gulf of Mexico is uniformly exposed to hydrocarbons, which are probably derived from primary production or from petroleum. Perhaps the ubiquitous population of hydrocarbon-oxidizing microorganisms and the low concentrations of hydrocarbons found in the Gulf are a reflection of a dynamic system whereby hydrocarbons derived from petroleum, plants, and animals are introduced into the Gulf and are continuously reduced to a very low baseline level through physical, chemical, and biological processes.

ACKNOWLEDGMENT

This paper previously appeared as Contribution No. 385 in Marine Science, The University of Texas.

REFERENCES CITED

Ahearn, D. G. and S. P. Myers, eds. 1973. The Microbial Degradation of Oil Pollutants. Publication #LSU-SG-73-01. Baton Rouge, Louisiana: Louisiana State University.

Barbier, M., D. Joly, A. Saliot, and D. Tourres. 1973. Hydrocarbons from sea water. Deep-Sea Research **29**:305-314.

Bartha, R. and R. M. Atlas. 1973. Biodegradation of oil in seawater: Limiting factors and artificial stimulation. *In* D. G. Ahearn and S. P. Myers, eds., Microbial Degradation of Oil Pollutants. Publication #LSU-SG-73-01. Baton Rouge, Louisiana: Louisiana State University.

Blumer, M. 1969. Oil pollution of the ocean. *In* D. Hoult, ed., Oil on the Sea. New York: Plenum Press.

——— and D. W. Snyder. 1965. Isoprenoid hydrocarbons in recent sediments: Presence of pristane and probable absence of phytane. Science **150**:1588-1599.

———, R. R. L. Guillard, and T. Chase. 1971. Hydrocarbons of marine phytoplankton. Marine Biology **8**:183-189.

Brent, C. R., H. P. Williams, W. A. Bergin, J. L. Tyvoll, and T. E. Myers. 1979. Organic carbon, inorganic carbon, and related variables in offshore oil production areas of the Northern Gulf of Mexico. Rice University Studies **65**(4):245-264.

Brown, R. A., T. D. Searl, J. J. Elliot, B. G. Phillips, D. E. Brandon, and P. H. Monaghan. 1973. Distribution of heavy hydrocarbons on some Atlantic Ocean waters. *In* Proceedings of Joint Conference on Prevention and Control of Oil Spills. Washington, D.C.: American Petroleum Institute, pp. 505-519.

Butler, M. J. A. and F. Berkes. 1972. Biological aspects of oil pollution in the marine environment, A review. Marine Science Center Manuscript Report No. 22.

Byrom, J. A. and S. Beastall. 1971. Microbial degradation of crude oil with particular emphasis on pollution. *In* Institute of Petroleum, Microbiology, 1971 (A Symposium). Amsterdam: Applied Science Publishers, pp. 73-86.

Colwell, R. R., J. D. Walker, and J. D. Nelson, Jr. 1973. Microbial ecology and the problem of petroleum degradation in Cheasapeake Bay. *In* D. G. Ahearn and S. P. Myers, eds., Microbial Degradation of Oil Pollutants. Publication #LSU-SG-73-01. Baton Rouge, Louisiana: Louisiana State University, pp. 185-197.

Crow, S. A., S. P. Myers, and D. G. Ahearn. 1974. Microbiological aspects of petroleum degradation in the aquatic environment. Bulletin de la Société Franco-japonaise d'Oceanographie **12**(2):95-112.

Davies, J. A. and D. E. Hughes. 1968. The biochemistry and microbiology of crude oil degradation. The Biological Effects of Oil Pollution on Littoral Communities. Field Studies Council, July, pp. 139-144.

Davis, J. B. 1967. Petroleum Microbiology. New York: Elsevier.

Fucik, K. W. and S. Z. El-Sayed. 1979. Effect of oil production and drilling operations on the ecology of phytoplankton in the OEI study area. Rice University Studies **65**(4):325-353.

Gholson, R. K., P. Guire, and J. Friede. 1972. Assessment of Biodegradation Potential for Controlling Oil Spills on the High Seas. Report to the U.S. Coast Guard 4110.1/3.1 NTIS AD-759 848.

Gunkel, W. 1968. Bacteriological investigation of oil-polluted sediments from the Cornish coast following the Torrey Canyon disaster. *In* The Biological Effects of Oil Pollution on Littoral Communities (Suppl. to Field Studies, vol. 2). Field Studies Council.

———. 1973. Distribution and abundance of oil-oxidizing bacteria in the North Sea. *In* D. G. Ahearn and S. P. Myers, eds., Microbial Degradation of Oil Pollutants. Publication #LSU-SG-73-01. Baton Rouge, Louisiana: Louisiana State University, pp. 127-139.

Jannasch, H. and C. Oppenheimer. 1962. Bacterial populations in turbid and clear waters in shallow marine bays. Journal of the Institute of Marine Science **8**:56-60.

Laseter, J. L. and E. J. Ledet. 1979. Hydrocarbons and free fatty acids associated with the air/water interface, sediments, and beaches of the Timbalier Bay and offshore Louisiana area. Rice University Studies **65**(4):265-286.

Levy, E. M. 1971. The presence of petroleum residues off the east coast of St. Lawrence and the St. Lawrence River. Water Research **5**:723-733.

Lytle, T., J. Lytle, and P. Parker. 1973. A geochemical study of a marsh environment. Gulf Research Reports **4**:214-232.

———, J. Lytle, and P. Parker. Personal communication.

McKenna, E. J. and R. E. Kallio. 1965. The biology of hydrocarbons. *In* Annual Review of Microbiology. Palo Alto, California: Annual Reviews, Inc. **19**:183-208.

Meinschein, W. G. 1959. Origin of petroleum. American Association of Petroleum Geologists Bulletin **43**(5):925-943.

———. 1969. Hydrocarbons—Saturated, unsaturated and aromatic. *In* Organic Geochemistry. Berlin: Springer-Verlag, pp. 339-356.

Menzies, R. J. 1973. A Handbook on Procedures and Methods Employed in the Offshore Ecology Investigation (OEI). Gulf Universities Research Consortium. Report No. 121.

Miget, R., C. Oppenheimer, H. Kator, J. L. Laseter, and E. J. Ledet. 1974. New sampling device for the recovery of petroleum hydrocarbons and fatty acids from aqueous surface films. Analytical Chemistry **46**(8):1154-1157, reprinted in this volume, pp. 265-275.

———, C. Oppenheimer, H. Kator, and P. LaRock. 1969. Microbial degradation of normal paraffin hydrocarbons in crude oils. *In* Proceedings of the Joint Conference on the Prevention and Control of Oil Spills. New York: American Petroleum Institute, pp. 327-331.

Myers, S. P. and C. G. Gunnerson. 1976. Hydrocarbons in the Ocean. MESA Special Report. Washington, D.C.: U.S. Department of Commerce.

Oppenheimer, C. 1965. Bacterial production of hydrocarbon-like materials. Zeitschrift für Allgemeine Mikrobologie **5**:284-307.

Parker, P. L., J. K. Winters, and J. Morgan. 1975. A baseline study of petroleum in the Gulf of Naples. Unpublished IDOE Manuscript.

Smith, P. V., Jr. 1952. Preliminary note on the origin of petroleum. Tulsa, Oklahoma: American Association of Petroleum Geologists.

Stone, Robert W. and C. E. ZoBell. 1952. Bacterial aspects of the origin of petroleum. Industrial and Engineering Chemistry **44**(11):2564-2567.

Stransky, K., M. Streibl, and V. Herout. 1967. On natural waxes. VI—Distribution of wax hydrocarbons in plants at different evolutionary levels. Collection of Czechslovakian Chemical Communications **32**:3213-3220.

Strickland, J. and T. Parsons. 1968. A Practical Handbook of Seawater Analysis. Fishery Research Board of Canada Bulletin.

Vorosilova, A. A. and E. V. Dianova. 1950, 1952. Quoted in Gunkel 1973, cited above.

Weete, J. D. 1972. Aliphatic hydrocarbons of the fungi. Phytochemistry **11**:1201-1205.

ZoBell, C. E. 1945. The role of bacteria in the formation and transformation of petroleum hydrocarbons. Science **102**(2650):365-369.

———. 1946a. Marine Microbiology. Waltham, Mass.: Chronica Botanica Co.

———. 1946b. Functions of bacteria in the formation and accumulation of petroleum. The Oil Weekly, February 18, pp. 1-7.

———. 1950. Assimilation of hydrocarbons by microorganisms. Advances in Enzymology **10**:443-486.

———. 1952. The part played by bacteria in petroleum formation. Journal of Sedimentary Petrology **22**(1):42-49.

———. 1969. Microbial modification of crude oil in the sea. Joint Conference on the Prevention and Control of Oil Spills. New York: American Petroleum Institute, pp. 317-326.

——— and J. F. Prokop. 1966. Microbial oxidation of mineral oils in Barataria Bay bottom deposits. Zeitschrift für Allgemeine Mikrobiologie **6**:143-162.

EFFECT OF OIL PRODUCTION AND DRILLING OPERATIONS ON THE ECOLOGY OF PHYTOPLANKTON IN THE OEI STUDY AREA

by Kenneth W. Fucik and Sayed Z. El-Sayed

ABSTRACT

Between June 1972 and January 1974, twelve cruises were made to study photosynthetic rates and the standing crop of phytoplankton off the Louisiana coast. Data from a production platform were compared with data from a control station 6 miles northeast of the platform, with regard to photosynthetic activity of phytoplankton, chlorophyll *a* as a measure of the standing crop, species composition, and nutrient salts.

Seasonal variations in all categories were observed at the platform and control station. Chlorophyll *a* values were lowest in November 1972 and January 1973. Highest standing crop values were recorded in April 1973, during the bloom of *Skeletonema costatum*.

The OEI study area is considered one of the most productive areas of phytoplankton in the Gulf. Our investigation showed no deleterious effects from the oil production platform.

INTRODUCTION

The Gulf of Mexico is one of our most valuable national resources. It provides energy, transportation, and recreation; but its productive continental shelves, noted for their economically valuable shrimp, oyster, and menhaden fisheries, are also the sites of thousands of offshore oil

---

Sayed El-Sayed is with the Department of Oceanography, Texas A&M University. At the time of this study, Kenneth Fucik was also with Texas A&M University.

fields, and the routes of heavy ship traffic.

In recent years, we have witnessed an increasing public awareness of the deterioration of the marine environment caused by dumping toxic pollutants in domestic and industrial wastes, and of the widespread threat of pollution caused by production and transport of oil and petroleum products. The S.S. *Torrey Canyon* oil spill disaster heightened public apprehension and sensitivity to the threat of oil in the marine environment; it also sharpened the conflict over the multiple uses of the Gulf's resources by oil and gas producers and the protectors of the environment.

Field observations of the effects of the actual oil spills on phytoplankton are limited and often contradictory. Soto et al. (1975) observed some kill among the phytoplankton following the *Torrey Canyon* disaster. Smith (1968), following the *Torrey Canyon* spill, observed that most of the stations that were studied showed healthy populations; at one station, however, there was a film of oil on the water and harmful effects were apparently suffered by members of the unicellular Prasinophyceae. The Santa Barbara blowout, according to Straughan (1973), also produced detectable harmful effects in phytoplankton productivity.

Besides the oil spill investigations (see Wheeler [1978] for review of recent literature), a few field and laboratory studies suggested that petroleum and its derived products could affect photosynthesis and growth of phytoplankton. For instance, the photosynthetic rate of marine phytoplankton could be stimulated or depressed by the level of hydrocarbon present and by means and duration of exposure. Galstoff et al. (1935), Mirnov and Lanskaya (1969), and Strand et al. (1971) reported that the growth of marine phytoplankton was retarded by the presence of oil in seawater. Gordon and Prouse (1973) studied the effect of low concentrations of three oils (Venezuelan crude oil, No. 2 fuel oil, and No. 6 fuel oil) on the photosynthetic activity of natural phytoplankton communities from Bedford Basin (Nova Scotia) and the open ocean between Halifax and Bermuda. Their results showed that while the hydrocarbon levels in the Bedford Basin area were sufficiently high to depress photosynthetic activity of the phytoplankton by a few percent, the levels of hydrocarbon present in the open ocean were not of sufficient magnitude to depress such activity. Pulich, Winters, and van Baalen (1974) also reported that soluble fractions of No. 2 fuel oil in concentrations of 0.04 to 0.40 μm/l reduced photosynthesis of the diatom *Thalassiosira pseudonana* and affected the photosynthetic activities of certain green and blue-green algae.

There seems to be a dearth of information, however, on the effects of oil drilling and production operations in marine ecosystems in which disastrous spills or blowouts have not occurred. We know very little about the chronic effects of low-level discharges of oil and gas into the marine environment, or about the ability of marine ecosystems to cope with slowly increasing background levels of oil.

The OEI was conducted between June 1972 and January 1974 in Timbalier Bay and the offshore region. Its main objective was to assess the cumulative ecological effects of 40 years of oil and gas drilling and production operations in that 400-square-mile area. Of major concern were the possible effects of prolonged low-level discharges of hydrocarbons and heavy metals (attributable to those operations) on the ecological health of the marine environment.

This paper addresses the effects of the prolonged low-level discharges of hydrocarbons and heavy metals on the phytoplankton populations off the Louisiana coast. The experimental strategy focused on a comparison between point source effects (in this case the Exxon Production Platform 54A, also known as HOR-54A) and an "unimpacted" (control) area experiencing the same natural variabilities as those of the "impacted" regions. While this study emphasized the intercomparability of the phytoplankton populations of the production platform region and the "control" (ambient) site, the occurrence of the 100-year maximum flood of the Mississippi River during this study produced some unexpected results, which will also be discussed.

MATERIALS AND METHODS

The locale chosen for the OEI contained a very high density of coastal and offshore production and marine pipeline operations, which had been functioning for 40 years. The production platform (hereafter referred to as the platform) was located in 20 m of water approximately 50 miles from the Southwest Pass of the Mississippi River. The control area (hereafter referred to as the control) was located six miles northeast of the platform (figure 1). Its selection as a control site was made on the basis of (1) minimal or no petroleum activity, and (2) its location upstream from the platform.

Twelve cruises aboard the R/Vs *Alaminos*, *Longhorn*, *Southwest Researcher*, and *Tursiops* were made between June 1972 and January 1974. The amount of chlorophyll *a*, which was used as an estimate of the phytoplankton standing crop, was determined spectrophotometrically (Strickland and Parsons 1970). The fluorometric method of Yentsch and Menzel (1963) with the modification of Holm-Hansen et al. (1965) was used for estimation of both chlorophyll *a* and phaeopigments (degradation products of chlorophyll).

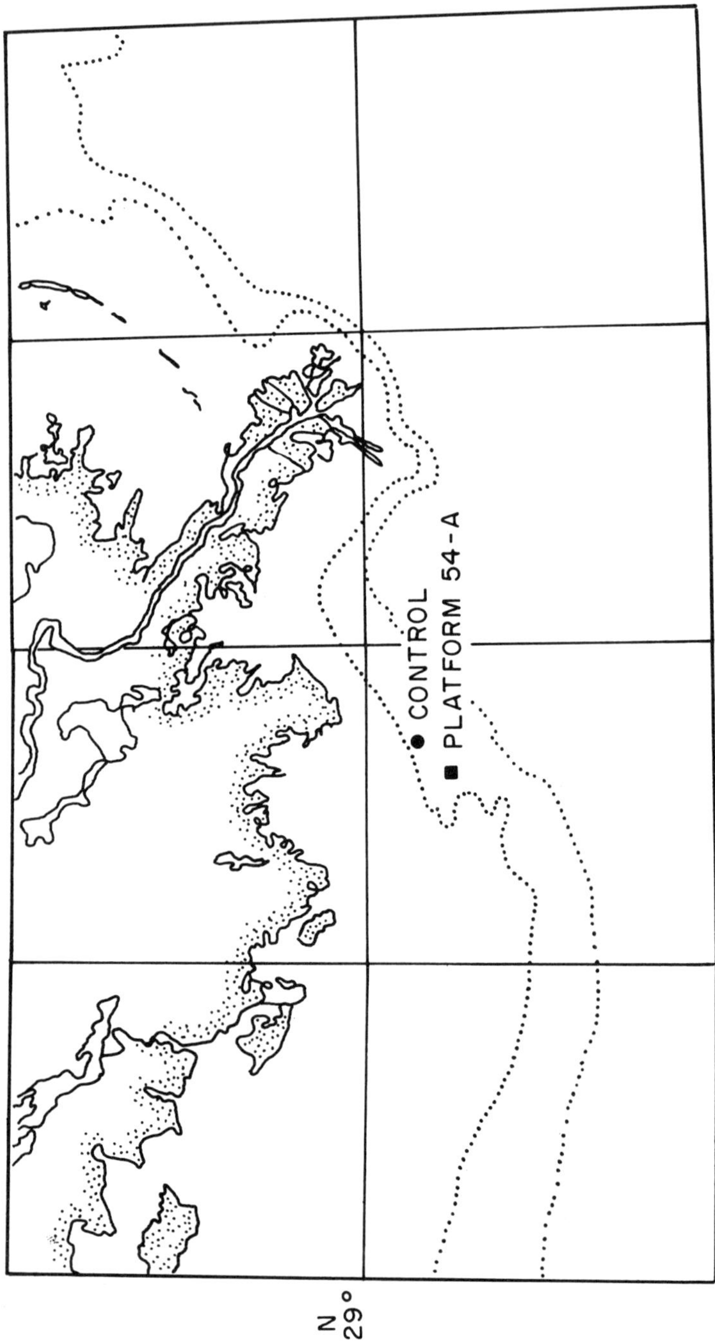

FIG. 1. LOCATION OF EXXON PLATFORM 54A AND THE CONTROL SITE OFF THE LOUISIANA COAST.

The photosynthetic activity of phytoplankton was determined by *in situ* experiments using the C^{14} uptake method of Steemann Nielsen (1952) as modified by Strickland and Parsons (1970). Measurements of submarine light penetration were made using a Secchi disc before Local Apparent Noon (LAN). The water samples were collected at 100%, 50%, 25%, 12%, and 1% of surface light intensity. These water samples were inoculated with C^{14}-labeled bicarbonate, lowered to the depths from which they were taken, and incubated between LAN and sunset. In cases where the euphotic depth (depth receiving 1% of surface light intensity) was very shallow, sampling depths were determined by the position of the halocline or thermocline. Occasionally, samples were taken below the euphotic zone to estimate chlorophyll presence and primary production.

During the R/V *Alaminos* cruises, inorganic nutrients (phosphates, silicates, and nitrates) were detected according to the methods outlined in Strickland and Parsons (1970) by using the *Technicon Auto-Analyzer*. When other research vessels were used, the water samples were frozen and shipped to the Texas A&M University campus for later analysis. The physical data, other than those taken on board the R/V *Alaminos* cruises, were furnished by The University of Texas Marine Science Institute and the Southwest Research Institute.

RESULTS

Phytoplankton standing crop

The pertinent physical, chemical, and biological data collected between June 1972 and January 1974 are given in El-Sayed's (1974) report to GURC.

Surface chlorophyll *a* at the platform ranged between 0.18 mg/m^3 and 8.44 mg/m^3 with a mean value of 1.51 mg/m^3. The corresponding values for the control site were 0.38, 10.08, and 1.87 mg/m^3, respectively. Integrated chlorophyll *a* values (i.e., integrated throughout the water column between surface and near bottom) at the two sampling sites showed trends similar to those observed for surface chlorophyll. These values ranged between 5.18 and 75.81 mg/m^2 at the platform (mean, 26.79 mg/m^2), and between 10.33 and 84.40 mg/m^2 (mean, 20.31 mg/m^2) at the control site.

The vertical distribution of chlorophyll *a* at both the platform and control site (figure 2) showed that except for June and September 1972, when the chlorophyll maximum layer showed the sub-surface maximums, and April 1973, when it showed the surface maximum, chlorophyll *a* was homogeneously distributed throughout the water column, thus indicating that the water columns were well mixed. The largest chlorophyll maximums were found at a depth of 11 m in June 1972 (12.85 mg/m^3),

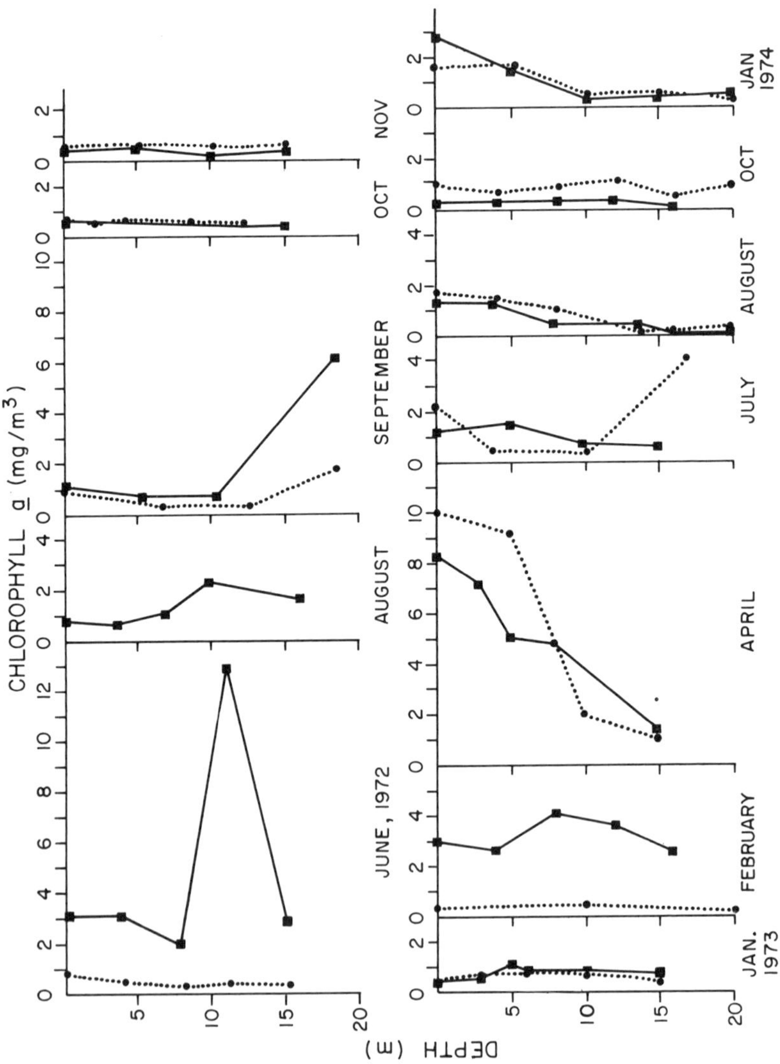

FIG. 2. THE VERTICAL DISTRIBUTION OF CHLOROPHYLL *a* (mg/m^3) at the platform (solid line) and control (dotted line) stations during the period June 1972-January 1974.

at 19 m in September 1972 (9.36 mg/m^3), and at the surface in April 1973.

The vertical distribution of chlorophyll at the control site displayed a pattern very similar to that found at the production platform. In June 1972 and July 1973, however, the vertical distributions of chlorophyll at the control site were markedly different from those at the platform. While the June 1972 vertical distribution showed a relatively homogeneous distribution of chlorophyll *a* at the control site, the distribution at the platform was characterized by a conspicuous maximum at 11 m. In July 1973 the chlorophyll maximum layer was located near the bottom of the control site, while at the platform, chlorophyll was homogeneously distributed throughout the water column, and no maximum concentration was discernible.

Phaeopigment distribution

The distribution of phaeopigments at both the platform and the control site (figure 3) presented a picture not very different from that of chlorophyll. The only noticeable variations occurred in April and July 1973. The high standing crops of phytoplankton that were recorded at the platform and control sites in April 1973 corresponded to very low phaeopigments at both localities. On the other hand, the rather low values of chlorophyll in July 1973 corresponded to high phaeopigment values recorded for that month.

The average surface phaeopigment value at the platform (1.06 mg/m^3) was nearly the same as that at the control site (1.02 mg/m^3). In integrated values, the average value at the platform (18.25 mg/m^2) was slightly higher than at the control (16.96 mg/m^2). Subsurface phaeopigment levels were generally low, except for the September 1972 and July 1973 values. With a few exceptions, however, subsurface values were generally higher than surface ones.

Primary production

Surface primary production was generally low during June and October 1972 and January 1973. During these months, productivity was less than 5 $mgC/m^3/hr$. A much higher value (12.04 $mgC/m^3/hr$) was obtained during November 1972. Between April 1973 and January 1974, surface primary production was conspicuously higher than during the preceding period. The average primary production was 26.04 $mgC/m^3/hr$ for surface water samples collected at the platform, and 27.03 $mgC/m^3/hr$ for those collected at the control site.

Average primary productivity values of the water column were slightly higher at the platform (101.04 $mgC/m^2/hr$) than at the control site (96.08 $mgC/m^2/hr$). These integrated primary production values

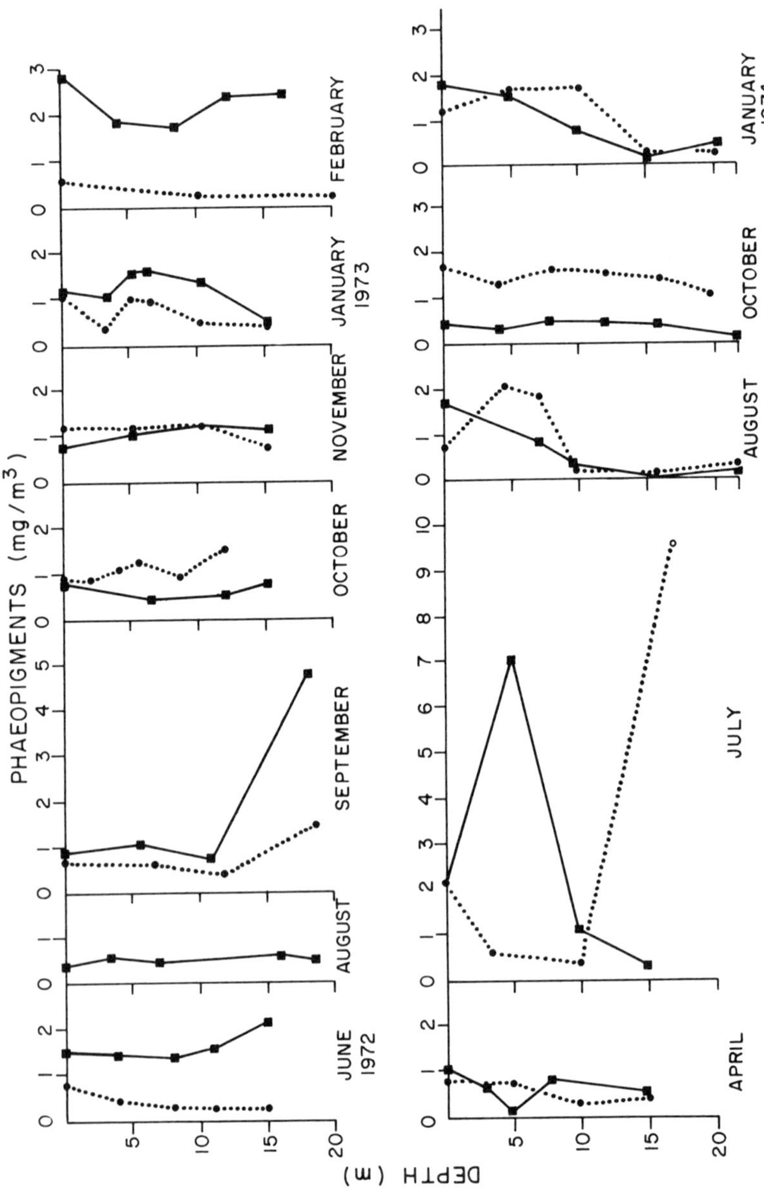

FIG. 3. THE VERTICAL DISTRIBUTION OF PHAEOPIGMENTS (mg/m^3) at the platform (solid line) and control (dotted line) stations during the period June 1972-January 1974.

showed trends similar to those exhibited by the surface water samples. Between June 1972 and January 1973, integrated values were much lower than the values obtained between April 1973 and January 1974. Primary production reached maximum levels during April 1973 at both the platform (283.89 mgC/m^2/hr) and the control site (247.57 mgC/m^2/hr). On the whole, the average values of primary production at the platform and the control site did not vary appreciably.

With very few exceptions (June 1972), primary production was maximized at the surface and decreased rapidly with depth (figure 4). Very low photosynthetic activities were recorded near the bottom of the platform and control site. However, a noticeable increase in the photosynthetic activity was found just above the bottom of the platform in September 1972.

Distribution of inorganic nutrients

The vertical distributions of silicates, phosphates, and nitrates showed, in general, similar trends at the production platform and the control site throughout the study period (figure 5). Major differences in the distribution of the inorganic nutrients were found in April and July 1973.

Seasonal variations of chlorophyll *a*, phaeopigments, and primary production at the production platform and the control site

Figure 6 shows the month-to-month variations in surface chlorophyll *a* at both the platform and the control site. After reaching a high surface chlorophyll value of 3.0 mg/m^3 in June 1972, the phytoplankton standing crop at the platform between August 1972 and January 1973 remained below 1.0 mg/m^3. The standing crop values seemed to reach their lowest ebb during November 1972 and January 1973. This was followed by a marked increase during February 1973, which culminated in the bloom of *Skeletonema costatum* in April 1973, resulting in the highest chlorophyll value recorded at the platform. In July and August 1973, the chlorophyll *a* values decreased to about one-fifth the preceding April values. A further decrease in October was followed by a higher value during January 1974. The monthly distributions of chlorophyll *a* at the control site, with some minor variations (for instance, June 1972 and February 1973), were similar to those at the platform.

The integrated chlorophyll *a* values at the platform (figure 7) showed that following the high value recorded during June 1972 there was a steady decline in the period between September 1972 and January 1973. A marked increase in the standing crop was noted in February 1973, with the phytoplankton density reaching a high value in April 1973, comparable to that of June 1972. A steady decline in the standing crop was

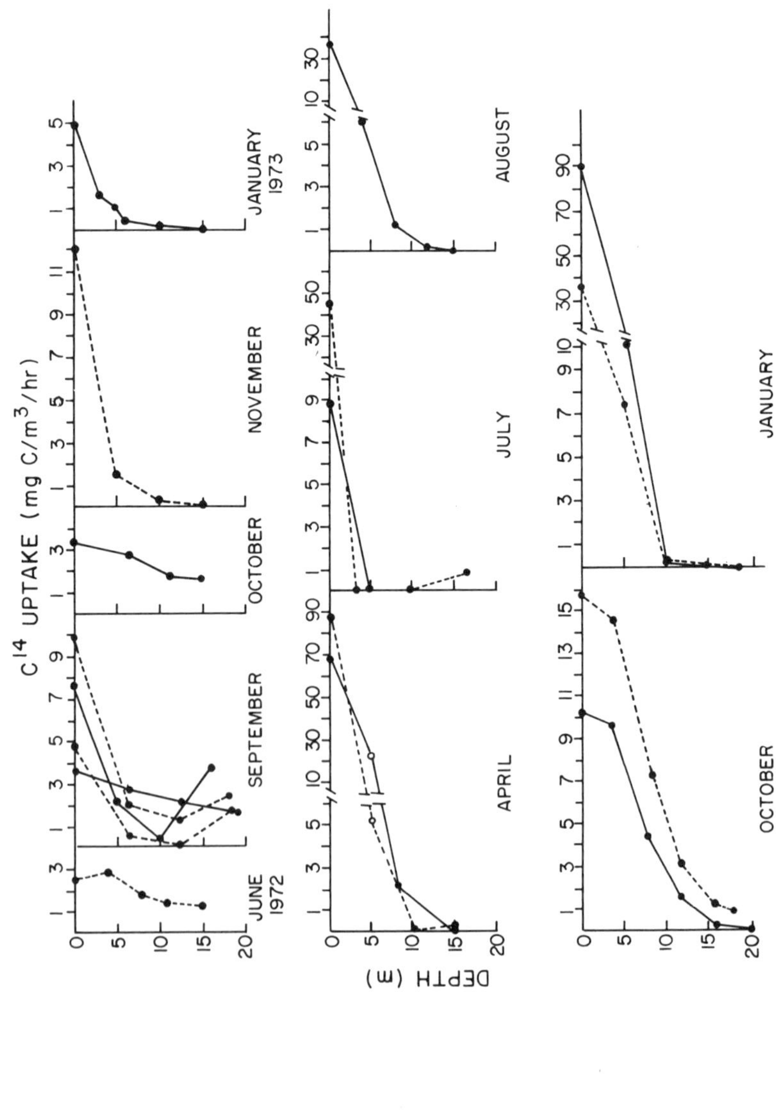

FIG. 4. THE VERTICAL DISTRIBUTION OF C^{14} UPTAKE (mgC/m^3/hr) at the platform (solid line) and control (dashed line) stations during the period June 1972-January 1974.

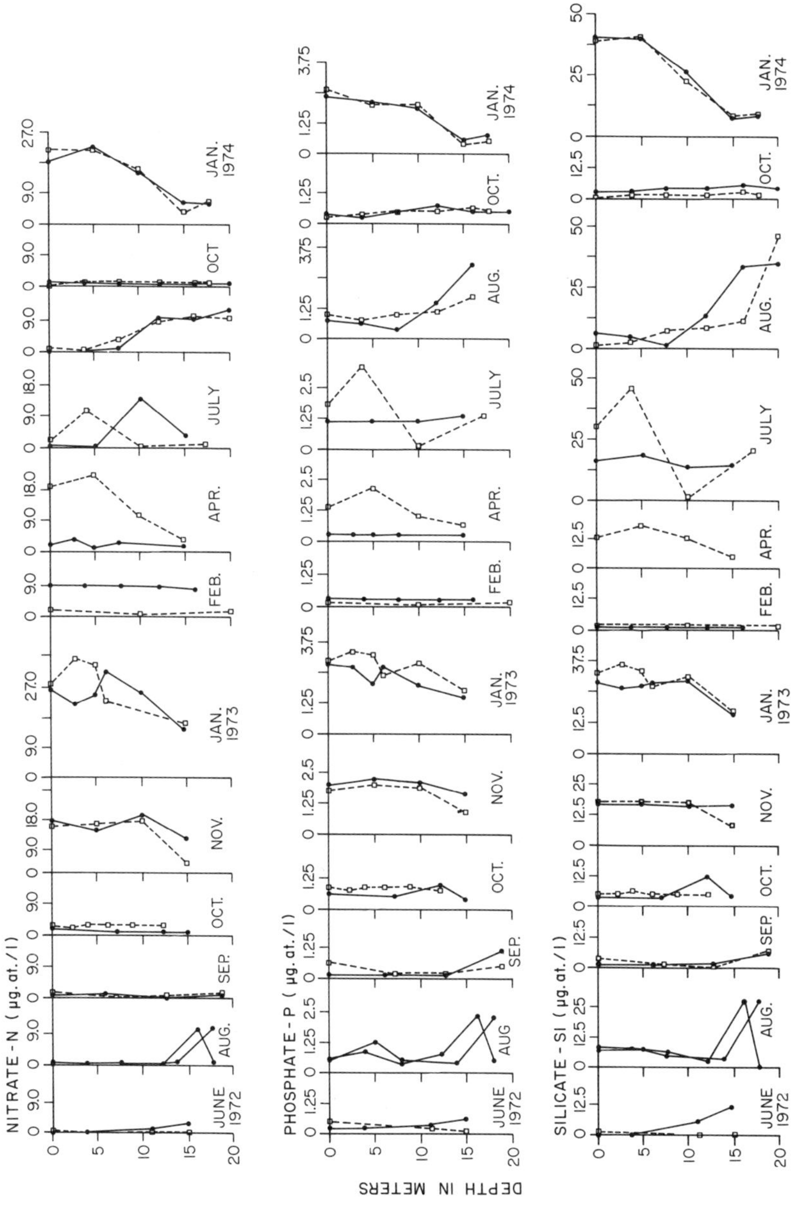

FIG. 5. THE VERTICAL DISTRIBUTION OF THE NUTRIENT SALTS in μg-at/l at the platform (solid line) and control (dashed line) stations during the period June 1972-January 1974.

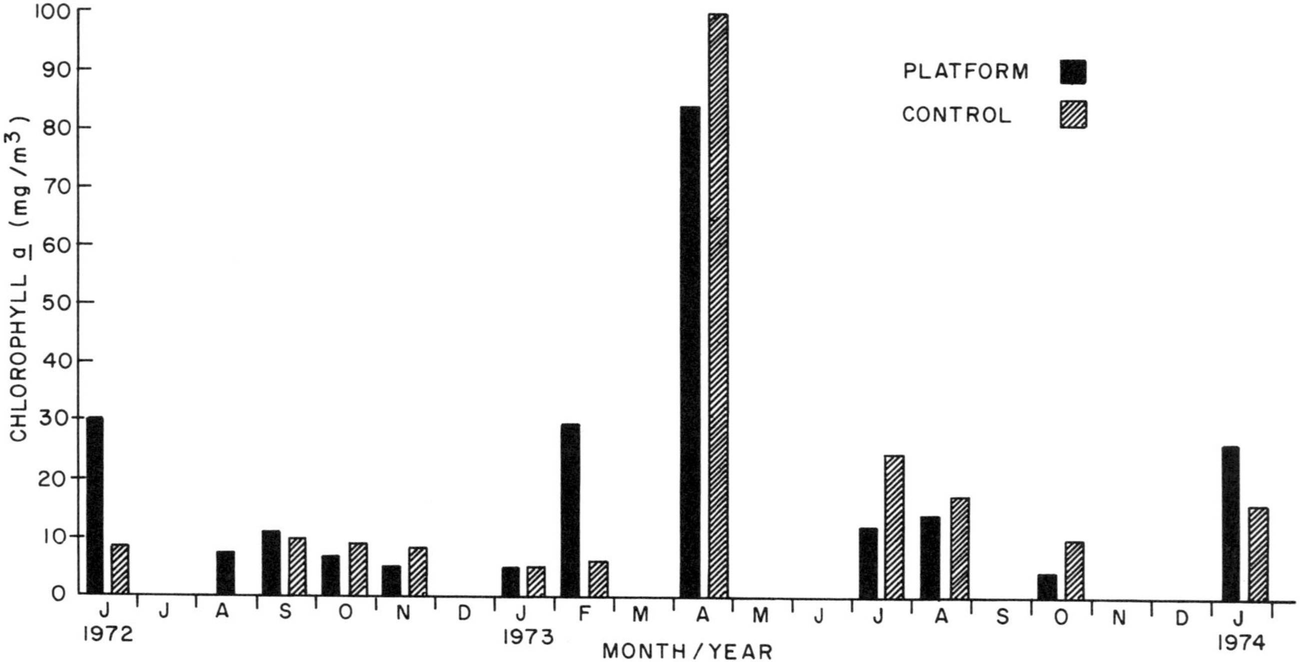

FIG. 6. THE MONTHLY VARIATIONS IN THE SURFACE PHYTOPLANKTON STANDING CROP (mg chl. *a*/m³) at the platform and control stations during the period June 1972-January 1974.

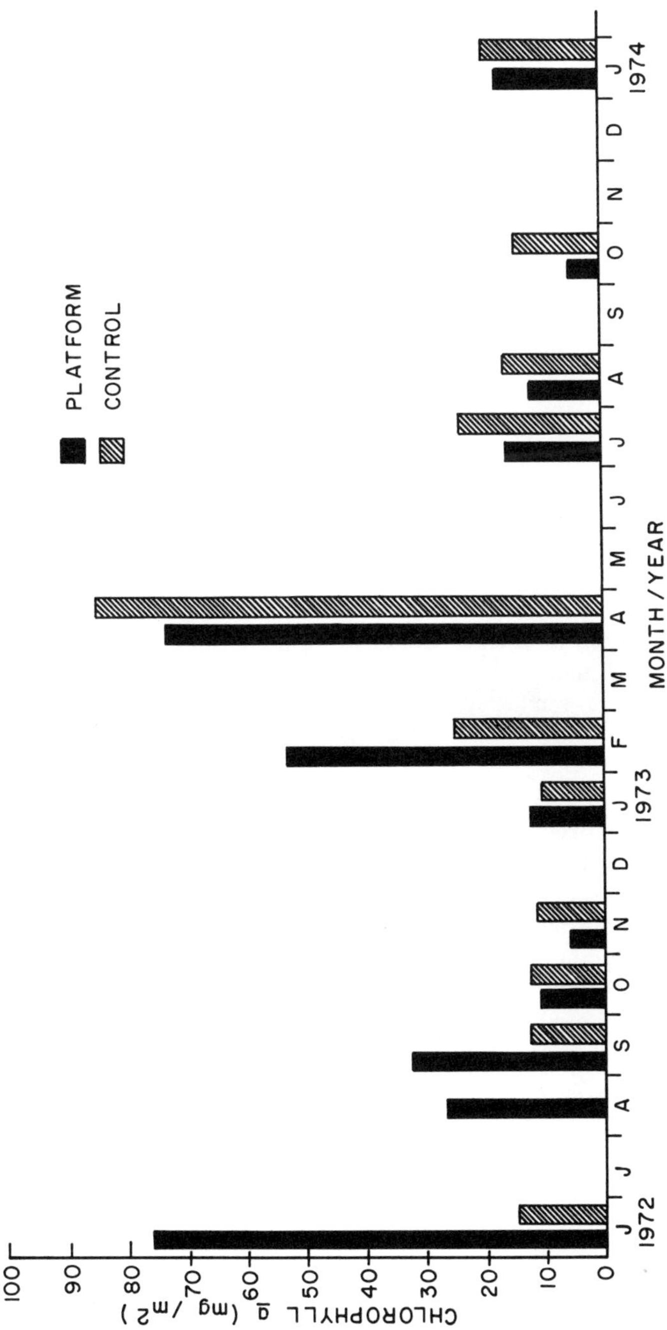

FIG. 7. THE MONTHLY VARIATION IN THE INTEGRATED PHYTOPLANKTON STANDING CROP (mg chl. *a*/m^2) at the platform and control stations during the period June 1972-January 1974.

noted the following July and August, with integrated chlorophyll reaching its lowest value during October. A slight increase in the standing crop was recorded from October 1973 to January 1974. In general, the monthly distributions of chlorophyll *a* at the control site closely followed those at the platform; notable exceptions were the values for June and September 1972 and February and October 1973.

Monthly variations of surface primary production at both the platform and the control site are plotted in figure 8. The photosynthetic rate at both sites was very low between June 1972 and January 1973. Very high C^{14} uptake values were recorded at both places in April 1973, during the bloom of *Skeletonema costatum*. Following the high April values was a marked decline in the photosynthetic activity of the phytoplankton at the platform in July, which was in marked contrast to the high primary production at the control site during the same month. The C^{14} uptake showed an increase in August at the platform (no productivity measurements were made at the control site during that month), and then a decrease in October. The highest primary productivity value during this study was encountered at the platform during January 1974 (90 $mgC/m^3/hr$).

Figure 9 shows the marked contrast between the generally low water column productivity values obtained between June 1972 and January 1973, and the high values recorded between April 1973 and January 1974. Note the remarkable similarity (except for July 1973) in the distribution of the monthly C^{14} uptake values between the platform and the control site.

Work on seasonal distribution of the nutrients showed that high levels generally occurred in winter and low levels occurred during the rest of the year (figures 10 and 11). The unusually heavy rains in the spring of 1973 resulted in an increase in the discharge of the Mississippi River in the spring and summer of that year, and phosphate and silicate concentrations were much higher in the summer of 1973 than in the previous year. Data collected on nitrates during the winter and fall cruises were not sufficient to allow generalizations to be made on their seasonal variability. Integrated nutrient values showed monthly distributions similar to those at the surface. Nutrient concentrations were consistently higher at the control site than at the platform.

The distribution of inorganic nutrients observed during the summer and fall lends support to the conclusion by Thomas and Simmons (1960) that the availability of nitrogen probably limits phytoplankton growth during these seasons. The present study also corroborates the findings of Riley (1937), and Thomas and Simmons (1960), who reported high concentrations of phosphates in the plume waters of the Mississippi River.

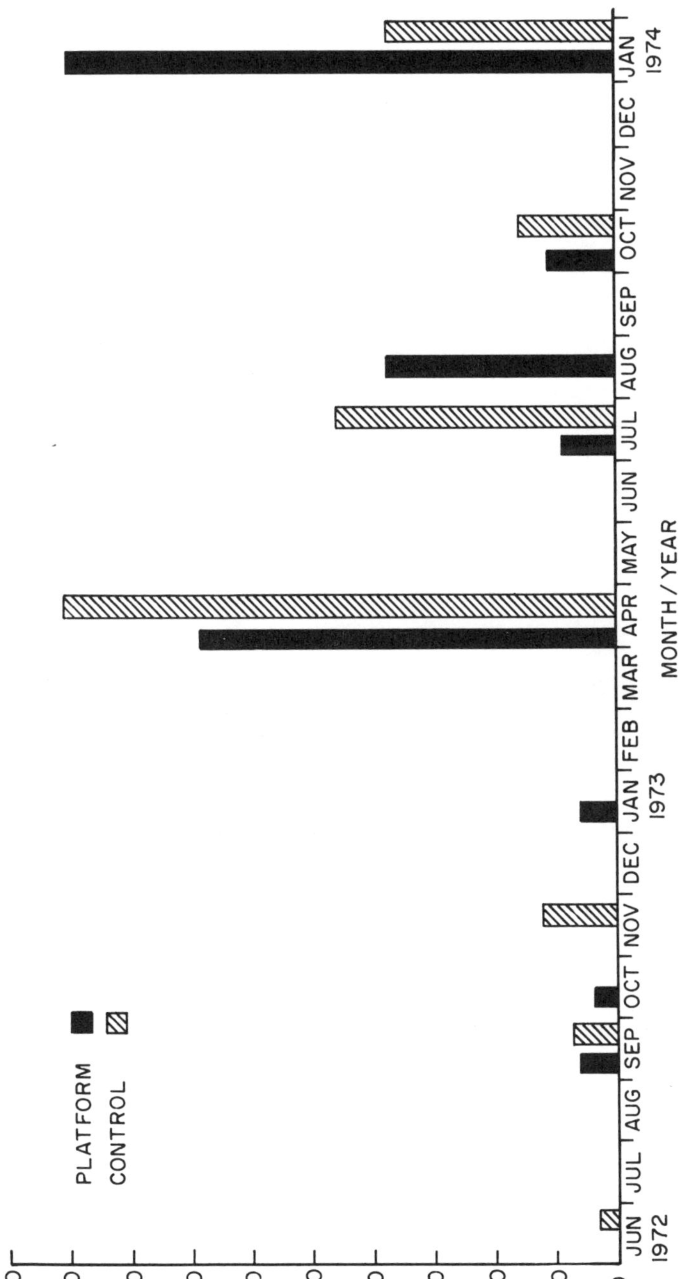

FIG. 8. THE MONTHLY VARIATION IN THE SURFACE C^{14} UPTAKE (mgC/m^3/hr) at the platform and control stations for the period June 1972-January 1974.

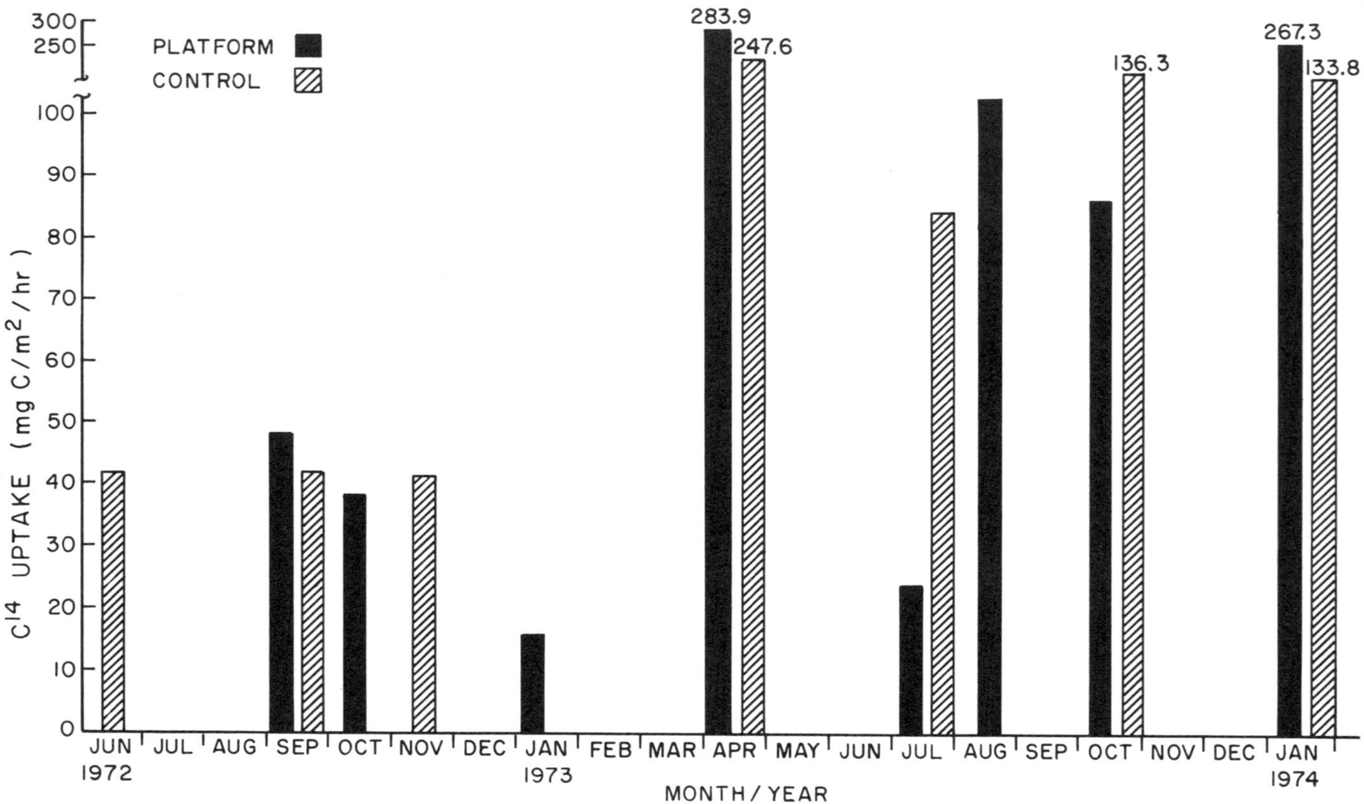

FIG. 9. THE SEASONAL VARIATIONS IN THE MONTHLY INTEGRATED C^{14} UPTAKE ($mgC/m^2/hr$) at the platform and control stations for the period June 1972-January 1974.

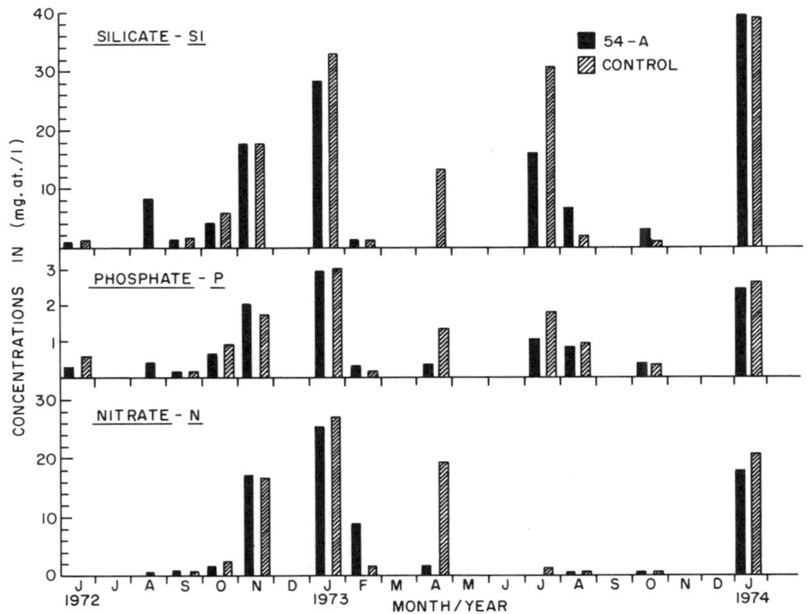

FIG. 10. MONTHLY VARIATIONS IN THE SURFACE NUTRIENT CONCENTRATIONS (mg-at/l) at the platform and control stations during the period June 1972-January 1974.

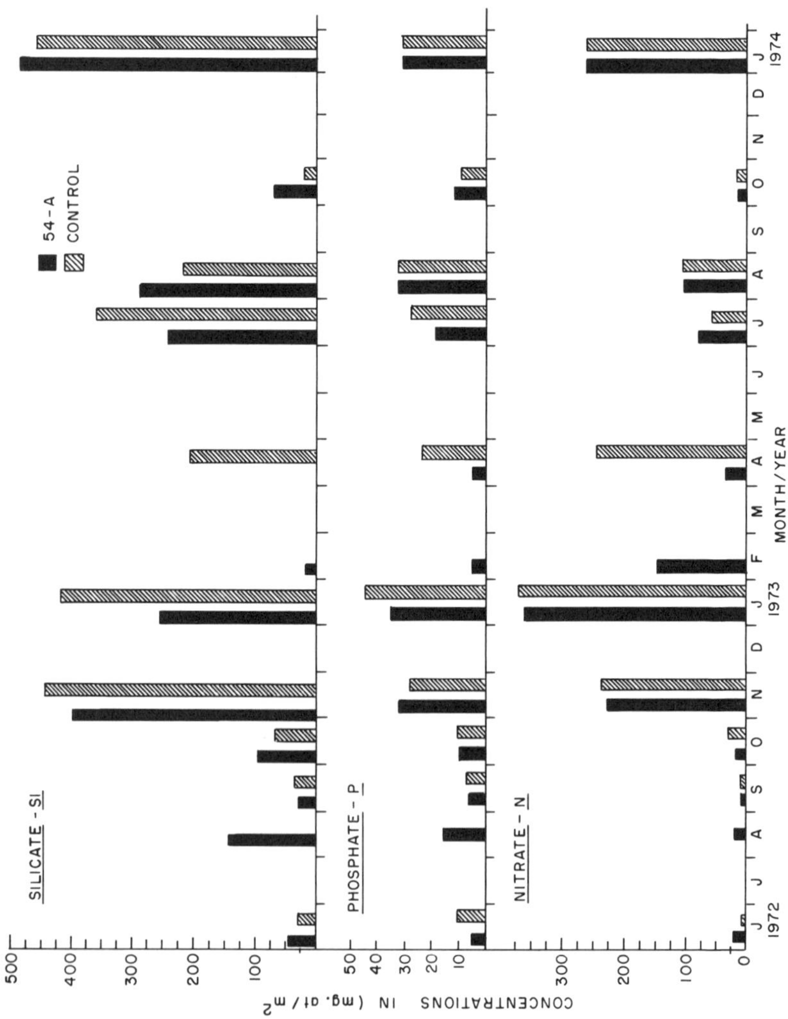

FIG. 11. SEASONAL VARIATIONS IN INTEGRATED NUTRIENT CONCENTRATIONS at the platform and control stations.

Species composition and relative abundance of phytoplankton

Throughout most of the study period, diatoms constituted over 90% of the phytoplankton populations in the OEI study area (table 1). The number of species of diatoms was always greater than that of the dinoflagellates. The dinoflagellates reached their greatest abundance during September 1972, when they contributed about 28% of the total population at the control site.

Diatom abundance was very pronounced at the platform in September 1972 and April 1973. During these months a single species, *Skeletonema costatum*, contributed between 75% and 97% of the total cell counts (in excess of 9×10^6 cells/liter in September 1972). The dominance of this species in the study area was apparent throughout much of the year, as it made up 40% of the total phytoplankton biomass in January 1973 and 74% in October 1973.

Other diatom species that made significant contributions to the standing crop at certain times of the year were the *Cyclotella-Melosira* complex (greater than 90% in July 1973); *Rhizosolenia fragillissima* (greater than 86% in August 1973); and to different degrees at various times, *Nitzschia pungens* and *N. subfraudulenta*. The most dominant dinoflagellate was *Exuviella compressa*, which was present during much of the year. *Gonyaulax monilata* and *G. turbynei* also showed periods of dinoflagellate dominance.

Phytoplankton numbers were lowest in October and November 1972; these values correlated well with the low chlorophyll *a* values recorded during the same time period. In general, the phytoplankton cell counts paralleled the estimates of chlorophyll *a*.

The composition of the phytoplankton populations at the platform and control sites was similar. The diatom populations encountered in this study were also similar to those reported by Simmons and Thomas (1962) in the eastern delta region. The diatom species encountered were nearly identical in both sampling programs, so the scheme presented in Simmons and Thomas' study for identifying and delineating water masses in the eastern delta region should also be valid for the region west of the river mouth. This scheme utilized the *Melosira-Cyclotella* complex to identify the river and plume river waters, and the *Nitzschia*, *Thalassiothrix*, *Thalassionema*, *Skeletonema*, *Chaetoceros*, and *Asterionella* associations to indicate the more saline Gulf waters. Our studies and those of Simmons and Thomas clearly indicate that, in general, the dinoflagellates do not significantly contribute to the phytoplankton standing crop or primary production in the study area.

TABLE 1. Species Composition and Percent Contribution of the Ten Most Abundant Species in the Phytoplankton Samples Collected at the Production Platform and the Control Site.

Month	Sampling Site	Species	Percent Contribution
June 24, 1972*	Platform	*Nitzschia pungens*	
		Chaetoceros curvisetus	
		Guinardia flaccida	
		Rhizosolenia delicatula	
		C. affinis	
		Thalassionema nitzschioides	
		C. laciniosus	
		C. decipiens	
		R. setigera	
		Bacteriastrum delicatulum	
June 24, 1972*	Control	*Nitzschia pungens*	
		Rhizosolenia delicatula	
		Chaetoceros affinis	
		C. lorenzianus	
		C. atlanticus	
		Thalassionema nitzschioides	
		C. messanensis	
		C. pseudocurvisetus	
		C. didymus	
		Bacteriastrum delicatulum	
August 2, 1972*	Platform	*Guinardia flaccida*	
		Nitzschia sp.	
		Rhizosolenia stolterfothii	
		R. styliformis	
		Thalassiosira oestrupii	
Sept. 20, 1972	Platform	*Skeletomena costatum*	94.4%
		Nitzschia pungens	< 1
		Thalassionema nitzschioides	< 1
		Rhizosolenia setigera	< 1
		R. fragilissima	< 1
		Thalassiothris mediterranea	< 1
		N. closterium	< 1
		R. delicatula	< 1
		Thalassiosira oestrupii	< 1
		Guinardia flaccida	< 1
Sept. 20, 1972	Control (surface)	*Rhizosolenia delicatula*	59%
		Gonyaulax monilata	15
		G. turbynei	4
		Chaetoceros pelagicus	4
		Skeletonema costatum	2
		G. polygramma	< 1
		Exuviaella compressa	< 1
		Ceratium fusus	< 1
		Thalassiosira oestrupii	< 1
		Thalassionema nitzschioides	1

*Samples were taken by a 35 μ net haul (vertical); no quantitative studies were made during these months.

TABLE 1 continued

Oct. 20, 1972	Platform (surface)	*Nitzschia closterium*	23%
		Rhizosolenia shrubsolei	18
		Thalassionema nitzschioides	12
		Peridinium trochoideum	10
		N. pungens	9
		N. sp.	6
		Guinardia flaccida	3
		Denticula elegans	3
Oct. 20, 1972	Control (surface)	*Rhizosolenia shurbsolei*	25%
		Denticula elegans	17
		Skeletonema costatum	17
		Thalassionema nitzschioides	9
		Melosira sp.	8
		Nitzschia closterium	3
		N. vitrea	3
		N. subfraudulenta	3
		R. setigera	2
		Gonyaulax turbynei	2
		Prorocentrum gracile	2
Nov. 24, 1972	Platform (surface)	*Synedra acus*	38%
		Nitzschia subfraudulenta	23
		Skeletonema costatum	9
		Thalassiosira eccentrica	5
		Melosira sp.	4
		Navicula hyalina	3
		Nitzschia sp.	2
		N. closterium	2
		Coscinodiscus centralis	1
		Pleurosigma naviculaceum	1
		Dictyocha fibula	1
Nov. 24, 1972	Control (surface)	*Thalassionema nitzschioides*	38%
		Thalassiosira eccentrica	17
		Melosira sp.	16
		Asterionella japonica	7
		Exuviaella compressa	5
		Coscinodiscus centralis	4
Jan. 10, 1973	Platform (surface)	*Skeletonema costatum*	40%
		Leptocylindrus danicus	11
		Rhizosolenia delicatula	9
		Asterionella glacialis	6
		Thalassiosira decipiens	4
		Eucampia cornuta	3
		Nitzschia subfraudulenta	3
		Chaetoceros affinis	3
		Thalassiosira aestivalis	2
		Cerataulina pelagica	2
		Hemiaulus chinensis	2

TABLE 1 continued

Jan. 11, 1973	Control (surface)	*Skeletonema costatum*	21%
		Rhizosolenia delicatula	10
		Melosira sp.	8
		Thalassiosira aestivalis	7
		Corethron criophilum	7
		Leptocylindrus danicus	6
		T. rotula	3
		T. eccentrica	3
		Hemiaulus chinensis	3
		Nitzschia closterium	2
April 9, 1973	Platform (surface)	*Skeletonema costatum*	82%
		Nitzschia pungens	8
		N. subfraudulenta	7
		Chaetoceros curvisetus	< 1
		C. decipiens	< 1
		Exuviaella compressa	< 1
		Leptocylindrus danicus	< 1
		Thalassionema nitzschioides	< 1
		Thalassiothrix mediterranea	< 1
		C. affinis	< 1
April 11, 1973	Control (surface)	*Skeletonema costatum*	75%
		Nitzschia subfraudulenta	8
		N. pungens	8
		Chaetoceros	3
		Leptocylindrus danicus	2
		Asterionella japonica	1
		C. atlanticus	< 1
		Exuviaella compressa	< 1
		Thalassionema nitzschioides	< 1
		Thalassiosira rotula	< 1
July 9, 1973	Platform (surface)	*Cyclotella striata*	>90%
		Thallassionema nitzschioides	< 1
		Cerataulina pelagica	< 1
		Melosira sp.	< 1
		Coscinodiscus radiatus	< 1
		Nitzschia subfraudulenta	< 1
		N. pungens	< 1
		Thalassiosira eccentrica	< 1
		Rhizosolenia alata	< 1
		R. calcar-avis	< 1
July 11, 1973	Control (surface)	*Cyclotella striata*	>90%
		Melosira sp.	< 1
		Coscinodiscus radiatus	< 1
		Pleurosigma naviculaceum	< 1
		Rhizosolenia shrubsolei	< 1
		Nitzschia pungens	< 1
		R. calcar-avis	< 1
		R. alata	< 1
		N. subfraudulenta	< 1
		Thalassiosira eccentrica	< 1

TABLE 1 continued

Oct. 1, 1973	Control (surface)	*Skeletonema costatum*	74%
		Chaetoceros compressus	7
		Rhizosolenia delicatula	7
		C. affinis	2
		Thalassiothrix mediterranea	2
		Nitzschia subfraudulenta	2
		R. stolterfothii	2
		Biddulphia mobiliensis	1
		R. setigera	1
Jan. 16, 1974	Platform (surface)	*Skeletonema costalum*	44%
		Leptocylindrus danicus	29
		Cyclotella striata	14
		Rhizosolenia delicatula	4
		Chaetoceros affinis	3
		Asterionella glacialis	2
		Cerataulina pelagica	1
		Thalassiosira sp.	< 1
		Nitzschia closterium	< 1
Jan. 15, 1974	Control (surface)	*Leptocylindrus danicus*	41%
		Cyclotella striata	38
		Skeletonema costatum	11
		Cerataulina pelagica	3
		Gonyaulax sp.	3
		Thalassiosira eccentrica	2
		Asterionella glacialis	1
		Exuviaella compressa	< 1
		Nitzschia closterium	< 1
Aug. 1, 1973	Platform (surface)	*Ceratulina pelagica*	95%
		Thalassionema nitzschioides	1
		Nitzschia closterium	< 1
		Thalassiosira sp.	< 1
		Rhizosolenia stolterfothii	< 1
		Nitzschia pungens	< 1
		Cyclotella sp.	< 1
		Nitzschia sp.	< 1
		Peridinium pellucidum	< 1
		Gonyaulux diegensis	1
Aug. 1, 1973	Platform (4 m)	*Cerataulina pelagica*	95%
		Rhizosolenia stolterfothii	2
		Thalassionema nitzschoides	< 1
		Nitzschia closterium	< 1
		Thalassiosira sp.	< 1
		Nitzschia pungens	< 1
		Rhizosolenia alata	< 1
		Cyclotella sp.	< 1
		N. sp.	< 1
		R. calcar-avis	< 1

TABLE 1 continued

Aug. 1, 1973	Platform (8 m)	*Cerataulina pelagica*	92%
		Rhizosolenia stolterfothii	5
		Thalassionema nitzschioides	2
		Nitzschia closterium	< 1
		N. pungens	< 1
		R. alata	< 1
		Thalassiosira sp.	< 1
		Gonyaulax fragilis	< 1
		Gonyaulax sp.	< 1
		Nitzschia sp.	< 1
Aug. 1, 1973	Platform 54A (12 m)	*Cerataulina pelagica*	58%
		Thalassiosira sp.	17
		Nitzschia pungens	17
		Thalassionema nitzschioides	4
		N. closterium	< 1
		N. sp.	< 1
Aug. 1, 1973	Platform (16 m)	*Cerataulina pelagica*	40%
		T. nitzschioides	20
		Cyclotella sp.	20
		Nitzschia sp.	20
Aug. 1, 1973	Platform (20 m)	*Cerataulina pelagica*	100%
Oct. 1, 1973	Platform (surface)	*Skeletonema costatum*	74%
		Rhizosolenia delicatula	6
		Nitzschia closterium	6
		R. calcar-avis	3
		Nitzschia subfraudulenta	2
		Chaetoceros lorenzianus	2
		N. pungens	< 1
		R. stolterfothii	< 1
		R. styliformis	< 1
		C. affinis	< 1
		Asterionella glacialis	< 1

DISCUSSION

The main objective of the OEI was to determine the cumulative effects of low-level discharge from offshore oil and gas operations on the ecological health of the marine ecosystem. To detect any effects, the investigation had to be carried out in an environment that had been subjected to prolonged oil and gas drilling and production. Therefore, a region that had experienced nearly 40 years of such operations was selected. Our experimental approach required that drilling and production operations be considered as possible "point sources" for discharges (i.e., Platform 54A), and that biological and chemical attributes be compared with those of "unimpacted" areas (i.e., the control site).

TABLE 2. Chlorophyll a (chl. a), Phaeopigments (Phaeo.), C^{14} uptake, and the Photosynthetic Index (P.I.) at the Platform and Control Sites during the Period June 1972 to January 1974.

		Min.	Max.	Mean	Std.Dev.	% Obs.
Surface chl. a (mg/m^3)	Platform	0.18	8.44	1.51	2.03	16
	Control	0.38	10.08	1.87	2.73	11
Int. chl. a (mg/m^2)	Platform	5.18	75.81	26.79	22.85	17
	Control	8.50	84.40	20.31	21.64	12
Surface Phaeo. (mg/m^3)	Platform	0.19	2.82	1.06	0.79	15
	Control	0.69	2.11	1.02	0.47	11
Int. Phaeo. (mg/m^2)	Platform	8.25	47.17	18.25	11.67	15
	Control	8.18	42.56	16.96	10.35	11
Surface C^{14} ($mgC/m^3/hr$)	Platform	3.42	67.92	26.04	22.98	9
	Control	2.50	87.98	27.03	31.05	8
Int. C^{14} ($mgC/m^2/hr$)	Platform	15.29	283.89	101.04	87.44	9
	Control	26.45	247.57	96.08	78.44	8
P.I. (mgC/mg chl/hr)	Platform	3.74	27.18	11.76	9.49	8
	Control	3.05	20.33	10.95	6.17	7

Table 2 summarizes the surface and integrated (i.e., throughout the water columns) data on the standing crop of phytoplankton, primary production, phaeopigments, and photosynthetic index (P.I.) at the production platform and the control site. There was close similarity between the levels of the parameters at both sites. Of special significance is the close correspondence between the photosynthetic index (calculated as the ratio of primary production to chlorophyll *a*) at the platform and the control site (11.76 and 10.95 mgC/mg/chl. *a*/hr, respectively). To the extent that the P.I. values reflect the state of ecological health of the phytoplankton populations (Strickland 1960), the closeness of these values suggests that the production platform has had no detectable harmful effects on the phytoplankton production in the OEI area. The uniformity observed between the vertical profiles of chlorophyll *a*, phaeopigments, primary production, and nutrients at both the platform and control sites tends to corroborate this finding. Moreover, the lack of variation in the above parameters between the platform and the control indicates that the two sites were influenced by the same factors.

There was no evidence to suggest that the platform operations added large amounts of sediments to the water columns. Increased sedimentation could have attenuated light penetration and decreased the amount of primary production in the water column. While the presence of a turbid layer near the bottom caused some concern in the early stages of this investigation, Griffin and Ripy (1973) found no connection between the origin of the turbid layer and the presence of the drilling platform in the study area.

Studies of the toxic metals (Brady and Montalvo 1973) and of barium (Chan and Hanor 1973), which were carried out as components of the OEI program, showed that concentrations of these elements were generally greater near the platform than at the control site. Nevertheless, the similarity in the phytoplankton growth between the two sites suggests that these chemicals had no inimical effect on phytoplankton production in the OEI area.

The results of our investigation would have been more significant if baseline studies had been undertaken prior to the construction of oil platforms in the Gulf of Mexico some 40 years ago. Lacking adequate recent baseline data for the OEI area, however, we found the data on phytoplankton production and species composition collected east of the Mississippi delta nearly 20 years ago by Thomas and Simmons (1960) and Simmons and Thomas (1962) to be very useful. Comparisons between the data sets show that the standing crop, primary production, and species composition of phytoplankton remain relatively unchanged. Further, variations in the seasonal distribution of the standing crop of phytoplankton and primary production were almost identical to those reported by Thomas and Simmons.

The single most important environmental event during the OEI study was the 100-year maximum flood of the Mississippi River, which occurred in April and June 1973. While there is no denial that the Mississippi runoff greatly influences the hydrography, nutrient chemistry, and biological productivity of the north central and western Gulf of Mexico (Fucik and El-Sayed, in preparation), one wonders whether the unusual flood in the spring of 1973 masked possible effects attributable to the offshore platforms. While the river flood waters markedly changed the phytoplankton patterns and greatly contributed to the increase in the phytoplankton standing crop and primary production in the study area, we doubt that the flood waters masked any platform effect. For one thing, the first effect of the flood waters was not observed until July 1973, and prior to this time, normal phytoplankton cycles were observed. The surface bloom observed in September 1972 and the massive bloom of *Skeletonema costatum* in April 1973 took place without influence from the flood waters or occurred just as the plume waters began to reach the area. The effect of the spring 1973 flood could still be detected in the OEI area as late as January 1974. During this month, primary productivity was an order of magnitude higher than it had been the previous year. How long beyond January 1974 these productivity levels remained above normal we do not know. However, Sklar (1976), in a

TABLE 3. Average Values of Chlorophyll *a* and C^{14} Uptake in the OEI Study Area Compared with the Gulf of Mexico.

Parameter	OEI Platform	OEI Control	Gulf of Mexico
Average chlorophyll *a* (surface) (mg/m^3)	1.51	1.87	0.20
Average chlorophyll *a* (integrated) (mg/m^2)	26.79	20.31	12.42
Average C^{14} uptake (surface) ($mgC/m^3/hr$)	26.04	27.03	17.5
Average C^{14} uptake (integrated) ($mgC/m^3/hr$)	101.04	96.08	6.63

separate study, measured phytoplankton production in the OEI area some time after the flood, and his data indicate that normal (i.e., pre-flood) conditions were being observed at least by November 1974.

Further evidence that production and drilling operations did not have harmful effects on the phytoplankton populations in the OEI region is attested to by the fact that the OEI area is one of the most productive regions in the Gulf of Mexico. This is shown in table 3, where the average surface and integrated chlorophyll *a* values measured in the OEI study are compared with those of the Gulf of Mexico (El-Sayed 1972).

It is clear from table 3 that in its surface phytoplankton standing crop, the OEI region is several times richer than the Gulf of Mexico; the integrated values, however, are only double those of the Gulf. The primary productivity data in the water columns obtained at both the production platform and at the control site are more than an order of magnitude higher than the primary productivity of the Gulf of Mexico.

ACKNOWLEDGMENTS

We thank Mr. Hussein Abdel Reheim for his assistance in nutrient analysis, Dr. J. T. Turner for his help with the collection of the field data, and Mr. W. J. Hyoun for the statistical analysis of the data collected.

Grateful acknowledgment is extended to the Petroleum Allied Industries for the financial support of this study and to Mr. Joe Tyson of Gulf Universities Research Consortium for his excellent cooperation and assistance throughout the investigation.

REFERENCES CITED

Brady, D. V. and J. G. Montalvo. 1973. Water column measurements—toxic metals (first year of the OEI). Gulf Universities Research Consortium. Report No. 131.

Chan, L. H. and J. Hanor. 1973. Input of barium into the marine environment as a result of offshore drilling operations. Gulf Universities Research Consortium. Report No. 131.

El-Sayed, S. Z. 1972. Primary productivity and standing crop of phytoplankton. *In* V. C. Bushnell, ed., Chemistry, Primary Productivity, and Benthic Algae of the Gulf of Mexico. New York: American Geographic Society, Folio 22, pp. 8-13.

———. 1974. Effect of oil production on the ecology of phytoplankton off the Louisiana coast. Gulf Universities Research Consortium. Appendix to Report No. 138.

Fucik, K. W. and S. Z. El-Sayed. Impact of the 1973 spring flood of the Mississippi River on phytoplankton productivity of offshore Louisiana. (In preparation)

Galtsoff, P. S., H. F. Prytherch, R. O. Smith, and V. Koehring. 1935. Effects of crude oil pollution on oysters in Louisiana waters. Bulletin of the Bureau of Fisheries **18**:143-210.

Gordon, D. C. and N. J. Prouse. 1973. The effects of three oils on marine phytoplankton photosynthesis. Marine Biology **22**:329-333.

Griffin, G. M. and B. J. Ripy. 1973. Horizontal and vertical distribution of turbidity on the Louisiana Shelf south of Timbalier Bay—August, 1972-July, 1973. Gulf Universities Research Consortium. Report No. 131.

Holm-Hansen, O., C. J. Lorenzen, R. W. Holmes, and J. D. H. Strickland. 1965. Fluorometric determination of chlorophyll. Journal du Conseil (Conseil Permanent International Pour L'Exploration de la Mer, Denmark) **30**:3-15.

Mirnov, O. G. and L. A. Lanskaya. 1969. Growth of marine microscopic algae in seawater contaminated with hydrocarbons. (In Russian) Biologiya Morya **17**:31-38.

Pulich, W. M., Jr., K. Winters, and C. van Baalen. 1974. The effects of a No. 2 fuel oil and two crude oils on the growth and photosynthesis of microalgae. Marine Biology (Berlin) **28**:87-94.

Riley, G. A. 1937. The significance of the Mississippi River drainage for biological conditions in the northern Gulf of Mexico. Journal of Marine Research **1**:60-74.

Simmons, E. G. and W. H. Thomas. 1962. Phytoplankton of the eastern Mississippi delta. Publications of the Institute of Marine Science, The University of Texas **8**:269-298.

Sklar, F. H. 1976. Primary productivity in the Mississippi delta bight near a shallow bay estuarine system in Louisiana. M.S. thesis, Louisiana State University.

Smith, J. E., ed. 1968. Torrey Canyon—Pollution and Marine Life. Report by the Plymouth Laboratory of the Marine Biological Association of the United Kingdom. Cambridge: Cambridge University Press.

Soto, C., J. A. Hellebust, and T. C. Hutchinson. 1975. Effect of naphthalene and aqueous crude oil extracts on the green flagellate *Chlamydomonas angulosa*. II. Photosynthesis and the uptake and release of naphthalene. Canadian Journal of Botany **53**:118-126.

Steemann Nielsen, E. 1952. The use of radioactive carbon (C^{14}) for measuring organic production in the sea. Journal du Conseil (Conseil Permanent International Pour L'Exploration de la Mer: Denmark) **18**:117-140.

Strand, J. A., W. L. Templeton, J. A. Lichatovich, and C. W. Apts. 1971. Development of toxicity test procedures for marine phytoplankton. *In* Proceedings of the Joint Conference on Prevention and Control of Oil Spills, January 15-17. Washington, D.C.: American Petroleum Institute, pp. 279-286.

Straughan, D. 1973. Biological studies of the Santa Barbara oil spill. Santa Barbara Channel Region revisited. AAPG Trip 3, 1973, Annual Meeting American Association of Petroleum Geologists, SEPM. SEG. pp. 4-16.

Strickland, J. D. H. 1960. Measuring the production of marine phytoplankton. Bulletin 122, Fisheries Research Board of Canada.

——— and T. R. Parsons. 1970. A Practical Handbook of Sea-Water Analysis. Ottawa, Canada: Fisheries Research Board, Queen's Printer.

Thomas, W. H. and E. G. Simmons. 1960. Phytoplankton production in the Mississippi delta. *In* Recent Sediment Northwestern Gulf of Mexico, 1951-1958. Tulsa, Oklahoma: American Association of Petroleum Geologists, pp. 103-116.

Yentsch, C. S. and D. W. Menzel. 1963. A method for the determination of phytoplankton chlorophyll and phaeophytin by fluorescence. Deep-Sea Research **10**:221-231.

Wheeler, R. B. 1978. The fate of petroleum in the marine environment. Exxon Production Research Company. Special Report, August.

SIGNIFICANCE OF DISTRIBUTION PATTERNS OF PLANKTONIC COPEPODS IN LOUISIANA COASTAL WATERS AND RELATIONSHIPS TO OIL DRILLING AND PRODUCTION

by James P. Marum

ABSTRACT

Zooplankton collections were made in the OEI area and in two additional Gulf of Mexico coastal areas to determine distribution patterns of holoplankton (single species copepod populations, diversity, biomass) and their relationship to long-term oil drilling and production. The plankton community in Timbalier Bay was found to consist of coastal-neritic and estuarine species, and a community in the inner shelf consisted of oceanic, slope-oceanic, shelf, and coastal-neritic species. Only limited faunal exchange appears to take place (except during severe spring floods) between Timbalier Bay and the adjoining shelf, because of a shallow sill, low tidal amplitude, and barrier islands.

In comparing the collections at the two oil producing platforms with their respective controls, no significant differences (F-test) were found between means of copepod single species populations (abundant species), diversity index values, or total biomass. When the data were subjected to two-way analysis of variance (stations $\times$ sampling dates), differences in counts (log-transformed) of abundant species between platforms and controls were never significant, while seasonal differences were always greater and usually significant. Seasonal variability, which simultaneously was affecting both platforms and controls, is a

---

At the time of this study J. P. Marum was at the Oceanography Department, Florida State University, Tallahassee, FL.

major controlling factor in zooplankton populations in this area. Copepod species composition of the OEI area is typical of coastal waters of the northern Gulf of Mexico, and apparently has not changed since limited records were made in 1952 and 1956. Offshore biomass values were high for neritic waters of the Gulf of Mexico (>0.3 g/m^3); studies in 1937 showed that this phenomenon is due to high primary productivity associated with Mississippi River runoff. These results and observations were interpreted as indicating that there have been no major changes in the quantity or composition of zooplankton in the OEI study areas as a result of long-term oil drilling and production. The hydrographic complexity of the area, however, could effectively mask localized short-term pollution damage, such as from an oil spill, to a portion of a zooplankton population.

INTRODUCTION

A zooplankton study was undertaken as part of the Offshore Ecology Investigation (OEI) in order to determine major distribution patterns of planktonic copepods and possible relationships to long-term oil drilling and production. Quantitative information is lacking on the extent and seasonality of faunal exchange between Gulf of Mexico coastal and estuarine waters, although it has long been known qualitatively that plankton zonation occurs in the Gulf of Mexico (Grice 1957) and other temperate and subtropical waters (Bowman 1971). Because coastal waters are characterized by rapid and large-scale physical and chemical variations, it is necessary to understand the normal distribution patterns for these parameters before one can fully evaluate pollution-caused stresses.

Low levels of pollution occurring over a number of years may adversely affect plankton populations in one or more of the following ways: increased numbers of the more resistant species, increased or decreased biomass, absence or decrease in numbers of the less resistant species, a shift in equitability toward uneven distribution, decrease in diversity index, and increase in patchiness of distribution. Long-term effects of pollutants on plankton (Odum et al. 1963) and fish (Bechtel and Copeland 1970) communities have been reported as occurring in some partially enclosed water bodies, such as in Galveston Bay near the Houston ship channel. Field evidence regarding the effects on plankton communities of chronic oil pollution in coastal waters is lacking. The study reported on herein was undertaken to provide a better understanding of this relationship.

METHODS AND MATERIALS

Locations of stations

A series of 20 stations was established (figure 1), spaced in such a way that a transect was formed connecting upper Timbalier Bay with the neritic waters 29 km from shore. Collections were made during eight sampling trips over an 18-month period in 1972-1974. One of the offshore stations (54A) and one of the bay stations (7) were within 100 m and 30 m, respectively, of oil production stations. An offshore control station (15A) and a bay control station (12) were established for faunal comparisons with the platform stations.

In Timbalier Bay, the oil platform was a tank battery located near the Philo Brice Islands. This platform was chosen because it was the only permanent platform in upper Timbalier Bay. Choosing a control station for Timbalier Bay was difficult, because of the great amount of oil industry activity throughout the bay. Capped oil and gas wells, and signs of previous drilling and production, were present at many places throughout the bay, and drilling rigs were frequently seen operating in the bay. Because it was virtually impossible to select a control station in the bay that had not been exposed to oil industry operations at some time during the previous twenty years, a control station was selected 9 km southeast of the platform, a location that appeared to have undergone little oil industry activity over the course of this study.

The offshore control station was 22 km northeast of the offshore platform. The platform was located approximately in the center of a group of six platforms occupying a 32 km^2 area. No platform was within 7 km of the control station.

To compare the zooplankton of the OEI study area with that of other Northern Gulf of Mexico areas having little or no oil exploration, six additional stations were established, three off the Texas coast sampled in July 1973, and three off the Mississippi coast sampled in October 1973. The Texas stations were 475, 350, and 250 km west of the OEI area, and the Mississippi stations were east of the Mississippi River along a transect extending toward the Chandeleur Islands, designed to be as close as possible to the OEI shelf transect in length, depth range, and salinity range.

Sample collection

Quantitative zooplankton collections were taken with 202 micron (#8) mesh nets with 0.6-meter mouth opening, and having a 5 to 1 net length-to-mouth diameter ratio. Flow meters (TSK) were centered in each net mouth. When opening-closing tows were taken for sampling discrete depths, General Oceanics opening-closing mechanisms (Niskin and Jones

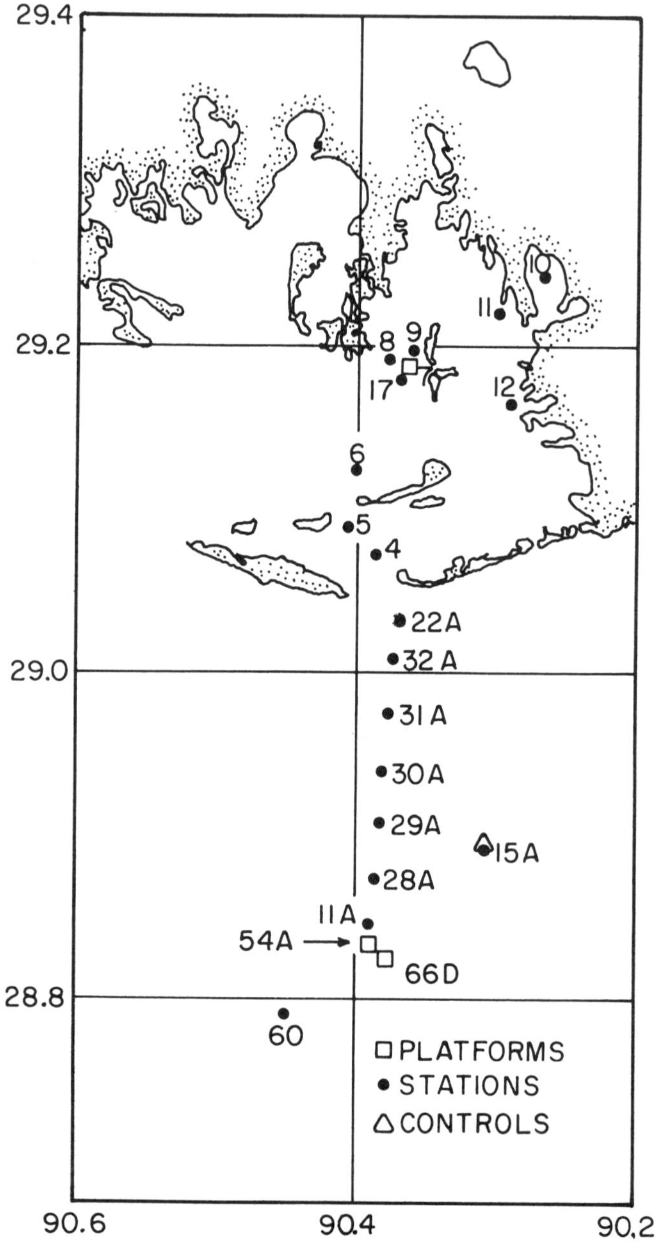

FIG. 1. LOCATIONS OF OEI ZOOPLANKTON STATIONS.

1963) were used with the nets. Some additional collections were taken with 116 micron (#12) mesh nets for determining net loss through the #8 mesh nets.

Attempts were made to take at least one sample over the entire water column at each station during each sampling trip. Because of inclement weather, however, particularly during winter months, sometimes the entire complement of samples was not obtained. Nearly complete complements of samples were collected on eight sampling trips, while the Timbalier Bay stations were sampled on a ninth trip.

Because Timbalier Bay is shallow (average depth, 1.8 m) and well-mixed, only horizontal surface tows were taken at the 10 bay stations. Because of large concentrations of ctenophores and detritus, tow times in the bay were limited to four minutes to avoid clogging. Approximately 50 m^3 of water were filtered in a four-minute tow.

Vertical tows over the entire water column were taken at offshore stations. Vertical tows consisted of a series of 7 to 10 round trip samplings of the water column while the ship was at anchor. With this procedure, sampling took place on the downward tow (with the net inverted) as well as on the upward tow. Seven to ten round trip tows were required to filter 50 m^3 of water. When sea states were high, vertical tows became unmanageable and oblique tows were used. With oblique tows, three depths were sampled, each for two minutes.

Opening-closing tows were taken at selected offshore stations to determine vertical distribution patterns. Three depths were sampled: surface, mid-depth, and bottom. All samples were preserved in 5% formalin buffered in sea water.

Laboratory methods

In the laboratory, samples were placed on #12 netting (116 μ), rinsed for five minutes with fresh water, and voided of interstitial water by allowing samples to drain on several pieces of absorbent paper toweling until more than 50% of the towel surface area beneath the plankton was dry after approximately three minutes. Samples were then weighed on a triple-beam balance. Because Timbalier Bay collections contained large amounts of detritus and ctenophore parts, biomass determinations on bay samples were limited to those few samples in which detritus and ctenophore parts were of minimal importance (determined by microscopic inspection).

Aliquots of the samples were taken with a modified Hopkins subsampler (Hopkins 1962) in preparation for counting. Each aliquot was 1/20 of the original volume. From 1/20 to 1/400 of a sample was counted, depending upon the size of the sample. Three identical aliquots were counted for each sample and the average count was used for determining numbers per cubic meter of water filtered. Depending upon the evenness of the distribution, 400 to 1500 (av. 750) copepods were counted. Approximately 10,000 additional individuals were examined in each sample in searching for rare species. Samples were counted and stored in 70% isopropyl alcohol.

Copepods were identified and counted at species level. Ninety-three copepod species were identified: 44 calanoids, 28 cyclopoids, 17 harpacticoids, and 4 caligoids (Marum 1974).

Name changes from previous reports were required in some cases because species once thought to be found worldwide are now considered to consist of distinct Atlantic and Pacific species (T. E. Bowman, personal communication). Thus, *Centropages fucatus* has become *C. velificatus* (Oliveira).

Bowman (1971) recognized three distinct species in the *Paracalanus parvus* group L: *P. parvus* (Claus), *P. indicus* (Wolfenden), and *P. quasimodo* (Bowman). He indicated that many of the specimens previously reported from tropical and subtropical waters labeled as *P. parvus* are either *P. indicus* or *P. quasimodo.* Both of these species were identified from the OEI collections, but both were counted together as *P. indicus,* the dominant species.

Some of the harpacticoid species were probably benthic harpacticoids. However, because each of these species occurred in more than one sample, I decided to include them as part of the plankton. In addition, several caligoid species (fish parasites) were counted as plankton. This is because they were found in more than one sample, and as Wilson (1932) indicates, are frequently freely swimming in the water column in search of new hosts.

Statistics and community indices

In preparation of statistical analyses, all counts were standardized to numbers of individuals per cubic meter. The numbers/m^3 were then logarithmically transformed prior to analysis of variance. Two-way analysis of variance was used to examine spatial and temporal variations along the transect and between oil platforms and controls. Linear and quadratic regression equations were developed and tested for significant inshore-offshore trends.

A method of grouping together species with similar inshore-offshore distribution patterns was developed (Marum 1974). I evaluated the data by applying the following criteria: (1) frequency of occurrence in Timbalier Bay and offshore, (2) stations at which maximum numbers occurred, and (3) presence of significant linear and quadratic trends.

Species diversity was determined by the Shannon-Weaver (1963) information content index. Despite some criticisms of this method

(Hurlbert 1971), the Shannon-Weaver index is useful because it is dimensionless, is nearly independent of sample size, expresses the relative importance of each species, and is widely used by marine ecologists. To correct for sample size, the Basharin (1959) correction was subtracted from the Shannon-Weaver value. Faunal affinity between samples was determined by the Morisita (1959) index.

Sources of error

Three types of potential error were investigated as part of this study: (1) laboratory subsampling (aliquoting) error, (2) error due to variability of replicate tows, and (3) organism loss through the 202-micron mesh net. Results of tests on reproducibility of the aliquoting procedure revealed low coefficients of variation (±10%) when 17 or more individuals were present in the counting tray, but dramatically increased variation (>40%) when there were fewer than 10 individuals counted per tray (Marum 1974). Although large errors from the aliquoting procedure may occur with the rarer species, these results are not thought to alter significantly the community structure analysis, because I found that species having low counting errors accounted for 98% of the total individuals.

Variability of replicate tows was determined for numbers of individuals of abundant species, diversity index, biomass, and faunal affinity index (table 1). Coefficients of variation (c.v.) for counts of abundant species ranged from 6% to 32% with an average of 16.5%. The average c.v. for diversity was 10% and the c.v. for biomass was also 10%. Furthermore, there was some evidence that replicate tow variability in the bay (stations 12, 4) was greater than variability offshore (station 54A), although data were limited. For example, c.v.'s of species counts in the bay ranged from 17% to 32%, while c.v.'s for offshore species counts ranged from 6% to 14%.

There have been reports of zooplankton loss through all mesh sizes except the very smallest (Hopkins 1966; Deevey 1971). The loss of organisms in estuarine and coastal waters can be large, because estuarine and coastal organisms are generally smaller than oceanic zooplankton. In order to estimate the loss of zooplankton through the 202 micron (#8) mesh, seventeen paired tows were taken with #8 mesh (202 μ) and #12 mesh (116 μ) during January 1974. The results are presented in table 2. It is evident that loss is greatest for those copepods having body volumes less than that of *Acartia tonsa*. *Acartia tonsa* is approximately 1 mm in length, and therefore those copepods less than 1 mm in length are probably not efficiently sampled by a #8 mesh net. Of those organisms not efficiently sampled, *Sapphirella* sp., *Oithona brevicornis*, and *Paracalanus crassirostris* are coastal species (i.e., they are present both in

TABLE 1. Variability of replicate tows. These data were obtained from 6 tows near offshore station 54A in July, 1973, 5 tows at bay station 12 in February, 1973, 4 tows at bay station 4 in July, 1973, and 3 tows at bay station 7 in June, 1973.

Parameter	$\bar{x}$	±	S.D.		C.V.	n
SPECIES						
Acartia tonsa	6095.	±	442.		7.	6
	2440.	±	783.		32.	5
	21700.	±	4958.		23.	4
Paracalanus indicus	380.	±	54.		14.	6
Centropages velificatus	164.	±	10.		6.	6
Paracalanus crassirostris	145.	±	25.		17.	4
Average			$\bar{x}$	=	16.5	
DIVERSITY INDEX (H')						
for n = 6	0.494	±	0.057		12.	6
for n = 5	0.259	±	0.043		17.	5
for n = 4	0.153	±	0.011		7.	4
for n = 3	0.453	±	0.022		5.	3
Average			$\bar{x}$	=	10.2	
BIOMASS						
for n = 6	0.329	±	0.024		7.	6
for n = 4	0.440	±	0.064		14.	4
Average			$\bar{x}$	=	10.5	
FAUNAL AFFINITY INDEX (Cλ)						
for n = 6[1]	0.9996	±	0.0004		0.04	15

[1] Cλ values were determined for all possible pairs of samples. Hence, the mean and standard deviation are computed from 15 values.

TABLE 2. Comparison of average catch of copepods from 116 μ (#12) and 202 μ (#6) mesh nets.

Species	#12	#8	Loss (°/o)	Relative volume[1]
Sapphirella sp.	3810	20	99.	7
Oithona brevicornis	1769	0	100.	10
Paracalanus crassirostris	6063	401	95.	15
Oithona nana	218	0	100.	20
Acartia tonsa	4890	3555	27.	100
Paracalanus indicus	511	405	21.	135
Temora turbinata	468	378	19.	214
Centropages velificatus	25.6	26.7	0.	349
Eucalanus pileatus	242	227	6.	688
Labidocera aestiva	79	59	25.	766
Undinula vulgaris	6.7	9.3	0.	2039

[1] Relative to *A. tonsa*, which equals 100. Volume values taken from Grice 1957, p. 60.

the bay and offshore). *Oithona nana,* a shelf species, does not regularly occur in Timbalier Bay and so it was lost only in offshore collections.

Samples with the #12 net showed that 33% of the total numbers were small (< 1 mm) species offshore, while 40% were small species in the bay. It is possible that losses of small species were even greater than indicated if loss occurred through the #12 mesh. Although loss of numbers influences diversity and faunal affinity measurements, this type of error affects accuracy only, and therefore comparisons between stations in this study remain valid. Furthermore, sampling with the smaller mesh sizes greatly increases the chances for errors of precision, because of the potential for clogging (Yentsch and Duxbury 1956).

RESULTS

Distribution of single species populations

Eighty-four copepod species were identified from the offshore collections, while 43 species were identified from Timbalier Bay collections. Of the 93 total species, 36 were present offshore but not in Timbalier Bay, nine were present in Timbalier Bay but not offshore, and 48 were present both offshore and in Timbalier Bay.

Offshore, the numerically abundant species were (in order of decreasing abundance): *Acartia tonsa*, *Temora turbinata*, *Paracalanus indicus*, *Paracalanus crassirostris*, *Eucalanus pileatus*, *Centropages velificatus*, *Oncaea venusta*, and *Paracalanus aculeatus*. In Timbalier Bay, the numerically abundant species were *Acartia tonsa*, *Paracalanus crassirostris*, *Pseudodiaptomus coronatus*, *Labidocera aestiva*, and *Tortanus setacaudatus*. Distribution by station and season is given in Marum (1974).

On the average, offshore samples contained three times the number of individuals as Timbalier Bay samples (7700/m^3 vs. 2600/m^3). Total numbers of copepods of single samples ranged from 32,000 to 52,000/m^3 offshore and from 11,000 to 69,000 in Timbalier Bay. Maximum abundances were observed in the warmer months. The average number of specimens occurring from April to September was five times greater than from October to March (12,500/m^3 vs. 2,500/m^3).

A persistent trend of copepod zonation was evident along the length of the transect. Species that were present in the upper part of Timbalier Bay were replaced at offshore stations with marine species. Zonation of zooplankton, particularly zonation along a salinity gradient, has been reported extensively in many previous investigations. In this study, using the method described in Marum (1974), the distribution of species along the inshore-offshore transect was found to occur in four general types of patterns: (1) species restricted to Timbalier Bay waters (estuarine species), (2) species present along the entire transect, with maximum numbers at the estuarine-marine transition zone (coastal species), (3) species present in lower Timbalier Bay and offshore, with maximum numbers offshore (shelf species), and (4) species restricted to the offshore region (oceanic species). The inshore-offshore trends (seasonal averages) of abundant species are shown in figure 2.

Thus, the copepod fauna of Timbalier Bay and the adjacent offshore region is composed of four more or less distinct groups of species, each group exhibiting distribution patterns similar to other members of the group but unlike distribution patterns of other groups (figure 2). Because these patterns agree with those of previous studies (Fleminger 1956; Bowman 1971), it is probably correct to say that these patterns will be consistently found in tropical and sub-tropical waters.

Each net collection was composed of copepod fauna of two or more of the groups, i.e., the upper part of Timbalier Bay contained estuarine and coastal faunas, the lower part of Timbalier Bay had estuarine, coastal, and shelf faunas, and the offshore region had coastal, shelf, and oceanic faunas. There is evidence in the literature (Bowman 1971; McGowan 1971) that such mixtures of species groups are unique to hydrographically unstable areas, particularly estuarine and neritic waters, and result from mixtures of waters of different origins. In the OEI area, mixtures of species groups were observed both on the horizontal axis and on the vertical axis.

Variations due to seasonal fluctuations changed the basic patterns. Table 3 summarizes variation over the inshore-offshore transect from October 1972 to October 1973. In April 1973 the influence of the spring floods was already apparent, but the greatest impact of the floods was observed in July 1973. Coastal species were present in such great numbers that they dominated even the offshore collections. By October 1973, however, the waters had again become well-mixed, and the horizontal zonation pattern was similar to that of the previous October.

It became apparent during the course of this study that vertical zonation was frequently more marked than horizontal zonation. Over a vertical distance of a few meters, large variations occurred in total number of copepods, single species populations, diversity, and relative percentages of the species groups. The degree of difference was directly related to the heterogeneity of the water column (table 4). For example, in October 1972 the offshore waters were vertically mixed, although surface waters contained twice as many copepods (10,000/m^3) as mid-depth waters (5,200/m^3), and five times as many as bottom waters (1,800/m^3). Diversities increased with depth, and thus numbers of in-

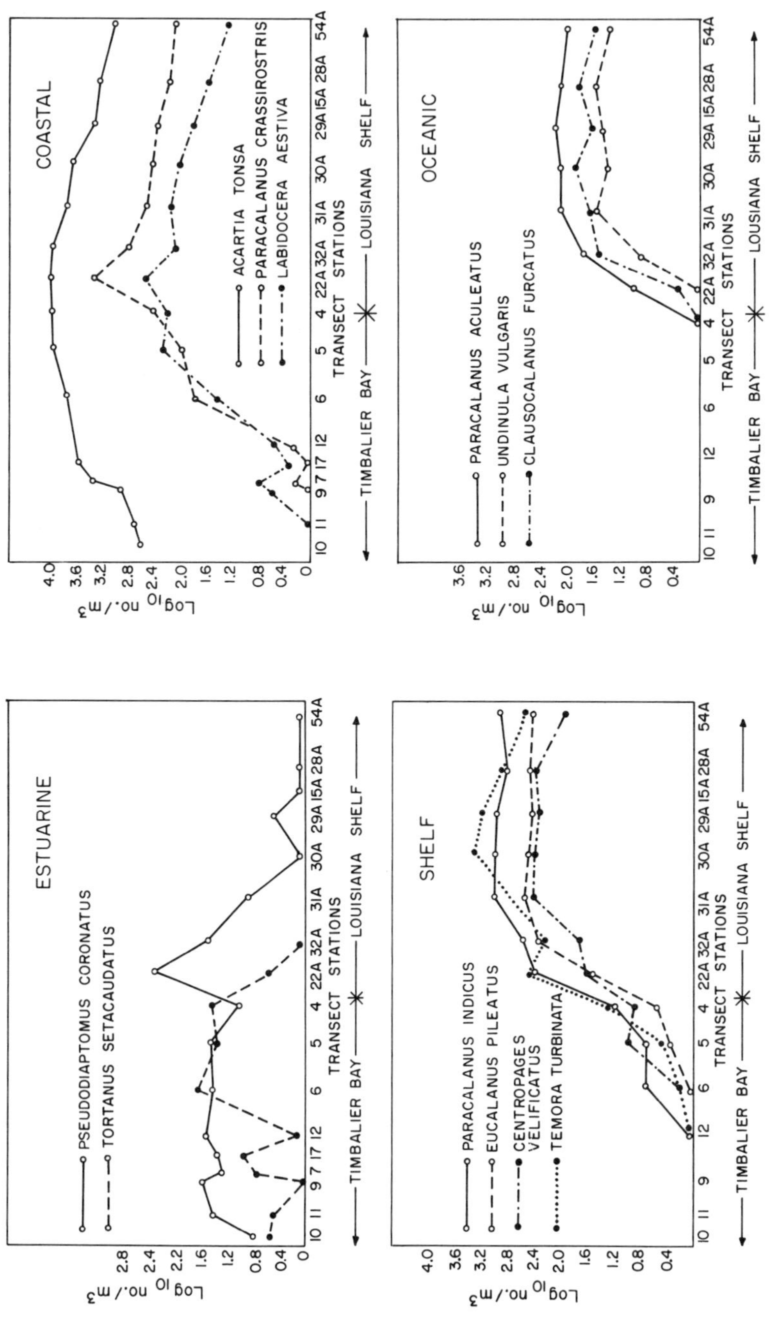

FIG. 2. DISTRIBUTION PATTERNS OF ABUNDANT SPECIES OF THE FOUR SPECIES GROUPS. Data averaged over all seasons.

TABLE 3. Seasonal variation of species groups at points along the transect.

Date	Species Group	Percent Composition Inner Bay	Outer Bay	6 km Offshore	21 km Offshore
Oct. 1972	Estuarine	11	-	-	-
	Coastal	80	83	53	1
	Shelf	9	17	42	76
	Oceanic	-	-	5	23
April 1973	Estuarine	1	1	-	-
	Coastal	99	92	71	49
	Shelf	-	6	24	45
	Oceanic	-	1	5	6
July 1973	Estuarine	-	1	-	-
	Coastal	100	98	98	96
	Shelf	-	-	1	3
	Oceanic	-	1	1	1
Oct. 1973	Estuarine	6	1	0	0
	Coastal	90	89	35	12
	Shelf	4	10	59	69
	Oceanic	-	-	6	19

TABLE 4. Variation of species groups with depth.

Date	Species Group	Present Composition Surface	Mid-Depth (7 m)	Bottom (14 m)
Oct. 1972	Coastal	3%	12%	2%
	Shelf	84%	76%	77%
	Oceanic	13%	12%	21%
July 1973	Coastal	98%	92%	1%
	Shelf	2%	7%	43%
	Oceanic	-	1%	56%

dividuals were inversely related to diversity values. The surface community was similar to the mid-depth community (similarity calculated by Morisita index), which was transitional with the bottom community. Exchange, therefore, was able to take place between surface and mid-depth communities. The mid-depth community was able to exchange only partially with the community at 14 meters.

A very different pattern of vertical zonation was observed in July 1973. Waters were highly stratified, a phenomenon that was reflected in the copepod communities. The surface and mid-depth communities were similar to each other, although dissimilar to the October patterns. The mid-depth community was unlike the bottom community. Therefore, exchange was able to take place between surface and mid-depth communities, but not between mid-depth and bottom communities. Surface and mid-depth waters contained large numbers of zooplankton, most of which were coastal species. In contrast, bottom waters had a much smaller population, composed primarily of oceanic and shelf species.

While a change from one community to another extended usually over a 16 to 25 km distance on the horizontal axis, the same change required only a few meters on the vertical axis. This reflects the hydrographic characteristics of the waters.

Species diversity

Copepod species diversity increased with distance from shore and decreased with distance into the bay. The average diversity of Timbalier Bay collections was 0.45 as compared with 1.62 for offshore collections. Lowest average diversities (both bay and offshore) were in July 1973 and highest in October 1972.

Offshore, diversity increased with depth and the difference between surface waters and bottom waters increased when spring floods carried large numbers of coastal species offshore. In October 1972, when offshore waters were well-mixed, diversities of offshore station 15A were 1.36 at the surface, 1.77 at mid-depth, and 1.83 at bottom. In July 1973, the equivalent values were 0.01, 0.32, and 1.91. The effect of the spring floods on copepod species diversity (H'), including subsequent "recovery" to a more homogeneous community, is shown graphically in figure 3.

Total zooplankton biomass (wet weight) values ranged from 0.66 g/m^3 in January 1974 to 2.57 g/m^3 in July 1973. Average values were lowest in December 1973 (0.179 g/m^3) and highest in July 1973 (0.775 g/m^3).

Biomass was infrequently determined for bay collections because of large concentrations of Ctenophora and detritus. Some Timbalier Bay values were determined, however, but only for those few samples in which the relative amount of detritus and ctenophore parts was insignificant as compared with the amount of non-ctenophore zooplankton. Values were in general lower than offshore biomass values. In addition, values in the upper bay were generally lower than values in the lower bay (stations 4, 5, and 6), and this related inversely to ctenophore biomass. Ctenophores are predators of other smaller zooplankton. A

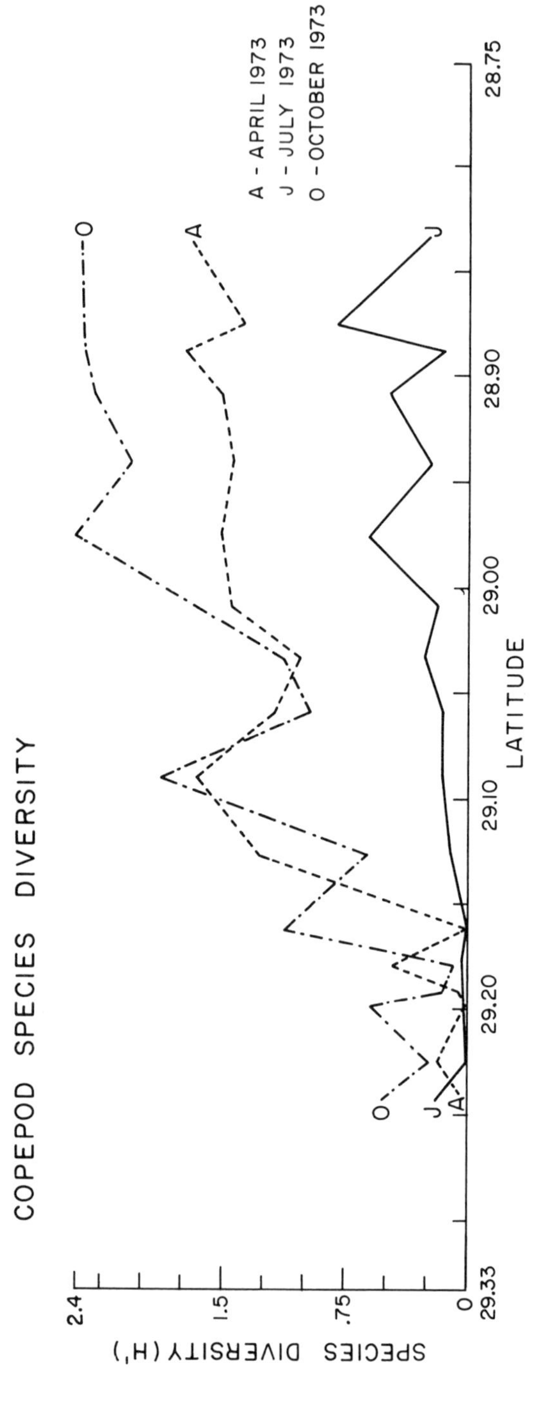

FIG. 3. SEASONAL VARIATION OF COPEPOD SPECIES DIVERSITY OVER THE LENGTH OF THE INSHORE-OFFSHORE TRANSECT.

summary of average biomass and diversity at platforms, reference stations, and other stations is shown in table 5.

Oil platform versus controls

The zooplankton community near the offshore platform (station 54A) was compared with that of the offshore control (station 15A) 10 km from the platform by analysis of variance of copepod single species populations, copepod diversity, and total zooplankton biomass. Species by species comparisons are in Marum (1974) for both bay and offshore comparisons. The eight most abundant species at the control station were the eight most abundant at the offshore platform. This closeness in ranking did not remain constant for species of lesser abundances. However, these differences in rankings are not thought to be due to the presence of the platform because (1) the variations were not consistent, and (2) errors in ranking often occur when one enumerates rare species, as I explained earlier (see also McGowan 1971). A total of 47 species was found at the control, 52 species were found at the platform; 43 of these were found at both locations.

The seasonal variations of abundant species are shown graphically in figure 4. Analyses of variance on log-transformed counts showed no significant differences between platform and control mean values. Seasonal variations were most pronounced and were significant for six of the abundant species. At least two of the species (*A. tonsa*, *T. turbinata*) showed large interaction terms, indicating clumped distribution patterns. But because the patches were not consistently present at either the plat-

TABLE 5. Average biomass and diversity at platforms, controls, and other stations.

Station	Biomass (g/m^3)	Diversity (H')
Offshore		
Platform (54A)	0.350 ± 0.133, n=9	1.60 ± 0.79, n=7
Control (15A)	0.344 ± 0.162, n=9	1.71 ± 0.78, n=7
Other OEI Stations[1]	0.436 ± 0.257, n=40	1.74 ± 0.59, n=37
Timbalier Bay[2]		
Platform (7)	0.090, n=2	0.331 ± 0.355, n=10
Control (12)	0.124, n=2	0.305 ± 0.358, n=10
Other Upper Bay (8,9,10,11,17)	0.065 ± 0.024, n=6	0.392 ± 0.322, n=39
Other Lower Bay (4,5,6)	0.233 ± 0.223, n=10	0.655 ± 0.567, n=25

[1]Stations 31A, 30A, 29A, 28A, 11A, and 60.

[2]Timbalier Bay biomass values are non-ctenophore biomass.

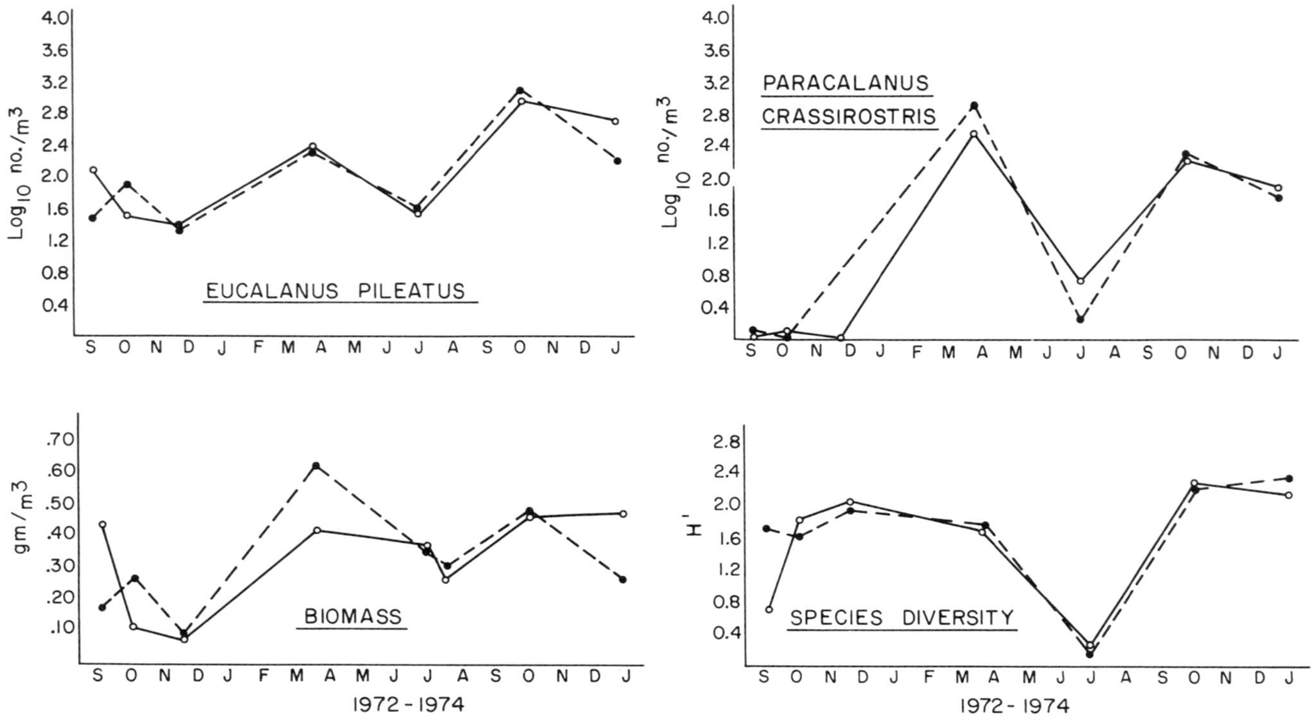

FIG. 4. SEASONAL VARIATIONS OF ABUNDANT COPEPODS, SPECIES DIVERSITY INDEX, AND TOTAL ZOOPLANKTON BIOMASS at offshore platform station 54A (solid line) and offshore control station 15A (dashed line).

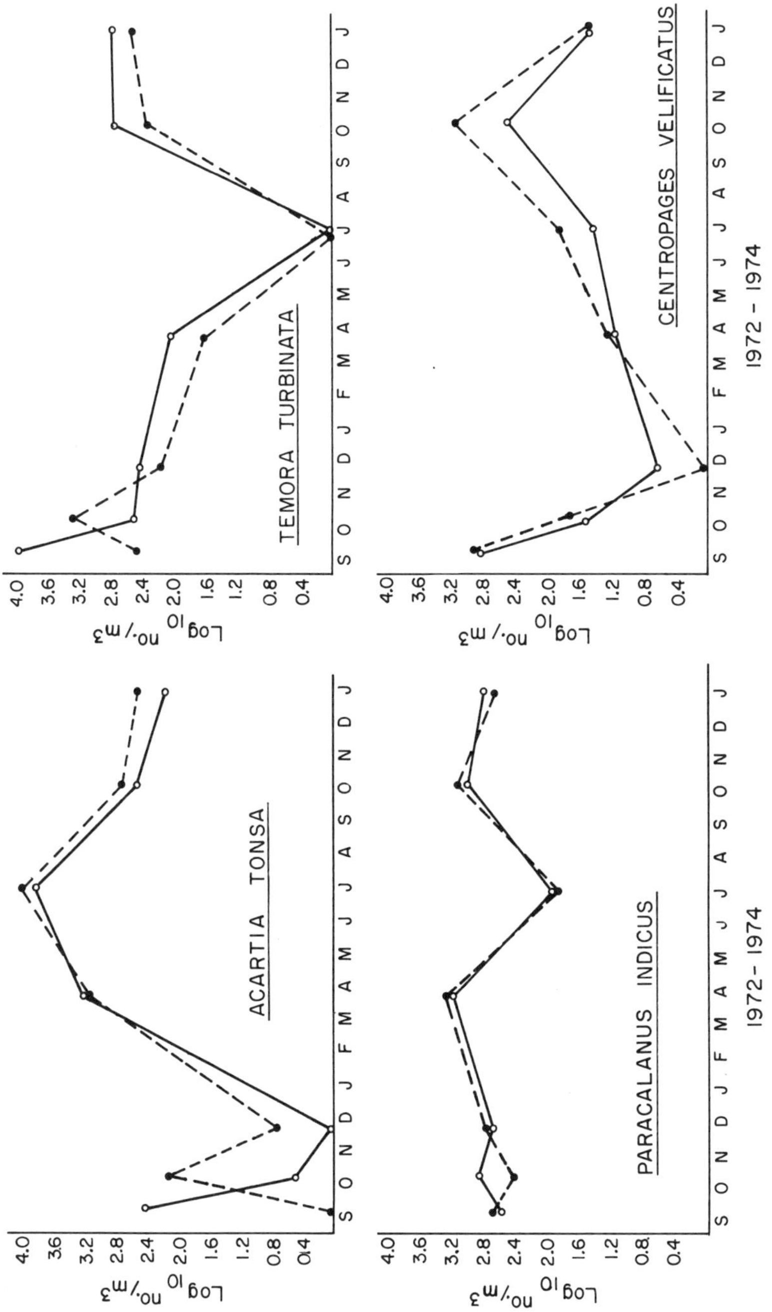

FIG. 4 continued.

form or the control, the mean values between the two stations were similar. Patchiness in single species populations has been observed in several previous studies (e.g., Wiebe et al. 1973) and are normal for estuarine and neritic waters where hydrographic and nutrient fluctuations are large.

Biomass values (wet weight) at Platform 54A and the control were also compared by analysis of variance; mean differences were not significant (figure 4). In addition, there were no significant seasonal differences over the months sampled. Biomass values of the OEI area were typically high for neritic waters in the direct influence of the Mississippi River (Bogdanov et al. 1968). The average biomass value is twice the Gulf shelf average of 1.17 g/m^3 (Arnold 1958) and is reported to be among the most productive in the Gulf (Riley 1937).

Seasonal changes in Shannon-Weaver diversity index are also shown in figure 4. An analysis of variance showed no difference between platform and control values. However, a significant seasonal trend was shown.

In Timbalier Bay, *Acartia tonsa* was by far the most numerically abundant, at both platform (Station 7) and control (Station 12). Of the five most abundant species at the control, four were among the most abundant at the platform. A total of 20 species was found at the control, 21 at the platform, with 15 of these at both locations.

The seasonal variations of *Acartia tonsa* and diversity are shown graphically in figure 5. Although the graphs from the three stations are somewhat similar to each other, there was considerable variability in results, although no significant differences between stations were shown by analysis of variance.

Faunal comparisons with non-drilling areas of the northern Gulf ot Mexico

In evaluating the results of faunal comparisons between the OEI area and non-drilling areas off Texas, it is important to note that samples were taken during July 1973, when the impact of the Mississippi and Atchafalaya River plumes was greatest. This fact makes comparisons of the OEI area with outside areas more difficult because another variable, the effect of the plume on zooplankton populations, has been introduced.

The Texas station closest to the OEI area (250 km west) showed high faunal similarity (Morisita index) with OEI stations, indicating that the plume effect extended as far west as 360 km from the Mississippi River. For the other two stations (350 and 475 km west of OEI) there was low faunal similarity with OEI stations. However, these differences are thought to be due to the plume effect and not to oil industry effects

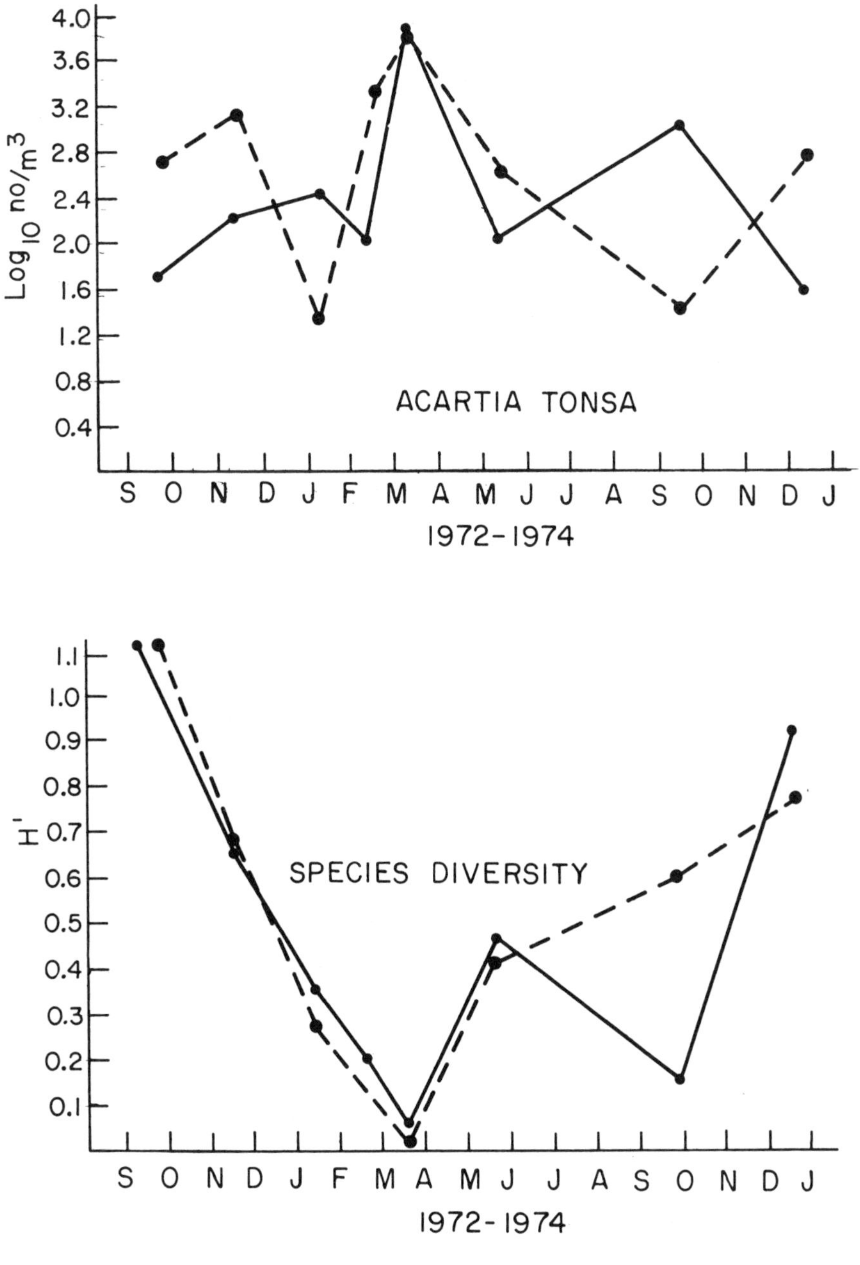

FIG. 5. SEASONAL VARIATIONS OF *A. TONSA* AND SPECIES DIVERSITY INDEX at Timbalier Bay platform station 7 (solid line) and Timbalier Bay control station 7 (dashed line).

because (1) the faunas of stations off the Texas coast were similar to the fauna found in the OEI area under non-stratified conditions, (2) the percentage of *Acartia tonsa* decreased with distance from the Mississippi River, *A. tonsa* being an indicator of low salinity water, (3) all three OEI stations showed similar values regardless of the proximity of an oil platform, and (4) all species found at Texas stations were also found in the OEI area.

DISCUSSION

Based on the faunal analyses, it appears that two separate copepod communities exist in the study area: a bay community composed of estuarine and coastal species, and an offshore community composed of coastal, shelf, and oceanic species. Between the two communities along the transect is a transition population composed of estuarine, coastal, shelf, and oceanic groups. The transition population is composed of species found either in upper Timbalier Bay or at the stations farthest offshore.

Shelf species exchange into lower Timbalier Bay throughout the year, but are prevented from exchanging into upper Timbalier Bay by Casse-tete and Calumet Islands. It is unlikely, however, that shelf species are able to propagate in the lower bay, but must depend upon tidal currents to carry them over the shallow sill. Oceanic species do not exchange into the bay. Estuarine species are generally restricted to Timbalier Bay, but are able to exchange with offshore surface waters during spring flood conditions. The movement of estuarine and coastal species offshore probably occurs every year. However, it is unlikely that the exchange in 1973 would have been so extensive were it not for the magnitude of the 1973 floods. The spring floods resulted in coastal species being moved offshore in such great numbers that they dominated the samples throughout the transect in July 1973. By October 1973, however, the waters had again become well-mixed, the communities had become more equitable, and the horizontal zonation pattern was similar to that of the previous October.

Communities of organisms that are in environmentally unstable areas are sometimes called *physically controlled communities*, where physical conditions fluctuate widely and the organisms are primarily adapted to stresses from the physical environment.

Community changes were much more pronounced through the water column offshore than over the inshore-offshore transect. While a change from one community to another extended usually over a 16 to 25 km distance on the horizontal axis, the same change required only a few meters on the vertical axis. This reflects the hydrographic characteristics of the waters. For example, a salinity change of 12 ppt over the horizontal axis covered 45 km, while a similar salinity change on the vertical axis occurred over a distance of a few meters.

Evidence presented shows that a wedge of shallow bottom water (underlying the surface layers) contains residual populations of oceanic species, which are probably present year-round regardless of surface phenomena. Therefore, it is not necessary to assume that oceanic incursions take place (e.g., from the Loop Current) in order for repopulation of surface waters to occur. A simple mechanism, involving only vertical mixing of the water column following breakdown of stratified conditions, is sufficient to explain the recovery of oceanic zooplankton fauna in surface waters. Thus, although this offshore ecosystem undergoes severe natural stresses, it is stable in the sense that the "resident" populations persist, if only in small numbers, in undisturbed wedges of water below the surface.

With regard to comparisons of fauna between oil platforms and reference areas, the evidence indicates that seasonal variations are substantially greater than variations that can be related to the presence of an oil platform or its associated activities. Seasonal fluctuations in the OEI area appear to be typical of coastal regions. Both the timing and the magnitude of the seasonal cycle in the OEI area are similar to other observations in the northern Gulf of Mexico, e.g., Cuzon du Rest (1963), Hopkins (1966), and Bogdanov et al. (1968).

A review of the species list from a sample collected in February 1952 in the OEI study area (28°54′N, 90°11′W) revealed that 20 of 23 calanoid copepods reported by Fleminger (1956) were found in spring collections from the OEI study area, and 22 of 23 were found in all OEI collections. Moreover, there were several species found in OEI collections that are not found in the 1952 collections, although this fact does not necessarily mean that they were not present in the area in 1952.

An attempt was made to compare the data of Cuzon du Rest (1963), collected east of the Mississippi River in 1959-1961, with the data from Timbalier Bay. Sub-area 3 in that study is similar in hydrography to Timbalier Bay, and there had been little or no oil drilling in sub-area 3 prior to the collections. Because of differences in collection sites, collection techniques, and presentation of the data, it was difficult to make rigid comparisons. Nevertheless, it is instructive to note that *Acartia tonsa* was the dominant copepod and was far more abundant than any other species. The list of other important copepod species from Cuzon du Rest was in reasonable agreement with that of Timbalier Bay collections. It should be pointed out that comparisons of OEI data with historical data are useful only as an indication that no gross effects have occurred in the copepod community as a result of the Louisiana oil drilling and production.

CONCLUSIONS

The totality of evidence from this study and the literature indicates that the zooplankton communities of the OEI study area (1) are similar to those of other northern Gulf of Mexico areas; and (2) have not changed over the past twenty years. In addition, there is no evidence of differences in zooplankton biomass or community structure between the waters of oil platforms and control waters. However, the OEI area is subject to extreme environmental fluctuations, primarily due to the runoff of the Mississippi River, which is carried in and out of the area by the complex patterns of tides and currents. If there were temporary changes in the zooplankton community structure caused by localized oil pollution, they could be effectively masked by the variability caused by the river.

ACKNOWLEDGMENTS

This work represents part of the fulfillment of the Ph.D. degree requirements at Florida State University. I thank the following persons for assistance during this investigation: Robert J. Menzies, for guidance and direction; James I. Jones, for loan of the equipment to make the collections; Thomas E. Bowman, for confirmation and identification of calanoid copepods; Frank Ferrari, for confirmation and identification of cyclopoid copepods; and Allen Z. Paul, for review of the manuscript. This work was supported by Gulf Universities Research Consortium grant number OB55HJM to Robert J. Menzies and James P. Marum.

REFERENCES CITED

Arnold, E. L. 1958. Gulf of Mexico plankton investigations, 1951-1953. U.S. Fish and Wildlife Special Scientific Report Fisheries **269**:1-53.

Basharin, G. B. 1959. On a statistical estimate for the entropy of a sequence of independent random variables. Theory Prolab. Applic. **4**:333-336.

Bechtel, T. J. and B. J. Copeland. 1970. Fish species diversity indices as indicators of pollution in Galveston Bay, Texas. Contributions in Marine Science, The University of Texas **15**:103-132.

Bogdanov, D. V., V. A. Sokolov, and N. S. Khromov. 1968. Regions of high biological and commercial productivity in the Gulf of Mexico and Caribbean Sea. Oceanology (USSR) **8**(3):371-381.

Bowman, T. E. 1971. The distribution of calanoid copepods off the southeastern United States between Cape Hatteras and Southern Florida. Smithsonian Contribution, Zoology **96**:1-58.

Cuzon du Rest, R. P. 1963. Distribution of the zooplankton in the salt marshes of southeastern Louisiana. Publications of the Institute of Marine Science, The University of Texas **9**:132-155.

Deevey, G. B. 1971. The annual cycle in quantity and composition of the zooplankton of the Sargasso Sea off Bermuda. I. The upper 500 M. Limnology and Oceanography **16**(2):219-240.

Fleminger, A. 1956. Taxonomic and distributional studies on the epiplanktonic calanoid copepods (Crustacea) of the Gulf of Mexico. Ph.D. dissertation, Harvard University.

Grice, George D. 1957. The copepods of the Florida west coast. Ph.D. dissertation, Florida State University, Tallahassee.

Hopkins, T. L. 1962. A zooplankton subsampler. Limnology and Oceanography **7**(3):424-426.

———. 1966. The plankton of the St. Andrew Bay System, Florida. Publications of the Institute of Marine Science, The University of Texas **11**:12-64.

Hurlbert, S. H. 1971. The nonconcept of species diversity: A critique and alternate parameters. Ecology **52**:577-586.

Marum, J. P. 1974. Spatial and temporal variations of zooplankton in relation to offshore oil drilling and estuarine-marine faunal exchange. Ph.D. dissertation, Florida State University, Tallahassee.

McGowan, J. A. 1971. Oceanic biogeography of the Pacific. *In* Funnell and Reidel, eds., The Micropaleontology of Oceans. Cambridge University Press, pp. 3-74.

Morisita, M. 1959. Measuring of interspecific association and similarity between communities. Memoirs of the Faculty of Science, Kyushu University, Series E (Biol.) **3**(1):65-80.

Niskin, S. and J. I. Jones. 1963. New collecting device for limnological and oceanographic research. *In* Proceedings of the 6th Conference on Great Lakes Research, University of Michigan, Great Lakes Research Division of the Institute of Science and Technology.

Odum, H. T., R. P. Cuzon du Rest, R. J. Beyers, and C. Allbaugh. 1963. Diurnal metabolism, total phosphorus, able anomaly, and zooplankton diversity of abnormal ecosystems of Texas. Publications of the Institute of Marine Science, The University of Texas **9**:404-453.

Riley, G. A. 1937. The significance of the Mississippi River on biological conditions in the Gulf of Mexico. Journal of Marine Research **1**:60-74.

Shannon, C. E. and W. Weaver. 1963. The mathematical theory of communication. Urbana: University of Illinois Press.

Wiebe, P. H., G. D. Grice, and E. Hoagland. 1973. Acid-iron waste as a factor affecting the distribution and abundance of zooplankton in the New York Bight. II. Spatial variations in the field and implications for monitoring studies. Estuarine and Coastal Marine Science **1**:51-64.

Wilson, C. B. 1932. The copepods of the Woods Hole region, Massachusetts. Bulletin of the U.S. National Museum **158**.

Yentsch, C. S. and A. C. Duxbury. 1956. Some of the factors affecting the calibration number of the Clarke-Bumpus quantitative plankton sampler. Limnology and Oceanography **1**(4):268-273.

THE BENTHIC MARINE ALGAE OF TIMBALIER BAY, LOUISIANA

by Harold J. Humm and Theresa M. Bert

ABSTRACT

A study of the benthic algae of Timbalier Bay, Louisiana, was made from June 1972 to September 1974, as part of an interdisciplinary study conducted by the Gulf Universities Research Consortium to determine the effect of the petroleum industry on the bay. Eighty-three species of benthic algae were recorded in the bay, divided among the four major groups as follows: Cyanophyta 26 (32%), Rhodophyta 17 (21%), Phaeophyta 12 (15%), Chlorophyta 28 (33%). Apparently the oil and gas wells in Timbalier Bay have not had a significant adverse effect upon the distribution or abundance of benthic algae during 35 years of operation.

INTRODUCTION

From June 1, 1972, to September 15, 1974, the Gulf Universities Research Consortium conducted an interdisciplinary study, known as the Offshore Ecology Investigation, of the ecological impact of the offshore oil and gas wells of the Louisiana coast. Field work was carried on in Timbalier Bay and among the offshore platforms covering an area of about 400 square miles and to depths of 30.5 meters. This report represents the results of a study of the benthic algae of Timbalier Bay, which has not been studied previously for these organisms.

At the time of this investigation, Harold Humm and Theresa Bert were at the Department of Marine Science, University of South Florida.

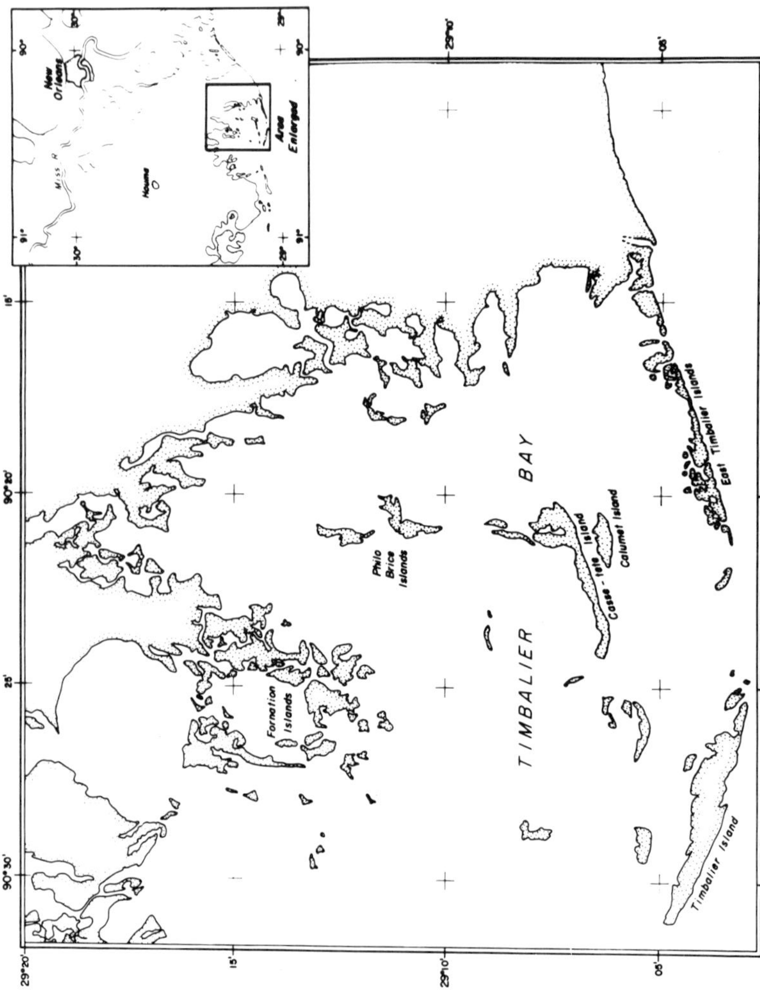

FIG. 1. MAP OF TIMBALIER BAY showing the location of the islands around which collections of algae were made. Inset shows the location of Timbalier Bay in relation to New Orleans and the Mississippi River.

THE HABITAT

Location

Timbalier Bay is the second largest bay west of the Mississippi delta on the Louisiana coast (figure 1). Its northern boundary is approximately at latitude 29°17 'N and its southern boundary at 29°03 'N. It is bounded on the east by marshlands and on the west by a diagonal line from Chinois Pass through the Fornation Islands to the western end of Timbalier Island at approximately 90°31 'W. The northern part of Timbalier Bay, an area of about 15 square miles, is known as Lake Raccourci, which was not included in this study. Separating Lake Raccourci from the southern part of Timbalier Bay are the Philo Brice Islands. Between these islands and the southern boundary of the bay are Casse-tete and Calumet Islands.

Salinity

Timbalier Bay contains somewhat brackish water, and most of the year has a decreasing salinity gradient from its boundary with the Gulf of Mexico between Timbalier and East Timbalier islands northward to Chinois Pass. Since the major source of fresh water for the bay is drainage from the vast marshes to the north, rather than large rivers, the salinity of Timbalier Bay is generally higher than other similar bays of coastal Louisiana (figure 2).

The salinity pattern varies from year to year (Gagliano 1970), but the lowest salinities usually occur in the spring, when land run-off and river discharge are highest, and highest salinities occur in late summer or fall, when land run-off is minimal, when sea level in the northern Gulf of Mexico is highest, and when southerly winds tend to blow Gulf water into the lower part of the bay (Walsh 1969).

Figure 2 shows the salinity pattern at various times for the years 1972 and 1973, indicating that the range from the Philo Brice Islands southward is 15-30 ppt most of the time, a range tolerated by many species of euryhaline marine algae.

Temperature

During the years 1972 and 1973, the surface water temperature in Timbalier Bay ranged from 12° C in February to 32° in July and August, a range of 20°. Figure 3 shows the surface water temperature variations latitudinally in Timbalier Bay at all seasons from October 1972 to July 1973.

Turbidity

Light transmission through the water of Timbalier Bay is generally

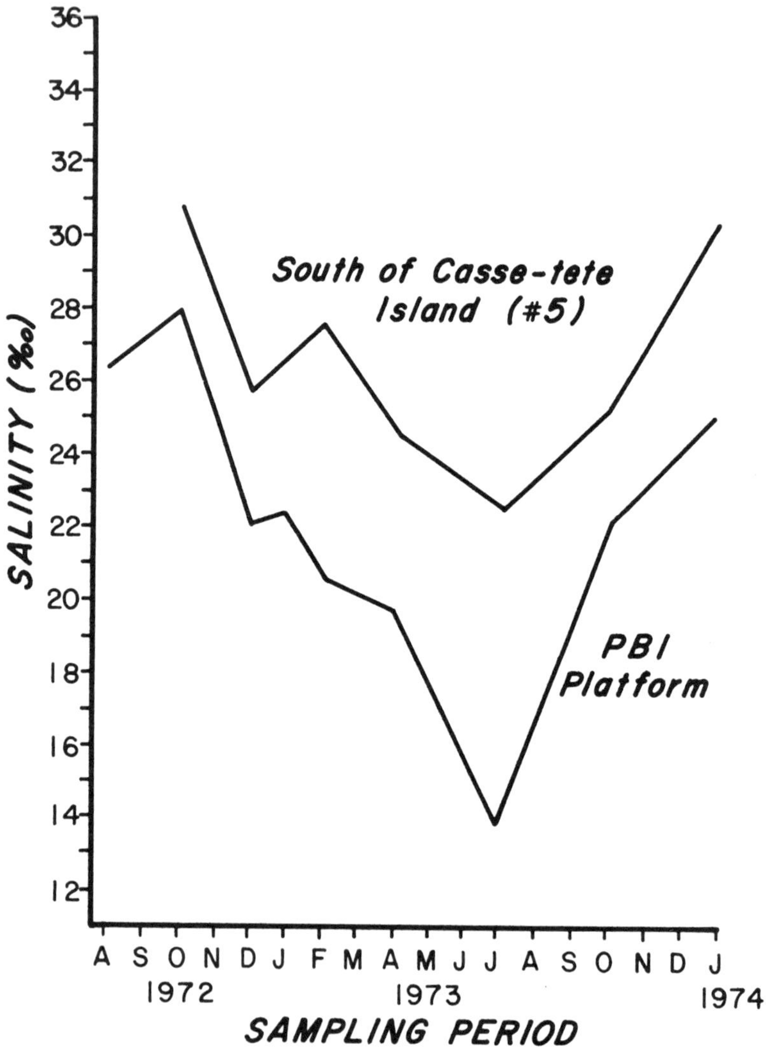

FIG. 2. SALINITY RECORDS IN TIMBALIER BAY for the period April 1972 to January 1974, from the Philo Brice Islands platform and from the south side of Casse-tete Island.

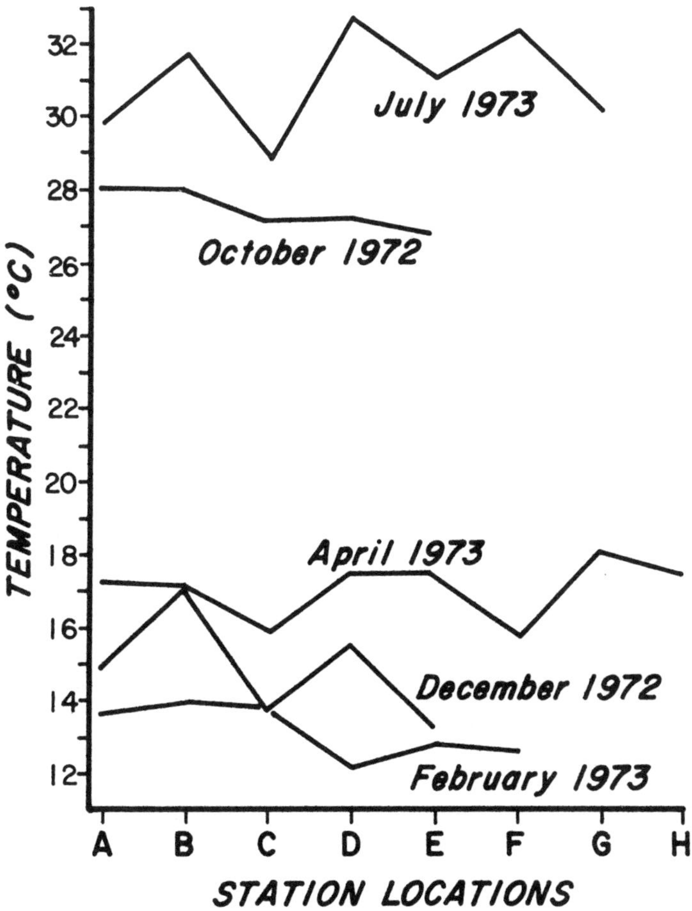

FIG. 3. WATER TEMPERATURE RECORDS at 5 to 7 stations in Timbalier Bay during five different months from October 1972 to July 1973.

so reduced that benthic algae are limited to a depth of from 10 cm to a few hundred mm below mean low water. The shallowness of the bay and the high proportion of silt and clay in its bottom sediments (Williams and Jones 1974) combine to keep the water turbid whenever the wind velocity exceeds 12 to 15 mph. In addition, boat traffic and shrimp trawling in the bay are a constant cause of resuspension of fine sediments. Periods of calm weather result in an increase in water clarity, but the duration of these periods is too short to permit benthic algae to colonize successfully at deeper levels.

Substrate

Since the benthic algae in Timbalier Bay are not able to grow deeper than 100 to 300 mm below mean low water, the principal natural substrates they colonize are the oyster shells and other molluscs, the pneumatophores of black mangroves, the stems of marsh grasses, the upper ends of annelid worm tubes, and the intertidal sediments in calm areas. Man-made structures added considerable substrate in the intertidal zone for benthic algae, especially rock break-waters, sea walls, pilings and timbers of oil platforms, channel markers, and buoys. A few species of benthic algae could survive and grow for long periods of time as loose, drifting plants in areas of low current velocity and reduced wave action.

Depth

The maximum depth of Timbalier Bay is about 1.8 to 2.1 m at mean low water. The depth of the northern half of the bay is 0.9 to 1.2 m and the southern half 1.2 to 1.5 m. Because the bay is shallow, water turbulence arising from prevailing winds keeps large amounts of silt and clay in suspension most of the time, and tidal currents distribute this turbidity rather evenly throughout the bay. The southernmost part of the bay, south of Brush and Casse-tete Islands, is slightly deeper, and the bottom sediments contain less silt and clay because of stronger tidal currents. For this reason the southern segment is generally less turbid than the rest of the bay, as shown by the latitudinal transparency variation data (Price 1974).

Tides and other sea level changes

The tides in Timbalier Bay are chiefly diurnal. The mean amplitude on the north side of Timbalier Island and around the Pelican Islands is about 365 mm. From the central part of the bay northward, the amplitude is progressively less. The maxima in amplitude occur in December and June at the approximate time of the winter and summer solstices. The minima occur in March and September in conjunction with the vernal and autumnal equinoxes. Lunar monthly maxima occur around new and full moon and minima during first and last quarters.

During the colder months of the year, mean sea level in Timbalier Bay is lower than that determined by celestial tidal forces because of prevailing north winds. During the warmer months of the year, the opposite is true.

Nutrients

Data on nutrient salts in the water of Timbalier Bay indicate that the quantities present most of the time are favorable to algal growth. The highly organic sediments of the bay and the abundance of non-living particulate organic matter in suspension contribute to rapid regeneration of nutrients by bacterial activity. The rate of regeneration may be of much greater importance to algal nutrition than the level of concentration of nutrients at any given time. The shallowness of the bay and the constant turbulence resulting from tidal currents assure that nutrients released from bottom sediments are quickly mixed throughout the water column.

Between the Philo Brice Islands and the southern margin of the bay, the area in which the benthic algae were best developed, orthophosphate values in the water ranged from 62 to 100 μg/l, ammonia values from 0.08 to 0.75 mg/l, except for a 4.5 reading one mile north of Casse-tete Island in February 1973; nitrite plus nitrate values ranged from 0.03 to 1.8, except for a reading of 5.8 in Little Timbalier Pass in February 1973 (Burchfield et al. 1974).

All nutrients were highest during winter and lowest during summer, with some fluctuations during spring and fall that may have been caused by a surge or bloom of a species of planktonic diatom.

ANNOTATED LIST OF SPECIES

Cyanophyta

Twenty-six species of blue-green algae have been recorded in Timbalier Bay. Of these, 24 species are considered marine (though they may occur in fresh water as well), and two species (*Anacystis montana* and *Microcoleus vaginatus*) are regarded as primarily fresh water algae, although they are often found in brackish water, especially off mouths of streams. There are about 32 marine species of blue-green algae in the Atlantic ocean and Gulf of Mexico *sensu stricta* Drouet (Drouet and Daily 1956; Drouet 1968, 1973, 1978). Thus, 75% of all known marine blue-greens from the Gulf and Atlantic Ocean have been found in Timbalier Bay.

Coccochloris elabens Drouet and Daily. Common throughout Timbalier Bay during the warmer months in the intertidal zone and somewhat below; attach to a variety of substrates, especially barnacles.

Coccochloris stagnina Drouet and Daily. Forms a bright blue-green gelatinous layer low in the intertidal zone on an oil well platform piling near the Philo Brice Islands; apparently not common in the bay.

Anacystis aeruginosa Drouet and Daily. Occasional colonies on or in dead barnacle shells in the intertidal zone on an oil well platform piling near Philo Brice Islands and on leaves of *Halodule* about two miles north of Little Timbalier Pass.

Anacystis dimidiata (Kützing) Drouet and Daily. Mixed with other blue-greens from a piling near the Philo Brice Islands, intertidal zone.

Anacystis marina Drouet and Daily. Common in the intertidal zone on pilings around the Philo Brice Islands in the form of small colonies.

Anacystis montana (Lightfoot) Drouet and Daily, *forma montana* Drouet and Daily. Intertidal on pilings around the Philo Brice Islands the year around.

Johannesbaptistia pellucida Taylor and Drouet. Found in the bay only during the winter months on the north side of Casse-tete Island among pneumatophores of black mangroves.

Gomphosphaeria aponina Kützing. In lower Timbalier Bay only; on *Halodule* leaves, oyster shells, and waterlogged wood.

Agmenellum thermale (Kützing) Drouet and Daily. Scattered colonies mixed with *Microcoleus lyngbyaceus* and other mat-forming blue-greens on pilings and driftwood around the Philo Brice Islands, and on *Halodule* leaves along the north side of Timbalier and East Timbalier Islands, summer.

Family Chamaesiphonaceae

Entophysalis conferta (Kützing) Drouet and Daily. Epiphytic on larger algae throughout the study area of the bay and on seagrass leaves along the south side of Casse-tete Island.

Entophysalis deusta Drouet and Daily. At all collecting sites on the surface of and boring into shells and other forms of limestone.

Order Hormogonales

Family Oscillatoriaceae

Spirulina subsalsa Oersted. Forms bright blue-green patches in algal mats, on the concave surface of dead shells, in dead barnacles, on and within hydroid and bryozoan tests especially around the Philo Brice Islands, but distributed throughout the bay.

Oscillatoria erythraea (Ehrenberg) Kützing. Planktonic in lower Timbalier Bay during summer periods of high salinity, when it formed patches of tiny flakes visible without magnification, which often adhered to pilings or other surfaces following contact.

Oscillatoria lutea C. Agardh. A constituent of algal mats on muddy sand in the intertidal zone and on pilings the year around throughout the bay.

Oscillatoria submembranacea Ardissone and Strafforello. An epiphyte of other algae in small tufts from the Philo Brice Islands to the south end of the bay.

Schizothrix arenaria (Berkeley) Gomont. Mixed with other blue-greens, especially on pilings, from the Philo Brice Islands southward, the year around.

Schizothrix calcicola (C. Agardh) Gomont. Probably the most widely distributed species of blue-green in the bay. Found in all collections the year around, both on the surface of all substrates and boring into limestone.

Schizothrix mexicana Gomont. On the old leaf bases of *Spartina*, south side of Calumet Island; probably widely distributed in the bay.

Schizothrix tenerrima (Gomont) Drouet. Intertidal on muddy sand among the aerial roots of black mangroves, north side of Casse-tete Island, typically with many trichomes in the sheath.

Porphyrosiphon kurzii (Zeller) Drouet. On intertidal sand, north side of Timbalier Island; summer months.

Porphyrosiphon notarisii (Meneghini) Kützing. Intertidal on stones, shells, wood, and mangrove roots during the warmer months from the Philo Brice Islands to the south end of the bay.

Microcoleus lyngbyaceus (Vaucher) Gomont. Probably the blue-green species of greatest biomass in the bay the year around. At all collection sites, usually in the form of a mat of which it is the principal constituent, but also forming loose skeins over seagrasses south of Casse-tete; summer.

Microcoleus vaginatus (Vaucher) Gomont. In the upper part of the bay north of the Philo Brice Islands where the water is more brackish, but after the Mississippi River flood of 1973, it was also found in the lower bay.

Family Nostocaceae

Calothrix crustacea Thuret. One of the most widely distributed of blue-greens. Abundant on pilings and woodwork where it forms a black band high in the intertidal zone; common also as an epiphyte on larger algae throughout the bay at all seasons.

Nostoc spumigena (Mertens) Drouet. A constituent of the blue-green film or mat on pilings around the Philo Brice Islands, intertidal the year around; on intertidal sand, north side of Timbalier Island.

Mastigocoleus testarum Lagerheim. Boring into the shells of dead barnacles and molluscs in lower Timbalier Bay the year around.

Rhodophyta

Class Rhodophyceae

Twenty-one species and one variety of Rhodophyta have been recorded for Timbalier Bay. The bay is not a favorable habitat for red algae because of the turbidity, salinity fluctuation, low water temperature of the winter months and rapid changes, and the limited amount of suitable substrata in the euphotic zone.

Order Bangiales

Family Bangiaceae

Goniotrichum alsidii (Zanardini) Howe. An epiphyte on the leaves of *Halodule* and on *Enteromorpha*, south side of Casse-tete Island, winter and spring 1972-73.

Erythrotrichia carnea (Dillwyn) J. Agardh. An epiphyte of *Halodule* leaves, *Gracilaria*, and *Enteromorpha*, the year around along the south side of Calumet and Casse-tete Islands and along the north side of East Timbalier Island.

Bangia fuscopurpurea (Dillwyn) Lyngbye. On *Halodule* leaves and aerial roots of black mangroves, southernmost part of Timbalier Bay, winter and spring.

Order Nemaliales

Family Acrochaetiaceae

Acrochaetium gracile Børgesen. On leaves of *Halodule*, south part of Timbalier Bay.

Acrochaetium sagraeanum (Montagne) Bornet. The most abundant species of *Acrochaetium* in lower Timbalier Bay on seagrass leaves and *Gracilaria*.

Acrochaetium virgatulum (Harvey) J. Agardh. Abundant on the leaves of *Halodule* during winter and spring along the south side of Casse-tete Island.

Order Gigartinales

Family Gracilariaceae

Gracilaria foliifera (Forsskal) Børgesen. Common on shells on the seagrass flats south of Casse-tete and Calumet Islands, winter and spring, occasional in summer.

Gracilaria verrucosa (Hudson) Papenfuss. On shells on the seagrass flats south of Casse-tete and Calumet Islands, from July to December 1973.

Gracilariopsis sjoestedtii (Kylin) Dawson. In great abundance as loose plants (hundreds of tons) on the seagrass flats along the south side of Calumet and Casse-tete Islands during the summer and fall of 1972, but in reduced quantity during 1973. These plants washed ashore in November and December. They appear in the spring as epiphytes of *Halodule* leaves but soon break off and continue growth as loose, drifting plants.

Family Hypneaceae

Hypnea musciformis (Wulfen) Lamouroux. Occasional as epiphytes of *Halodule* leaves or as larger, loose plants along the south side of Casse-tete and Calumet Islands from spring to early winter.

Order Ceramiales
Family Rhodomelaceae

Polysiphonia echinata Harvey. Occasional as an epiphyte of *Halodule* along the south side of Casse-tete and Calumet Islands, spring.

Polysiphonia hemisphaerica Areschoug. Occasional on shells and pilings low in the intertidal zone from the Philo Brice Islands to the southern end of the bay, year around.

Polysiphonia subtilissima Montagne. On pilings, shells, *Spartina* stems, and black mangrove aerial roots throughout the bay the year around. This species is one of the most euryhaline of the marine red algae since it occurs in salinity as low as 10 ppt.

Bostrichia radicans Montagne. On intertidal woodwork, aerial roots of black mangroves, around the base of *Spartina* stems, and in mats on mud, north side of Casse-tete Island and generally distributed around the fringes of salt marshes throughout the bay.

Bostrichia radicans forma moniliforme Post. On aerial roots of black mangroves along the north side of Casse-tete Island and on woodwork of an oil well platform, Philo Brice Islands.

Bostrichia tenella (Vahl) J. Agardh. On aerial roots of black mangroves along the north side of Casse-tete Island.

Lophosiphonia subadunca (Kützing) Falkenberg. On woodwork low in the intertidal zone, oil well platform near Philo Brice Islands, and on leaves of *Halodule*, south side of Casse-tete Island; rare.

Laurencia poitei (Lamouroux) Howe. One plant obtained by

dredging in upper Timbalier Bay, fall of 1973; a few plants on leaves of *Halodule*, south side of Casse-tete Island, winter.

Phaeophyta

Class Phaeophyceae

The brown algae are represented in Timbalier Bay by only 10 species that grow there and two additional species that occasionally drift into the bay from the Gulf. Of the four major groups of benthic algae, the Phaeophyta have the fewest fresh water species and few species that tolerate brackish water. In general, the brown algae of Timbalier Bay are species that do best during the cooler months of the year and tolerate reduced salinities.

Order Ectocarpales

Family Ectocarpaceae

Ectocarpus dasycarpus Kuckuck. On barnacle and oyster shells attached to well platform pilings low in the intertidal zone, central Timbalier Bay around the Philo Brice Islands and in the Bayou Lafourche boat channel, winter and spring. This species is the most tolerant of reduced salinity of any of the brown algae recorded for the bay.

Ectocarpus elachistaeformis Heydrich. An epiphyte during winter and spring on *Enteromorpha* and *Cladophora* from pilings around the Philo Brice Islands and southward in the bay.

Ectocarpus intermedius Kützing. On shells and seagrass leaves along the south side of Casse-tete and Calumet islands winter and spring. This species has been referred to as *E. confervoides* (Roth) LeJolis in much of the current literature.

Giffordia mitchaelliae (Harvey) Hamel. In lower Timbalier Bay on barnacles from pilings along the channel leading to Bayou Lafourche and on leaves of *Halodule* along the south side of Casse-tete and Calumet Islands and the north side of Timbalier Island. Present the year around but especially in spring.

Giffordia rallsiae (Vickers) Taylor. An epiphyte of *Halodule* along the south side of Casse-tete Island, year around except during periods of low salinity, but most common in spring.

Bachelotia antillarum (Grunow) Gerloff. Occasional on shells from the seagrass flats along the south side of Casse-tete Island, spring.

Phaeostroma pusillum Howe and Hoyt. Epiphytic on *Sphacelaria* from pilings around the Philo Brice Islands, winter.

Order Sphacelariales
Family Sphacelariaceae

Sphacelaria furcigera Kützing. Low intertidal on pilings around the Philo Brice Islands, especially on barnacles.

Order Dictyosiphonales
Family Striariaceae

Hummia onusta (Kützing) Fiore. Found only as an epiphyte on the leaves of *Halodule* in Timbalier Bay between Casse-tete and the Timbalier Islands. This dimorphic species was misunderstood until studied in culture by Fiore (1969, 1975). In the past, the gametophyte was known as *Myriotrichia subcorymbosa* (Holden) Blomquist and as *Farlowiella onusta* (Kützing) Kornmann. In both cases it had been originally described as *Ectocarpus*, and Taylor (1957, 1960) lists it as *E. subcorymbosus*. The gametophyte is found year around and is only a few mm in length. The sporophyte, previously known as *Stictyosiphon subsimplex* Holden, is several cm tall. It appears in late fall and disappears in spring.

Order Fucales
Family Striariaceae

Myriotrichia repens Hauck. On leaves of *Halodule* just inside Little Timbalier Pass and along the north side of East Timbalier Island in the extreme southern part of the bay, summer. This species may occur only sporadically in the bay during summer months.

Family Sargassaceae

Sargassum fluitans Børgesen and *Sargassum natans* (L.) Meyer. These two species of *Sargassum* are strictly pelagic, never attached. Their origin and center of distribution is the Sargasso Sea, but they often develop in considerable abundance in the Gulf of Mexico. During periods of abundance in the Gulf, both of these species are occasionally blown into Timbalier Bay and are to be found temporarily drifting in the bay or washed ashore.

Chlorophyta

Chlorophyceae

The green algae are better represented in Timbalier Bay (28 species) than any of the other three major groups. The reason is that the genera *Enteromorpha*, *Ulva*, and *Rhizoclonium* comprise species that tolerate a wide range of salinity and temperature. In addition, they tolerate turbidity and are able to grow in the intertidal zone. Species of these genera constitute half of all green algae known for the bay.

Order Tetrasporales
Family Palmellaceae

Pseudotetraspora antillarum Howe. Gelatinous colonies of this species were found adhering to *Enteromorpha* around the Philo Brice Islands in July.

Order Ulotrichales
Family Ulotrichaceae

Ulothrix flacca (Dillwyn) Thuret. Mixed with *Cladophora* and *Enteromorpha* from pilings along the channel leading to Bayou Lafourche and in drift algae on the seagrass beds along the south side of Casse-tete Island, winter and spring.

Stichococcus marinus (Wille) Hazen. An epiphyte on *Enteromorpha* from the seagrass beds in the southern part of the bay. This species is adapted to estuaries and salt marshes.

Family Chaetophoraceae

Pseudoendoclonium marinum (Reinke) Aleem and Schulz. Forms a green coating on barnacles and woodwork in the intertidal zone in the central and lower bay the year around.

Ulvella lens Crouan. An epiphyte on *Polysiphonia subtilissima* and probably other algae from the seagrass beds along the north side of East Timbalier Island, year around.

Family Gomontiaceae

Gomontia polyrhiza (Lagerheim) Bornet and Flahault. Boring into old mollusc shells in lower Timbalier Bay, year around.

Order Ulvales
Family Ulvaceae

Enteromorpha clathrata (Roth) J. Agardh. One of the most common species in the central and southern parts of the bay the year around, especially spring.

Enteromorpha compressa (L.) Greville. Common in at least the southern part of the bay, winter and spring.

Enteromorpha erecta (Lyngbye) J. Agardh. On pilings of the channel leading to Bayou Lafourche in the southeastern part of the bay and in other open, exposed places, winter and spring.

Enteromorpha flexuosa (Wulfen) J. Agardh. A common species on all substrata in the lower intertidal zone throughout the bay, year around, especially spring.

Enteromorpha intestinalis (L.) Link. Throughout the bay in the intertidal zone from December to April.

Enteromorpha lingulata J. Agardh. On pilings along the channel leading to Bayou Lafourche in the southeastern part of Timbalier Bay during winter and spring and probably the year around in forms difficult to recognize.

Entermorpha plumosa Kützing. Common throughout Timbalier Bay on various substrata, especially barnacles, in the intertidal zone. The species is best developed during winter and spring, but is present the year around.

Enteromorpha prolifera (Müller) J. Agardh. On pilings, shells, and rocks, low in the intertidal zone around the Philo Brice Islands and southward in the bay the year around.

Enteromorpha salina Kützing, variety *polyclados* Kützing. An epiphyte on seagrass leaves in the southern part of the bay, summer and fall, and probably the year around.

Ulva curvata (Kützing) Detoni. Common throughout the bay, especially in spring, but not abundant.

Family Monostromataceae

Monostroma oxyspermum (Kützing) Doty. On roots of black mangroves along the north side of Casse-tete Island and around the base of *Spartina* stems in salt marshes throughout the bay, winter and spring.

Order Cladophorales
Family Cladophoraceae

Chaetomorpha brachygona Harvey. Common as an epiphyte on *Halodule* leaves along the south side of Casse-tete Island, summer and fall.

Chaetomorpha linum (Muller) Kützing. In loose masses on the seagrass flats along the south side of Casse-tete and Calumet Islands and at other stations in the central and southern part of the bay.

Chaetomorpha minima Collins and Harvey. On *Halodule* leaves, south side of Casse-tete Island, summer and fall.

Rhizoclonium hookeri Kützing. On aerial roots of black mangroves and mixed with the blue-green mat around the base of *Spartina* stems along the edge of salt marshes, Philo Brice Islands and elsewhere in the bay.

Rhizoclonium kerneri Stockmayer. On pilings just below the blue-green band, Philo Brice Islands and elsewhere in the central and southern parts of the bay, and around the base of black mangrove aerial roots.

Rhizoclonium kochianum Kützing. Mixed with other loose algae on the seagrass flats along the south side of Casse-tete and Calumet Islands, year around.

Rhizoclonium riparium (Roth) Harvey. Throughout Timbalier Bay on virtually all substrates in the intertidal zone and loose in salt marshes and on the seagrass beds.

Rhizoclonium tortuosum Kützing. Forms a central band on aerial roots of black mangroves on the north side of Casse-tete Island and on oyster shells and waterlogged planks in the intertidal zone. Some plants of this species had rows of empty cells each with a lateral protuberance, apparently pores through which gametes or spores had escaped.

Cladophora dalmatica Kützing. On shells in Little Timbalier Pass near the eastern end of Timbalier Island.

Cladophora delicatula Montagne. Abundant as an epiphyte of *Halodule* and many larger algae along the south side of Casse-tete and Calumet Islands, year around.

Cladophora gracilis (Griffiths) Kützing. Around the bases of *Spartina* at the edge of salt marshes around the Philo Brice Islands and elsewhere in the lower part of the bay.

DISCUSSION

The relative richness of the benthic algal flora of Timbalier Bay is indicated by a comparison with other areas of the Gulf and nearby areas as summarized in table 1. The algal diversity of Timbalier Bay, 83 known species, is noteworthy in comparison to records of only 53 around the

TABLE 1 -- COMPARISON OF THE BENTHIC ALGAE OF TIMBALIER BAY WITH OTHER HABITATS AND AREAS OF THE GULF OF MEXICO AND NEARBY REGIONS

Locality	Cyanophyta		Rhodophyta		Phaeophyta		Chlorophyta		Total Species	Reference
	Species	Percent	Species	Percent	Species	Percent	Species	Percent		
Timbalier Bay	26	32	17	21	12	15	28	33	83	This paper
Louisiana Offshore	22	17	38	29	23	17	49	37	132	Mueller 1976; Kapraun 1974
Chandeleur Island	4	7	26	49	11	21	12	23	53	Humm/Darnell 1959 Mullahy 1959
Mississippi Sound	16	25	24	37	13	20	12	18	65	Humm/Caylor 1957
Port Aransas Area	16*	16	51	51	12	12	21	21	100	Edwards/Kapraun 1973
Anclote Estuary	20	14	59	42	19	13	42	30	140	Ballantine/Humm 1975; Hamm/Humm 1976
Alacran Reef	21	11	79	40	29	15	69	35	198	Kim 1964
Biscayne Bay	32	10	151	47	43	13	96	30	322	Humm 1964, 1976
Gulf Region "D"	21	16	42	32	24	19	43	33	130	Earle 1972
Gulf of Mexico	26	4	325	53	82	13	173	28	617	Earle 1972

*Based on present author's estimate since Edwards and Kapraun (1973) did not include Cyanophyta.

Chandeleur Islands, 65 in Mississippi Sound, and 100 for the Port Aransas, Texas, area. In fact, 63% of all species recorded for the Louisiana coast have been found in Timbalier Bay.

The estuarine habitats of the localities given in table 1 are characterized by the high percentage of Cyanophyta, ranging from 14% in the Anclote River estuary (at Tarpon Springs, Florida) to 32% in Timbalier Bay. The high salinity habitats are characterized by a high percentage of Rhodophyta in the following order: the Gulf of Mexico as a whole, the Port Aransas area, the Chandeleur Islands (east of the Mississippi delta), and Biscayne Bay at Miami. Only 5% of all the Rhodophyta known for the Gulf of Mexico have been found in Timbalier Bay, but nearly all the Cyanophyta were found. The total number of species of Cyanophyta recorded previously for the entire Gulf of Mexico (Earle 1972) and in this work for Timbalier Bay are the same, 26 species. However, several species are known for the Gulf that have not been found in the bay, and vice versa.

There are about 10 species of algae missing from the Timbalier Bay records whose known distribution is such that it seems they should grow in the bay. These species include the following:

Vaucheria (Xanthophyta, Vaucheriales) is a genus specifically adapted to the environment of the bay. Several species of *Vaucheria* are expected to be present within the bay. Reproduction in *Vaucheria*, essential to species determination, is usually restricted to the winter-spring season in the Gulf of Mexico.

Caloglossa leprieurii (Rhodophyta, Ceramiales) is another species that is highly adapted to bay conditions. It probably is present on the aerial roots of black mangroves, on intertidal woodwork, around the bases of *Spartina*.

Enteromorpha linza (Chlorophyta, Ulvales) is a winter-spring species in the Gulf of Mexico that is sufficiently tolerant to grow in lower Timbalier Bay.

Ectocarpus siliculosus (Phaeophyta, Ectocarpales) is a winter-spring species in the northern Gulf of Mexico and should occur in lower Timbalier Bay.

Fosliella (Rhodophyta, Cryptonemiales) is a common epiphyte on seagrass leaves. However, it was not found on leaves of *Halodule* in southern Timbalier Bay. It may invade the lower bay during favorable seasons, however, and from one to three species are to be expected.

Ceramium (Rhodophyta, Ceramiales) is a common epiphyte of seagrass leaves. Although not yet found on the seagrass leaves in southern Timbalier Bay, *C. byssoideum* and *C. fastigiatum* are to be expected there during favorable seasons.

Intertidal zonation

Although a detailed study of intertidal zonation was not made in Timbalier Bay, some general observations were made. Two species of algae mark the highest intertidal band in the bay, the blue-green *Calothrix crustacea*, and the green *Pseudoendoclonium marinum*. In the southern part of the bay, *Pseudoendoclonium* appears to form a slightly higher band where shading occurs, as on stems of the salt marsh grass, *Spartina*. It also occurs under bridges and on the north side of pilings. In exposed places, *Calothrix* is the highest alga, forming a band known as the "black zone" (Stephenson and Stephenson 1952), which has often been mistaken as a high tide stain caused by an oil spill.

Below the *Calothrix-Pseudoendoclonium* zone, *Microcoleus lyngbyaceus* is the dominant alga on treated pilings and woodwork. It tolerates the toxic compounds that leach from creosoted pilings better than any other species, with the result that it may be the only alga on the newer or more heavily treated intertidal wood.

On substrate suitable for its attachment, *Bostrichia radicans* forms the second highest band. *Bostrichia* does not tolerate creosote or other treated wood nor does it attach to smooth surfaces.

Below the *Bostrichia* or *Microcoleus* band, a green zone is present consisting of tufts of *Enteromorpha* and *Ulva*. They will grow on treated pilings that have been in the water for some time. A variety of blue-greens is also present beneath *Enteromorpha* and *Ulva* or between the tufts as well as upon the barnacles and oysters that are present at this level.

Below the green band is a fourth but poorly developed zone dominated by red algae as scattered plants or patches. *Polysiphonia hemisphaerica* is a common constituent of this band in the southern part of the bay from Philo Brice Islands southward. North of Philo Brice, *Polysiphonia subtilissima* is common in this low intertidal position.

In southern Timbalier Bay aerial roots of black mangroves exhibit a distinct intertidal zonation involving several species of algae. On the upper part of the tallest roots of the black mangrove, *Pseudoendoclonium marinum* formed a dark green band or patches 25 to 50 mm wide. Just below this zone near the central part of most roots, *Rhizoclonium tortuosum* formed a dense gray-green fuzz over a vertical band 8 to 12 mm high, along with *Entophysalis conferta* as an epiphyte. Below this band and downward to the soil surface, *Rhizoclonium riparium* was the dominant alga in a narrow zone that included several species of blue-greens and small patches of *Rhizoclonium kerneri*.

CONCLUSIONS

The distribution of the benthic algae in Timbalier Bay is apparently a product of the natural environment, which has not been significantly affected by the activity of the petroleum industry except possibly locally and temporarily. Oil spills soon spread widely in the bay so that areas distant from the wells are affected about as much as those close to the spill. In order to compare, qualitatively, the algal flora of Timbalier Bay with other similar bays of the Louisiana coast in which there is no petroleum activity, collections were made along the Louisiana coast from Timbalier Bay westward to Sabine Pass and the Texas border, including collections at various seasons in Atchafalaya Bay, Cote Blanche Bay, Vermilion Bay, and along the Gulf coast. Only two species were found that were not recorded from Timbalier Bay.

REFERENCES CITED

Ballantine, D. and H. J. Humm. 1975. Benthic algae of the Anclote estuary. I. Epiphytes of seagrass leaves. Florida Scientist **38**(3):150-162.

Burchfield, H. P., R. J. Wheeler, and W. Subra. 1974. Nutrient analysis of water and sediment in Timbalier Bay, Louisiana, and offshore, August 1972—January 1974, parts I (1 & 2) and II. Final Report to Gulf Universities Research Consortium from Gulf South Research Institute, New Iberia, La.

Drouet, F. 1968. Revision of the Classification of the Oscillatoriaceae. Monograph 15, Academy of Natural Science, Philadelphia.

———. 1973. Revision of the Nostocaceae with Cylindrical Trichomes. New York: Hafner Publishing Co.

———. 1978. Revision of the Nostocaceae with Constricted Trichomes. Vaduz, Liechtenstein: J. Cramer.

——— and W. A. Daily. 1956. Revision of the coccoid Myxophyceae. Butler University Botanical Studies **12**:1-218.

Earle, S. A. 1972. Benthic algae and seagrasses. *In* S. Z. El-Sayed et al., eds., Serial Atlas of the Marine Environment. Folio 22, Chemistry, Primary Productivity, and Benthic Algae of the Gulf of Mexico. New York: American Geographic Society, pp. 15-18, plate 6.

Edwards, P. and D. Kapraun. 1973. Benthic marine algal ecology in the Port Aransas, Texas, area. Contributions in Marine Science, The University of Texas **17**:15-52.

Fiore, J. 1969. Life history studies of Phaeophyta from the Atlantic coast of the United States. Ph.D. dissertation, Duke University.

———. 1975. A new generic name for *Farlowiella onusta* (Phaeophyta). Taxon **24**:497-498.

Gagliano, S. M., H. J. Kwon, and J. L. Beck. 1970. Salinity regimes in Louisiana estuaries. Report No. 2, U.S. Corps of Engineers, New Orleans District.

Hamm, D. and H. J. Humm. 1976. Benthic algae of the Anclote estuary. II.Bottom-dwelling species. Florida Scientist **39**(4):209-229.

Humm, H. J. 1964. Epiphytes of the sea grass, *Thalassia testudinum*. Florida Bulletin of Marine Science **14**:306-341.

———. 1976. The benthic algae of Biscayne Bay. *In* E. H. Man et al., eds., Biscayne Bay Symposium I, University of Miami Sea Grant Special Report No. 5, pp. 71-93.

——— and R. L. Caylor. 1957. The summer marine flora of Mississippi Sound. Publications of the Institute of Marine Science, The University of Texas **4**:228-264.

——— and R. M. Darnell. 1959. A collection of marine algae from the Chandeleur Islands. Publications of the Institute of Marine Science, The University of Texas **6**:265-276.

Kapraun, D. F. 1974. Seasonal periodicity and spatial distribution of benthic marine algae in Louisiana. Contributions in Marine Science, The University of Texas **18**:140-167.

Kim, C. S. 1964. Marine algae of Alacran Reef, Southern Gulf of Mexico. Ph.D. dissertation, Duke University.

Muller, T. M. 1976. The distribution and seasonality of marine algae of coastal Louisiana and the adjacent offshore continental shelf. M.S. thesis, University of South Florida.

Mullahy, J. H. 1959. Preliminary survey of the algal flora of the Chandeleur Islands. Proceedings of the Louisiana Academy of Science **22**:62-68.

Price, K. C. 1974. Analysis of water quality data, Timbalier Bay, Louisiana, August 1972-January 1974. Report to Gulf Universities Research Consortium from Gulf South Research Institute, New Iberia, La. GSR1 Project No. N4-573-03.

Stephenson, T. A. and A. Stephenson. 1952. Life between tide marks in North America. II. Northern Florida and the Carolinas. Journal of Ecology **40**:1-49.

Taylor, W. R. 1957. Marine Algae of the Northeastern Coast of North America. Ann Arbor: University of Michigan Press.

———. 1960. Marine Algae of the Eastern Tropical and Subtropical Coasts of the Americas. Ann Arbor: University of Michigan Press.

Walsh, D. 1969. Characteristic pattern of river outflow in the Mississippi delta. Ph.D. dissertation, Texas A&M University.

Williams, S. E. and J. I. Jones. 1974. The distribution and origin of bottom sediments in Timbalier Bay, Louisiana. Gulf Universities Research Consortium. Report No. 138.

BENTHIC MOLLUSCAN AND CRUSTACEAN COMMUNITIES IN LOUISIANA

by Douglas H. Farrell

ABSTRACT

This investigation examines species diversity, faunal similarities, and biomass relations of benthic communities in Timbalier Bay and offshore Louisiana. The bay and shelf regions were faunistically distinct with limited species exchange. Benthic fauna was temporally variable with many species apparently cyclic in occurrence. Seasonality was observed throughout the study area; maximum biomass and maximum species density occurred during spring. Biomass in the bay was not significantly different among different sites, but biomass of the shallow shelf was significantly higher than deeper offshore stations. A marked oxygen diminution occurred in the offshore region during July 1973 and was apparently limiting to biomass and most benthic species. The low dissolved oxygen levels apparently resulted from natural causes, probably the unusually high Mississippi River flood during the previous spring. Sampling was designed to examine possible long-term deleterious effects of oil platforms on the benthic environment, but no abnormal environmental stress could be attributed to the petroleum activity.

INTRODUCTION

Bordering the Gulf of Mexico, Louisiana contains 29,500 km^2 of

---

At the time of this investigation, Douglas Farrell was at the Department of Oceanography, Florida State University.

coastal marshes, which include about 13,700 km^2 of estuaries (Perret et al. 1971a). Extensive fisheries activities in these estuaries and on the adjacent continental shelf have resulted in Louisiana's position among the leading states in commercial fishery products—it is the leading producer on the Gulf. Louisiana tidelands and coastal waters support a multimillion dollar industry as well as thriving sports fishing activities, both of which depend on the "health" of coastal resources. Many commercial species use the estuaries and marshes as nursery grounds, and they migrate through the various biotopes of the estuary and shelf regions (Rounsefell 1966; McHugh 1966; Gunter 1967; Christmas 1973). Sport fisheries are not subject to regularly maintained statistics, but Gunter estimates that sport fishermen annually catch 100 million pounds or more of estuarine fishes in the Gulf. Despite the economic importance of the marine and estuarine resources of the Louisiana coast, most biological investigations have been limited to commercially important species. There is an obvious need for data on primary productivity, biomass, and marine and estuarine communities from the Louisiana coast.

Recent concern about pollution has prompted many investigations of marine resources around the United States. Perret (1971a) summarizes pollution sources in Louisiana from domestic and industrial wastes, dredging, and filling, but only briefly mentions petroleum industry activity. The impact of producing wells, associated petroleum industry activity, and long-term effects on the environment have only recently been studied (St. Amant 1971, 1972).

Timbalier Bay has been the subject of some biological study. The Louisiana Wildlife and Fisheries Commission emphasized the commercially important species in their assessment of Louisiana estuaries (Perret et al. 1971b). Waldron (1963), in the only benthic study, divided the bay into two biocoenoses based on distribution of Foraminifera. Several faunal studies of Louisiana were systematic in nature (Behre 1950; Causey 1953; Dawson 1966; Deichmann 1946; Harry 1942; Hoff 1943; Sprague 1950; and Willis 1942). Parker (1956, 1960) attempted to establish "faunal assemblages" on the distribution of living and dead mollusc shells.

My investigation was one aspect of a multi-discipline study of Timbalier Bay and the adjacent continental shelf. This region has been subject to active, sustained petroleum production for at least thirty years. The main objective of the present study was analysis of the epibenthic communities of estuary and shelf regions, with respect to petroleum industry activity and its long-term effect on benthic communities.

METHODS AND MATERIALS

Sampling methodology

Qualitative samples were collected with a small biological trawl (SBT). The triangular frame was fitted with a 505 mesh bag 2.4 m in length and 0.75 m on a leading edge. The trawl was towed for five minutes at a speed of one knot at selected stations. Quantitative samples were collected with a 1.5 m^2 Van Veen grab, which penetrated to at least 10 cm into the substrate.

All samples were washed aboard ship in a 505 sieve to reduce the size to a more easily handled bulk. Samples were then preserved in a 10% seawater formalin buffered with hexamethyleneamine and transported to the laboratory, where all benthic fauna were removed. The material was then sorted into major taxa and transferred to 70% isopropyl alcohol. Replicate Van Veen samples were collected whenever possible.

Sampling Stations

Before a comprehensive study of Timbalier Bay and offshore benthos could be initiated, an exploratory sampling program was necessary to delineate various biotopes and to define factors influencing distribution of benthic invertebrates in the area. Exploration began in August 1972 and concluded in October 1972, with pooled efforts of several investigators in a "biotope survey" of the near-shore region. Based on the surveys and historical data (Barrett et al. 1971a, 1971b), sampling sites were established in Timbalier Bay and adjacent near-shore regions (figure 1, table 1). Bay sampling stations were distributed over the general area with the principal experimental station near the Philo Brice platform (station 7) and the main control located near Wrong Name Pass (station 12). Selection of the control site was based on the lack of oil industry activity within the area and similarity to the platform site in substrate type and salinity range.

In the offshore region, producing platform 54A (see figure 1) was selected as the platform site. This platform is 25 km south of Timbalier Bay in about 20 m of water. Because of substrate differences, the control site used by the phytoplankton study and several other simultaneous investigations was unsatisfactory. However, one site examined during the biotope survey closely resembled the platform in substrate, depth, and fauna. This site was selected as the initial benthic control. Other sites located between the platform and shore were sampled whenever possible to obtain data for species exchange analysis. During July 1973, several

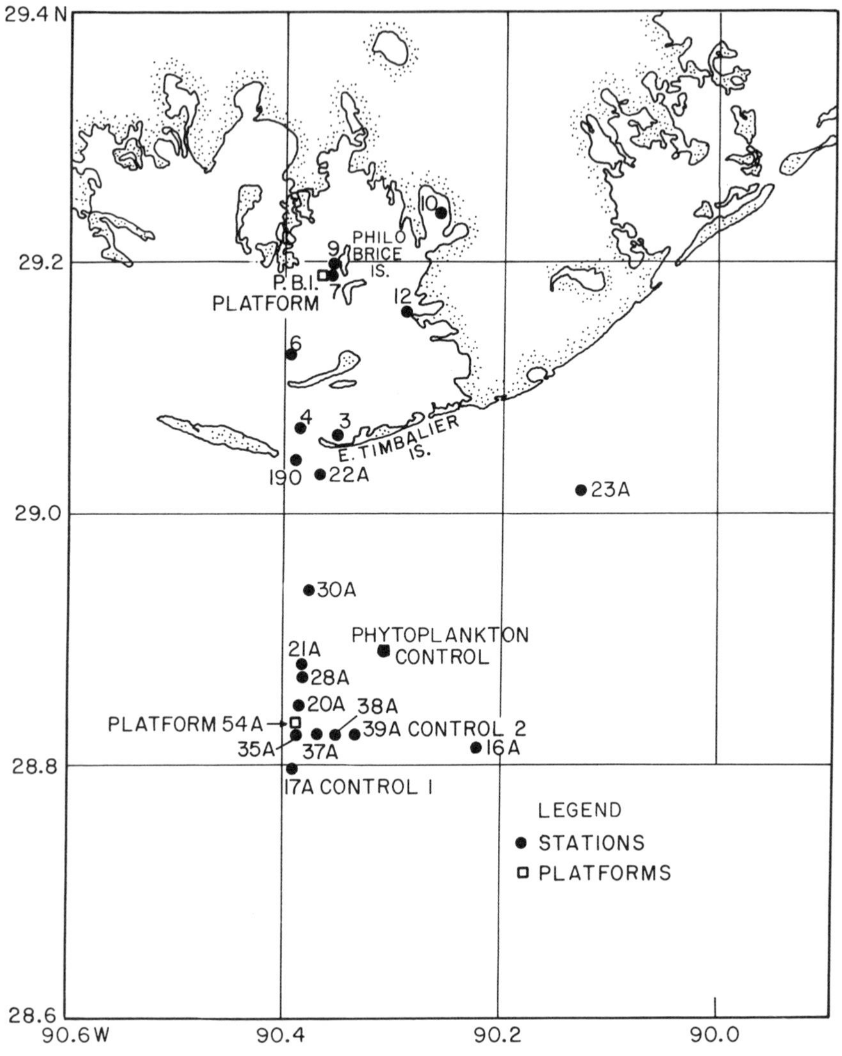

FIG. 1. STATION LOCATIONS IN TIMBALIER BAY, LOUISIANA, AND THE ADJACENT NEAR-SHORE REGION.

additional sampling stations were included as part of a cuttings survey in the area of the "oil well patch," and station 39A was retained as a second benthic control.

Analytical procedures

All crustacean and molluscan species were determined to the lowest possible taxon.

Table 1. General physical characteristics, location and sampling data for Timbalier Bay and nearshore stations, nearshore stations indicated by "A" following station number. Van Veen samples indicated by "x"; small biological trawl samples indicated by "y".

Station Number	Depth (m)	Bottom Salinity range o/oo	Substrate % sand/% mud	N. Latitude W. Longitude	Sampling Dates F72	W73	Sp73	Su73	F73	S74
3	1.7	24.8-29.3	95.2/4.6	$29^{\circ}03.72'$ $90^{\circ}21.26'$	y	x	x	x		
4	1.2	26.2-34.8	99.7/0.2	$29^{\circ}04.16'$ $90^{\circ}23.35'$		x	xy	x		y
6	1.6	18.3-27.3	39.0/60.7	$29^{\circ}07.59'$ $90^{\circ}23.96'$		x	x	x	x	
7 (Platform)	2.0	15.1-26.8	18.8/71.6	$29^{\circ}11.47'$ $90^{\circ}21.62'$		x	xy	x	x	y
9	1.2	13.4-19.0	35.3/53.5	$29^{\circ}11.88'$ $90^{\circ}21.43'$		x	x	x		
10	1.6	10.2-17.4	4.4/93.8	$29^{\circ}14.50'$ $90^{\circ}15.50'$		x	x	x	x	
12 (Control	1.8	17.1-33.3	18.3/81.5	$29^{\circ}09.70'$ $90^{\circ}17.50'$		x	xy	x	x	y
190	3.0	31.8	- -	$29^{\circ}02.60'$ $90^{\circ}23.50'$	xy					
16A	25.6	-	66.3/32.9	$28^{\circ}49.00'$ $90^{\circ}13.30'$	x					
17A (Control)	18.6	31.4-36.1	98.9/1.1	$28^{\circ}47.97'$ $90^{\circ}23.58'$	x	x	xy	x	y	xy
20A	20.1	31.2	49.6/50.2	$28^{\circ}49.96'$ $90^{\circ}23.41'$	x					
21A	18.7	30.7	34.3/64.9	$28^{\circ}52.90'$ $90^{\circ}23.05'$	x					
22A	5.1	23.9-35.7	82.2/17.4	$29^{\circ}01.90'$ $90^{\circ}22.18'$	x	x	xy	x	x	xy
28A	14.1	34.2-34.7	33.7/66.3	$28^{\circ}50.44'$ $90^{\circ}23.14'$			x		x	
30A	14.3	33.2-36.1	7.9/92.1	$28^{\circ}56.42'$ $90^{\circ}22.73'$			x	y	x	
35A (Platform)	20.1	31.2-36.1	97.0/2.9	$28^{\circ}49.96'$ $90^{\circ}23.41'$	x	x	xy	x	x	xy
37A	19.6	35.4	55.2/44.7	$28^{\circ}49.83'$ $90^{\circ}22.30'$				x	x	
38A	19.4	35.2	82.2/17.4	$28^{\circ}49.75'$ $90^{\circ}21.18'$				x	x	
39A (Control #2)	20.0	28.2-35.1	85.6/14.3	$28^{\circ}49.67'$ $90^{\circ}20.06'$				x	x	xy
Phytoplankton Control	20.1	34.73	12.8/72.2	$28^{\circ}53.5'$ $90^{\circ}18.5'$	y					

Biomass (wet weight) of benthic fauna, exclusive of meiofauna, was measured by volumetric displacement techniques for smaller specimens and direct weighing for larger organisms. For the Mollusca, correction factors were determined for the shell.

Species diversity was defined according to the methods of Shannon and Weaver (1963). The Basharin (1959) correction for sample size was included, and Pielou's (1966) "evenness" (J ′) component was also calculated.

The Wilhm (1967) modification of the Shannon-Weaver formula was used to determine biomass diversity. Faunal affinities between pairs of stations and samples were quantified with the Morisita (1959) index of similarity. Both the standard method and nonparametric applications of the similarity index are used to compare stations and biotopes.

Benthic biomass from the Van Veen samples was examined using two different designs of analysis of variance (ANOV). A two-way design ANOV was used on bay samples with seasons and stations as sources of variations, and a factorial two-way design was used for offshore biomass data.

Water quality data for Timbalier Bay stations and logistical support for bay work were furnished by Gulf South Research Institute (GSRI). Substrate samples were collected in cooperation with GSRI for interstitial nutrient analysis. Hydrography, water mass data, and logistical support for offshore collections were furnished by Southwest Research Institute (SWRI). Additional offshore support during September 1972 and July 1973 was furnished by the R/V *Longhorn,* The University of Texas. Most of the data on substrate type and distribution were furnished by Dr. J. I. Jones and Mr. Sam Williams. Total carbon and organic carbon substrate data for benthic stations were furnished by Dr. Charles Brent.

RESULTS

Data on physical parameters were normally taken from instantaneous measurements at time of sampling, and range of variation of these parameters between sampling periods is unknown. Sampling was based on quarter-annual intervals, and measurements presented here should approach seasonal variation. Unless otherwise noted, values of temperature, salinity, and dissolved oxygen represent bottom measurements and were collected as near the bottom as possible, depending on height of surface waves.

Salinity

The bay is a shallow (1-2 m depth), semi-enclosed basin with no major input of fresh water. Mean bottom salinities at bay stations ranged from 14.4 ppt at Little Lake (station 10) to 29.0 ppt near East Timbalier Pass (station 4), and mean bottom salinites at the platform and control sites were 20.5 ppt and 23.9 ppt, respectively. Station 22A near East Timbalier Pass exhibited the greatest range in salinity among near-shore stations (23.9-35.7 ppt). Salinities increased seaward and were most stable at station 17A (31.7-36.1 ppt). Salinities of bottom water at deeper stations were lowest when there was evidence of mixing between bottom and surface waters, during fall 1972.

Dissolved oxygen

Within the bay, dissolved oxygen (DO) levels followed a predictable pattern, with a maximum of 11.3 mg/l observed during February 1973 and a minimum of 3.6 mg/l occurring in July 1973. Dissolved oxygen in bottom water of near-shore stations reached a mean high of 7.3 mg/l during January 1973 and a mean low of 1.4 mg/l during July 1973. The low DO readings were apparently affecting the entire offshore study area during July sampling, and at some stations bottom water DO measurements as low as 0.05 mg/l were recorded. Fish kills were observed during this period. Although quantification was not possible, large numbers of dead, floating fish were seen at station 22A on July 11 and in East Timbalier Pass near bay station 4 on July 16. Seven of the fish at the bay site were identified as the Gulf menhaden *Brevoortia patronus*. Timbalier Bay apparently was not affected by this phenomenon. DO levels remained comparatively high and the two sampling sites nearest the bay mouth, stations 3 and 4, had DO levels of 7.4 mg/l and 8.0 mg/l respectively.

Temperature

While bay temperatures showed great seasonal variation, there was little difference among stations. Average bottom temperatures varied from 12.7° C in February 1973 to 29.0° C during July 1973. Among the near-shore stations, the shallowest, 22A, had the greatest seasonal variation in temperature (13.9° C - 24.2° C), and the offshore platform station, 35A, showed the least variation with a minimum of 19.0° C during February 1973 and a maximum of 24.9° C during July 1973.

Sediments

General substrate types are given in table 1. Bay stations ranged from nearly pure sand (99.65%) at station 4 to the very muddy sediments (93.83%) at station 10 in Little Lake. The platform station and control site in the bay both had sandy mud substrates. Among the offshore stations, the platform station and the controls had predominantly sandy bottoms, as did the station nearest to shore (22A). Intervening stations (20A, 21A, 28A, 30A), as well as the phytoplankton control station, were predominantly mud. With the exception of offshore platform station 35, total carbon content was correlated with mud content of the substrate. Only bay platform station 7 had an inorganic carbon level (1.69 mg/g) higher than 1.0 mg/g. Organic carbon content ranged from 0.26 mg/g at bay station 7 to 19.00 mg/g at bay station 10.

Other physical factors

Currents in the bay appeared to be tidal and reached noticeable velocities only in the passes. Southwest Research Institute reports wind, tides, and regional Gulf of Mexico currents as primary driving forces of currents in the offshore sampling area. Direction of bottom currents appears to be seasonal, and net transport was to the northwest, with an average velocity of 0.21 knots. I observed high turbidity of the bottom water in the form of a "turbid layer," as did other members of a diving crew in the near-shore area. Presence of the layer was confirmed by ship's transmissivity readings, and continued examination proved the layer to be temporally and spatially variable, reaching an apparent maximum density in April 1973. Griffin and Ripy (1974) reported the apparent source of the turbidity as the Mississippi River and not the existent substrate in the sampling area.

BIOLOGICAL RESULTS

Twenty-six Van Veen samples and 8 SBT samples were collected from the eight Timbalier Bay stations, and 58 Van Veen samples and 9 SBT samples from 12 sites were collected offshore (figure 1). Sampling was concentrated at stations 35A (platform 54A), 17A (benthic control 1), and 22A throughout the sampling phase. Station 39A (benthic control 2) was included during the second year. Samples were collected from other stations to furnish data from different biotopes whenever time permitted.

One hundred and one species of epibenthic molluscs and crustaceans were collected from the bay and offshore waters (tables 2 and 3). Crustaceans numbered 65 species, including 39 species of decapods, 19 amphipods, 4 cumaceans, 3 isopods, and 1 each of stomatopods and tanaids. There were 22 species of pelecypods and 14 gastropods. All groups were more numerous in terms of number of species offshore, where 78 species were taken compared to 55 from the bay.

There was little evidence of sessile flora or fauna at any of the sampling stations. The lack of vegetation apparently resulted from the high turbidity of the bottom water, which reduced the penetration of light at the bottom, and from the general lack of solid substratum. One bay SBT haul contained several small pieces of a benthic green alga, which probably originated from the vicinity of Casse-tete and adjacent islands. At stations 17A, 35A, and 39A occasional small pieces of branching ectoprocts were collected, usually attached to worm tubes.

TABLE 2. Species list of epibenthic invertebrates collected in Timbalier Bay and Offshore Louisiana, 1972-1974.

	Bay	Offshore
Phylum Mollusca		
Class Gastropoda		
Anachis avara		X
A. obesa	X	
Cantharus cancellarus		X
Epitonium krebsi		X
E. rupiola	X	
Murex fulvescens		X
Nassarius acutus	X	X
N. vibex	X	X
Olivia sayana		X
Polinices duplicatus	X	X
Sinum perspectivum		X
Terebra dislocata	X	X
Thais haemastoma	X	
Turbonilla sp.		X
Class Pelecypoda		
Argopecten irradians		X
Anadonta ovalis	X	
A. transversa	X	X
Chione cancellata		X
C. grus		X
C. clenchi		X
Corbula sp.		X
Diplodonta punctata	X	X
Docinus discus	X	
Donax tumidus	X	
Ensis minor	X	
Macoma sp.	X	
M. mitchelli	X	
Mercenaria campechiensis	X	X
Mulina lateralis	X	X
Noetia ponderosa		X
Nucula proxima		X
Nuculuna concentrica	X	X
Solen viridis		X
Tellidora cristata	X	
Tellina alternatis	X	
T. iris	X	X
Phylum Arthropoda		
Class Crustacea		
Order Stomatopoda		
Squilla empusa	X	X
Order Cumacea		
Cyclaspis sp.	X	X
Eudorella monodon	X	
Leucon sp.	X	
Oxyurostylis smithi	X	X
Order Tanaidacea		
Leptochelia sp.		X
Order Isopoda		
Ancinus depressus	X	X
Edotea montosa	X	X

(continued next page)

Munna sp.		X
Order Amphipoda		
Acanthohaustoruis sp.	X	
Ampelisca holmesi	X	X
A. vadorum		X
Batea catharinensis	X	X
Caprella sp.		X
Cerapus tubularis		X
Corophium acherusicum	X	X
Elasmopus rapax	X	
Erichthonius brasiliensis	X	X
Gitanopsis tortugae		X
Hemigena minuta		X
Lembos sp.		X
Listrella clymenella	X	X
Microprotopus raneyi		X
Monoculoides nyei	X	X
Orchestia platensis	X	
Protohaustorius sp.	X	X
Stenothoe minuta	X	X
Synchelidium sp.	X	X
Order Decapoda		
Albunea paretti		X
Alpheus sp.		X
Calappa sulcata		X
Callinectes sapidus	X	X
C. similis	X	X
Clibanarius vittatus	X	
Euceramus praelongus		X
Hepatus sp.		X
Hexapanopeus angustifrons	X	
H. paulensis		X
Isocheles wurdmanni	X	X
Labina dubia	X	
Latreutes fucorum	X	X
L. parvulus	X	X
Leptochela serratorbita		X
Metaporhapis calcarata		X
Neopanope texana		X
Ogyrides alphaerostris		X
O. limicola	X	X
Ovalipes guadulpensis	X	X
Paguridae	X	
Paguristes hummi		X
Pagurus longicarpus	X	
P. pollicaris	X	X
Penaeus aztecus	X	X
Persephona punctata		X
P. sp.		X
Petrochirus diogenes		X
Pinnixa cristata		X
P. sp.		X
Portunus gibbesi		X
Processa hemphilli		X
Sicyonia brevirostris		X
S. dorsalis		X
Trachypenaeus constrictus	X	X
Upogebia sp.		X
Xanthidae	X	

Table 3. Molluscan and crustacean biostandards from bay and offshore stations in faunal exchange analysis.

	Number	Rank
BAY STATION 3		
Mulinia lateralis (Say)	67	10
Nassarius vibex (Say)	14	9
Callinectes similis Williams	5	8
Ampelisca holmesi Pearse	4	6.5
Pagurus longicarpus Say	4	6.5
Ensis minor Dall	3	5
Penaeus aztecus Ives	2	4
Oxyurostylis smithi Calman	1	2
Clibanarius vittatus (Bosc)	1	2
Callinectes sapidus Rathbun	1	2
BAY STATION 4		
Mulinia lateralis (Say)	71	33
Callinectes similis Williams	69	32
Tellina iris Say	68	31
Synchelidium sp.	62	30
Pagurus longicarpus Say	56	29
Monoculoides nyei Shoemaker	54	28
Ampelisca holmesi Pearse	29	27
Anachis obesa Adams	13	26
Corophium acherusicum Costa	12	25
Latreutes parvulus (Stimpson)	11	24
Ovalipes guadulpensis (Saussure)	11	23
Hexapanopeus angustifrons (Benedict and Rathbun)	7	22
Nassarius vibex (Say)	5	21
Elasmopus rapax Costa	4	17
Acanthohaustorius sp.	4	17
Isocheles wurdmanni Stimpson	3	17
Mercenaria campechiensis (Gmelin)	3	17
Oxyurostylis smithi Calman	3	17
Trachypenaeus constrictus (Stimpson)	3	17
Pagurus pollicaris Say	3	17
Anadara transversa Say	2	10.5
Ancinus depressus (Say)	2	10.5
Erichthonius brasiliensis (Dana)	2	10.5
Stenothoe minuta Holmes	2	10.5
Protohaustorius sp.	2	10.5
Callinectes sapidus Rathbun	2	10.5
Donax tumidus Philippi	1	4
Terebra dislocata Say	1	4
Edotea montosa Stimpson	1	4
Libina dubia H. Milne Edwards	1	4
Squilla empusa Say	1	4
Anadara ovalis Bruguiere	1	4
Nassarius acutus (Say)	1	4

TABLE 3 continued.

BAY STATION 7, PRODUCING PLATFORM		
Mulinia lateralis (Say)	163	22
Ogyrides limicola Williams	73	21
Oxyurostylis smithi Calman	24	20
Macoma mitchilli Dall	18	19
Ensis minor Dall	17	18
Ampelisca holmesi Pearse	9	17
Batea catharinensis Muller	5	16
Eudorella monodon Calman	4	13.5
Ovalipes quadulpensis (Saussure)	4	13.5
Penaeus aztecus Ives	4	13.5
Callinectes similis Williams	4	13.5
Synchelidium sp.	3	11
Diplodonta punctata Say	2	7.5
Monoculoides nyei Shoemaker	2	7.5
Leucon sp.	2	7.5
Mercenaria campechiensis (Gmelin)	2	7.5
Stenothoe minuta Holmes	1	3.5
Cyclaspis sp.	1	3.5
Latreutes parvulus (Stimpson)	1	3.5
Docinus discus (Reeve)	1	3.5
Epitonium rupicola Kirtz	1	3.5
Polinices duplicata Say	1	3.5

BAY STATION 12, CONTROL		
Mulinia lateralis (Say)	315	16
Macoma mitchilli Dall	49	15
Nassarius vibex (Say)	18	14
Ogyrides limicola Williams	6	13
Nuculana concentrica Say	5	12
Eudorella mondon Calman	4	11
Oxyurostylis smithi Calman	3	9
Ampelisca holmesi Pearse	3	9
Paguridae	3	9
Thias haemostoma L.	2	7
Xanthidae	2	6
Orchestia platensis Kroyer	1	3
Callinectes similis Williams	1	3
Penaeus aztecus Ives	1	3
Listriella clymenella Mills	1	3
Batea catharinensis Muller	1	3

TABLE 3 continued.

OFFSHORE STATION 22A		
Mulinia lateralis (Say)	3415	31
Synchelidium sp.	195	30
Monoculoides nyei Shoemaker	84	28
Trachypenaeus constrictus (Stimpson)	84	28
Tellina iris Say	84	28
Oxyurostylis smithi Calman	63	26
Latreutes parvulus (Stimpson)	23	25
Nassarius acutus (Say)	18	24
Nassarius vibex (Say)	17	23
Callinectes similis Williams	17	23
Ogyrides limicola Williams	17	23
Albunea paretti Guerin	14	21
Stenothoe minuta Holmesi	12	20
Penaeus aztecus Ives	10	18.5
Isocheles wurdmanni Stimpson	10	18.5
Protohaustorius sp.	9	17
Ancinus depressus (Say)	8	16
Polinices duplicatus Say	6	15
Corophium acherusicum Costa	5	14
Pagurus pollicaris Say	4	12.5
Persephona punctata Stimpson	4	12.5
Leptochela serratorbita Bate	3	10.5
Listrella clymenella Mills	3	10.5
Diplodonta punctata Say	2	7
Pinnixa cristata Rathbun	2	7
Oliva sayana Ravenel	2	7
Batea catharinensis Muller	2	7
Callinectes sapidus Rathbun	2	7
Caprella sp.	1	2.5
Edotea montosa (Stimpson)	1	2.5
Epitonium krebsi (Morch)	1	2.5
Anadara transversa Say	1	2.5

OFFSHORE STATION 30A		
Corbula sp.	302	14
Anadara transversa Say	44	13
Diplodonta punctata Say	16	12
Nuculana concentrica Say	13	11
Nucula promixa Say	6	10
Sicyonia dorsalis Kingsley	3	9
Neotia ponderosa Say	2	7
Solen viridis Say	2	7
Mercenaria campechiensis (Gmelin)	2	7
Neopanope texana (Stimpson)	1	3
Tellina iris Say	1	3
Squilla empusa Say	1	3
Chione cancellata L.	1	3
Sinum perspectivum Say	1	3

TABLE 3 continued.

OFFSHORE STATION 17A, BENTHIC CONTROL #1		
Oxyurostylis smithi Calman	181	47
Monoculoides nyei Shoemaker	177	46
Stenothoe minuta Holmes	161	45
Synchelidium sp.	126	44
Lembos sp.	121	43
Erichthonius brasiliensis (Dana)	83	42
Latreutes parvulus (Stimpson)	72	41
Cyclaspis sp.	64	40
Leptochela serratorbita Bate	63	39
Gitanopsis tortugae Shoemaker	33	38
Euceramus praelongus Stimpson	27	37
Ovalipes guadulpensis (Saussure)	19	35
Albunea paretti Guerin	19	35
Processa hemphilli Manning and Chase	19	35
Paguristes hummi Wass	14	33
Nassarius acutus (Say)	13	31
Sicyonia dorsalis Kingsley	13	31
Hexapanopeus paulensis (Stimpson)	10	30
Cerapus tubularis (Say)	8	28
Hemigena minuta Mayer	8	28
Tellina iris Say	7	27
Trachypenaeus constrictus (Stimpson)	6	25
Ampelisca vadorum Mills	6	25
Portunus gibbesi (Stimpson)	6	25
Oliva sayana Ravenel	5	22.5
Diplodonta punctata (Say)	5	22.5
Sicyonia brevirostris Stimpson	4	20
Hexapanopeus angustifrons (Benedict and Rathbun)	4	20
Anadara transversa (Say)	4	20
Cantharus cancellarus (Conrad)	2	17.5
Neotia ponderosa (Say)	2	17.5
Munna sp.	1	8.5
Mulinia lateralis (Say)	1	8.5
Ogyrides alphaerostris (Kingsley)	1	8.5
Penaeus aztecus Ives	1	8.5
Alpheus sp. B	1	8.5
Metaphorhapis calcarata (Say)	1	8.5
Hepatus sp.	1	8.5
Petrochirus diogenes (L.)	1	8.5
Squilla empusa Say	1	8.5
Argopecten irradians (Lamarck)	1	8.5
Terebra dislocata Say	1	8.5
Turbonilla sp.	1	8.5
Anachis avara (Say)	1	8.5
Solen viridis Say	1	8.5
Murex fulvescens Sowerby	1	8.5
Chione grus (Holmes)	1	8.5

TABLE 3 continued.

OFFSHORE STATION 39A, BENTHIC CONTROL #2		
Latreutes parvulus (Stimpson)	154	27
Anadara transversa (Say)	47	26
Sicyonia dorsalis Kingsley	30	25
Trachypenaeus constrictus (Stimpson)	27	24
Euceramus praelongus Stimpson	18	23
Diplodonta punctata (Say)	16	22
Stenothoe minuta Holmes	15	21
Calappa sulcata Rathbun	8	20
Ampelisca vadorum Mills	7	18
Oxyurostylis smithi Calman	7	18
Sicyonia brevirostris Stimpson	7	18
Portunus gibbesi (Stimpson)	6	16
Hexapanopeus angustifrons (Benedict and Rathbun)	5	15
Leptochela serratorbita Bate	4	14
Erichthonius brasiliensis (Dana)	3	13
Synchelidium sp.	2	10.5
Hepatus sp.	2	10.5
Corbula sp.	2	10.5
Chione clenchi Pulley	2	10.5
Monoculoides nyei Shoemaker	1	4.5
Persephona (?) sp.	1	4.5
Nuculana concentrica Say	1	4.5
Argopecten irradians (Lamarck)	1	4.5
Noetia ponderosa (Say)	1	4.5
Squilla empusa Say	1	4.5
Processa hemphilli Manning and Chase	1	4.5
Oliva sayana Ravenel	1	4.5

Biomass

Data were sufficient to indicate a great variability in standing crop in the bay. Standard deviations were large, approaching or exceeding one-half the range at most stations (figures 2 and 3). Using analysis of variance methods to test bay biomass data, there were no significant differences between stations at one collecting period at the $F_{.1}$ level, but the spring biomass was significantly higher than other dates at the $F_{.05}$ level of significance.

Using a factorial analysis of variance, I tested offshore biomass data for differences in stations, dates, and station-date interaction. The most complete set of data was available from stations 35A, 17A, and 22A (figures 4 and 5). Dates, stations, and station-date interaction were all significantly different at the $F_{.005}$ level. By partitioning the degrees of freedom and sums of squares, I found station 22A to have a significantly higher biomass ($F_{.005}$) than stations 17A and 35A. The latter two stations

TABLE 3 continued.

OFFSHORE STATION 35A, PRODUCING PLATFORM		
Oxyurostylis smithi Calman	82	32
Synchelidium sp.	70	31
Latreutes parvulus (Stimpson)	68	30
Ampelisca vadorum Mills	50	29
Monoculoides nyei Shoemaker	36	28
Stenothoe minuta Holmes	33	27
Leptochela serratorbita Bate	19	26
Lembos sp.	16	25
Euceramus praelongus Stimpson	15	24
Processa hemphilli Manning and Chase	13	23
Sicyonia dorsalis Kingsley	10	22
Cyclaspis sp.	9	20.5
Diplodonta punctata (Say)	9	20.5
Gitanopsis tortugae Shoemaker	7	18.5
Tellina iris Say	7	18.5
Albunea paretti Guerin	6	16
Listriella clumenella Mills	6	16
Trachypenaeus constrictus (Stimpson)	6	16
Solen viridis Say	4	13.5
Portunus gibbesi (Stimpson)	4	13.5
Oliva sayana Ravenel	3	11
Alphaeus sp. A	3	11
Erichthonius brasiliensis (Dana)	3	11
Hexapanopeus angustifrons (Benedict and Rathbun)	2	8
H. paulensis Rathbun	2	8
Cantharus cancellarus (Conrad)	2	8
Noetia ponderosa (Say)	1	3.5
Paguristes hummi Wass	1	3.5
Sicyonia brevirostris Stimpson	1	3.5
Terebra dislocata Say	1	3.5
Ovalipes guadulpensis (Saussure)	1	3.5
Ampelisca holmesi Pearse	1	3.5

were then compared and while the stations were not significantly different in biomass, a difference in dates ($F_{.005}$) and date-station interaction ($F_{.05}$) was apparent.

Inspection of the data reveals consistent diminution in biomass values in the July 1973 sampling when the dissolved oxygen minimum was observed offshore. Comparison of values before and after July 1973 indicates that the drop in biomass was highly significant (at $F_{.005}$). The significant station-date interaction apparently results from station 17A being more severely affected by the O_2 minimum than station 35A.

Muddy stations were compared with 35A and 17A, and no significant difference was found between stations (at $F_{.1}$). Station 39A had a significantly greater biomass ($F_{.005}$) when compared with other

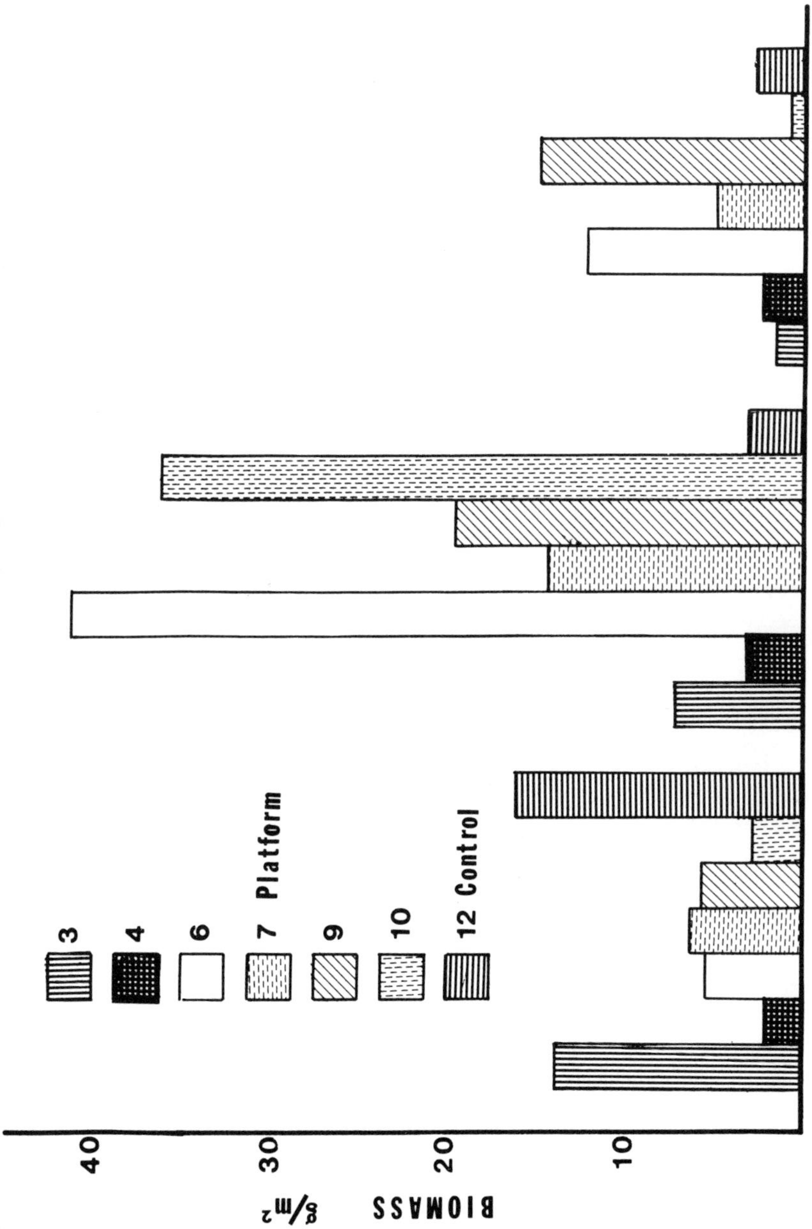

FIG. 2. BENTHIC BIOMASS FROM TIMBALIER BAY STATIONS, February 1973-July 1973.

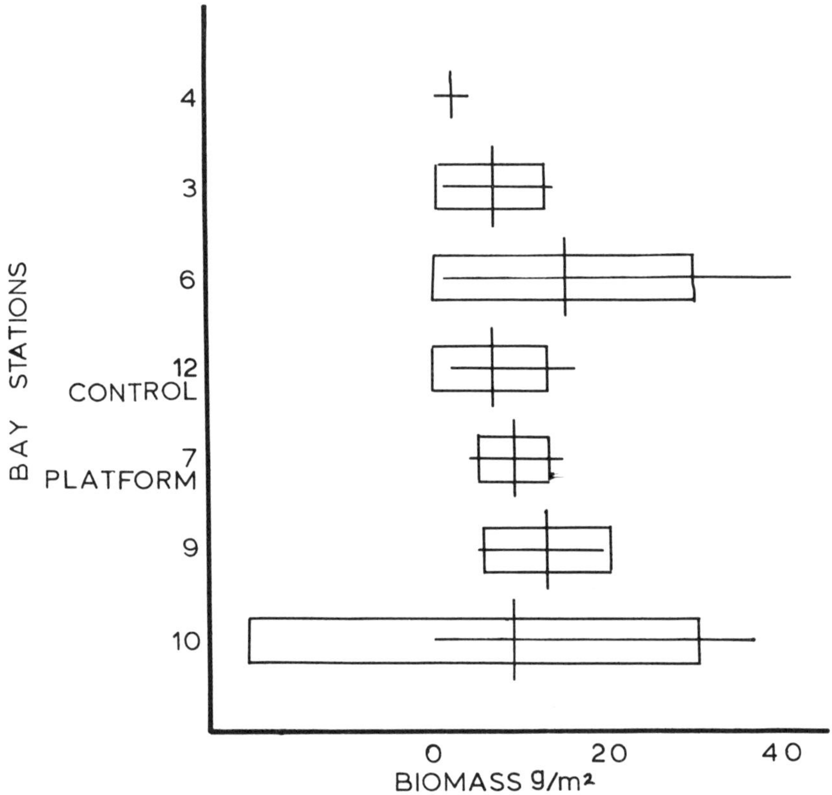

FIG. 3. TIMBALIER BAY BIOMASS DATA FROM TOTAL SAMPLES. Range indicated by horizontal line, mean indicated by vertical line, two standard deviations indicated by horizontal bar.

stations, but 39A was not sampled before the O_2 minimum and the effect on the station is unknown.

Faunal diversity and affinities

The Shannon-Weaver diversity index and Morisita index of similarity were used to analyze the molluscan and crustacean benthic communities. All benthic molluscs and crustaceans identified were used in the calculations. To eliminate sampling error and natural variation from calculation of index measurements, samples were pooled by station. Total Van Veen samples were combined for each station, and the SBT samples pooled. For faunal indices of offshore muddy substrates, station 30A was used, and separate calculations were included on total samples from all mud stations.

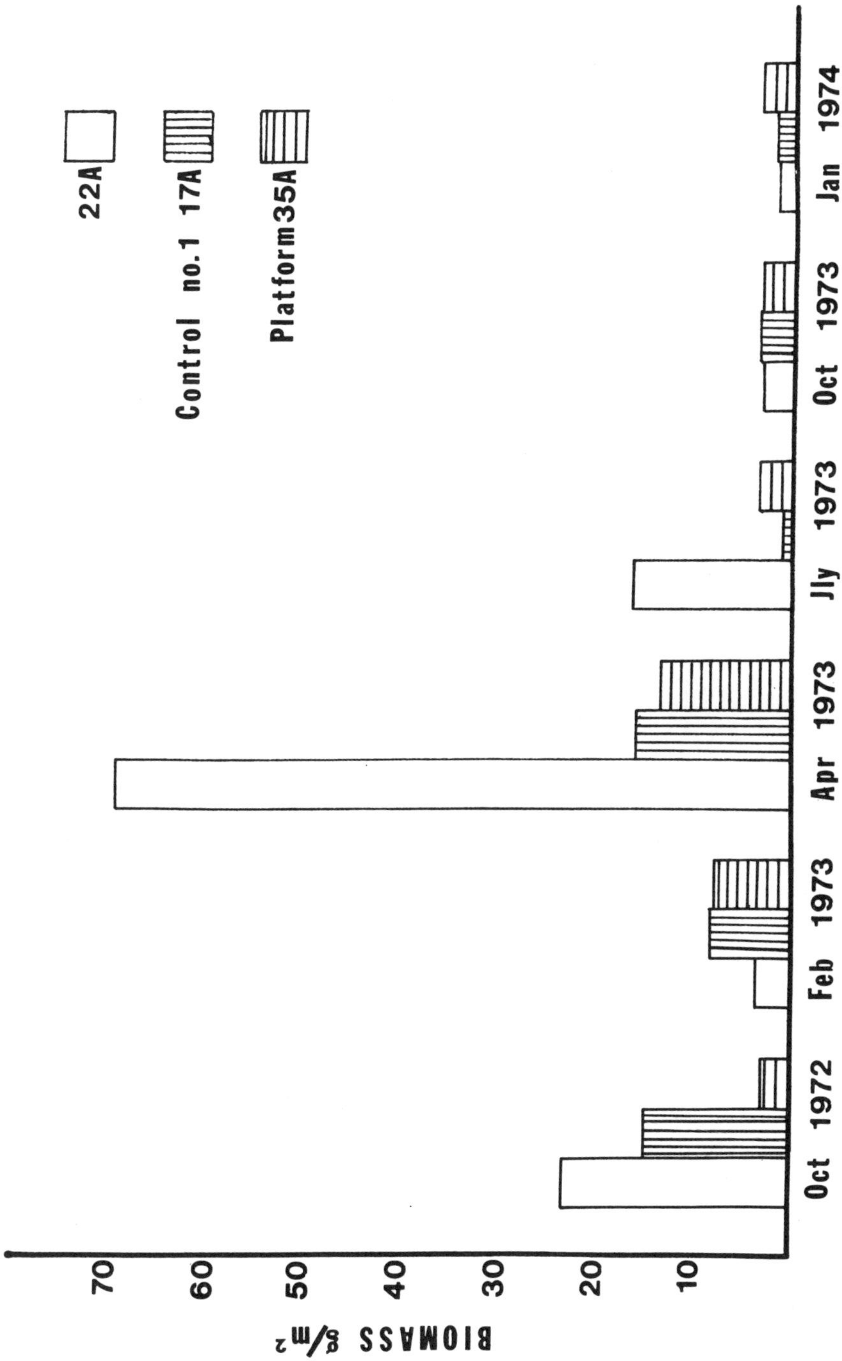

FIG. 4. BENTHIC BIOMASS FROM OFFSHORE STATIONS, October 1972-January 1974.

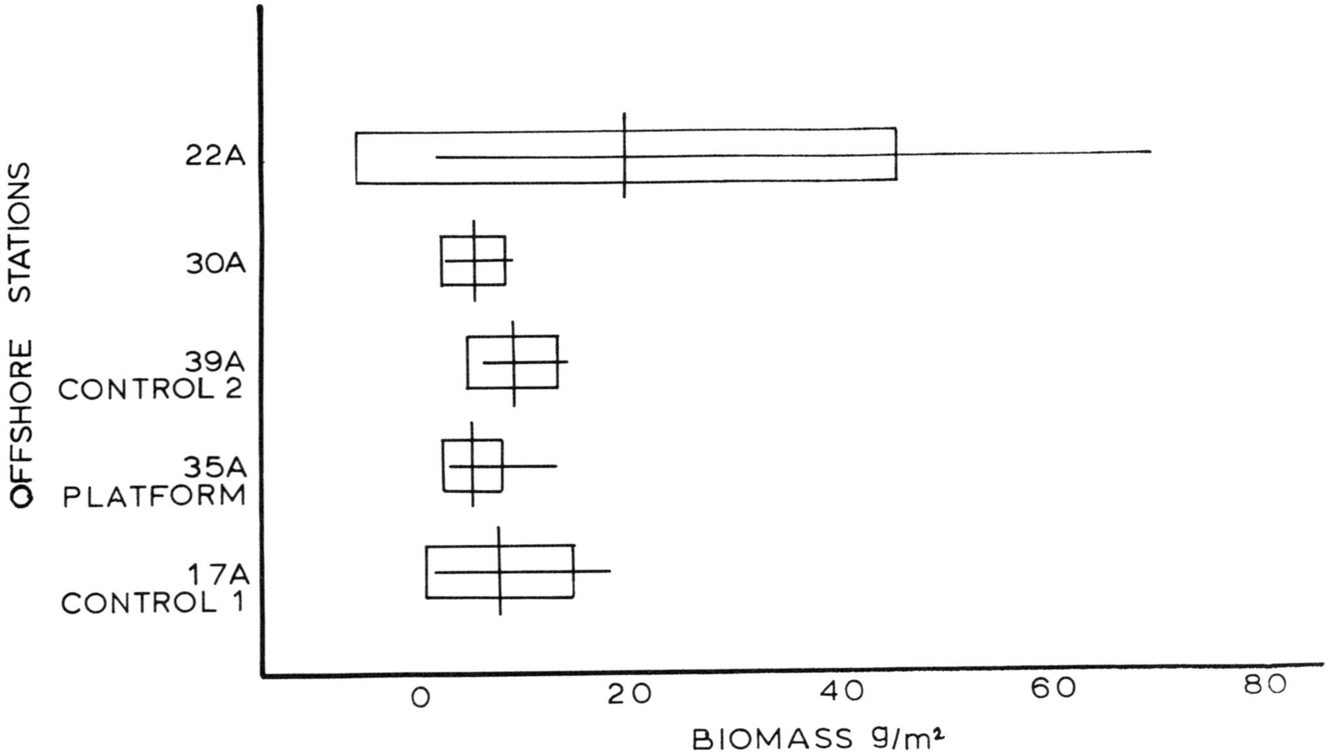

FIG. 5. OFFSHORE BIOMASS DATA FROM TOTAL SAMPLES. Range indicated by horizontal line, mean indicated by vertical line, two standard deviations indicated by horizontal bar.

Species diversity among bay Van Veen samples (table 4) was low (0.324-1.085), but overall diversity increased with increasing salinity (figure 6). The platform station was slightly more diverse (0.727) than the control (0.578). Diversity of SBT samples was higher than Van Veen samples, but, with the exception of station 4, species diversity remained relatively low. For SBT samples, diversity at the platform station (1.918) remained higher than the control (0.921). Evenness (J ′) of bay Van Veen samples ranged from 0.26 to 0.52, and J ′ of SBT samples ranged from 0.35 to 0.73. Another measure of diversity deals with distribution of biomass among taxa within a sample (figure 6). This measure of biomass diversity for bay stations ranged from 0.558 to 1.737, with values of 1.236 for platform station 7 and 1.031 for control station 12.

Diversity of near-shore stations (table 4) is relatively high at the platform and control site 1. Diversity of Van Veen samples was slightly higher at the platform (2.448) than at control 1 (2.340), and slightly lower at the platform (2.576 to 2.861) for SBT samples. Values from the second control were about 80% of the diversity of the platform, and diversity of station 22A (1.160 Van Veen, 0.861 SBT) was much lower than other offshore, sandy substrates. Values from muddy substrates were moderately low with Van Veen diversity of station 30A (0.947) exceeding SBT diversity (0.753); Van Veen diversity of pooled muddy stations (1.744) exceeded pooled SBT samples (0.849). Species equitability

Table 4. Species diversity for total Van Veen samples and total SBT samples. H'=sample diversity, H_{max}=Maximum possible diversity, J'=measure of species equitability.

Station Number	Van Veen Samples			SBT Samples		
	H'	H_{max}	J'	H'	H_{max}	J'
3	0.411	1.609	0.26	1.426	1.946	0.73
4	1.023	2.397	0.42	2.455	3.367	0.73
6	0.977	2.565	0.38	-	-	-
7 (Platform)	0.727	1.386	0.52	1.918	3.091	0.62
12 (Control)	0.578	1.386	0.42	0.921	2.639	0.35
9	0.5229	1.099	0.48	-	-	-
10	0.324	1.099	0.29	-	-	-
22A	1.160	3.044	0.38	0.923	2.398	0.39
30A	0.947	1.609	0.59	0.753	2.197	0.34
39A (Control #2)	1.899	2.302	0.82	2.066	3.091	0.67
35A (Platform)	2.448	2.996	0.82	2.576	3.296	0.78
17A (Control #1)	2.340	2.944	0.79	2.861	3.737	0.72
Muddy Substrates	1.744	2.485	0.70	0.849	2.773	0.31

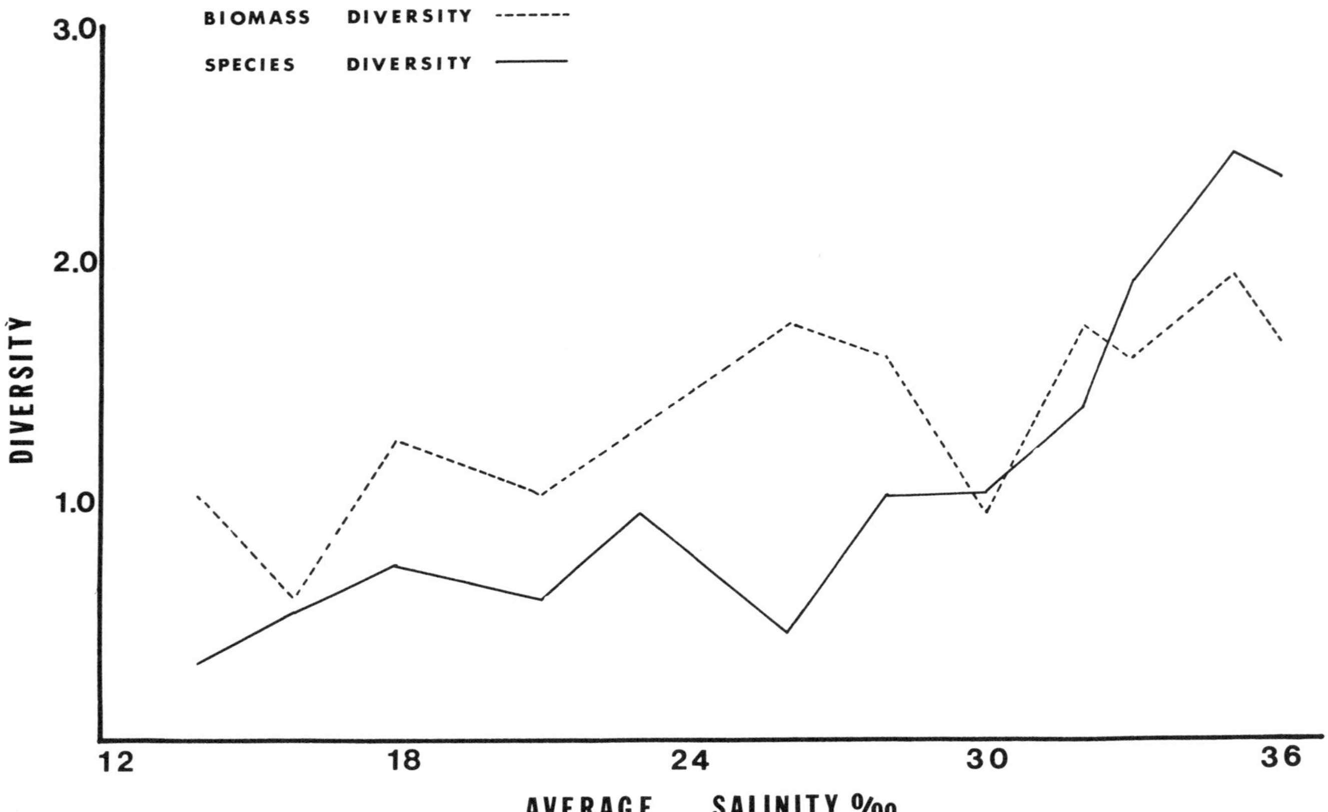

FIG. 6. SPECIES DIVERSITY AND BIOMASS DIVERSITY from Van Veen samples over salinity range of study area.

Table 5. Morisita-Ono index of faunal affinity values for total Van Veen samples.

Station No.	190	10	12 Control	9	7 Platform	6	3
4	.175	.003	.036	.036	.036	.050	.037
3	.001	.081	.987	.962	.953	.980	
6	.000	.085	.979	.965	.962		
7 Platform	.000	.078	.990	.996			
9	.000	.077	.975				
12 Control	.000	.081					
10	.000						

(J ′) values for offshore stations ranged from 0.38 to 0.82 for Van Veen samples and 0.26 to 0.78 for SBT samples. Biomass diversity from offshore stations ranged from 0.923 to 1.927 at the platform station to 1.629 at benthic control 1.

The general trends of species diversity and biomass diversity over the full salinity range of the study area are depicted by figure 6. With some variation, species diversity increased with increasing salinities. Biomass diversity does not appear to follow the same trend. Using analysis of variance to test for difference between bay and offshore diversities, I found no distinction between areas in biomass diversity, even at $F_{.1}$. Offshore, however, species diversity was significantly higher than bay diversity ($F_{.005}$).

Values for the Morisita index of faunal affinity range from 0.0 to about 1.0, with affinity between sample pairs increasing with increasing index values. Results of affinity calculations on bay Van Veen samples (table 5) indicate a very high correlation among stations 3, 6, 7, 9, and 12, index values ranging from 0.996 to 0.953 among these sampling sites. Stations 10, 4, and 190 showed little or no correlation to other stations or among themselves. Similarity of SBT samples follows the same trend as affinities of Van Veen samples (table 6).

Affinities among Van Veen samples were relatively low for the nearshore stations. The highest value was found between the platform station and control 1 (0.449), and the next highest between the platform and control 2 (0.350). These values were only a little lower than expected. Affinities among five replicate samples at station 17A varied from 0.59 to 1.00, indicating that distribution was not as homogeneous as that found at station 22, where agreement among replicate samples was

Table 6. Morisita index values for SBT stations from Bay and offshore Louisiana. Upper value from standard method, lower value, in parentheses, from nonparametric method.

Station No.	Offshore Control 17A	Offshore Platform 35A	39A	30A	22A	4	3	Bay Control 12
7 Bay Platform	.046 (.187)	.113 (.206)	.011 (.096)	.001 (.041)	.831 (.401)	.473 (.409	.854 (.606)	.355 (.896)
12 Bay Control	.008 (.105)	.013 (.105)	.002 (.049)	.001 (.139)	.902 (.507)	.355 (.381)	.890 (.687)	
3	.000 (.030)	.000 (.030)	.000 (.029)	.000 (.000)	.912 (.482)	.447 (.633)		
4	.378 (.354)	.380 (.350)	.086 (1.34)	.004 (.005)	.388 (.682)			
22A	.052 (.540)	.054 (.566)	.018 (.329)	.001 (.000)				
30A	.004 (.009)	.006 (.116)	.046 (.435)					
39A	.421 (.401)	.768 (.602)						
35A Offshore Platform	.9072 (.897)							

greater than 1.00. Affinity values from SBT samples indicate a high degree of faunal similarity between the platform station and control 1, and only moderate faunal affinity between the platform station and control 2. Among remaining stations, faunal relationship is low or nonexistent.

In addition to the standard Morisita index of faunal affinity, affinity values were calculated on the basis of species rank instead of number of individuals (table 6). Rank values were assigned to species in order of numerical dominance. When more than one species occurred with the same frequency, an average of the rank values was assigned. Because very large numbers of individuals at some stations may completely dominate the samples, the importance of some rarer species in transitional areas may be lost. If the species are ranked in order of numerical dominance, only the degree of domination is diminished. Using the rank value in the Morisita index results in a nonparametric application of faunal affinity. With this method, stations that agreed in most faunal components retained high affinity values. Station pairs that had high affinity values under the standard method, but were faunistically different except for one or two dominant species, had lower affinity values with the ranking method. Likewise, index values increased between station pairs with the same rarer species, but different dominant forms.

DISCUSSION

Maximum standing crop was observed during spring at most bay stations, and maximum density of molluscs and crustacean species occurred at this time. The lack of significant differences in biomass among various bay stations may be a function of low sample representation. However, while station 4 appeared lower than other stations in maximum observed biomass, this site did not exhibit the range of seasonal variation shown by other stations. This stability was the apparent reason that station 4 was not significantly different from other bay stations. Biomass of the offshore area apparently varies with seasons and depth. Highest biomass observed at any site (69 g/m^2) occurred in the muddy sand of the inner shelf (station 22A), and the clam *Mulinia lateralis* contributed over 90% of the weight. Since no significant differences were indicated among other stations sampled before the dissolved oxygen minima, substrate was apparently not a major factor in biomass distribution. Among stations sampled before July 1973, spring samples appear to represent biomass maxima.

Comparative biomass data from the northern Gulf of Mexico prior to this study were essentially nonexistent. I collected three Van Veen samples from medium sand in Mississippi Sound at a depth of 4 m, averaging 21.13 g/m^2. Gunter (1967) reported Petersen grab samples from "around the mouth of the Mississippi River" to have "drained" weights of 3.5 g from 3.5 to 4 fathoms and 0.8 g at depths greater than 4 fathoms. Fauchald (1971) found a standing crop off California of 300-800 g/m^2 in an echuroid worm (*Listhiolobus pelodes*) community in the vicinity of oil industry activity. Some fifteen years earlier, Barnard, Hartman, and Jones (1958) had reported biomass values up to 2,000 g/m^2 in the *Listhiolobus* community in the same area prior to the Santa Barbara oil spill of 1969. Since no benthic samples had been taken from this area, it is impossible to determine whether the decrease in biomass was the result of the oil spill or the result of natural variability of the benthic community. Barnard, Hartman, and Jones reported benthic biomass from several areas of the California shelf ranging from about 150 g/m^2 to about 400 g/m^2. Wigley and McIntyre (1964) studied the continental shelf off Massachusetts and found wet weights from 24.30 to 81.55 g/m^2 in depths of 40 to 58 m. By comparison, standing crop in Louisiana waters appears low, particularly for one of the most productive areas in the world.

Before species diversity is discussed, habitat diversity of the study area should be examined. The "level bottom" environment of Timbalier Bay and the adjacent shallow shelf varied in sediment type only from mud to fine sand; medium and coarse sand components were lacking. Except for shallow, isolated areas in the bay, there is no benthic flora, and little evidence of branching ectoprocts, sponges, or other sessile fauna. On the eastern side of the Mississippi River these floral and faunal components serve as habitat for numerous associated species. In Mississippi Sound, Farrell (1970) reported over 30 species of benthic amphipods, and total collections contained over 50 species. The present study includes only 16 benthic species. The amphipod components missing from Louisiana collections are largely those associated with either medium to coarse sand or benthic flora. Habitat diversity of the Louisiana substrate is low and can acount for low species diversity when compared with other areas. Shannon-Weaver diversity includes species richness and equitability as factors in the calculated diversity value. The species equitability measure "J" was low for most bay stations and the shallowest offshore stations, indicating an oligomixic community numerically dominated by one or two species.

Use of species diversity as a means of detecting pollution in aquatic environments has been demonstrated as an effective and sensitive method (Wilhm 1972; Wilhm and Dorris 1966, 1968; Boesch 1972, 1973). The assumption is made that more sensitive fauna will be reduced or eliminated from the area of abnormal stress. Lower species diversity values of the community reflect the loss of species.

Within the scope of this study, there were no significant differences between the bay platform station and the other central bay stations in standing crop or species diversity. As a result of the July 1973 oxygen minima, the offshore situation was not as clear cut. However, present data indicate that there were no real differences in biomass or standing crop between the platform and offshore control 1. In both the bay and offshore areas, there was no indication of long-term detrimental effects on the benthos by oil industry activity.

Dissolved Oxygen Minima

The exact cause of the dissolved oxygen minima is a matter of speculation. Brongersma-Sanders (1957), in a survey of mass mortality at sea, includes lack of oxygen in isolated bottom water and in association with noxious waterbloom among known causes for marine fish kills. During the summer months the envelope of bottom water on the Louisiana shelf was isolated; a source of eutrophication was present in the abnormally high flood of the Mississippi River; and high phaeopigments were reported during July near the bottom by investigators studying primary productivity (El Sayed 1974). From the evidence at hand, I believe that the offshore Louisiana oxygen minima resulted from high respiration and high BOD from mortality of an abnormally large phytoplankton bloom. The high phaeopigment level during July sampling tends to support this contention. Of course, mortality of benthic species would only contribute to the BOD. Whatever the cause, the oxygen minima were the most important stress factor on offshore benthos during the sampling phase of this study.

Community relationships

Carriker (1967) reviews various arguments concerning reality of benthic biocoenosis as a biological unit and applications to estuaries, and points out that an estuary is not simply an area of integration between fresh and salt water, but possesses unique faunal components. Whether an estuary is represented by a single community divided into a series of sub-communities (Barnard 1970) or a series of communities contributing to a major faunal zone is a matter of semantics. Throughout the Louisiana Bay and offshore study area, several changes in substrate were evident, as well as the gradient in salinity. From a standpoint of salinity ranges, the entire study area would have to be classified as estuarine. Even the most saline station (17A) ranged from 31.4-36.1 ppt. For simplicity the various ecological units in this study were considered benthic communities.

None of the Louisiana sampling sites could be characterized by a molluscan or crustacean species throughout the entire sampling phase.

Biocoenoses in this study appear to have a dynamic stability in which dominance of principal species is probably cyclic. Exceptions would be expected in the event of abnormal stress, such as the DO minima in the near-shore region. Fauna best able to cope with low oxygen levels would have a survival advantage, and dominance would shift.

Another factor to be considered in examining communities is continua of distribution. Mills (1971) observed that different benthic species that have population maxima at the same site may have radically different distribution patterns. During this study several euryhaline crustacea were found ranging from the bay to sandy offshore stations where apparent population maxima occurred. These crustaceans were among the numerically dominant fauna at control site 1 (17A) and the platform station, but were sufficiently eurybiotic to span other biotopes.

Examination of physical parameters of all stations in the bay and near-shore regions shows two distinct biotopes within each region and a transitional area in the vicinity of the bay mouth (figure 7):

1. Mesohaline bay, muddy substrate, minimum temperature below 15° C, low energy;
2. Polyhaline bay, sandy mud substrate, minimum temperature below 15° C, low energy;
3. Polyhaline bay mouth (transitional area), fine sand-muddy sand substrate, minimum temperature below 15° C, moderate energy;
4. Euryhaline near-shore, sandy mud-mud substrate, minimum temperature about 15° C, low energy;
5. Euryhaline near-shore, muddy sand-sand substrate, minimum temperature above 15° C, low energy.

The standard Morisita-Ono index of faunal affinity applied to samples throughout the study area helps define the real biotopes and evaluate interchange among stations. Station 10 in Little Lake was dominated by the clam *Macoma mitchilli* at spring sampling, and showed only minor exchange with any other stations. Central Bay stations 6, 7, 9, and 12 all have very high faunal similarity when examined on the basis of total Van Veen samples. All show, more or less, a biocoenosis of oligomixity with domination by the pelecypod *Mulinia lateralis*. Composition of less abundant species differed among some stations, but the pelecypod *Macoma mitchilli,* decapod *Ogyrides limicola,* amphipod *Ampelisca holmesi,* and the cumaceans *Oxyurostylis smithi* and *Eudorella mondon* were among the most common species from the area.

Station 3 agrees with the central bay in dominance of *M. lateralis* but differs in less abundant species. Station 3 was located on sand flats

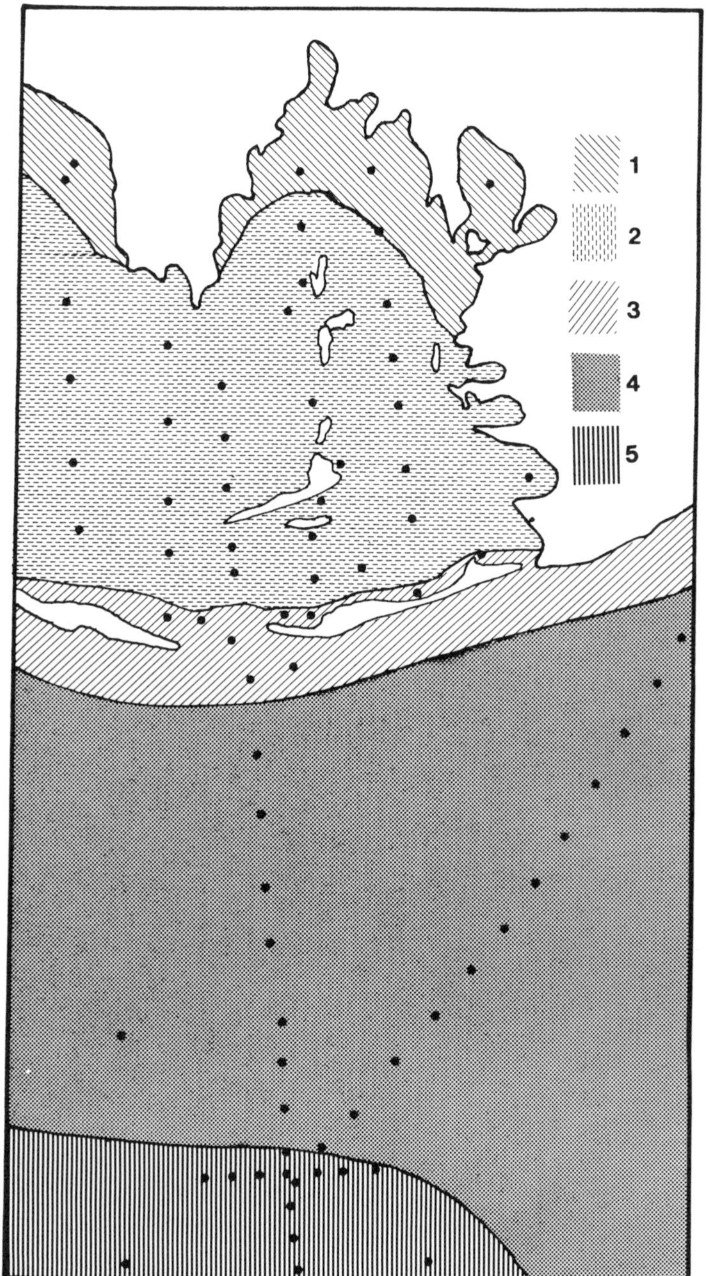

FIG. 7. IDEALIZED BIOTOPE DISTRIBUTION OF STUDY AREA based on general physical parameters. 1 - mesohaline Bay; 2 - polyhaline Bay, sandy mud substrate; 3 - polyhaline transitional area, sand substrate; 4 - euryhaline offshore, sandy mud substrates; 5 - euryhaline offshore, sandy substrate. Points represent locations of data acquisition.

along the inside of E. Timbalier Island. It was more comparable, ecologically, to station 4 than to other stations in the bay. Station 4 near the pass into Timbalier Bay represents the transition area, exemplified by low to medium low affinities in comparison to all other stations. Diversity is relatively high, because there is no clear-cut dominance. The clams *Tellina iris* and *Mulinia lateralis,* amphipods *Synchelidium* sp. and *Monoculoides nyei,* and decapods *Pagurus longicarpus* and *Callinectes similis* were equally abundant. Rather than just representing range extremes of bay or offshore species, station 4 fauna is largely euryhaline and estuarine species that are represented in both offshore and bay samples. Also included are several apparent endemic forms that probably prefer the higher energy environment.

Despite the location on the shallow shelf, station 22A had very high affinities with the central bay stations and no real correlation with other offshore sites with the standard Morisita index. *Mulinia lateralis* reached its apparent population peak at this station, and on the basis of this species station 22A would be lumped into the bay biocoenosis. Inspection of species components showed that station 22A is really a transition station and represents the range extremity of many offshore and bay species. The Morisita index, using species rank, indicates that similarity to other offshore stations was equal to bay stations if dominance of *M. lateralis* was reduced.

Stations 35A, 17A, and 39A are offshore sites with sand to muddy sand substrates and less variability among the physical parameters. The cumacean *Oxyurostylis smithi*; amphipods *Synchelidium* sp., *Monoculoides nyei,* and *Stenothoe minuta*; and the caridean shrimp *Latreutes parvulus* and *Leptochela serratorbita* were the dominant species in this area.

Between the deep stations, which are characterized by sand substrate, and the shallow shelf lies a distinct mud-sandy mud biotope represented by stations 21A, 28A, and 30A. Since SBT data are available from station 30A, this site was used to represent offshore muddy bottom communities. The dominant species, *Corbula* sp. (Pelecypoda), was encountered only rarely in other biotopes.

As was previously mentioned, both standard and nonparametric application of the affinity index were used to test similarity among SBT stations throughout the study area. Values between station-pairs were different only when the stations agreed in one or two dominant species but did not agree in the majority of the less common species. To depict faunal exchange between a bay station and other sites, platform station 7 was selected, since it was the most brackish SBT sampling site (figure 8). With the ranked Morisita index, station 7 shows high similarity only to station 12, moderately high affinity with bay station 4 and shallow shelf

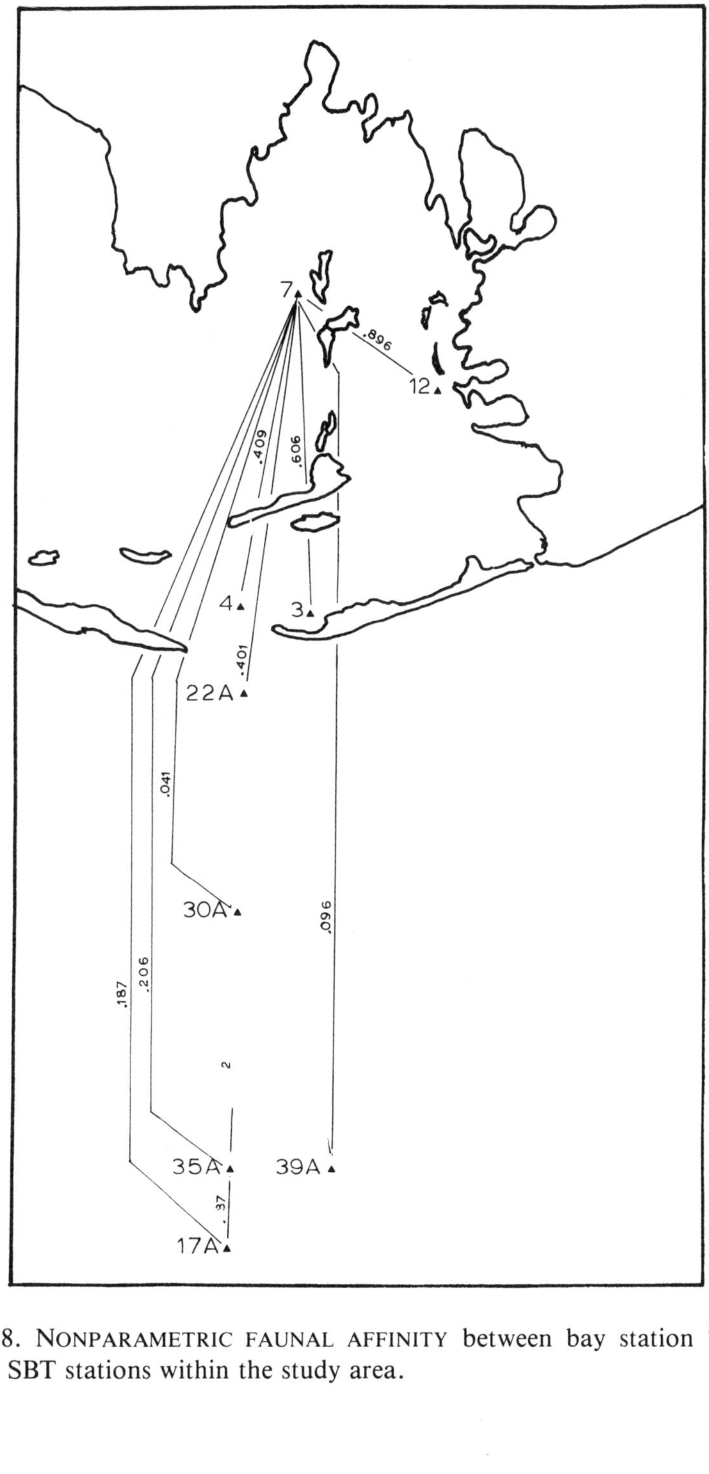

FIG. 8. NONPARAMETRIC FAUNAL AFFINITY between bay station 7 and other SBT stations within the study area.

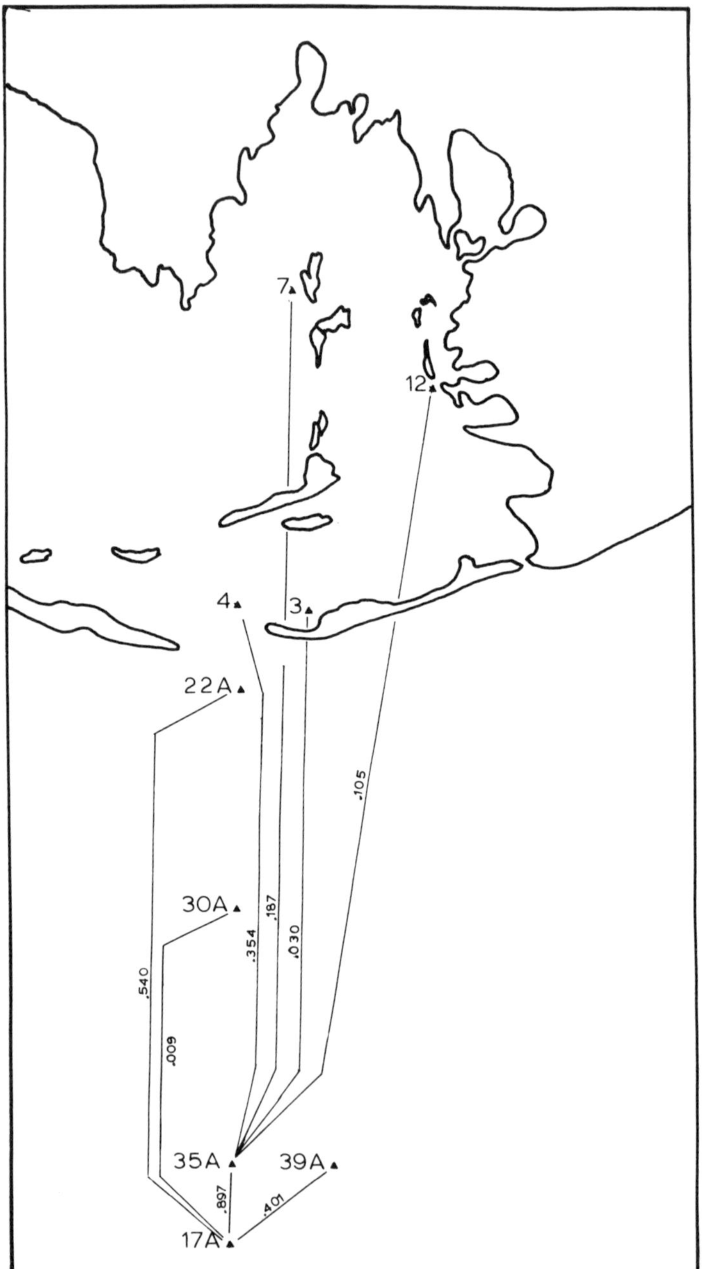

FIG. 9. NONPARAMETRIC FAUNAL AFFINITY between offshore station 17A and other SBT stations within the study area.

site 22A, low affinity to the deep sandy substrates, and no similarity to the offshore muddy bottoms.

For offshore stations, control 1 (17A) was selected for comparison with other stations because this site was the most saline and the most distant from shore (figure 9). Station 17A showed a high affinity with station 35A, a moderate affinity with offshore stations 39A and 22A, and a low affinity with station 4 at the mouth of the bay.

Despite the fact that salinity ranges and species composition indicated that the entire area was estuarine, the sampling area can be divided into two major faunal zones, bay and offshore, with at least two biocoenoses in each zone. The area in the vicinity of East Timbalier Pass could be called a fifth or transitional community. To gain further insight into species exchange between the zones, the various species were categorized according to estuarine habitat (figure 10). Migrants, largely commercial shrimp and crabs, included only a few species and ranged

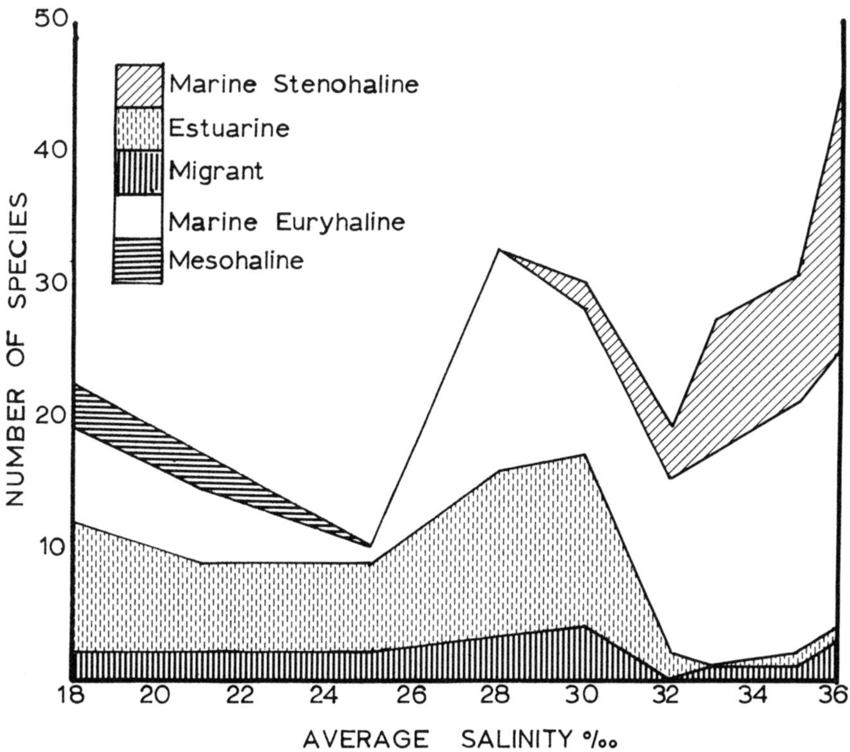

FIG. 10. CUMULATIVE NUMBER OF SPECIES for SBT stations, species classified by ecological habitat.

over the entire sampling area. Mesohaline species were not collected beyond bay stations 7 and 12, and the remaining bay fauna (estuarine species) barely ranged beyond the transition area, offshore station 22A. The low, but distinct, faunal affinity between the deeper offshore areas and bay stations 7 and 12 resulted from invasion of marine euryhaline species—pelecypod, *Diplodonta punctata*; cumaceans, *Oxyurostylis smithi* and *Cyclaspis* sp.; amphipods, *Monoculoides nyei*, *Synchelidium* sp., and *Stenothoe minuta*; and the decapod *Latreutes parvulus*.

While the effect of substrate on species distribution was pronounced on the shelf, bottom type was apparently not as critical in the bay. For example, the fossorial amphipods, *Synchelidium* sp. and *Monoculoides nyei*, were limited entirely to sandy substrates at offshore stations, but in Timbalier Bay they were collected in substrates ranging from sand to sandy mud. McLusky (1968) found that local distribution of the amphipod *Corophium volutator* was substrate-dependent within its optimal salinity range, but in more brackish water, salinity was the factor controlling distribution. Similar factors may have contributed to spatial distribution of marine euryhaline species in Timbalier Bay.

ACKNOWLEDGMENTS

I wish to thank Dr. Robert Menzies for his guidance throughout the duration of this project. Thanks are also due Drs. Jim Jones, Ruby Krishnamurti, and Winston Menzel of the Oceanography Department of the University of Florida for their invaluable advice, and Drs. Van Belle and Metter for assistance with statistical aspects of this problem. Drs. James Marum, Allen Paul, and Robert Avent contributed much help through lengthy discussions and constructive criticism. This project was funded by the Gulf Universities Research Consortium, Contract No. OB69JM, and was submitted in partial fulfillment of the requirements for the degree of Doctor of Philosophy.

REFERENCES CITED

Barnard, J. Laurens. 1970. Benthic ecology of Bahia de San Quintin, Baja California. Smithsonian Contributions to Zoology, No. 44.

———, O. Hartman, and G. F. Jones. 1958. Benthic biology of the mainland shelf of southern California. *In* Oceanographic Survey of the Continental Shelf Area of Southern California. Allan Hancock Foundation, University of Southern California, pp. 265-430.

Barrett, B. B., J. W. Tarver, W. R. Latapie, J. F. Pollard, W. R. Mock, G. B. Adkins, W. J. Gaidry, C. J. White, and J. S. Mathis. 1971a. Phase 2, hydrology. *In* W. S. Perret, Coordinator, Cooperative Gulf of Mexico Estuarine Inventory and Study, Louisiana. Louisiana Wildlife and Fisheries Commission.

———. 1971b. Phase 3, sedimentology. *In* W. S. Perret, Coordinator, Cooperative Gulf of Mexico Estuarine Inventory and Study, Louisiana. Louisiana Wildlife and Fisheries Commission.

Basharin, G. P. 1959. On a statistical estimate for the entropy of a sequence of independent random variables. Theory of Probability and its Applications **4**:333-336.

Behre, E. H. 1950. Annotated list of the fauna of the Grand Isle region. Occasional Papers, Marine Laboratory, Louisiana State University **6**:1-66.

Boesch, D. F. 1972. Species diversity of marine macrobenthos in the Virginia area. Chesapeake Science **13**(3):206-211.

———. 1973. Classification and community structure of macrobenthos in the Hampton Roads area, Virginia. Marine Biology **21**:226-244.

Brongersma-Sanders, M. 1957. Mass mortality in the sea. *In* J. W. Hedgpeth, ed., Treatise on Marine Ecology and Paleoecology. Vol. I, pp. 941-1010. Geological Society of America Memo **67**.

Carriker, Melbourne R. 1967. Ecology of estuarine benthic invertebrates: a perspective. *In* George R. Lauff, ed., Estuaries. No. 83, American Association for the Advancement of Science, pp. 442-487.

Causey, D. 1953. Parasitic Copepoda from Grand Isle, Louisiana. Occasional Papers, Marine Laboratory, Louisiana State University **7**:1-18.

Christmas, J. Y., ed. 1973. Cooperative Gulf of Mexico estuarine inventory and study, Mississippi. Gulf Coast Research Laboratory, State of Mississippi.

Dawson, C. E. 1966. Additions to known marine fauna of Grand Isle, Louisiana. Proceedings, Louisiana Academy of Science **21**:175-180.

Deichmann, E. 1946. A new species of *Thyone* s.s. from the Gulf of Mexico *Thyone mexicana* sp. nov. Occasional Papers, Marine Laboratory, Louisiana State University **4**:1-4.

El-Sayed, S. Z. 1974. Effect of oil production on the ecology of phytoplankton off the Louisiana coast. Gulf Universities Research Consortium. Appendix to Report No. 138.

Farrell, D. H. 1970. Ecology and seasonal abundance of littoral amphipods from Mississippi. Unpublished M.S. thesis, Mississippi State University.

Fauchald, K. 1971. Benthic fauna in Santa Barbara channel following the January 1969 oil spill. *In* Dale Straughan, ed., Biological and Oceanographic Survey of Santa Barbara Channel Oil Spill, Vol. 1. University of Southern California.

Griffin, G. M. and B. S. Ripy. 1974. Turbidity, suspended sediment concentrations, clay mineralogy of suspended sediments, and the origin of the turbid near-bottom water layer—Louisiana shelf south of Timbalier Bay—August 1972-January 1974—with comments on a process model for turbid layer transport. Gulf Universities Research Consortium. Appendix to Report No. 138.

Gunter, G. 1967. Some relationships of estuaries to the fisheries of the Gulf of Mexico. *In* George H. Lauff, ed., Estuaries. No. 83, American Association for the Advancement of Science, pp. 621-638.

Harry, H. W. 1942. List of Mollusca of Grand Isle, Louisiana, recorded from the Louisiana State University Marine Laboratory 1929-1941. Occasional Papers, Marine Laboratory, Louisiana State University **1**:1-13.

Hoff, C. C. 1943. The description of a new ostracod of the genus *Potamocypris* from Grand Isle, Louisiana, and records of ostracods from Mississippi and Louisiana. Occasional Papers, Marine Laboratory, Louisiana State University **3**:1-11.

McHugh, J. L. 1966. Management of estuarine fisheries. American Fishery Society Publication **3**:133-154.

McLusky, D. S. 1968. Some effects of salinity on distribution and abundance of

Corophium volutator in the Ythan estuary. Journal of the Marine Biological Association, U.K. **48**:442-454.

Mills, E. L. 1971. Views of the community concept, with comments on continua and the role of instability in some marine benthic communities. Vie et Milieu, Troisième Symposium Europeen de Biologie Marine. Supp. **22**:145-153.

Morisita, M. 1959. Measuring of interspecific association and similarity between communities. Memoirs of the Faculty of Science, Kyushu University, Series E **3**:65-80.

Parker, Robert H. 1956. Macroinvertebrate assemblages as indicators of sedimentary environments in east Mississippi delta region. American Association of Petroleum Geologists Bulletin **40**(2):295-376.

———. 1960. Ecology and distributional patterns of marine macroinvertebrates, northern Gulf of Mexico. Recent Sediments, Northwest Gulf of Mexico, 1951-1958. Tulsa, Oklahoma: American Association of Petroleum Geologists.

Perret, W. S., W. R. Latapie, J. F. Pollard, W. R. Mock, G. B. Adkins, W. J. Gaidry, and C. J. White. 1971a. Phase I, area description. *In* W. S. Perret, Coordinator, Comparative Gulf of Mexico Estuarine Inventory and Study, Louisiana. New Orleans: Louisiana Wildlife and Fisheries Commission, pp. 3-27.

———. 1971b. Fishes and invertebrates collected in trawl and seine samples in Louisiana estuaries, phase 4, sec. 1. *In* W. S. Perret, Coordinator, Cooperative Gulf of Mexico Estuarine Inventory and Study, Louisiana. New Orleans: Louisiana Wildlife and Fisheries Commission, pp. 41-105.

Pielou, E. C. 1966. The measurement of diversity in different types of biological collections. Journal of Theoretical Biology **13**:131-144.

Rounsefell, G. A. 1966. Realism in the management of estuaries. Marine Research Bulletin **1**:1-13.

St. Amant, L. S. 1971. Impact of oil on the Gulf Coast. Transactions of the Thirty-Sixth North American Wildlife and Natural Resources Conference, pp. 206-219.

———. 1972. The petroleum industry as it affects marine and estuarine ecology. Journal of Petroleum Technology **24**(4):385-392.

Shannon, C. E. and W. Weaver. 1963. The Mathematical Theory of Communication. Urbana: University of Illinois Press.

Sprague, V. 1950. Notes on three Microsporidian parasites of Decapod Crustacea of Louisiana coastal waters. Occasional Papers, Marine Laboratory, Louisiana State University **5**:1-8.

Waldron, Robert P. 1963. A seasonal ecological study of Foraminifera from Timbalier Bay, Louisiana. Gulf Research Reports **1**(4):132-188.

Wigley, R. L. and A. D. McIntyre. 1964. Some quantitative comparisons of offshore meriobenthos and macrobenthos south of Martha's Vineyard. Limnology and Oceanography **9**:483-493.

Wilhm, J. L. 1967. Use of biomass units in Shannon's formula. Ecology **49**(1):153-156.

———. 1972. Graphic and mathematical analyses of biotic communities in polluted streams. Annual Review of Entomology **17**:223-252.

——— and T. C. Dorris. 1966. Species diversity of benthic macroinvertebrates in a stream receiving domestic and oil refinery effluents. American Midland Naturalist **76**(2):427-449.

——— and T. C. Dorris. 1968. Biological parameters for water quality criteria. Bioscience **18**(6):477-481.

Willis, E. R. 1942. Some mud shrimps of the Louisiana coast. Occasional Papers, Marine Laboratory, Louisiana State University **2**:1-6.

CHECKLIST OF THE MARINE ALGAE ON THE OFFSHORE OIL PLATFORMS OF LOUISIANA

by Theresa M. Bert and Harold J. Humm

ABSTRACT

A survey was conducted over a period of a year for the marine algae present on eight oil production platforms located in the offshore waters of Louisiana. A total of 120 species was recorded, bringing the total known for Louisiana to 180. Of the 120 species collected 18% are Cyanophyta, 31% Rhodophyta, 16% Phaeophyta, and 35% Chlorophyta. Two floristic provinces were noted with regard to algal occurrences on these platforms: a coastal province characterized by shallow water of reduced salinity with fewer species especially of browns and reds, and an offshore province characterized by deeper, clearer water and a large number of species especially of browns and reds.

INTRODUCTION

Prior to the initiation of this work, nearly all studies of the marine flora of Louisiana had been limited to bays or near-shore areas. An exception was Kapraun's (1974) collections from submerged offshore reefs. He reported 31 species new to the Louisiana coast and commented on their distribution, habitat, and reproduction.

Early reports of the marine algae of Louisiana included a list of species from the vicinity of the Gulf Biological Station at Cameron (Cocks 1907), and two lists from the Grand Isle area (Behre 1950; Flint 1951). Taylor (1954) mentioned a few species from Louisiana and in-

---

At the time of this investigation, Theresa Bert and Harold Humm were at the Department of Marine Science, University of South Florida.

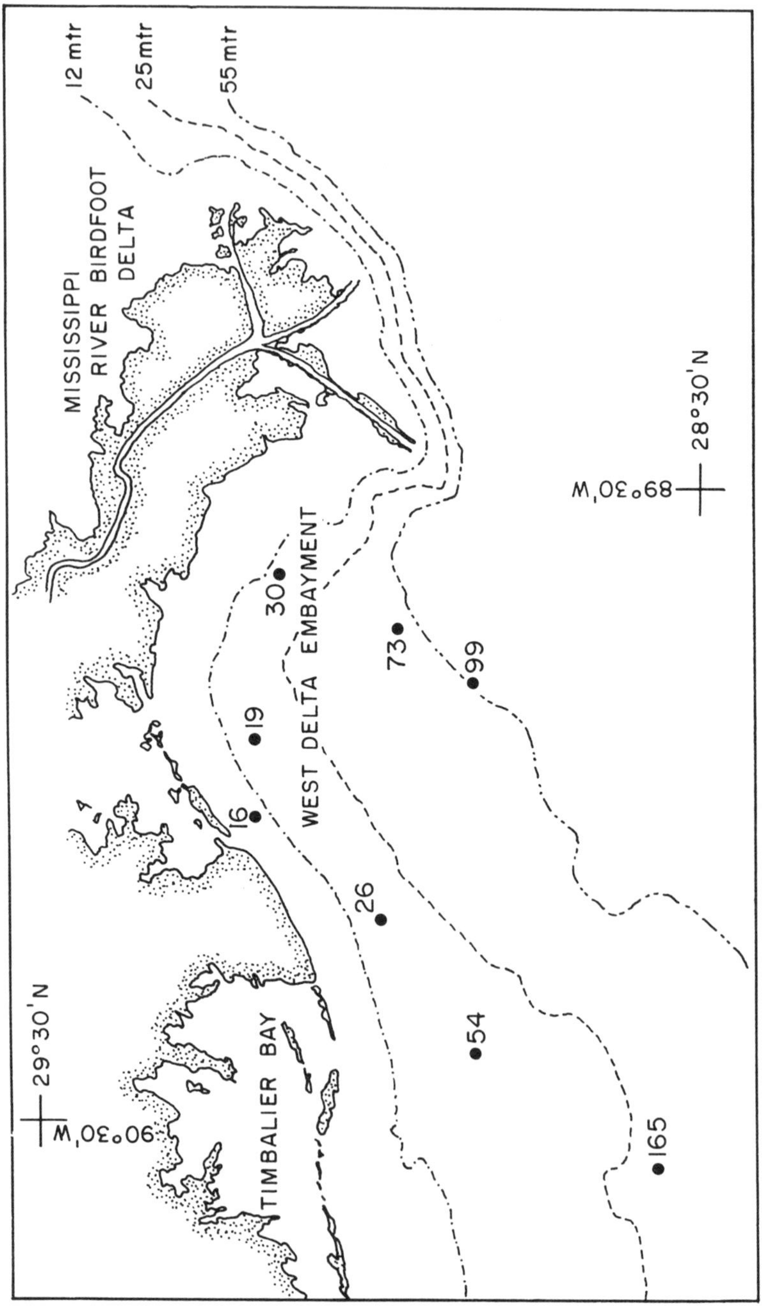

FIG. 1. ALGAL COLLECTION SITES IN THE CONTINENTAL SHELF OF LOUISIANA.

cluded these and a few additional species in his 1960 book. Humm and Darnell (1959) and Mullahy (1959) published articles on collections from the Chandeleur Islands. Drouet (1968, 1973, 1978) and Drouet and Daily (1956) listed the Cyanophyta that they and others had collected from the state. Earle (1969), in her study of the Phaeophyta of the eastern Gulf of Mexico, included the area east of Grand Isle.

The purpose of this paper is to present the results of an algal survey conducted over a period of a year from the offshore petroleum platform pilings. It is hoped that these results will be of use in future investigations to determine whether or not future petroleum activities have any adverse effect on these plant species. Ecological relationships between the algal species and abiotic parameters will be presented elsewhere.

MATERIALS AND METHODS

Seven field trips were made to eight offshore platform stations in Louisiana from July 17, 1973, to July 23, 1974 (figure 1). Petroleum platforms (stations) were selected to reach the widest possible range of water characteristics. Stations were chosen along two transects, one running in close proximity and perpendicular to the Mississippi River discharge (Stations 30, 73, and 99) and the second running away from the influence of the river discharge (16, 26, 54, and 165). Station 19 was added near shore within the west delta embayment.

Samples were taken from the pilings by use of SCUBA. Scrapings of 0.01 m^2 surface area were taken at regular intervals from the surface to the lowest level of the algal zone usually at about 9.2 m. Samples were generally taken from the southwestern side of the piling where the algal growth was richest and from the shaded northern side. Samples were fixed in 5% sea water formalin and brought to the laboratory for identification.

RESULTS

A total of 120 species of algae was collected from these platforms, of which 22 (18%) belong to Cyanophyta, 37 (31%) Rhodophyta, 19 (16%) Phaeophyta, and 42 (35%) Chlorophyta (table 1). These species bring the total of known species of marine algae from Louisiana to 180. Station 54, located in 21.6 m water depth, was the richest station, with 76 species present, followed by Station 165, in 27.7 m, with 63 species. The number of species collected from the remaining six stations ranged from 40 to 53.

TABLE 1

Occurrence of Marine Algae on Oil Platforms in Offshore Waters of Louisiana

Species	Station Number of Oil Platforms							
	16	19	26	30	54	73	99	165
Phylum Cyanophyta (22 species)								
Order Coccogonales								
Agmenellum thermale		X			X			X
Anacystis aeruginosa		X			X		X	X
A. dimidiata	X			X	X		X	X
A. montana	X			X	X		X	X
Coccochloris elebens					X		X	X
C. stagnina	X							
Entophysalis conferta	X	X	X	X	X	X	X	X
E. deusta	X	X	X	X	X	X	X	X
Johannesbaptistia pellucida	X	X	X	X	X	X	X	X
Order Hormogonales								
Calothrix crustacea					X	X		
Mastigocoleus testarum					X		X	X
Microcoleus lyngbyaceus	X	X	X	X	X		X	X
M. vaginatus	X							
Oscillatoria erythraea	X			X				
O. lutea	X	X	X	X	X		X	X
O. submembranacea	X	X	X	X	X			X
Porphyrosiphon notarisii		X	X	X	X		X	X
Schizothrix arenaria	X	X	X	X	X	X		
S. calcicola	X	X	X	X	X	X	X	X
S. mexicana	X	X	X	X	X	X	X	X
S. tenerrima			X		X			
Spirulina subsalsa	X	X	X	X	X	X		

TABLE 1 (continued)

Species	Station Number of Oil Platforms							
	16	19	26	30	54	73	99	165
Phylum Rhodophyta (37 species)								
Order Bangiales								
Asterocytis ramosa					X			
Bangia fusco purpurea	X	X	X	X	X	X	X	X
Erythrocladia recondita								X
E. subintegra	X				X	X	X	X
E. vagabunda		X	X				X	
Erythrotrichia carnea	X	X	X	X	X	X	X	X
Goniotrichum alsidii					X	X	X	X
Order Nemalionales								
Acrochaetium antillarum					X			
A. dufourii						X		X
A. flexuosum	X			X	X	X		X
A. gracile	X	X	X	X	X	X	X	X
A. hoytii	X							
A. sagraeanum							X	
A. sancti-thomae					X			X
A. seriatum					X	X	X	X
A. thuretii	X				X	X		X
A. trifilum	X				X			X
Order Gelidiales								
Pterocladia americana								X
Order Cryptonemiales								
Fosliella atlantica					X	X	X	X
F. farinosa								X
F. f. var. *solmsiana*								X
Jania adherens								X
Order Gigartinales								
Gracilaria foliifera			X		X			

TABLE 1 (continued)

Species	Station Number of Oil Platforms							
	16	19	26	30	54	73	99	165
Phylum Rhodophyta (continued)								
Order Ceramiales								
Antithamnion cruciatum					X			X
Ceramium byssoideum								X
C. fastigiatum								X
C. f. f. *flaccida*	X			X		X	X	
Herposiphonia secunda	X	X	X		X		X	X
Laurencia papillosa						X		
Lophosiphonia cristata								X
L. saccorhiza					X			
L. subadunca				X	X			
Polysiphonia havanensis								X
P. hemisphaerica			X		X	X	X	X
P. howeii								X
P. subtilissima	X	X	X	X	X	X	X	X
Spermothamnion investiens					X			
Phylum Phaeophyta (19 species)								
Order Ectocarpales								
Acinetospora crinita			X		X			X
Bachelotia antillarum					X			
Ectocarpus elachistaeformis	X	X	X	X	X	X	X	X
E. intermedius	X	X	X	X	X	X	X	X
E. variabilis	X	X	X		X			
Giffordia conifera		X				X		
G. indica		X	X		X	X		X
G. mitchelliae	X	X	X	X	X	X		X
G. rallsiae					X			X
Herponema tortugense					X			X
Phaeostroma pusillum	X			X		X	X	
Spongonema tomentosum						X		

TABLE 1 (continued)

Species	Station Number of Oil Platforms							
	16	19	26	30	54	73	99	165
Phylum Phaeophyta (continued)								
Order Sphacelariales								
Sphacelaria furcigera	X	X	X	X	X	X	X	X
S. tribuloides						X		
Order Chordariales								
Ascocyclus orbicularis							X	
Myrionema strangulans	X						X	
Order Dictyosiphonales								
Hummia onusta	X			X		X	X	
Order Fucales								
Sargassum fluitans				X	X	X	X	X
S. natans				X	X	X	X	X
Phylum Chlorophyta (42 species)								
Order Tetrasporales								
Pseudotetraspora antillarum	X	X						
Order Ulotrichales								
Gomontia polyrhiza					X		X	X
Stichococcus marinus		X	X		X		X	
Order Ulvales								
Enteromorpha chaetomorphoides	X	X	X				X	
E. clathrata	X	X	X	X		X	X	
E. compressa	X	X	X	X	X	X	X	
E. erecta					X			
E. flexuosa	X	X	X	X	X	X	X	X
E. intestinalis		X			X	X		
E. lingulata	X	X	X					
E. plumosa							X	
E. prolifera							X	
E. salina	X				X			
Order Chaetophorales								
Entocladia viridis					X	X		X
Phaeophila dendroides	X	X						
Protoderma marinum	X		X	X	X			X
Ulvella lens		X			X		X	

TABLE 1 (continued)

Species	Station Number of Oil Platforms							
	16	19	26	30	54	73	99	165
Phylum Chlorophyta (continued)								
Order Cladophorales								
Chaetomorpha brachygona			X		X			
C. gracilis							X	
C. linum					X			X
C. minima	X	X	X	X	X	X	X	X
Cladophora albida	X	X	X					
C. brachyclona			X			X	X	
C. catenata					X			
C. constricta								X
C. dalmatica	X	X	X	X	X	X	X	X
C. delicatula	X	X	X	X		X	X	
C. flexuosa						X		
C. gracilis			X		X		X	
C. repens				X		X	X	
C. sericea			X		X			
C. vagabunda			X		X			
Rhizoclonium hookeri	X			X				
R. kochianum			X		X	X		
R. riparum	X	X	X					
Order Siphonocladales								
Cladophoropsis membranacea					X			
Order Dasycladales								
Batophora oerstedii						X		
Order Siphonales								
Bryopsis hypnoides	X		X					
B. pennata			X		X	X		X
Derbesia lamourouxii					X			X
D. vaucheriaeformis	X	X	X	X	X	X	X	X
Ostreobium quekellii	X	X	X	X	X	X	X	X
Number of Species	52	43	49	40	76	50	53	63

Analysis of the species composition indicated two groups, based primarily on water depth: the coastal stations 16, 19, 26, and 30, which were located in water depths of 11.6 to 16.0 m, and an offshore group of four stations, 54, 73, 99, and 165, which were located in water depths of 21.6 to 58.0 m. The offshore province averaged 60 species as compared to 46 from the coastal province; this difference was largely the result of an increase in the number of species of reds and browns at the offshore stations.

DISCUSSION

Comparisons of the algal species composition with pre-oil production activity is impossible because of the near absence of suitable solid substrate for algal attachment in this area before the platforms were constructed. The pilings, supporting hundreds of oil platforms, constitute a substrate of major importance to the occurrence and distribution of marine algae. In addition to benthic algae, these pilings also support large populations of sessile and free-living invertebrates and serve as a base of the food sources for herbivorous and carnivorous fish.

The influence of the run-off of the Mississippi River is evident from the smaller species total of those shallow water stations located within the plume. The lower species totals are probably the result of lowered surface salinities and increased turbidity, which decreases the light penetration. The presence of a nepheloid area near the bottom limited algal growth at this depth (Oetking et al. 1974).

The knowledge of the species composition of marine algae on these pilings may be of value in assessing the effects of an oil spill or blowout in the event that one should occur within this area.

ACKNOWLEDGMENTS

This study was supported by research grant OB 70 HJM from the Gulf Universities Research Consortium. We are grateful to Curtis Kruer, Bruce DeGroot, and Bob Ernest for assisting in the field work, and to the Exxon Corporation for providing transportation and lodging while we conducted the field work.

REFERENCES CITED

Behre, E. H. 1950. Annotated list of the fauna of the Grand Isle Region, 1928-1946. Occasional Papers of the Marine Laboratory, Louisiana State University **6**:1-66.

Cocks, R. S. 1907. Flora of the Gulf biologic station. Bulletin of the Louisiana Gulf Biologic Station (Cameron) **7**:1-42.

Drouet, F. 1968. Revision of the Classification of the Oscillatoriaceae. Monograph 15. Philadelphia, Academy of Natural Sciences.

———. 1973. Revision of the Nostocaceae with Cylindrical Trichomes. New York: Hafner Publishing Co.

———. 1978. Revision of the Nostocaceae with Constricted Trichomes. Vaduz, Liechtenstein: J. Cramer.

——— and W. A. Daily. 1956. Revision of the Coccoid Myxophyceae. Butler University Botanical Studies **12**:1-222.

Earle, S. A. 1969. Phaeophyta of the eastern Gulf of Mexico. Phycologia **7**:71-254.

Flint, L. H. 1951. Some winter red algae of Louisiana. Proceedings of the Louisiana Academy of Sciences **14**:34-36

Humm, H. J. and R. M. Darnell. 1959. A collection of marine algae from the Chandeleur Islands. Publications of the Institute of Marine Science, The University of Texas **6**:265-276.

Kapraun, D. F. 1974. Seasonal periodicity and spatial distribution of benthic marine algae in Louisiana. Contributions in Marine Science, The University of Texas **18**:140-167.

Mullahy, J. H. 1959. Preliminary survey of the algal flora of the Chandeleur Islands. Proceedings of the Louisiana Academy of Sciences **23**:62-68.

Oetking, P., R. Back, R. Watson, and C. Merks. 1974. Hydrography on the nearshore continental shelf of South Central Louisiana. Gulf Universities Research Consortium. Appendix to Report No. 138.

Taylor, W. R. 1954. Sketch of the character of the marine algal vegetation of the shores of the Gulf of Mexico. *In* P. S. Galtsoff, ed., Gulf of Mexico, its Origin, Waters, and Marine Life. Fishery Bulletin 89, U.S. Fish and Wildlife Service **55**:177-192.

THE DISTRIBUTION OF RECENT FORAMINIFERA IN TIMBALIER BAY, LOUISIANA

by Christopher L. Ostrom

ABSTRACT

The species composition and distribution of 26 species of the living and dead Recent benthic foraminiferal faunas of Timbalier Bay, Louisiana, were determined. There are four species complexes: (1) species endemic to marshes, (2) open Gulf species, (3) bay species, and (4) barrier island marsh species. Predominant species were *Ammonia beccarii* varieties *parkinsoniana* and *tepida, Elphidium gunteri, E. matagordanum, E. poeyanum, E. vadescens*, *Ammotium salsum*, *Ammobaculites dilatatus*, and *Quinqueloculina tenagos*.

Low total populations and standing crops of benthic Foraminifera correlate with sediments exhibiting high percentages of sand and low percentages of both organic material and calcium carbonate. Low species diversities were related to extremes in salinity and sediment type, as well as to the barrier island marsh environment. Of the parameters studied, salinity and sediment type were the most important factors limiting the distribution of foraminiferal species within Timbalier Bay. Patterns of species distribution did not appear to be related to the location of oil and gas facilities. Comparisons with a previous study of Timbalier Bay indicated that little change has occurred in the foraminiferal fauna over the past 15 years.

At the time of this investigation, Christopher Ostrom was with the Department of Oceanography, Florida State University, Tallahassee, FL.

INTRODUCTION

Several investigations of the ecology and distribution of Recent benthic Foraminifera have been completed within the past 30 years, especially since the introduction by Walton (1952) of an efficient and reliable method of staining living Foraminifera. Distribution of selected taxa of benthic Foraminifera have been correlated with numerous environmental parameters and combinations of parameters, including grain size and organic content of the sediment, salinity, temperature, dissolved oxygen concentration, availability of food, depth, and bottom topography. A few studies have demonstrated that species diversity, abundance of individuals, and the presence of certain species of Foraminifera can be used as indicators of specific environmental conditions. Because it has been shown that benthic Foraminifera may be sensitive to changes in environment, the distribution and ecology of benthic Foraminifera have been measured in the northern Gulf Coast Region, including Timbalier Bay, to determine the effects, if any, of oil and gas production there.

Any study of Recent Foraminifera in the northern Gulf of Mexico must include reference to the early classic taxonomic works of d'Orbigny (1839), Brady on the Challenger collections (1884), and Cushman (1922), and to texts by Phleger (1960) and Murray (1973), which summarize and integrate much of the principal work on the ecology of Recent Foraminifera. The results of some previous investigations suggest that it is possible that impacts from oil and gas development activities in Timbalier Bay might be reflected in patterns of distribution of selected taxa of benthic Foraminifera. For example, Morin (1971) observed that the normal distribution of one species was extended toward the source of the oil spill following the 1969 Santa Barbara oil spill. Watkins (1961) and Bandy et al. (1964a, 1964b, 1964c) demonstrated that sewage pollution in southern California was indicated by the absence of living Foraminifera directly beneath sewage outfalls, few Foraminifera where the sewage was heavily concentrated, and flourishing populations where sewage was diluted. Clark (1971) showed that total populations reflected the effects of an aquaculture outfall in Clam Bay, Nova Scotia.

Investigations of the distribution of benthic Foraminifera, and of the relationship between patterns of species distribution and measured environmental parameters, have been conducted in coastal areas adjacent to the Mississippi River Delta and along the coastlines of Texas and Alabama. Lowman (1949), Parker et al. (1953), Phleger (1954, 1956), Bandy (1956), and Lamb (1972) found patterns of change in the distribution of foraminiferal taxa that were associated with changes in salinity. Phleger (1955) and Lankford (1959) both recognized faunal assemblages in the eastern Mississippi Delta region that were correlated with bottom water-masses of variable physical and chemical composition. Lehman (1957) used statistical methods to show that distribution patterns of Foraminifera in Matagorda Bay, Texas, were the result of a complex interaction of many ecological factors. Salinity, organic content of the sediment, the extent and nature of water currents, and turbulence were the more important parameters limiting the distribution, while temperature, depth, and turbidity were thought to have little effect. Phleger and Lankford (1957) examined seasonal occurrences of living benthic Foraminifera in several southwest Texas bays, and found that living populations in lower bays were similar during all seasons, whereas in upper bays the largest living populations occurred in November and January.

An earlier ecological study of the Foraminifera of Timbalier Bay was made by Waldron (1957). Twenty-three species were found within the bay. Waldron was able to divide them into two groups, an upper bay facies characterized by fluctuating salinities and a lower bay facies with relatively high constant salinities. Two stations with a sandy bottom had unique faunas. Waldron stated that salinity and grain size of the sediment influenced foraminiferal distribution, but pH, Eh, alkalinity, and the concentration of sulphates, phosphates, nitrates, and carbohydrates had no apparent effect.

The objectives of my investigation of the distribution and ecology of Recent benthic Foraminifera collected from Timbalier Bay were: (1) to provide a detailed description of the foraminiferal distribution, (2) to determine whether or not patterns of species distribution and species diversity may be related to measured changes in environmental parameters, and (3) to compare the current foraminiferal fauna of Timbalier Bay with the fauna described by Waldron approximately 15 years earlier.

MATERIALS AND METHODS

Selection of stations

The location and description of the stations sampled are shown in table 1 and figure 1. The locations of two oil fields in Timbalier Bay (figure 2) were taken from Barrett (1971) and the general circulation pattern from Waldron (1957). The sand in the area of Station 2 was believed to be derived from an eroded natural levee (Waldron 1957). Station 3 was located on an oyster reef, which had a high calcium carbonate content. Stations 8-12 had high sand contents.

Stations 18, 20, and 22 were sampled during a preliminary OEI study of Timbalier Bay. Stations 1-12 were selected to form a north-

TABLE 1 -- STATION LOCATIONS

Station Number	Date Sampled	Latitude	Longitude	Remarks
1	9/72	N 29° 13.50'	W 90° 18.50'	
2	9/72	N 29° 12.50'	W 90° 18.50'	On eroded natural levee, thus high sand content.
3	9/72	N 29° 11.50'	W 90° 18.50'	On oyster reef, thus high shell hash content.
4	9/72	N 29° 09.50'	W 90° 18.50'	
5	9/72	N 29° 08.50'	W 90° 18.50'	
6	9/72	N 29° 07.50'	W 90° 18.50'	
7	9/72	N 29° 06.50'	W 90° 18.50'	
8	9/72	N 29° 05.50'	W 90° 18.50'	Near oil rigs.
9	9/72	N 29° 04.50'	W 90° 18.50'	
10	9/72	N 29° 04.22'	W 90° 18.50'	
11	9/72	N 29° 04.22'	W 90° 18.50'	Near beach.
12	9/72	N 29° 04.22'	W 90° 18.50'	In marsh environment.
13	10/73	N 29° 08.43'	W 90° 17.88'	Replicate of Waldron's (1957) station.
14	10/73	N 29° 07.44'	W 90° 14.93'	
15	10/73	N 29° 05.77'	W 90° 20.29'	
16	10/73	N 29° 06.22'	W 90° 22.95'	
17	10/73	N 29° 07.88'	W 90° 22.21'	Replicate of Waldron's (1957) station.
18	8/72	N 29° 09.50'	W 90° 20.95'	
19	10/73	N 29° 10.00'	W 90° 21.00'	
20	8/72	N 29° 10.60'	W 90° 20.40'	
21	10/73	N 29° 11.11'	W 90° 21.19'	Replicate of Waldron's (1957) station.
22	8/72	N 29° 11.50'	W 90° 21.44'	
23	10/73	N 29° 11.70'	W 90° 23.10'	Replicate of Waldron's (1957) station. In oyster reef.
24	10/73	N 29° 15.00'	W 90° 21.70'	Replicate of Waldron's (1957) station.
25	10/73	N 29° 14.20'	W 90° 19.30'	Replicate of Waldron's (1957) station.
26	10/73	N 29° 11.30'	W 90° 17.35'	Replicate of Waldron's (1957) station.
27	10/73	N 29° 11.14'	W 90° 19.06'	Replicate of Waldron's (1957) station. In oyster reef.
		Philo Brice Platform is located at: N 29° 11.50'	W 90° 21.65'	
28	9/72	100 m. off Philo Brice Platform at a heading of 020° magnetic		
29	9/72	463 m. off Philo Brice Platform at a heading of 020° magnetic		
30	9/72	925 m. off Philo Brice Platform at a heading of 020° magnetic		
31	9/72	100 m. off Philo Brice Platform at a heading of 110° magnetic		
32	9/72	463 m. off Philo Brice Platform at a heading of 110° magnetic		
33	9/72	925 m. off Philo Brice Platform at a heading of 110° magnetic		
34	9/72	100 m. off Philo Brice Platform at a heading of 200° magnetic		
35	9/72	463 m. off Philo Brice Platform at a heading of 200° magnetic		
36	9/72	925 m. off Philo Brice Platform at a heading of 200° magnetic		
37	9/72	925 m. off Philo Brice Platform at a heading of 290° magnetic		

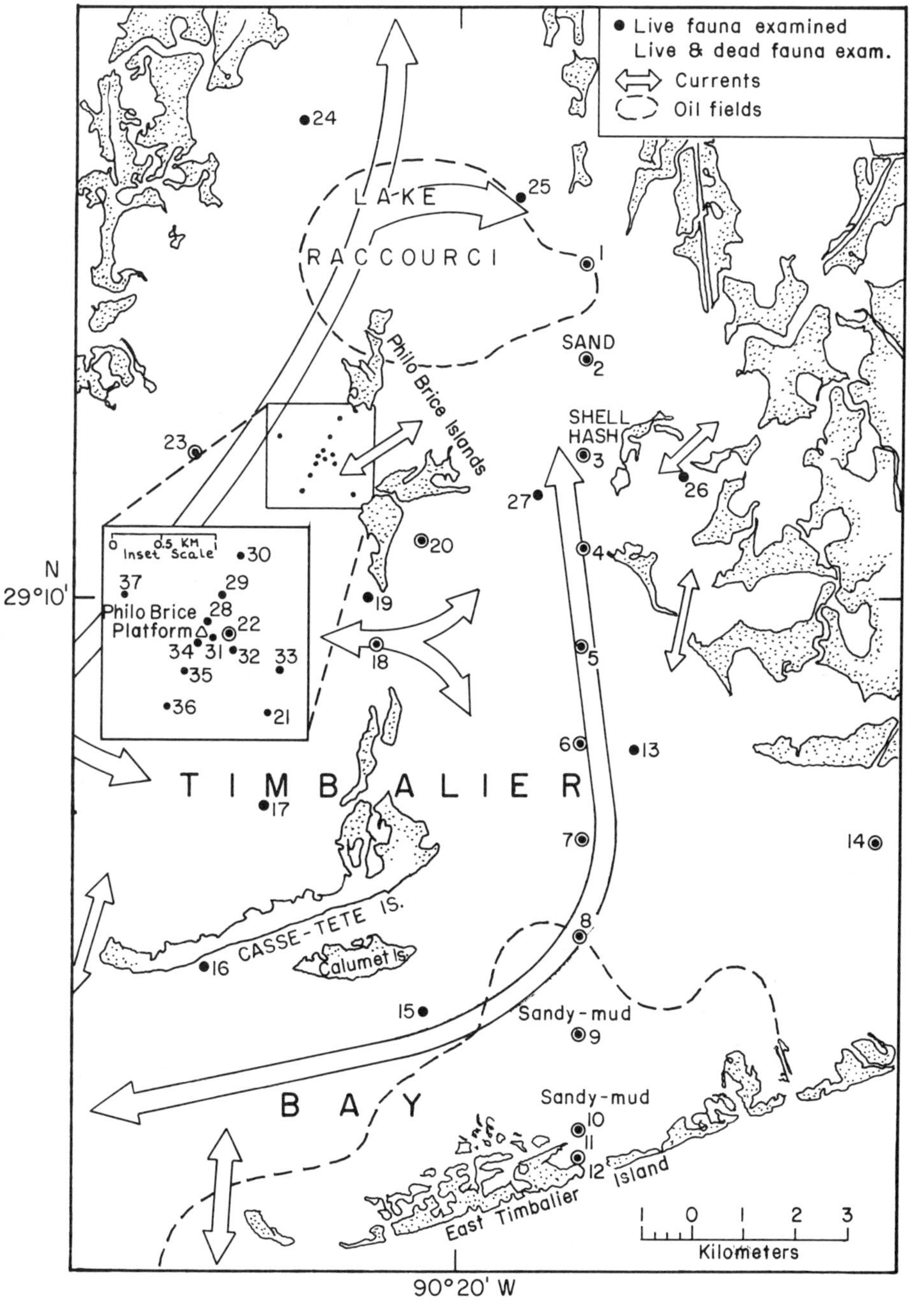

FIG. 1. STATION LOCATIONS.

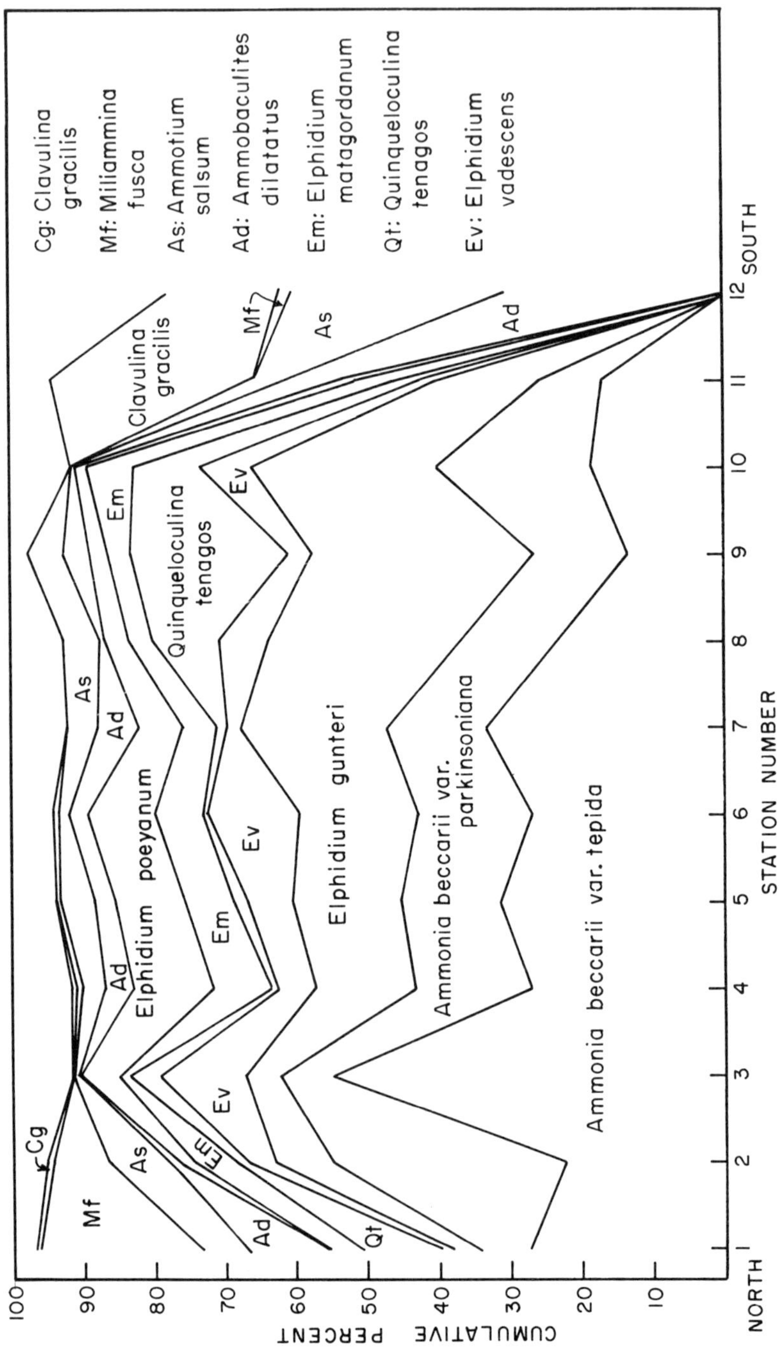

FIG. 2. CUMULATIVE GRAPH OF PERCENTAGES OF FORAMINIFERAL SPECIES at stations along a north-south transect of Timbalier Bay.

south transect of the entire bay. Stations 28-37 were chosen to form four short transects originating at the Philo Brice Island Platform. Stations 13, 17, 21, and 23-37 were established to coincide with those taken during Waldron's earlier study. Stations 14-16 and 19 were selected to increase the coverage of the sampling to include the entire area of the bay.

Collection and treatment of samples

Populations of benthic Foraminifera were analyzed from sediment samples obtained from cores collected by divers as described by Williams and Jones (1974). Each core sample from Stations 13-17, 19, 21, and 23-27 was prepared for foraminiferal analysis by removing the top 1 cm of sediment and overlying water. A solution was prepared of 1 g rose Bengal and 1 liter of 50% ethanol, which was added to the sediment to preserve and stain the living Foraminifera. This procedure was a modification of the staining technique described by Walton (1952).

Core samples from Stations 1-12, 18, 20, 22, and 28-37 were used for both Foraminifera and sediment analyses and were therefore treated differently. After the layer of water directly above the sediment was collected, the top 5 cm of sediment were extruded and cut lengthwise into four equal sections. One of the sections was added to the layer of water and was mixed with the rose Bengal-ethanol solution. The remaining three sections of sediment were used for the sediment analyses described by Williams and Jones (1974).

All samples were washed through a 63 μ sieve to remove the silts and clays. Samples from Stations 13, 17, 21, and 23-27, established during Waldron's study, were subsequently washed through a 149 μ sieve to separate the Foraminifera into two size fractions. The larger size fraction, greater than 149 μ, was the same as that used by Waldron.

After drying, the Foraminifera were separated from the denser components of the sediment by flotation with carbon tetrachloride. After flotation, the remaining sediment was examined with a binocular microscope and the residual Foraminifera were collected by hand with the help of a fine camel's hair brush. The Foraminifera were dried and stored in labeled envelopes.

Analysis of Foraminifera

A random sample of 300 dead specimens, and all living specimens from Stations 1-12, 14, 18, 20, 22, and 23 were selected for identification. Only the living Foraminifera were identified from the remaining stations. Tentative species identifications were made from the descriptions and illustrations presented by Andersen (1953, 1961, and 1971), Bock (1967 and 1971), Kornfeld (1931), Phleger and Parker (1951), Todd and Brönnimann (1957), and Warren (1957). Verification of species identifications was made by comparison with type specimens at the Smithsonian Institution in Washington, D. C., under the guidance of Ms. Ruth Todd.

The internal diameter of the core liner used in collecting the samples was 5.7 cm, thus the surface area of the sediment samples analyzed from Stations 13-17, 19, 21, and 23-27 was 25.5 cm². The surface area of sediment samples from Stations 1-12, 18, 20, 22, and 28-27 was 6.4 cm². Studies by Boltovskoy (1966) and Brooks (1967) demonstrated that while a substantial number of Foraminifera live beneath the sediment surface, the greatest proportion of living benthic Foraminifera are found within the top 1 cm of sediment. Because of the difference in surface area sampled, the data generated from the 6.4 cm² sample were multiplied by four to compare with the surface area sampled with the larger corer.

Total populations were only estimated for the samples having a surface area of 6.4 cm² and depth of 5 cm at Stations 1-12, 14, 18, 22, and 23, and were expressed as the number of living and dead specimens per 32 ml of sediment. Total populations were estimated as follows: After all the living, and 300 dead, specimens were removed for identification purposes, the remaining Foraminifera were divided into eight equal portions by using a sediment splitter. The specimens in one of the portions were counted, multiplied by eight, then added to the number of living and dead specimens previously removed to give the estimated total population. The accuracy of the splitting technique was evaluated by comparing the number of specimens in splits from the same sample. The percentage of error in three such comparisons ranged from 7% to 20%.

Distribution of individual species throughout the bay was analyzed by examination of the relative and absolute abundances of the species at each station. Relative abundance of a species refers to the percentage of either the standing crop or the total population that is contributed by the species at a station. Absolute abundance of a species refers to the percentage of the total number of either living or dead specimens of that species from all the stations that were found at any given station.

Measured environmental parameters

In situ physical and chemical measurements made by other workers in conjunction with the OEI, and considered in the present study, included: (1) salinity measured to within ± 0.3 parts per thousand; (2) temperature measured to within ± 0.5° C; (3) dissolved oxygen (ppm) measured to within ± 1% of the observed value; and (4) sediment characteristics including: (a) grain size according to weight percentages at 0.25 phi size intervals; (b) percentage of organic matter by ignition loss;

and (c) percentage of calcium carbonate by weight loss after acidification. The methods used to measure these parameters are described in the OEI handbook (Menzies 1973).

RESULTS AND DISCUSSION

Most abundant species

Twenty-six species of benthic Foraminifera were common in Timbalier Bay. The relative percentages of living and dead specimens of each of the 26 species have been calculated for each station (tables 2 and 3).

A few species predominated. Foraminiferal species comprising greater than 1% of the total number of living and dead specimens identified are listed in table 4. Only 11 taxa comprise greater than 1%, and 3 greater than 10% of the dead population. Only 10 taxa comprise greater than 1%, and 3 greater than 10% of the living population. Specimens of *Ammonia beccarii* formed 40.9% of the dead foraminiferal population in Timbalier Bay, and 59.1% of the live population. *A. beccarii* and species of *Elphidium* were also dominant in Copano Bay, Aransas Bay, Mesquite Bay, San Antonio Bay (Parker et al. 1953), and Matagorda Bay (Lehman 1957) along the Texas coast; in Mobile Bay, Alabama (Lamb 1972); and in Barataria Bay, Louisiana (Lowman 1949).

A cumulative graph shows the relative abundances of the dominant foraminiferal forms identified from the dead populations from stations along a north-south transect of Timbalier Bay (figure 2). It illustrates the major faunal trends within Timbalier Bay. The upper portion of the bay, in the Lake Raccourci area (Station 1), is characterized by an abundance of arenaceous forms such as *M. fusca* and species of *Ammotium*. These species are also abundant, along with *C. gracilis*, at Station 12, the marsh station at East Timbalier Island. The middle portion of Timbalier Bay (Stations 3-10) is characterized by an abundance of calcareous forms such as *A. beccarii* varieties *tepida* and *parkinsoniana*, *Q. tenagos*, and species of *Elphidium*. Stations 2 and 11 represent transition zones between the middle bay and the Lake Raccourci and East Timbalier Island marsh areas. Station 3, located on an oyster reef, displayed an unusually high relative percentage of *A. beccarii* var. *tepida*. Relative percentages of both *E. gunteri* and *Q. tenagos* tend to increase toward the lower portion of Timbalier Bay (in Stations 7-10), while *A. beccarii* var. *tepida*, *E. poeyanum*, *E. vadescens*, and *E. matagordanum* tend to decrease in abundance in the lower bay.

Groups of species

Analysis of the distributions of relative and absolute abundances of

TABLE 2

Relative Percentages of Dead Benthic Foraminifera

LOCATION	Timbalier Bay, Louisiana																					
STATION	1	2a	2b	3	4	5	6	7	8	9	10	11	12	13	14	15	16	17a	17b	18	19	20
LIVING POPULATION (per 25cc)	316	224	472	684	92	192	656	228	216	280	32	44	208	256	26	44	71	329	262	164	293	152
Ammobaculites crassus	2.5	3.6	1.3	0	0	0	0	0	0	0	0	9.1	13.5	0	0	0	5.6	0	0	0	0	7.9
Ammobaculites dilatatus	0	0	0	0	0	2.1	0	1.8	0	1.4	0	9.1	0	0	0	2.3	0	0	0	0	0	0
Ammobaculites exiguus	0	0	0	0	0	0	0	0	1.9	0	0	18.2	5.8	0	0	0	0	0.3	0	0	0	0
Ammonia beccarii var. *parkinsoniana*	19.0	21.4	35.0	9.4	26.1	2.1	16.5	3.5	7.4	5.7	0	0	0	14.5	3.8	0	8.5	1.8	2.7	0	2.4	5.3
Ammonia beccarii var. *tepida*	51.9	50.0	38.1	80.1	21.7	45.8	47.6	26.3	31.5	15.7	37.5	45.5	0	7.8	19.2	6.8	12.7	5.5	16.0	43.9	4.1	60.5
Ammonia pauciloculata	0	1.8	0.6	4.7	0	0	0	0	0	0	0	0	0	0	0	2.3	2.8	0	0	2.4	1.0	0
Ammotium fragile	0	0	0	0	0	0	0	0	0	1.4	0	0	0	0	0	0	0	0	0	0	0	0
Ammotium salsum	11.4	12.5	10.7	0	0	12.5	0.6	3.5	14.8	5.7	0	9.1	55.8	1.2	3.8	9.1	0	7.6	5.7	4.9	1.0	2.6
Arenoparella mexicana	0	0	0	0	0	2.1	0	0	0	0	0	0	0	0	0	0	0	0	0	0	0	0
Bolivina lowmani	0	0	0	0	0	0	2.4	1.8	3.7	1.4	0	0	0	0	0	0	0	0	0.4	7.3	0	0
Bolivina striatula	0	0	0	2.9	0	2.1	8.5	1.8	11.1	2.9	0	0	0	0.8	0	0	1.4	0.3	1.5	9.8	0.7	5.3
Buliminella elegantissima	0	0	0	0	0	2.1	1.8	1.8	1.9	1.4	0	0	0	0	0	4.5	0	0	0	0	0	0
Elphidium gunteri	0	0	2.5	0.6	8.6	8.3	1.8	0	0	48.5	12.5	9.1	0	8.2	3.8	0	8.4	2.4	0.8	7.3	0	5.3
Elphidium matagordanum	3.8	3.6	8.8	0.6	8.7	18.7	4.3	36.8	18.5	7.1	0	0	11.5	2.0	0	0	2.8	0	8.4	22.0	1.4	5.3
Elphidium poeyanum	0	1.8	2.5	1.2	30.4	2.1	4.9	5.3	3.7	0	0	0	0	0	23.1	2.3	4.2	1.2	0.4	2.4	1.7	7.9
Elphidium vadescens	1.3	0	0	0	0	0	8.5	1.8	1.9	0	50.0	0	0	53.5	3.8	4.5	8.5	77.2	53.8	0	33.8	0
Gaudryina exilis	0	0	0	0	0	0	0.6	3.5	1.9	1.4	0	0	1.9	0	0	6.8	0	0	0	0	0	0
Haplophragmoides wilberti	0	0	0	0	0	0	0	0	0	1.4	0	0	0	0	0	0	0	0	0	0	0	0
Miliammina fusca	5.1	1.8	0	0	0	0	0	0	0	0	0	0	1.9	2.7	26.9	2.3	2.8	0	0	0	1.4	0
Palmerinella gardenislandensis	0	0	0	0	0	0	0.6	10.5	1.9	4.3	0	0	0	2.3	0	13.6	21.1	3.0	9.5	0	0.7	0
Quinqueloculina tenagos	3.8	3.6	0	0.6	4.3	2.1	1.2	0	0	0	0	0	0	7.0	15.4	45.5	16.9	0.6	0.8	0	51.9	0
Tiphotrocha comprimata	0	0	0	0	0	0	0	0	0	1.4	0	0	3.8	0	0	0	0	0	0	0	0	0
Trochammina inflata	1.3	0	0	0	0	0	0	0	0	0	0	0	0	0	0	0	4.2	0	0	0	0	0

TABLE 3

Relative Percentages of Living Benthic Foraminifera

LOCATION	Timbalier Bay, Louisiana																		
STATION	20	21	22	23	24	25	26	27a	27b	28	29	30	31	32	33	34	35	36	37
LIVING POPULATION (per 25cc)	152	98	72	346	461	474	125	41	253	0	92	428	40	776	360	228	76	112	148
Ammobaculites crassus	7.9	0	0	0	0	0	0	0	0	0	0	0	0	0	0	0	0	0	0
Ammobaculites dilatatus	0	3.1	11.1	0	0	0	0	0	0	0	0	0	0	0	0	0	0	0	0
Ammonia beccarii var. *parkinsoniana*	5.3	1.0	5.6	4.6	66.4	17.5	60.0	2.4	15.0	0	34.8	28.0	0	23.7	25.6	3.5	36.8	14.3	43.2
Ammonia beccarii var. *tepida*	60.5	3.1	16.7	52.3	32.8	70.9	16.0	34.1	71.0	0	43.5	57.0	50.0	61.3	55.6	7.0	42.1	39.3	29.7
Ammonia pauciloculata	0	0	0	0.6	0.2	0.2	2.4	2.4	0	0	0	0	0	0	1.1	1.8	0	3.6	2.7
Ammonia rolshauseni	0	0	0	0	0	0.2	1.6	4.9	0	0	0	0	0	0	0	0	0	0	0
Ammotium salsum	2.6	19.4	0	0.9	0	1.3	0.8	14.6	0.4	0	0	0	0	0	0	1.8	0	0	2.7
Bolivina lowmani	0	0	0	0	0	0	0	0	0	0	0	0	0	0.5	0	1.8	0	3.6	0
Bolivina striatula	5.3	0	5.6	2.0	0	0	0	0	0	0	8.7	5.6	0	7.2	1.1	54.4	5.3	7.1	2.7
Elphidium gunteri	5.3	4.1	16.7	0.6	0.7	0.8	1.6	0	4.3	0	0	1.9	0	1.0	6.7	5.3	5.3	10.7	8.1
Elphidium limosum	0	0	0	0.3	0	0	0	0	0	0	0	0	0	0	0	0	0	0	0
Elphidium matagordanum	5.3	0	11.1	0	0	0	0.8	0	0	0	0	0.9	20.0	0	0	3.5	5.3	3.6	0
Elphidium poeyanum	7.9	0	11.1	1.2	0	0.4	1.6	0	0	0	0	2.8	0	2.1	1.1	17.5	0	7.1	0
Elphidium vadescens	0	68.4	0	28.0	0	6.1	14.4	29.3	7.5	0	8.7	2.8	30.0	2.6	7.8	0	0	7.1	0
Gaudryina exilis	0	0	0	0	0	0	0	0	0	0	0	0.9	0	0.5	0	0	0	0	2.7
Haplophragmoides wilberti	0	0	0	0	0	0	0	0	0	0	0	0	0	0	0	0	0	0	2.7
Miliammina fusca	0	1.0	5.6	0.3	0	0	0	0	0	0	0	0	0	0	0	0	0	0	2.7
Palmerinella gardenislandensis	0	0	0	0.3	0	0	0.8	4.9	0.4	0	0	0	0	0	0	0	0	0	0
Quinqueloculina lamarkiana	0	0	0	0	0	0	0	0	0	0	0	0	0	0	0	0	0	3.6	0
Quinqueloculina tenagos	0	0	0	9.0	0	2.5	0	4.9	0.8	0	4.3	0	0	1.0	0	1.8	0	0	2.7
Trochammina inflata	0	0	5.6	0	0	0	0	0	0	0	0	0	0	0	1.1	1.8	5.3	0	0

TABLE 4 -- RELATIVE PERCENTAGES OF SPECIES FROM LIVING AND DEAD FORAMINIFERAL POPULATIONS IN TIMBALIER BAY

DEAD FAUNA		LIVING FAUNA	
Species	Percent of Total Number of Dead Specimens Identified	Species	Percent of Total Number of Living Specimens Identified
Ammonia beccarii var. *tepida*	27.5	*Ammonia beccarii* var. *tepida*	41.9
Elphidium gunteri	16.1	*A. beccarii* var. *parkinsoniana*	17.2
Ammonia beccarii var. *parkinsoniana*	13.4	*Elphidium vadescens*	10.7
Elphidium vadescens	6.0	*Bolivina striatula*	4.6
Elphidium poeyanum	5.4	*Ammotium salsum*	4.3
Elphidium matagordanum	5.3	*Elphidium matagordanum*	4.2
Quinqueloculina tenagos	5.2	*Elphidium gunteri*	4.1
Ammotium salsum	4.9	*Quinqueloculina tenagos*	3.7
Ammobaculites dilatatus	4.4	*Elphidium poeyanum*	2.7
Clavulina gracilis	2.5	*Palmerinella gardenislandensis*	1.0
Miliammina fusca	2.3		
All Others	1.0	All Others	1.0

the species identified in the Timbalier Bay samples, combined with information concerning the distribution of these species from previous studies, revealed that there were four species complexes represented in the present study. The species composition, relative abundance of species, and geographic distribution characterizing each complex are described below.

(1) Species endemic to marshes:

Ammoastuta inepta	*Miliammina fusca*
Arenoparella mexicana	*Tiphotrocha comprimata*
Haplophragmoides species	*Trochammina inflata*
Jadammina polystoma	

In earlier studies (Parker et al. 1953; Phleger 1965) these species were commonly reported from marsh environments. In the present study, specimens of these species were most abundant in stations near marshes and occurred in relatively low numbers in the bay. Presumably those species present within the bay were transported there from marsh areas. One species from this group, *M. fusca*, was a major component of both the living and dead foraminiferal fauna in areas of the bay adjacent to marshes. The probable explanation for the abundance of *M. fusca* in some bay stations is that living populations of this species have radiated into the bay environment from neighboring marshes.

(2) Open Gulf species:

Ammobaculites exiguus	*Buliminella* sp. cf. *B. bassendorfensis*
Ammonia pauciloculata	*Buliminella elegantissima*
Ammonia rolshauseni	*Gaudryina exilis*
Bolivina lowmani	*Quinqueloculina lamarckiana*
Bolivina striatula	

Earlier studies (Phleger 1951; Parker et al. 1953; Phleger 1954; Bandy 1954, 1956) demonstrated that these species were common, and often abundant, in the open Gulf, and occasionally occurred in small numbers in the southernmost portions of bays in the northern Gulf region. In the present study these species were commonly observed in low frequencies in the middle and lower portions of Timbalier Bay. One species from this *group, B. striatula,* was a major component of the living foraminiferal fauna at several middle and lower bay stations. Dead specimens of *B. striatula*, however, were not abundant in any of the bay stations. It is generally accepted by those who have conducted studies in this area that living populations of some open Gulf species on occasion, such as during periods of low fresh water inflow, radiate into Timbalier Bay.

(3) Barrier island marsh species:

One sample was collected directly from a marsh area on East Timbalier Island, Station 12, and another from a beach adjacent to the barrier island marsh, Station 11. The species *Clavulina gracilis* was found in abundance only at Stations 11 and 12, and characterized the barrier island marsh foraminiferal fauna. *Ammotium* and *Ammobaculites* species were also abundant in the barrier island marsh. The only calcareous species represented at Station 12 was *Elphidium matagordanum*, which was previously reported from a barrier island marsh by Parker.

(4) Bay species:

Ammobaculites species	*Elphidium gunteri*
Ammotium species	*E. matagordanum*
Ammonia beccarii var. *parkinsoniana*	*E. poeyanum*
Ammonia beccarii var. *tepida*	*E. vadescens*

These species were the major components of the foraminiferal fauna throughout Timbalier Bay. Evidence from earlier foraminiferal studies in the northern Gulf of Mexico, especially from Parker's, indicated that these bay species form two ecologic groups. One group, comprising mainly variants of *Ammobaculites* and *Ammotium*, was generally restricted to the brackish water environments of estuaries and marshes, and was perhaps endemic to these areas. The second group, composed mainly of *Ammonia* and *Elphidium*, was abundant in the open Gulf as well as estuaries. Parker hypothesized that *E. gunteri* and varieties of *A. beccarii* were originally offshore species that invaded bays and estuaries during periods of low fresh-water runoff. Phleger's observation that low salinities restricted these species to the area outside of the barrier islands along the Mississippi coast supported this hypothesis.

Groups of stations with similar faunas

Percentage of similarity values were calculated using the method described by Sanders (1960) from comparisons of the relative abundances of dead foraminiferal species in Stations 1-12, 14, 18, 20, 22, and 23. The percentage of similarity values were arranged in a trellis diagram to show groups of stations with similar faunas (figure 3). The species composition of these groups of stations was re-examined to determine those species of Foraminifera that characterized the groups. The five faunal groups from within Timbalier Bay are:

(1) *Miliammina fusca* fauna

A foraminiferal fauna characterized by an abundance of *M. fusca* and large numbers of *Ammotium* and *Ammobaculites* species was reported from Station 1, in the eastern portion of Lake Raccourci. This fauna was not observed at Stations 24 and 25 in the Lake Raccourci area. Waldron recorded large numbers of *M. fusca* from Coal Tar Bay, a marshy area just north of Station 1. It is believed that the *M. fusca*

	3	20	22	23	4	18	5	7	6	8	14	10	9	1	2	11	12
3																	
20	65																
22	69	84															
23	65	72	72														
4	54	76	77	69													
18	57	78	74	66	75												
5	60	83	83	77	85	81											
7	58	79	81	74	75	77	85										
6	60	80	77	79	83	77	87	78									
8	53	68	75	71	70	68	75	74	72								
14	45	60	64	62	69	64	69	64	72	72							
10	46	60	65	65	65	62	66	63	68	79	87						
9	40	54	61	59	58	53	61	64	58	81	74	69					
1	46	53	52	54	56	49	56	57	51	55	49	47	49				
2	44	60	55	58	67	63	65	61	63	60	58	54	54	59			
11	40	55	56	55	58	49	61	61	56	58	52	57	61	52	53		
12	2	9	7	8	12	6	10	13	6	7	5	1	9	20	14	31	

KEY % similarity: 0-30 | 31-50 | 51-70 | 71-100

FIG. 3. STATIONS WITH SIMILAR FAUNA.

fauna is characteristic of low salinity marshes and adjacent bay areas along the Lake Raccourci shoreline.

(2) Oyster reef fauna

Station 3 was located on an oyster reef, which exhibited a high percentage of shell hash in the sediment. The assemblage at Station 3 was characterized by unusually high percentages of *Ammonia beccarii* var. *tepida*, and large numbers of *A. beccarii* var. *parkinsoniana* and *Elphidium* species. Other stations exhibiting large amounts of shell hash in the sediment, Stations 23 and 27B, also displayed unusually large proportions of living *A. beccarii* var. *tepida*. Oyster reefs seem to be characterized by high relative percentages of *A. beccarii* var. *tepida*.

(3) Bay fauna

Stations 4-8, 18, 20, and 22, which represented the major portion of Timbalier Bay, were characterized by a group of species considered to be the typical bay fauna. The bay fauna was composed of high relative percentages of *A. beccarii* varieties *tepida* and *parkinsoniana*, moderate relative percentages of *Ammotium salsum*, *Ammobaculites dilatatus*, *Elphidium gunteri*, *E. matagordanum*, *E. poeyanum*, and *E. vadescens*, and low relative percentages of the typical marsh and offshore species.

(4) *Elphidium gunteri* fauna

Stations 8, 9, and 10 were represented by a foraminiferal fauna that was similar to the bay fauna, but was differentiated from the typical bay fauna by a greater proportion of *E. gunteri*, and a low proportion of arenaceous species. The fauna at Station 8 was similar to the bay fauna, and the station apparently represented a transition zone between the two faunal groups.

(5) Barrier island marsh fauna

The barrier island marsh station located on East Timbalier Island, Station 12, was represented by a foraminiferal fauna composed of large numbers of *Clavulina gracilis*, an abundance of other arenaceous species, and a small number of one calcareous species, *E. matagordanum*. Station 11, located on the northern edge of East Timbalier Island, exhibited a large number of *C. gracilis*, but also a large number of calcareous species, indicating that it represented a transition zone between the bay and barrier island marsh faunas.

Species diversity

Measures of species diversity were calculated to serve as indicators of environmental variability within Timbalier Bay. The mean S value (number of species per station) for the dead foraminiferal data was 17, with a range of 11-23 species per station and a standard deviation of 3. The number of species at each station that formed 5% or more of the total population ranged from 5 to 7 species, with a mean of 6 and a standard deviation of 0.5. The mean S value for the living Foraminifera was 8, with a range of 0-13 and a standard deviation of 3. The number of species representing 5% or more of the standing crop averaged 4 per station, with a range of 0-9 and a standard deviation of 1.5. A trend of decreasing species richness for both the living and dead fauna was apparent in the upper bay and Lake Raccourci area, and in the southernmost portion of the bay.

Shannon-Weaver H' values (Shannon and Weaver 1963) were calculated from the living and nonliving faunal data to reflect species

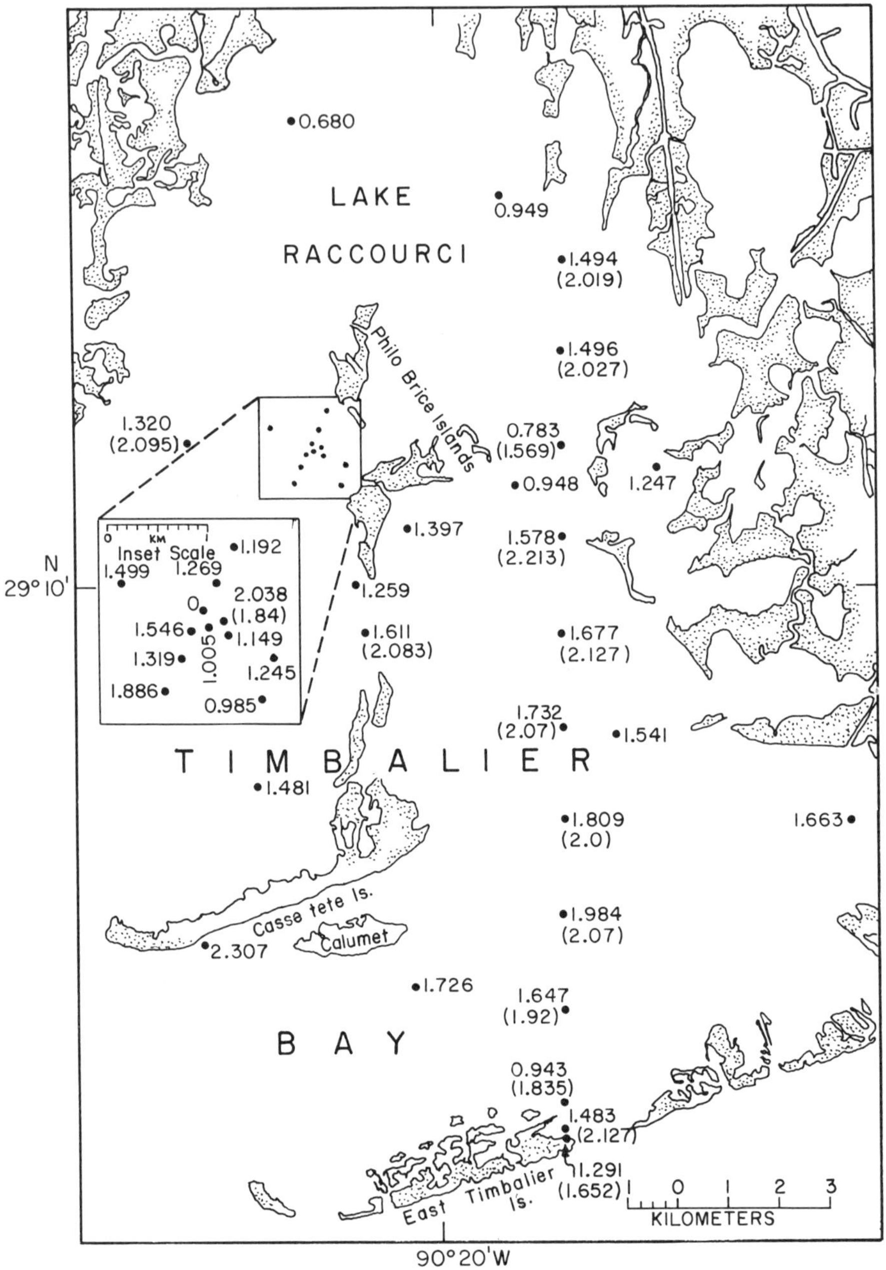

FIG. 4. SHANNON-WEAVER (H ′) INDEX OF SPECIES DIVERSITY VALUES for the living and dead benthic foraminiferal fauna (upper figure denotes living and lower figure denotes dead Foraminifera).

evenness (figure 4). Areas having relatively low H′ values, thus dominance by a few species, in both the living and nonliving faunas were: (1) the oyster reef represented by Station 3, (2) the barrier island marsh represented by Station 12 and the adjacent area represented by Station 10, and (3) some stations near the Philo Brice Island Production Platform. The Lake Raccourci area, Stations 24 and 25, also displayed lower diversity in the living fauna.

Comparisons between patterns of species distribution and measured environmental parameters

Waldron concluded that the distribution of species of benthic Foraminifera in Timbalier Bay was limited by salinity, sediment type, and water movements. Environmental parameters measured in Timbalier Bay during the OEI included salinity, temperature, dissolved oxygen, and sediment characteristics including grain size, percentage of calcium carbonate, and percentage of organic material.

Temperatures varied seasonally from a low of 8.5° C in January 1973 to a high of 39° C during the summer months. There were no areal temperature gradients. Dissolved oxygen concentrations (DO) ranged from 2.0 to 13.0 mg/l, varied inversely with the temperature, and showed no lateral trends throughout the bottom water of the bay. Because DO and temperature measurements show no areal gradients within the bay, I conclude that these parameters do not limit the distribution of the species observed during this study.

The grain size, percentage of calcium carbonate, and percentage of organic material data for Stations 1-12 are shown in table 5, along with the total populations expressed as the average number of specimens per species.

Figure 5 shows a correlation between sediment characteristics, foraminiferal faunas, and numbers of Foraminifera. Sandy sediments were associated with lower concentrations of organic material and smaller Foraminifera counts. Sediments with higher percentages of silt and clay had correspondingly higher concentrations of organic material, and higher Foraminifera counts. Cluster analysis, based upon the 10 most abundant species, produced a dendrogram display with three main branches that corresponded to the three sediment types—sandy, silt and clay, and shell hash.

Salinities fluctuated seasonally and exhibited an increasing trend from the northern to southern portion of the bay. Seasonal changes in salinity values (figure 6) ranged from 12 to 26 ppt in the upper portion of the bay (north of the Philo Brice Islands), from 15 to 29 ppt in the middle portion (south of the Philo Brice Islands and north of Casse-tete Island), and from 24 to 31 ppt in the lower portion (south of Casse-tete

TABLE 5 -- RELATIONSHIP BETWEEN SEDIMENT CHARACTERISTICS AND POPULATIONS OF BENTHIC FORAMINIFERA

Station	% $\geq$ 2mm (shell hash)	.037mm $\leq$ % $\leq$ 2mm (sand)	% $\leq$ .037mm (silt & clay)	Percent CO_3	Percent Organic Material	Total Population Living and Dead Average Number Per Species
1	1	38	61	7.40	6.50	1,621
2A	1	61	39	6.55	4.99	1,382
2B	1	50	49	6.85	5.91	
3	44	23	34	72.32	5.29	778
4	1	59	40	7.60	4.81	1,059
5	1	74	25	4.36	2.72	839
6	1	54	45	7.05	4.81	1,845
7	2	55	43	7.55	4.26	902
8	1	62	36	7.00	3.60	1,721
9	1	96	4	4.40	1.79	613
10	1	84	15	4.50	2.10	435
11	1	85	14	4.45	2.39	200
12	1	82	17	5.74	3.09	116

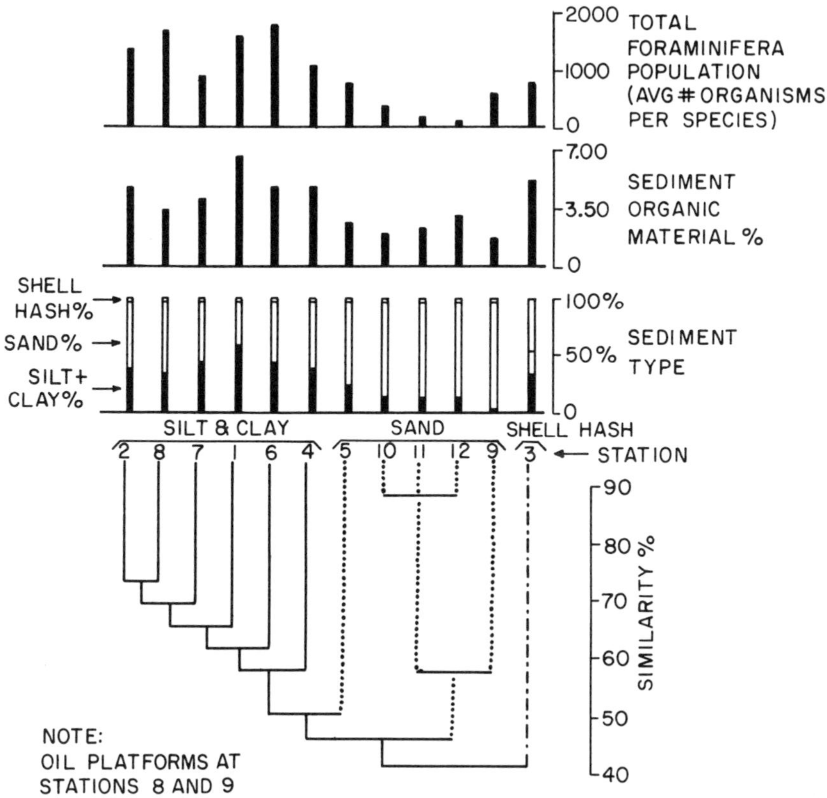

FIG. 5. RELATIONSHIP BETWEEN FAUNAL GROUPS, TOTAL POPULATIONS, AND SEDIMENT CHARACTERISTICS. (The lowered values for Station 7 were the result of a partial loss of sample.)

Island). The north-south salinity gradient varied from 4 ppt in September 1972 to 12 ppt in July 1973. The gradient appears to vary inversely with the average salinity of the bay.

Because the sediment types observed in Timbalier Bay follow a north-south trend that generally coincides with the north-south trend in the salinity gradient, it is difficult to determine the relative significance of salinity in limiting the distribution of most of the species observed. It is apparent from the data on dead Foraminifera, however, that the relative abundance of *Elphidium gunteri* increases toward the southernmost portion of the bay, over a series of stations having a similar sediment type (i.e., Stations 2, 4, 6, 7, and 8). Lamb's study of Mobile, Alabama, indicated that salinity was the main controlling factor in the

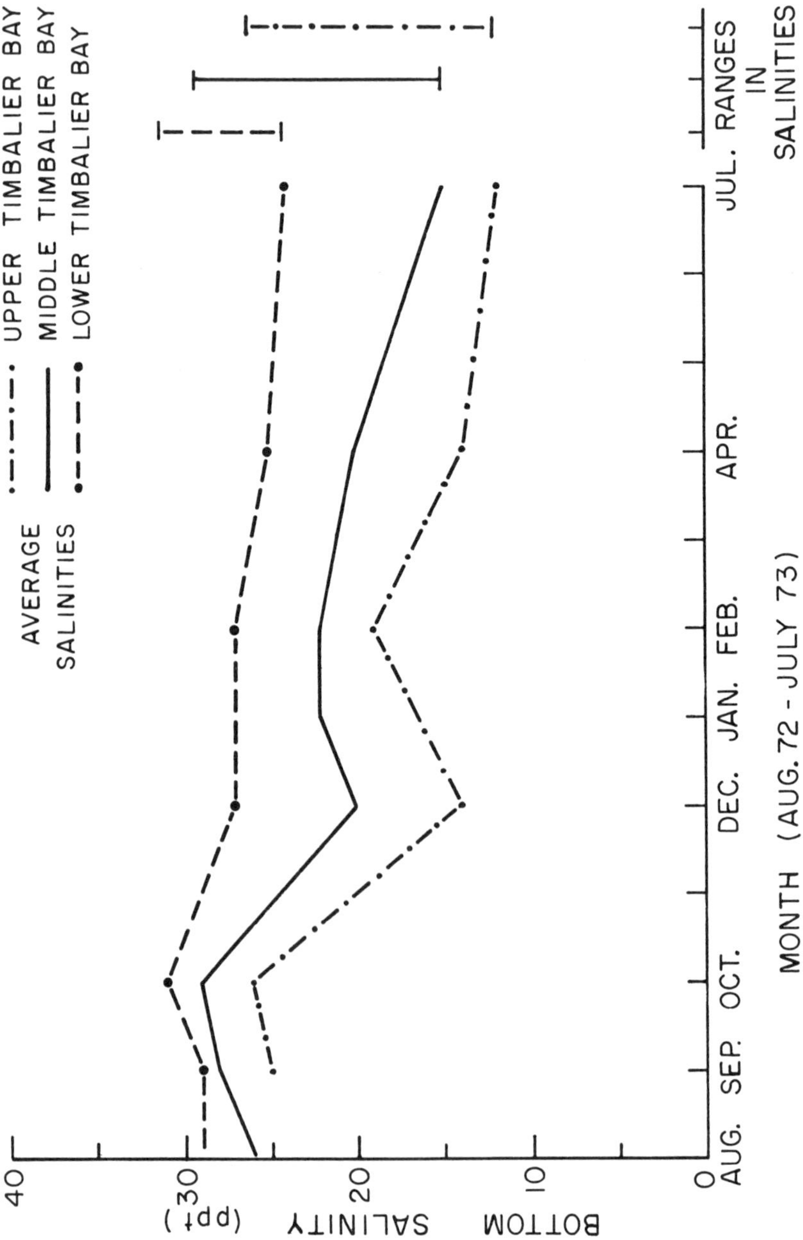

FIG. 6. SALINITY GRADIENT AND RANGES IN TIMBALIER BAY.

distribution of foraminiferal species. The foraminiferal species and range in salinity observed in the lower two-thirds of Mobile Bay were quite similar to those observed in Timbalier Bay.

In summary, the faunal and environmental data collected during the present study, combined with evidence from studies by Waldron and Lamb, strongly suggest that sediment characteristics and salinity were the major environmental parameters controlling the distribution of benthic Foraminifera in Timbalier Bay.

Comparison with previous study

Seasonal fluctuations of living populations of benthic Foraminifera in Stations 13, 17, 21, and 23-27 were analyzed in Waldron's earlier study. These stations were resampled and the living Foraminifera were analyzed from the same sediment size fraction (> 149 μ) examined by Waldron. The Morisita-Ono Index of Similarity between communities (Morisita 1959) was used to compare Waldron's results with the data from this study (table 6). The faunas observed during the two studies at Stations 24 and 27 were quite similar. The faunas observed during the two studies at Stations 25, 13, 23, 26, and 17 reflected some change, but were generally within the range of similarity values measured between groups of stations from the same regions of the bay in the present study (see figure 7, p. 469).

ACKNOWLEDGMENTS

This work was supported by Gulf Universities Research Consortium grant number OG58JM to James I. Jones.

I wish to express my deep gratitude to James I. Jones and R. J. Menzies for their guidance throughout this investigation. I thank Ruth Todd for verifying species identifications; and Donald J. Reish, S. G. Appan, and Karen Braun for their assistance in preparing the manuscript.

REFERENCES CITED

Andersen, H. V. 1953. Two new species of *Haplophragmoides* from the Louisiana coast. Contributions, Cushman Foundation for Foraminiferal Research **4**(1):21-22.

———. 1961. Foraminifera of the mudlumps, lower Mississippi River Delta. Louisiana Geological Survey, Geological Bulletin **35**(2):1-208.

———. 1971. Key to Cenozoic foraminiferal families and genera of the Gulf Coastal Plain Province. School of Geoscience, Louisiana State University, Miscellaneous Publication 71-2.

TABLE 6

Morisita-Ono Index Values Comparing Foraminiferal Faunas

Observed in Waldron's and Present Study

Station Number	Number of Living Specimens: Waldron's Data	Number of Living Specimens: Present Data	Similarity	Comments
24	196	462	0.88	No change in foraminiferal faunas.
27	62	67	0.88	No change in foraminiferal faunas.
25	90	474	0.61	Some change in foraminiferal faunas.
13	86	58	0.60	Some change in foraminiferal faunas.
21	116	4	0.01	Small numbers of living specimens, thus data not reliable.
23	122	20	0.76	Small numbers of living specimens, thus data not reliable.
26	18	88	0.68	Small numbers of living specimens, thus data not reliable.
17	13	24	0.71	Small numbers of living specimens, thus data not reliable.

FIG. 7. FAUNAL SIMILARITY 1957 AND 1972.

Bandy, O. L. 1954. Distribution of some shallow-water Foraminifera in the Gulf of Mexico. Geological Survey Professional Paper 254-F.

———. 1956. Ecology of Foraminifera in the northeastern Gulf of Mexico. Geological Survey Professional Paper 274-G.

———, J. C. Ingle, and J. M. Resig. 1964a. Foraminiferal trends, Laguna Beach outfall area, California. Limnology and Oceanography **9**:112-123.

———, J. C. Ingle and J. M. Resig. 1964b. Foraminifera, Los Angeles County outfall area, California. Limnology and Oceanography **9**:124-137.

———, J. C. Ingle, and J. M. Resig. 1964c. Modification of foraminiferal distribution by the Orange County outfall, California. Ocean Science and Ocean Engineering **1**:54-76.

Barrett, B. B. 1971. Cooperative Gulf of Mexico estuarine inventory and study, Louisiana. Phase II, hydrology, and Phase III, sedimentology. Louisiana Wildlife and Fisheries Commission.

Bock, W. D. 1967. Monthly Variation in the Foraminiferal Biofacies on Thalassia and Sediment in the Big Pine Key Area, Florida. Ph.D. dissertation, University of Miami.

———. 1971. A handbook of the benthonic Foraminifera of Florida Bay and adjacent waters. Memoir 1: Miami Geological Society. A symposium on recent south Florida Foraminifera **3**:1-72.

Boltovskoy, E. 1966. Depth at which Foraminifera can survive in sediments. Contributions, Cushman Foundation for Foraminiferal Research **17**(2):43-45.

Brady, H. B. 1884. Report on the Foraminifera. *In* Voyage of HMS Challenger. Volume 9, pp. 1-814.

Brooks, A. L. 1967. Standing crop, vertical distribution, and morphometrics of *Ammonia beccarii* (L). Limnology and Oceanography **12**:667-684.

Clark, D. F. 1971. Effects of aquaculture outfall on benthic Foraminifera in Clam Bay, Nova Scotia. Marine Sediments **7**(2):76-84.

Cushman, J. A. 1922. The Foraminifera of the Atlantic Ocean. U.S. National Museum Bulletin **104**, part 3, p. 83.

Kornfeld, M. M. 1931. Recent littoral Foraminifera from Texas and Louisiana. Contributions of the Department of Geology, Stanford University **1**(3):77-101.

Lamb, G. M. 1972. Distribution of Holocene Foraminifera in Mobile Bay and the effect of salinity changes. Geological Survey of Alabama, Circular 82.

Lankford, R. R. 1959. Distribution and ecology of Foraminifera from east Mississippi Delta margin. American Association of Petroleum Geologists, Bulletin **43**(9):2068-2099.

Lehman, E. P. 1957. Statistical study of Texas Gulf Coast recent foraminiferal species. Micropaleontology **3**:325-356.

Lowman, S. W. 1949. Sedimentary facies in the Gulf coast. American Association of Petroleum Geologists, Bulletin **33**(12):1939-1997.

Menzies, R. J. 1973. A handbook on procedures and methods employed in the Offshore Ecology Investigation (OEI). Gulf Universities Research Consortium. Report No. 121.

Morin, R. W. 1971. Offshore foraminiferal populations. *In* R. L. Kolpack, ed., Biological and Oceanographical Survey of the Santa Barbara Channel Oil Spill, 1969-1970. Vol. II. Physical, Chemical and Geological Studies. Allan Hancock Foundation, University of Southern California.

Morisita, M. 1959. Measuring interspecific association and similarity between communities. Memoirs of the Faculty of Science, Kyushu University, Series E (Biol.) **3**(1):65-80.

Murray, J. W. 1973. Distribution and Ecology of Living Benthic Foraminiferids. New York: Crane, Russak and Company, Inc.

Orbigny, Alcide D. d'. 1839. Foraminifères. *In* Ramon De la Sagra, ed., Histoire Physique, Politique, et Naturelle de l'île de Cuba. Paris, pp. 1-224.

Parker, F. L., F. B. Phleger and J. F. Peirson. 1953. Ecology of Foraminifera from San Antonio Bay and environs, southwest Texas. Cushman Foundation for Foraminiferal Research, Special Publication **2**:1-75.

Phleger, F. B. 1951. Ecology of Foraminifera, northwest Gulf of Mexico. Part I. Foraminifera distribution. Geological Society of America, Memoir **46**:1-88.

———. 1954. Ecology of Foraminifera and associated microorganisms from Mississippi Sound and environs. American Association of Petroleum Geologists, Bulletin **38**(4):584-647.

———. 1955. Ecology of Foraminifera in southeastern Mississippi Delta area. American Association of Petroleum Geologists, Bulletin **39**(5):712-752.

———. 1956. Significance of living foraminiferal populations along the central Texas coast. Contributions, Cushman Foundation for Foraminiferal Research **9**(4):106-151.

———. 1960. Ecology and Distribution of Recent Foraminifera. Baltimore: Johns Hopkins Press.

———. 1965. Patterns of marsh Foraminifera, Galveston Bay, Texas. Limnology and Oceanography **10**, Supplement R:169-184.

——— and R. R. Lankford. 1957. Seasonal occurrences of living benthic Foraminifera in some Texas bays. Contributions, Cushman Foundation for Foraminiferal Research **8**(3):93-105.

——— and F. L. Parker. 1951. Ecology of Foraminifera, northwest Gulf of Mexico. Part II. Foraminiferal species. Geological Society of America, Memoir **46**:1-64.

Sanders, H. L. 1960. Benthic Studies in Buzzard's Bay. III. The Structure of the Soft Bottom Community. Limnology and Oceanography **5**:138-153.

Shannon, C. E. and W. Weaver. 1963. The Mathematical Theory of Communication. Urbana: University of Illinois Press.

Todd, R. and P. Brönnimann. 1957. Recent Foraminifera and Thecamoebina from the eastern Gulf of Paria. Cushman Foundation for Foraminiferal Research, Special Publication 3.

Waldron, R. P. 1957. A seasonal ecological study of Foraminifera from Timbalier Bay, Louisiana. Coastal Studies Institute, Louisiana State University. Technical Report 16, Part B, pp. 132-188.

Walton, W. R. 1952. Techniques for the recognition of living Foraminifera. Contributions, Cushman Foundation for Foraminiferal Research **3**(2):56-60.

Warren, A. D. 1957. Foraminifera of the Buras-Scofield bayou region, southeast Louisiana. Contributions, Cushman Foundation for Foraminiferal Research **8**(1):29-40.

Watkins, J. G. 1961. Foraminiferal ecology around the Orange County, California, ocean sewer outfall. Micropaleontology **7**:199-206.

Williams, S. E. and J. I. Jones. 1974. The distribution and origin of bottom sediments in Timbalier Bay, Louisiana. Unpublished manuscript.

OIL PRODUCTION AND ECOLOGY OF THE LITTORAL POLYCHAETA OF TIMBALIER BAY

by Henry Kritzler

ABSTRACT

A seasonal quantitative study was made of the polychaetous annelids from Timbalier Bay in 1973. Collections were analyzed from samples taken near oil drilling and production platforms as well as from two sites located some distance from any oil activities. The species from each sample were identified and the results analyzed according to species diversity, faunal similarities, and biomass. Eighty-two species of polychaetes were identified, raising the total known species from Louisiana to 117. *Spiochaetopterus c. oculatus* accounted for over 80% of the specimen total. The polychaete fauna was diverse with many secondary dominants and many species present in small numbers. Ambient stations always had the highest species and specimen total, biomass, and species diversity, while the lowest values for these parameters were recorded from either the oil drilling station or the production station. The index of similarity showed a close relationship between the oil drilling station and one ambient station, which was probably the result of the high numbers of *Spiochaetopterus* at these two stations. Cluster analysis separated two groups of stations; one group linked the production station with ambient station AB-2; the second group linked the drilling station with ambient station AB-1 and one production station. On the basis of these data, there is no in-

Henry Kritzler is with the Department of Oceanography, Florida State University, Tallahassee, FL.

dication that the long-term oil activities in Timbalier Bay have had any adverse effect on the benthic polychaete communities.

INTRODUCTION

The primary objective of this study was to assess the effects of long-term oil drilling and production on the composition of the benthic polychaetous annelid fauna of estuarine Timbalier Bay. The study was planned in such a way that the results could constitute a scientifically sound baseline against which any possible past or future effects of oil activities could be measured. Since virtually all previous studies on polychaetes from the Gulf of Mexico have been based on qualitative collections, I hoped to contribute to knowledge not only of this group in a bay environment but also of the quantitative occurrences of these organisms.

Prior to this study only 53 species had been reported from Louisiana, the majority from the vicinity of Grand Isle. Warren (1942) recorded three species and later Behre (1950) added 28 species from Grand Isle. Hartman (1951, 1954) summarized the knowledge of the polychaetes from the Gulf of Mexico; she reported 16 species from Louisiana. In the first quantitative study of the marine benthos of Louisiana, Parker (1956) reported 29 species from the eastern Mississippi delta region including a new species, *Cossura delta*, later described by Reish (1958). The number of species of polychaetes known from Louisiana totaled 53 at the outset of this study of Timbalier Bay.

The polychaetes were chosen for study because Timbalier Bay lies within the Mississippi River delta, and its substrate consists exclusively of fine sand, silt, and clay (Barrett 1971). Accordingly, it was expected that the macrobenthos would be dominated numerically and in biomass by deposit-feeding polychaetes. This has generally been found to be the case in fine sediments elsewhere (Blegvad 1928; Sanders 1958; Lie 1968; Rhodes and Young 1970; O'Connor 1972). Since assemblages of polychaete species have been shown to serve as indicators of sewage and industrial pollution (Reish 1955, 1957, 1959), it was thought that the data might also show the effects of long-term oil drilling and production if, indeed, such occur.

MATERIALS AND METHODS

Collection

Sixteen replicate samples were taken from 29 stations for polychaetes in Timbalier Bay over a period of 17 months from September 1972

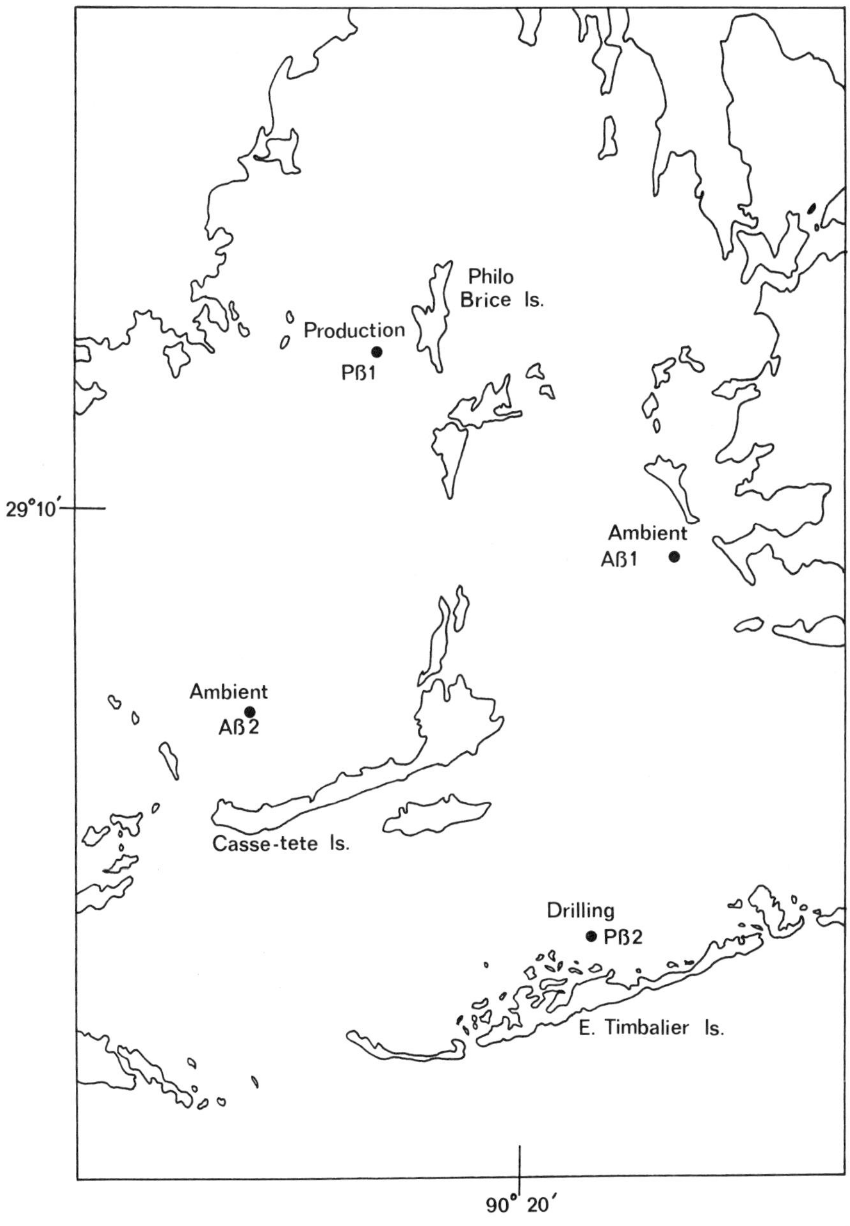

FIG. 1. STATION LOCATIONS.

through January 1974. Samples were taken with a plastic tube, which sampled an area of 625 cm^2 to a depth of 10 cm. The sampler was connected to an air hose that lifted the material into the boat, where it was washed through a 0.5 mm mesh sieve; the material retained on the sieve was preserved (Kritzler et al. 1974). Four stations were completely analyzed, two of which were located near oil operations (producing PB-1 and drilling PB-2) and two of which were used as representative of ambient or natural environmental conditions (AB-1 and AB-2) (figure 1). Data from four stations in winter, spring, and summer 1973 collections were analyzed and two from fall 1973 (PB-1 and AB-2). Since each sample covered a surface area of 625 cm^2, the combined surface area of the 16 replicates was 1.0 m^2.

Data analyses

The polychaetes were identified to species insofar as possible. Biomass was estimated by volumetry and converted to dry weight. The data were analyzed for species diversity and faunal similarities, and were clustered for index of similarities. Species diversity (H ′) was calculated according to the method defined by Shannon and Weaver (1963). This index measures both species index and evenness. The numerical value obtained is of particular use in environmental studies in comparing the results of a set of stations. The higher the value, the cleaner, healthier, or more natural the environment; in contrast, a low number indicates a dominance by one or two species with very few others present, which, in turn, may indicate a stressed or polluted environment.

Faunal affinities or similarities ($C\lambda$) between two communities (or stations) were measured by the Morisita-Ono index (Morisita 1959; Ono 1961). The values for $C\lambda$ will vary from 0 to 1.0. A numerical value of 1.0 indicates identical communities, and 0 indicates no species in common. A dendrogram was drawn from calculations based on the minimal faunal percentages (Sanders 1958), which were subjected to Mountford's (1962) clustering method.

RESULTS

The quantities of the polychaetes at the four stations analyzed are summarized in tables 1 and 2 according to species, station, date, number of species and specimens, biomass, and species diversity. The results of Morisita-Ono index of faunal similarity for all possible pairs of samples are represented in a trellis diagram in figure 2 (p. 482). A dendrogram showing the similarities and differences of the 14 stations is given in figure 3 (p. 483).

TABLE 1

Polychaetous Annelids in Timbalier Bay, Louisiana, Collected in Winter and Spring 1973

Species	Winter 1973				Spring 1973			
	Production PB-1	Ambient AB-1	Drilling PB-2	Ambient AB-2	Production PB-1	Ambient AB-1	Drilling PB-2	Ambient AB-2
Spiochaetopterus c. oculatus	113	4210	1872	70	208	4316	3283	56
**Paraeurythoe americana*	14	82	16	5	22	237	58	1
**Myriochele bioculata*	5	23	8	176	50	6		23
**Harmothoe trimaculata*	12	154	8	3	116	3	2	48
Lumbrineris parvapedata	44	45	37	39	68	44	7	8
**Heteromastus filiformis*	24	18	9		6	21	68	19
**Notomastus latericeus*	6	9	3	45	2	1	19	
**Prionospio sexoculata*	21		1		37	22	1	2
Lumbrineris bassi	16	10	13	2	31	32	8	15
**Melinna maculata*					30	10		
**Paraprionospio pinnata*	8	2			24	8		
**Glycinde polygnatha*	12	14	1	2	10	24	15	3
Clymenella t. calida		1	23	3		2	20	
Cossura delta	3	15	1		2	11	1	
**Euclymene* sp.	4	3	1	2	15	14	5	
**Polydora socialis*	1	2	6		2	2	5	13
**Magelona cincta*		4	2	12	12	3	1	2
**Cirriformia tentaculata*		3	4	12				
**Aedicira pallida*		3	5	4			3	10
Gyptis capensis	1	1	9	1	3	3	3	1
**Capitella capitata*		1		9				
**Mystides borealis*	1						9	
Neanthes succinea		2	3	3			9	1
**Asychis urceolata*		8	2		1			
**Harmothoe lunulata*		3		5	2	9		3
**Aricidea fauveli*			8	7				
Pectinaria gouldii					8	2		
**Arabella iricolor*	2	2	7					
**Sigambra tentaculata*	2	3		4	2	6	3	4

TABLE 1 (continued)

Species	Winter 1973				Spring 1973			
	Production PB-1	Ambient AB-1	Drilling PB-2	Ambient AB-2	Production PB-1	Ambient AB-1	Drilling PB-2	Ambient AB-2
**Neanthes arenaceodentata*			4			1	5	1
**Diopatra c. cuprea*	1		4	2	3		3	
Onuphis e. oculata			4					
**Ancistrosyllis hamata*	1		1	4		3		1
Maldane sarsi		1			3	2	4	1
**Spiophanes bombyx*	3			1				3
**Nereiphylla paretti*		3						1
**Malaceros fuliginosa*			3					
Semiodera roberti		1		3				1
Nerine agilis		1		3		1		
Owenia fusiformis		1			1	1	3	
**Eteone heteropoda*						2	3	
**Nephtys bucera*	1	1			1	2		3
**Bhawaia goodei*	1	1		2	2	1		1
Glycera americana	2		9				6	
**Prionospio cirrifera*		2		1				1
**Eumida sanguinea*		2						
**Tharyx monilaris*		1		2			1	
**Anaitides maculata*				2				
Lumbrineris erecta				2				
Cirratulus cirratus					2			
Glycera rouxi					2	1	1	
**Sabella microphthalma*						2		
**Cabira brevicirrus*								2
**Melinnexis* sp.	1			1				
**Streblospio benedicti*	1	1						
**Dorvillea rudolphi*		1	1					
**Ophiodromus* sp.		1						
**Poecilochaetus* sp.		1						
**Notomastus hemipodus*		1						

TABLE 1 (continued)

Species	Winter 1973				Spring 1973			
	Production PB-1	Ambient AB-1	Drilling PB-2	Ambient AB-2	Production PB-1	Ambient AB-1	Drilling PB-2	Ambient AB-2
**Armandia polyophthalma*		1						
Ninoe n. gracilis		1						
**Megalomma bioculata*		1						
**Dasybranchus lumbricoides*			1					
**Sthenelais boa*			1					
**Pilargis verrucosa*				1				
**Scoloplos rubra*				1				
**Haploscoloplos robustus*				1				
**Drilonereis magna*					17	1		
**Magelona longicornis*					1			1
Sthenelais articulata							1	
**Pista mirabilis*								1
**Aglaophamus verrillii*								1
Number of species	26	42	31	33	30	32	28	29
Number of specimens	300	4690	2067	433	692	4940	3544	211
Biomass (g/m^2 dry wt.)	0.29	5.79	2.94	0.86	0.91	5.34	4.57	0.77
Species diversity (H')	4.56	4.85	3.93	5.44	4.74	3.88	3.43	5.60

*denotes previously unreported species

TABLE 2

Polychaetous Annelids in Timbalier Bay, Louisiana, Collected in Summer and Fall 1973

Species	Summer 1973				Fall 1973	
	Production PB-1	Ambient AB-1	Drilling PB-2	Ambient AB-2	Production PB-1	Ambient AB-2
Spiochaetopterus c. oculatus	114	2553	1708	79	516	1113
**Harmothoe trimaculata*	48	175		1	61	16
**Paraeurythoe americana*	37	130	1	1	76	1
**Prionospio sexoculata*	120	5	29	11	32	6
Lumbrineris parvapedata	25	48	46	36	45	58
**Melinna maculata*	24	46		2	1	3
**Notomastus latericeus*		2	14	42		42
**Glycinde polygnatha*	15	41	15	1	9	5
Lumbrineris bassi	33	26	40	33	30	25
**Paraprionospio pinnata*	32	2	3	6	7	2
**Euclymene* sp.	7	29		4		
Cossura delta	9	8	27			
**Heteromastus filiformis*	8	2	1	5	23	7
**Aedicira pallida*	2	3	2	12	9	20
**Myriochele bioculata*	6			20		6
**Sigambra tentaculata*	6	1	8	4	19	3
**Magelona cincta*	6	5	1	13	3	13
**Harmothoe lunulata*				13		6
**Polydora socialis*		7	2	2		10
Clymenella t. calida		9		10	2	3
**Asychis urceolata*		9				
**Gyptis capensis*	1		6	4	4	1
Owenia fusiformis		2		5		4
Maldane sarsi		5				
Neanthes succinea	4	2	3	3		2
**Nainereis laevigata*						4
**Prionospio cirrifera*				4		
**Magelona longicornis*	1		2	3	1	3
**Diopatra c. cincta*	3	3	1		1	

TABLE 2 (continued)

Species	Summer 1973				Fall 1973	
	Production PB-1	Ambient AB-1	Drilling PB-2	Ambient AB-2	Production PB-1	Ambient AB-2
**Drilonereis magna*		2			3	
**Ancistosyllis hamata*	3	1		1	2	2
**Spiophanes bombyx*				3		1
Glycera americana	1	3			1	
Pectinaria gouldii		3				1
**Aricidea fauveli*		3		1		1
**Ammotrypane aulogaster*						3
**Sthenelais boa*				2		
Cirratulus cirratus	2				1	
**Laonice* sp.				2		
**Nainereis setosa*				2		
**Bhawania goodei*				1	1	
Diopatra c. cuprea		1	1			1
**Tharyx monilaris*		1	1	1		
**Neanthes arenaceodentata*		1		1		
**Loimia viridis*						1
**Myriowenia* sp.				1		1
**Syllidia armata*		1				
**Exogone* sp.		1				
Cabira brevicirrus						1
Glycera rouxi		1				
**Malaceros fuliginosa*				1		
**Cirriformia tentaculata*			1			
**Dorvillea rudolphi*				1		
**Mystides borealis*			1			
Number of species	24	35	22	36	23	32
Number of specimens	508	3132	1918	327	848	1365
Biomass (g/m^2 dry wt.)	0.26	3.30	2.18	0.20	0.70	1.55
Species diversity (H')	3.69	4.22	2.91	6.04	3.26	4.57

*denotes previously unreported species

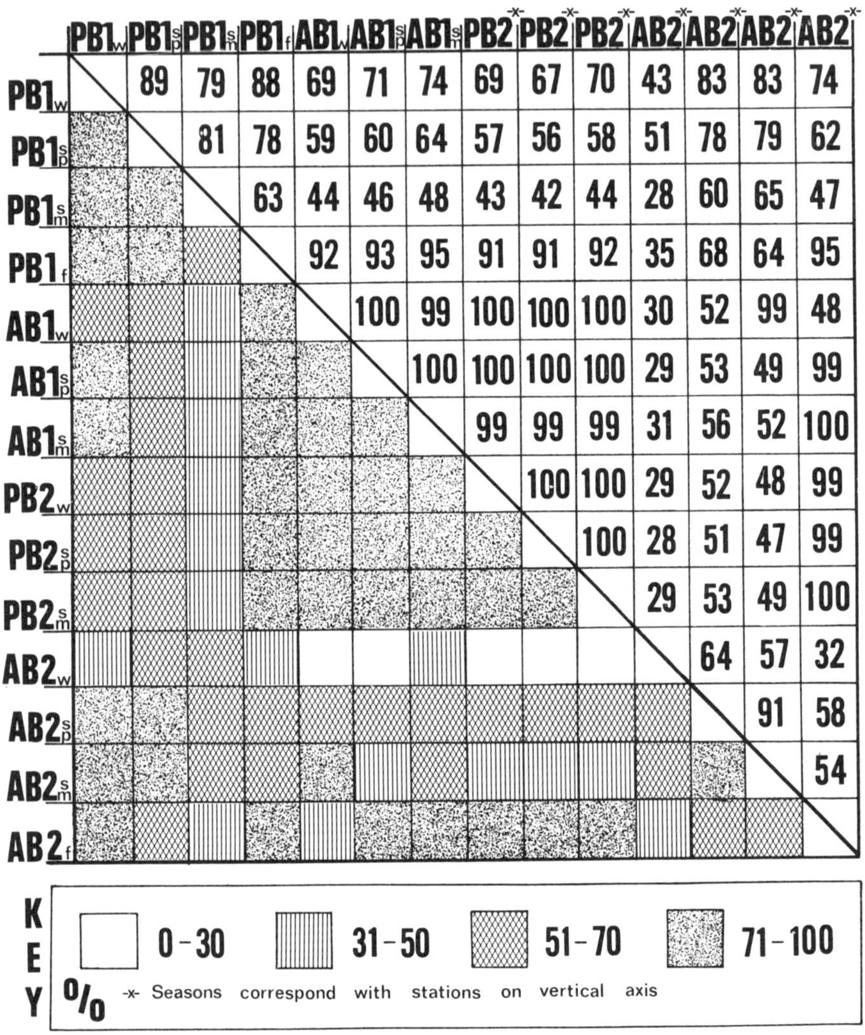

FIG. 2. TRELLIS DIAGRAM SHOWING MORISITA-ONO INDEX OF FAUNAL AFFINITY (×100) for the polychaete data from Timbalier Bay.

A total of 82 species in 33 families of polychaetes was collected in Timbalier Bay (tables 1 and 2). Nearly 25,000 specimens of polychaetes were collected, of which over 80% belong to a single species, *Spiochaetopterus c. oculatus*. Nearly one-half of the species were represented by 5 or fewer specimens; 5 species were encountered at every station and collection period. The more numerous species are listed in table 3 together with the total number of specimens collected.

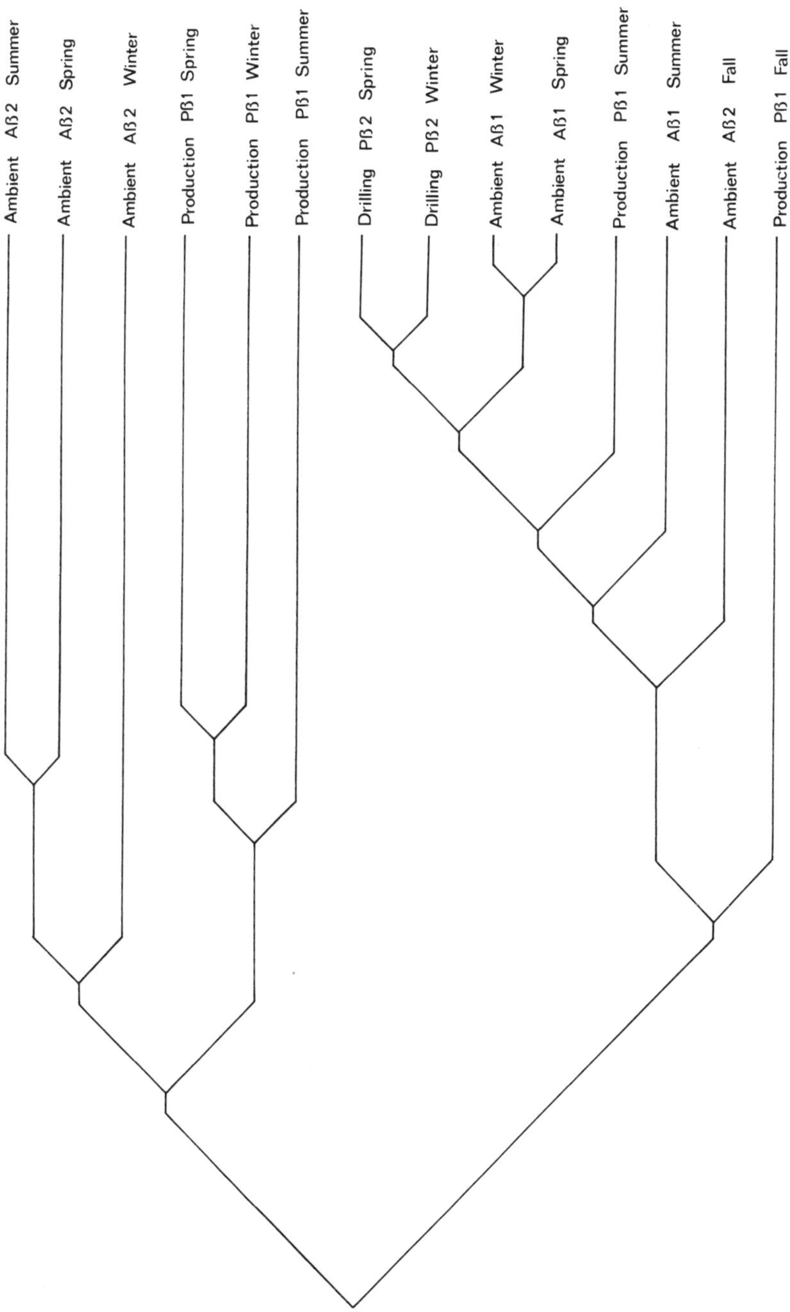

FIG. 3. DENDROGRAM SHOWING THE RESULTS OF A CLUSTER ANALYSIS of the polychaete data from 14 stations.

Spiochaetopterus c. oculatus, *Paraeurythoe americana*, *Lumbrineris parvapedata*, *L. bassi*, and *Glycinde polygnatha* were present at all stations from all collecting periods; these 5 species accounted for over 90% of the specimens collected. The results indicate a rich and diverse polychaete fauna, which is dominated by *Spiochaetopterus* and 4 others plus 14 secondary dominant species (table 3).

The number of species present per station ranged from 22 to 42 with a median of 30. Ambient station AB-1 had the highest average with 36, and production station PB-1 the lowest with 26 species. The number of specimens collected per station ranged from 211 to 4940 with a median of 1783. Ambient station AB-1 had the highest average with 4254 specimens, and production station PB-1 the lowest with 500. Polychaete biomass ranged from 0.2 to 5.79 g/m^2 (dry weight) with a median of 2.12. Ambient station AB-1 had the highest average biomass with 3.81 g/m^2 (dry weight), and production station PB-1 had the lowest with 0.54. Species diversity ranged from 2.91 to 6.04 with a median of 4.04. Ambient station AB-2 had the highest species diversity with 5.41, and drilling station PB-2 had the lowest with 3.42.

Seasonal differences in the various parameters measured were noted, but what occurred at one station during a particular season did not necessarily occur at the others. At the production station PB-1 the highest values were measured during spring for number of species, biomass, and species diversity, but the number of specimens was lowest at this time. The same parameters were highest during the winter collection period at ambient station AB-1. All parameters were more or less similar throughout the four collecting periods at ambient station AB-2. At drilling station PB-2 the number of species and species diversity were low during the summer and the number of specimens increased in the fall. When the various parameters between stations were compared, ambient station AB-1 was highest in biomass at all collecting periods and highest for the number of specimens present for two collections. Ambient station AB-2 was highest in numbers of specimens collected three of the four seasons. Drilling station PB-2 was lowest in species diversity from winter through summer and lowest in number of species present in the spring and summer collections.

When the number of species, specimens, biomass, and species diversity are averaged by collection period, a seasonal pattern is evident. All parameters were highest either during the winter or spring period with a decided drop during the summer period, which generally extended into the fall at the two stations analyzed.

The results of the Morisita-Ono index of affinity (× 100) are represented by the trellis diagram in figure 2. The polychaete fauna is shown to be very similar in the three collections from ambient station

TABLE 3

Dominant Species of Benthic Polychaetes as Measured by Percent Occurrence

Timbalier Bay, 1973

Species	Station and Collection Period														Total Number of Specimens	Percent Occurrence
	PB-1				AB-1			AB-2				PB-2				
	W	Sp	Sum	F	W	Sp	Sum	W	Sp	Sum	F	W	Sp	Sum		
Spiochaetopterus oculatus	X	X	X	X	X	X	X	X	X	X	X	X	X	X	21,211	100
Paraeurythoe americana	X	X	X	X	X	X	X	X	X	X	X	X	X	X	681	100
Lumbrineris parvapedata	X	X	X	X	X	X	X	X	X	X	X	X	X	X	550	100
L. bassi	X	X	X	X	X	X	X	X	X	X	X	X	X	X	314	100
Glycinde polygnatha	X	X	X	X	X	X	X	X	X	X	X	X	X	X	167	100
Harmothoe trimaculata	X	X	X	X	X	X	X	X	X	X	X	X	X		747	93
Heteromastus filiformis	X	X	X	X	X	X	X		X	X	X	X	X	X	211	93
Magelona cincta		X	X	X	X	X	X	X	X	X	X	X	X	X	77	93
Sigambra tentaculata	X	X	X	X	X	X	X	X	X	X	X		X	X	65	93
Gyptis capensis	X	X	X	X	X	X		X	X	X	X	X	X	X	38	93
Notomastus latericeus	X	X			X	X	X	X	X	X	X	X	X	X	187	86
Prionospio sexoculata	X	X	X	X		X	X		X	X	X	X	X	X	281	86
Polydora socialis	X	X			X	X	X	X	X	X	X		X	X	52	79
Aedicira pallida			X	X	X		X	X	X	X	X	X	X	X	73	79
Euclymene sp.	X	X	X		X	X	X	X		X		X	X		84	71
Myriochele bioculata	X	X	X		X	X		X	X	X	X	X			373	71
Neanthes succinea			X		X		X	X	X	X	X	X	X	X	32	71
Paraprionospio pinnata	X	X	X	X	X	X	X			X	X			X	90	71
Ancistrosyllis hamata	X		X	X		X	X	X	X	X	X	X			19	71

AB-1 and drilling station PB-2 as well as in the fall sample from ambient station AB-2. The index of affinity was above .95 in comparisons between these stations, indicating essentially identical polychaete faunas. Close affinity was shown between seasonal samples taken at a station (i.e., production station PB-1). The essential similarity of polychaete faunas at these stations is the result of the overwhelming dominance of *Spiochaetopterus* plus similar populations of *Lumbrineris parvapedata* and *L. bassi*.

The samples taken at ambient station AB-2 had a reduced affinity with samples taken at the same site at different seasons and a very low affinity for the polychaete fauna at other stations. Examination of the raw data included in Kritzler (1974) indicates that the separation of the winter sample at ambient station AB-2 is the result of a shifting in numbers of the dominant species. Ambient station AB-2, winter, was characterized by a large number of *Myriochele bioculata* and *Notomastus latericeus* and by the absence of *Heteromastus filiformis* as well as reduced numbers of *Harmothoe trimaculata* and *Paraeurythoe americana*.

The average index of similarity value for all possible sample pairs (91) was .69, and this figure became .75 when the indices for ambient station AB-2 for winter were excluded from the computation. This high average index value demonstrates a similar polychaete community at these stations in 1973 regardless of the presence or absence of oil production activity.

The results of the polychaete sample clustering based on Sanders's (1960) index of affinity is diagrammed in figure 3. There were two major separations of the station data for polychaetes on the basis of clustering. One group was limited to the winter, spring, and summer collections and was further separated according to location (production station PB-1 and ambient station AB-2). The second major cluster essentially gave a staircase effect after the initial winter and spring separation of drilling station PB-2 from ambient station AB-1. In the latter case, these two stations are dominated by *Spiochaetopterus*, which varied considerably in number of specimens present. *Spiochaetopterus* does not dominate the benthic fauna at production station PB-1 or ambient station AB-2 and the distinct separation, on the basis of location, is related to the presence or absence of secondary dominants at these stations.

DISCUSSION

The number of known polychaetous annelids known from Louisiana marine waters has been increased from 53 to 117 as the result of this quantitative study. It is reasonable to assume that this number will be increased further by additional quantitative studies especially from offshore waters. The previously unreported species of polychaetes are indicated by an asterisk in tables 1 and 2.

The polychaete assemblage from these four stations sampled in 1973 is basically one association dominated by *Spiochaetopterus* and additional species of polychaetes. This is especially true of ambient station AB-1 and drilling station PB-2 in which *Spiochaetopterus* is present in numbers from 1708 to 4316 per m^2. While *Spiochaetopterus* is not dominant at the other two stations (AB-2 and PB-1), it is still present in considerable numbers. The appearance of other dominants at these stations accounts for the separation of faunal similarities between these two groups of stations as indicated in figures 2 and 3. Since nothing is known concerning the reproduction and settlement of *Spiochaetopterus*, it is impossible to ascertain whether the large populations of this species are continuous in time or represent a single age group following a particularly good larval settlement.

The quantitative occurrence of the polychaetes at these four stations from the 1973 collections was analyzed statistically in three ways; these were species diversity (Shannon-Weaver), faunal similarities (Morisita-Ono), and station clustering. Since *Spiochaetopterus* was so abundant at two of the stations, the species diversities and faunal similarities were calculated with and without this species data. A factor of $\times$ 5 was applied to the crustacean and mollusc data in order to equate the data to 1 m^2 to coincide with the polychaete data; the species diversity and faunal similarities were calculated with and without this factor. Much of the data indicates a close relationship between ambient station AB-1 and drilling station PB-2. As indicated by the faunal similarities, cluster analysis, and biomass data from these two sites, the similarity of these two stations is probably the result of the numerical dominance by *Spiochaetopterus*.

The influence of the record 1973 flood waters of the Mississippi River may account for the differences between the summer collections and the winter-spring collections at all stations as shown by species diversity. Lower values were measured for the summer and fall collections. Since the water is shallow in Timbalier Bay, the higher temperatures during the summer may have also played a role in the reduction of the polychaete population. Whatever the cause of the reduction in the population, however, it was not of sufficient severity as to cause massive mortalities, which, in turn, could lead to a different assemblage of organisms.

The relatively large numbers of polychaete species and specimens found at these four stations in 1973 in Timbalier Bay suggest that the bay is a healthy one. In the array of species reported herein, we find repeated many of the genera of polychaetes reported by Reish and Winter (1954) in Alamitos Bay, by Barnard and Reish (1959) in Newport Bay, and by Reish (1959) from the unpolluted portions of Los Angeles-Long Beach Harbors, all localities from southern California. This is true also when this species list is compared to that of Sanders (1960) from Buzzard's Bay, Massachusetts. It seems likely that the array of polychaetes in Timbalier Bay is more or less typical of what one may expect to find present in such an area with silts, clays, and sand substrates. No exclusion of species at any one station was noted in Timbalier Bay such as what Reish (1959) found in those portions of Los Angeles-Long Beach Harbors that were heavily polluted with refinery wastes. However, there may be a difference between oil refinery wastes and those resulting from drilling and production of oil. If the ambient stations are representative of an unaffected area, one is led to the conclusion that long-term oil drilling and production in Timbalier Bay have not exerted demonstrated stress on the polychaete assemblages. A more complete answer to this question must wait until the remaining samples taken in this study are analyzed and a parallel study is conducted in a bay of similar characteristics in the Gulf of Mexico but free of oil activities.

CONCLUSIONS

The polychaete fauna in Timbalier Bay consists of about 30 different species per square meter. *Spiochaetopterus* dominates the polychaetes in both number of specimens and biomass at the oil drilling station and one of the ambient ones. Several species, including *Spiochaetopterus*, dominate the benthos at the production platform and at the second ambient station. The polychaete species present within Timbalier Bay seem to be typical of a healthy, unpolluted fine sediment benthic environment. The spatial and temporal variation in species composition, species diversity, faunal similarities, and biomass, which were demonstrated, cannot be correlated with the spatial distribution of oil drilling and production activities, which have occurred in the bay for the past 35 years.

REFERENCES CITED

Barnard, J. L. and D. J. Reish. 1959. Ecology of amphipoda and polychaeta of Newport Bay, California. Allan Hancock Foundation, University of Southern California. Occasional Paper No. 21.

Barrett, B. B. 1971. Cooperative Gulf of Mexico estuarine inventory and study, Louisiana. Phase III. Sedimentology. Louisiana Wildlife and Fisheries Commission, pp. 68-1911.

Behre, E. H. 1950. Annotated list of the fauna of the Grand Isle region. Occasional Papers Marine Laboratory, Louisiana State University, No. 6.

Blegvad, H. 1928. Quantitative investigations of bottom invertebrates in the Lonfjord 1910-1927 with special reference to plaice food. Report Danish Biological Station, Ministry of Fisheries **34**:35-52.

Hartman, O. 1951. The littoral marine annelids of the Gulf of Mexico. Publications of the Institute of Marine Sciences, The University of Texas **2**:7-124.

———. 1954. Polychaetous annelids of the Gulf of Mexico. *In* P. S. Galtsoff, ed., Gulf of Mexico, its Origin, Waters and Marine Life. Fishery Bulletin 89, U.S. Fish and Wildlife Service **55**:413-417.

Kritzler, H. 1974. Oil production and polychaetous annelids in a Louisiana estuary. Gulf Universities Research Consortium. Appendix to Report No. 138.

———, R. M. Hiskey, and P. J. Thomas. 1974. A system for the quantitative sampling of shallow water benthos. Internationale Revue der Gesamten Hydrobiologie **59**:621-627.

Lie, U. 1968. A quantitative study of benthic infauna in Puget Sound, Washington, U.S.A. in 1963-1964. Fiskeridirektorateis Skrifter, Serie Havuntersokelser, Norway **14**:229-556.

Morisita, M. 1959. Measuring of interspecific association and similarity between communities. Memoirs of the Faculty of Science, Kyushu University, Ser. E **3**:65-80.

Mountford, M. D. 1962. An index of similarity and its adaptation to classification problems. *In* P. W. Murphy, ed., Progress in Soil Zoology, pp. 43-50.

O'Connor, J. S. 1972. The benthic macrofauna of Moriches Bay, New York. Biological Bulletin **142**:84-102.

Ono, Y. 1961. An ecological study of the brachyuran community on Tomiolca Bay, Amakusa, Kyushu. Records of Oceanographic Works in Japan. Special Report No. 5, pp. 199-210.

Parker, R. H. 1956. Macro-invertebrate assemblages as indicators of sedimentary environments in east Mississippi delta region. American Association of Petroleum Geologists, Bulletin **40**:295-376.

Reish, D. J. 1955. The relation of polychaetous annelids to harbor pollution. U.S. Public Health Reports **70**:1168-1171.

———. 1957. The relationship of the polychaetous annelid *Capitella capitata* (Fabricius) to waste discharges of biological origin. U.S. Public Health Service Publication No. 208, pp. 195-200.

———. 1958. Description of a new species of *Cossura*. Journal of the Washington Academy of Sciences **48**:53-55.

———. 1959. An ecological study of pollution in Los Angeles-Long Beach Harbors, California. Allan Hancock Foundation, University of Southern California. Occasional Paper No. 22.

——— and H. A. Winter. 1954. The ecology of Alamitos Bay, California, with special reference to pollution. California Fish and Game **40**:105-121.

Rhodes, D. C. and D. K. Young. 1970. The influence of deposit-feeding organisms on sediment stability and community trophic structure. Journal of Marine Research **28**:150-178.

Sanders, H. L. 1958. Benthic studies in Buzzards Bay. I. Animal-sediment relationships. Limnology and Oceanography **3**:245-258.

———. 1960. Benthic studies in Buzzards Bay. III. The structure of the soft-bottom community. Limnology and Oceanography **5**:138-153.

Shannon, C. E. and W. Weaver. 1963. The Mathematical Theory of Communication. Urbana: University of Illinois Press.

Warren, W. S. 1942. A survey of the annelid worms of the Grand Isle region. Bioscience **13**:39-46.

DISTRIBUTION AND ECOLOGY OF THE LITTORAL FORAMINIFERA OF TIMBALIER BAY, LOUISIANA

by John Robert Inabinet and A. Geoffrey Fish

ABSTRACT

The living littoral Foraminifera of Timbalier Bay, Louisiana, were studied from September 1972 to May 1973, to determine species composition, the distribution and abundance with respect to environmental factors, and effects of activities of the oil industry of the bay upon Foraminifera. Four study areas were selected. A Mud Experimental Station (MES-1) was located close to producing wells in the northern bay, and a Mud Control Station (MCS-3) was isolated from oil production. A Sand Experimental Station (SES-2) was located close to oil production activities in the southern bay, and a Sand Control Station (SCS-4) was chosen remote from such activities. Data were collected from the high, mid, and low-tide level of each station and included temperature, salinity, pH, orthophosphate, nitrate, nitrite, sulfate, heavy metals, organic carbon, hydrocarbon, sediment size, and foraminiferal fauna. Species diversity analyses were performed on the foraminiferal data.

Thirty-nine species from 12 families were identified. The largest populations were measured in September with declines in January and May. The Mud Experimental Station supported larger numbers than MCS-3. In contrast, more Foraminifera were collected from SCS-4 than from SES-2. Foraminifera populations were divisible into a sand group, located at SES-2 and SCS-4, a mud group from MES-1 and MCS-3, and a group

---

At the time of this investigation, J. R. Inabinet and Geoffrey Fish were with the Department of Biology, University of Southern Mississippi, Hattiesburg, MS.

common to both environments. Foraminiferal distribution and abundance seemed related to seasonal temperature changes, salinity, sediment size, and the unusual precipitation of May 1973. Species diversity was reduced at MES-1 and SES-2 with respect to their control stations, indicating that Foraminifera may be experiencing some environmental stress; however, the effects were believed minimal.

INTRODUCTION

The objectives of the present study were to determine the species of living Foraminifera inhabiting selected littoral sites in Timbalier Bay, Louisiana; to relate the distribution and abundance of species to environmental factors; and to determine what effects, if any, the offshore oil industry had on the distribution and abundance of the species.

The early foraminiferal research was dominated by morphologic and taxonomic studies. In the early twentieth century, Foraminifera were recognized as economically important in exploration for petroleum, and numerous paleontological studies resulted in which Foraminifera were used for identification of geologic age and ancient environments. In the past 30 years, increasing interest in Foraminifera as living organisms has arisen. Numerous papers have been published concerning the distribution, abundance, and ecology of modern benthic Foraminifera (Boltovskoy 1966; Bradshaw 1968; Buzas 1969; Kornfeld 1931; Lankford 1959; Lehmann 1957; Parker et al. 1953; Phleger 1954; Waldron 1963; Warren 1956). Few of these studies have dealt with the intertidal zone, where environmental parameters are subject to sharp diurnal and seasonal variations (Moore 1958).

Little is known concerning the effects man has upon the distribution and abundance of benthic Foraminifera. Few papers deal with this problem (Bandy et al. 1964; Seiglie 1968; Watkins 1961), and literature concerning the effects of the oil industry on the foraminiferal ecology of Timbalier Bay is nonexistent; therefore, this bay, which has been historically associated with activities of the oil industry, is well suited for such a study. While Waldron carried out a seasonal ecological study of the Foraminifera of Timbalier Bay, he did not address the possible effects of the oil production activities on this group of organisms.

MATERIALS AND METHODS

Description of study area

Timbalier Bay extends from longitude 90°13′ to 90°29′ N and

latitude 29°03′ to 29°19′ W (figure 1). The maximum width is approximately 23 km and the length is 27 km from north to south. The bay averages approximately 1.85 m in depth. There is little natural fresh water influx other than local rainfall. Saline waters enter the bay through Grand Pass Timbalier, which connects the bay with the Gulf of Mexico. There is one tide each 24 hours with an average tidal range of about 0.3

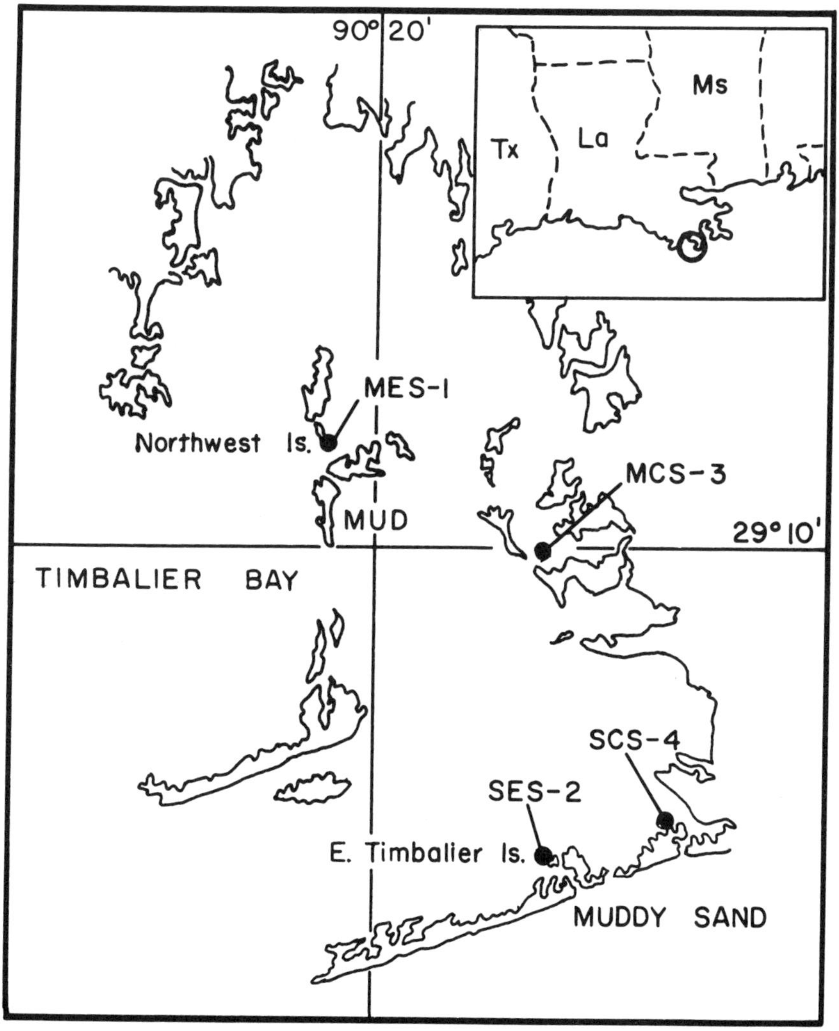

FIG. 1. THE LOCATION OF SAMPLING STATIONS WITHIN TIMBALIER BAY, LOUISIANA.

m. Many of the islands within the bay are remnants of natural levees. Most are low-lying land areas and marshes composed of muddy-sand sediments. The southern boundary of the bay is composed of natural sand barrier islands (Waldron 1963).

Four sites were chosen for study within the bay. Station 1 was established on the southeastern tip of Northwest Island (figure 1) at latitude 29°11.73′ N, and longitude 90°20.93′ W. The study area was a narrow beach of fine silt and mud, sporadically covered with crushed shell material. The site was located within sight of several active oil producing platforms. This station was designated the Mud Experimental Station (MES-1).

Station 2 was located on East Timbalier Island, approximately 68 m northeast of a Gulf Oil Company installation at latitude 29°04.8′ N and longitude 90°17.20′ W. The sampling area was an isolated sand beach bordered by *Spartina* and within sight of a productive oil platform and natural gas wells. This station will be referred to as the Sand Experimental Station (SES-2).

Station 3 was situated on the southwestern tip of a small low-lying island located at latitude 29°09.80′ N and longitude 90°17.18′ W. The sample site was an isolated mud beach, bounded by *Spartina*. This station was located as far from active oil production areas as possible and served as the control site for MES-1. It was designated as the Mud Control Station (MCS-3).

Station 4 was established on a narrow sand island located at latitude 29°05.65′ N and longitude 90°14.59′ W. This station was as isolated as possible from active oil production sites. It has been designated as the control site for SES-2 and was referred to as SCS-4.

Sampling techniques

Twenty-two equal-size faunal samples were collected during the period from September 19, 1972, to May 19, 1973. The dates of the collection trips were as follows: September 19-20, 1972; January 31-February 2, 1973; and May 18-19, 1973. Samples were collected from the high-, mid-, and low-water marks at each study site. Samples were taken using a plastic hand-operated coring tube, with inside diameter of 4 cm and length of 45 cm. Each core was 30 cm long and had a volume of 376.8 ml. The cores were divided into upper and lower halves and preserved in 70% isopropyl alcohol.

The sediment temperature at 5 and 25 cm was recorded from the low-, mid-, and high-water marks at each station and from the adjacent sea water. Water samples for the determination of salinity and pH were collected from the surface and from the 5 and 25 cm depths at the mid-water mark using a 50 ml glass syringe fitted with a long stainless steel needle whose end was plugged. Eight small holes were drilled in the shaft approximately 2 mm from the top of the plug. Salinity determinations were made with a Goldberg Refractometer.

Samples for the determination of orthophosphates, nitrates, nitrites, and sulfates were taken from the mid-water mark by digging a hole for water to flow into and filling the respective containers from the hole. The containers were then placed on ice. Samples for carbon analysis were taken from the sediment at the 5 cm and 25 cm depths from the low-, mid-, and high-water marks. Hydrocarbon samples were taken from the sediment at 5 cm and 25 cm depths at the mid-water mark. These samples were obtained by digging a hole approximately 45 cm in depth at the appropriate tide level and taking a small horizontal core at the desired depth. Core samples for sediment size analysis were collected during the January trip from the low-, mid-, and high-water marks at each station.

In the laboratory, an aliquot of 100 ml was taken from each faunal sample and washed through a U.S. standard sieve number 200. This aliquot was treated with rose Bengal stain for recognition of living Foraminifera (Walton 1952). After staining, samples were washed through U.S. screen sieves numbers 80, 100, and 200 to divide the sample into subsamples to facilitate identification. The finer fraction contained the smaller specimens, probably juveniles, which were unidentified. Therefore, only the organisms from the coarser fraction were identified (Parker et al. 1953).

Samples from mud and sand stations were treated differently. Little sediment was left on the sieves after washing mud samples, but large amounts of organic debris remained. This debris was removed by slowly decanting from a beaker onto a standard screen sieve No. 200. The decanted portion was examined under a binocular dissecting scope for loss of Foraminifera (Warren 1956). Sand samples contained less organic debris, but large amounts of sediment remained in the sieves. The washed residue was placed in a Boisseau extraction apparatus (Hulings and Gray 1971) and filtered at a constant flow rate for 20 minutes to remove most of the Foraminifera. Organic debris was removed by the decanting.

The washed residue of each sample was placed in a Petri dish whose bottom had been divided into 63 1-cm squares. Samples were examined under a binocular dissecting scope and all living Foraminifera were counted and identified. The proportion of dead Foraminifera was also estimated for each sample.

Sediment size analysis was performed using a modification of the method outlined by Hulings and Gray. Analysis of orthophosphate, sulfate, nitrate, nitrite, and carbon samples was conducted by Dr. C. R.

Brent of the University of Southern Mississippi, and the analysis of hydrocarbon samples was performed by Dr. J. L. Laseter of the University of New Orleans.

Species diversity indices were calculated from the data collected at each sampling site by using the formula proposed by Margalef (1956):

$$\bar{d} = \Sigma \left({}^{ni}/_{n}\right) \log_2 \left({}^{ni}/_{n}\right)$$

Where d = diversity index

ni = the number of individuals in the Ith species

n = the total number of individuals in a community

This diversity index was chosen because it takes into account the relative abundance of the different species, and it is relatively independent of sample size (Warren 1971).

RESULTS

Physical and chemical data

The temperature range of the interstitial water, 12° C to 35° C, differed slightly from that of the adjacent sea water, 12° C to 36° C (table 1). The maximum temperature of interstitial waters occurred from the high-water mark of SES-2 on September 20, 1972. The sea water maximum was also recorded on this date. The minimum temperature of interstitial water was from low-water, MES-1, on January 31, 1973. The minimum sea water temperature was also recorded on this date. The temperature range was greatest at the high-water mark and least at the low-water mark for all locations. Temperature data for the mid-water beach level are included in table 1.

The salinity of the interstitial water generally exceeded that of the adjacent sea water (table 1). Maximum salinities were recorded from the high-water mark of stations; minimal salinities were usually at the low-water mark. The salinity of the two southern stations always exceeded that of the two northern stations.

Orthophosphate levels were highest in January and lowest in September. Sulfate levels were generally highest in September and lowest in May. In general, the two southern stations contained higher levels of sulfates than the two northern stations (table 1).

Because of instrument failure, pH data were recorded only for September. Values of pH were lower for the interstitial water than for the adjacent sea water (table 1).

Nitrate and nitrite levels were so low as to be barely detectable. Because of their low values, only the September and January samples were anlayzed (table 1).

Organic and inorganic carbon levels demonstrated a general trend to

Table 1

Temperature, salinity, pH, phosphate, sulfate, nitrate, and nitrite at 5 and 25 cms from the mid-water mark of each station, 1972-1973.

Station	Date	Depth cm	Beach Level	Temp °C	Sal. °/oo	pH	PO_4 mg/l	SO_3 mg/l	NO_3 mg/l	NO_2 mg/l
MES-1	9-19-72	5	high	31.0	24.98	7.6	0.025	1580	0.0748	0.0732
		5	mid	30.0	24.98	8.3				
		25	mid	29.5						
			sea-water	29.0	25.52	8.3				
SES-2	9-20-72	5	high	35.0	41.63	7.6	0.000	3425	0.0088	0.0264
		5	mid	32.0						
		25	mid	31.5						
			sea-water	31.5	26.50					
MCS-3	9-19-72	5	high	34.5			0.200	1480	0.0264	0.0033
		5	mid	32.0	30.53					
		25	mid	29.5	27.76	7.3				
			sea-water	32.2	26.50					
SCS-4	9-20-72	5	high	31.0			2.075	1350	0.000	0.000
		5	mid	34.0	33.31	7.3				
		25	mid	32.0	33.31	7.4				
			sea-water	33.5	26.50	8.5				
MES-1	1-31-73	5	mid	12.0	21.11					
		25	mid	12.5	27.76		8.64	800	0.044	0.001
			sea-water	11.8	21.67					
SES-2	2-2-73	5	mid	14.0	27.22					
		25	mid	13.8	28.32		4.16	1300	0.020	0.003
			sea-water	14.0	25.54					
MCS-3	1/31/73	5	mid	13.8						
		25	mid	13.5			6.48	850	0.030	0.0001
			sea-water	12.5	19.43					
SCS-4	2/1/73	5	mid	20.0	39.98					
		25	mid	16.8	31.65		3.94	956	0.017	0.002
			sea-water	20.5	27.76					
MES-1	5-18-73	5	high	22.6	17.22		1.40	625		
		5	mid	22.9						
		25	mid	22.5						
			sea-water	24.0	15.00					
SES-2	5-19-73	5	high	22.5	12.71		1.68	1400		
		5	mid	23.4						
		25	mid	23.2						
			sea-water	24.7	17.22					
MCS-3	5-18-73	5	high	22.5	17.22		0.14	550		
		5	mid	22.3						
		25	mid	22.5						
			sea-water	24.6	15.00					
SCS-4	5-19-73	5	high	22.8			0.33	11800		
		5	mid	22.4						
		25	mid	22.0						
			sea-water	23.3	19.99					

decrease from September to January, followed by a slight increase in May. Both levels were found to be higher at the two northern stations than at the two southern stations (table 2).

Hydrocarbon data were available only for September (table 3). More complete data are reported elsewhere (Oetking et al. 1974).

The sediment analysis indicated that the two northern stations have an average composition of 92.1% mud, while the average composition of the two southern stations was 78.8% sand (table 4). The mean grain size of the two southern sand stations was 0.114 mm, which is listed as very fine sand in the Wentworth (1922) size scale.

Biological data

A total of 3,458 living specimens of Foraminifera was collected. Largest populations were found in September and smallest populations in May (figure 2).

Station MES-1, located near active oil production sites, supported larger numbers of living Foraminifera than the control beach, MCS-3, except in September. More living Foraminifera were taken from the control sand beach, SCS-4, than from the sand beach located near active oil production sites, SES-2.

Thirty-nine species from 12 families and 20 genera inhabited the four beaches. Of these, 37 species from 12 families and 18 genera were living at the time of collection. The remaining two species, representing two families and two genera, were found only as dead specimens (tables 5-7).

Living Foraminifera were recorded from both the upper and lower 15 cm of each faunal core, for both sand and mud beaches. In most cases, larger numbers were observed in the upper 15 cm. In mud beaches, much greater numbers of organisms were collected from the upper 15 cm than from the lower 15 cm, whereas the number of organisms living at both depth ranges was much more equal for sand beaches.

Diversity index data

The calculated diversity indices for the four study sites were similar and remained relatively constant throughout the study (table 8). Species diversity indices ranged from 2.19 to 2.43 (average 2.28) at stations located near active oil production sites and 2.51 to 2.70 (average 2.62) at control stations.

DISCUSSION

Thirty-nine species of Foraminifera were identified from the littoral

(text continues p. 505)

TABLE 2

Monthly sediment carbon analysis at 5 and 25 cms from the mid-water mark of each station.

Station	Date	Depth cm	Total ppt	Organic ppt	Inorganic ppt
MES-1	9-19-72	5	25.40	24.16	1.24
		25	19.78	19.34	0.44
SES-2	9-20-72	5	1.57	1.47	0.10
		25	33.84	30.16	3.68
MCS-3	9-19-72	5	33.20	31.60	1.60
		25	31.80	30.56	1.24
SCS-4	9-20-72	5	1.70	1.51	0.19
		25	1.97	1.65	0.32
MES-1	1-31-73	5	18.94	18.14	0.80
		25	16.92	16.52	0.40
SES-2	2-2-73	5	0.98	0.87	0.16
		25	1.98	1.81	0.17
MCS-3	1-31-73	5	31.28	30.88	0.40
		25	38.70	37.90	0.80
SCS-4	2-1-73	5	4.77	4.20	0.57
		25	4.43	4.05	0.38
MES-1	5-18-73	5	16.44	16.04	0.40
		25	23.96	23.96	0.80
SES-2	5-19-73	5	2.51	2.35	0.16
		25	1.42	1.32	0.10
MCS-3	5-18-73	5	31.60	31.00	0.60
		25	41.80	41.22	0.58
SCS-4	5-19-73	5	2.13	1.97	0.16
		25	2.52	2.42	0.10

TABLE 3

Hydrocarbon analysis of 0-15 and 15-30 cm depths from the four sampling stations for September.

Station Number	Depth cm	Total Lipids mg	Hexane Eluate mg	Benzene Eluate mg	MeOH Eluate mg
MES-1	0-15	9.0	0.1	3.1	1.1
	15-30	1.2	0.0	0.8	1.1
SES-2	0-15	1.9	0.5	0.8	0.6
	15-30	-	-	-	-
MCS-3	0-15	3.9	0.5	0.9	2.8
	15-30	0.9	0.0	0.7	0.8
SCS-4	0-15	2.0	0.1	0.7	1.5
	15-30	-	-	-	-
Outboard Motor Fuel	-	0.6	0.0	2.1	1.7

TABLE 4

The percent mud and sand and the mean grain size from 0-15 and 15-30 cm depths at the mid-water mark of the sampling stations.

Station Number	Depth cm	% Mud	% Sand	Mean Grain Size
MES-1	0-15	88.88	11.12	
	15-30	83.78	16.22	
SES-2	0-15	13.82	86.18	0.106
	15-30	10.70	89.30	0.102
MCS-3	0-15	98.35	1.65	
	15-30	97.48	2.52	
SCS-4	0-15	14.31	85.69	0.148
	15-30	45.97	54.03	0.101

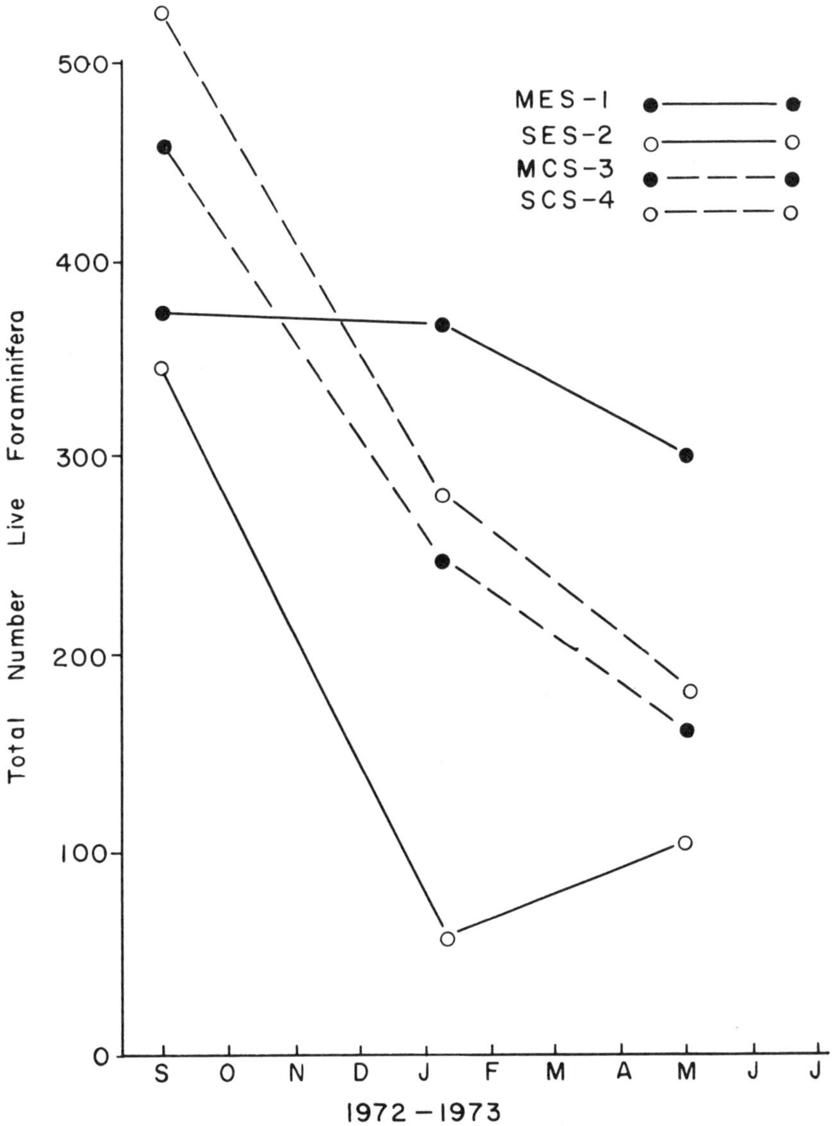

FIG. 2. NUMBER OF LIVE FORAMINIFERA RECORDED from Stations One, Two, Three, and Four for September 1972 and January and May 1973.

TABLE 5

DISTRIBUTION OF FORAMINIFERA IN LITTORAL SEDIMENT SAMPLES
FROM TIMBALIER BAY, SEPTEMBER 1972

Station	MES-1						MCS-3						SES-2						SCS-4					
Sample	1	2	3	4	5	6	7	8	9	10	11	12	13	14	15	16	17	18	19	20	21	22	23	24
Total Population 0.149 mm	52	198	56	36	19	53	63	32	131	153	76	46	82	17	48	45	68	129	41	105	90	54	136	122
Ammoastuta inepta							2	11	3	3	25	5												
Ammotium dilatatum		.6											1		2	3		.9	2	1	3	2		.8
Ammotium salsum	2	1	4				23	9	5	1	6		1		2			2		2	9	2	2	
Ammobaculites pseudospiralis																3					1			
Arenoparella mexicana	2	22	8	39	26	39		14	4	21	18	33							2				2	
Bolivina lowmani		.6																						
Cornuspira sp.	4						5		3															
Elphidium delicatulum		.6																						
E. discoidale									.9									2			1			.8
E. gunteri	17	20	35	3			3	6	14	27	7	5	31	40	26	35	21	32	34	35	24	27	18	22
E. mexicanum	2									1		2									1			
E. matagordanum	2								.9	1			3		9	5	13	6	5	4	7	7	22	18
E. poeyanum		2					3		3	.7			5				5	4	7	3	5	6	26	24
Epistominella vitrea									2									.9						
Hanzawaia strattoni																								.3
Haplophragmoides wilberti	4					4	2			.7		5						.9						
H. manilaensis	2				5																			
Miliammina fusca		.6						3	2	.7											3			3
Nonionella atlantica																		4		10	1		2	.8
N. opima																		3		1				.8
Quinqueloculina compta		1							.9		3		3			3	5	3		4	3	4	4	5
Q. funafutiensis																						2		3
Q. lamarckiana															6			2	2				2	
Q. rhodiensis															2			2			1	2		
Q. wiesneri																	2	.9			1			.8
Ammonia beccarii var. A	11	25	29		5		17	3	19	14			28	20	15	25	20	30	20	19	22	20	9	10
Ammonia beccarii var. B	17	4	4		11		8	3	27	9	2	10	28	40	34	28	30	9	27	21	17	29	15	12
Trochammina comprimata	4	2	6	6	5	45	10	6	4	.7	2	12			2		2						2	
T. inflata	30	20	14	53	47	12	27	46	15	20	37	29								1				
T. macrescens		1								.7	2													

TABLE 6

DISTRIBUTION OF FORAMINIFERA IN LITTORAL SEDIMENT SAMPLES
FROM TIMBALIER BAY, JANUARY 1973

Station	MES-1						MCS-3						SES-2						SCS-4					
Sample	1	2	3	4	5	6	7	8	9	10	11	12	13	14	15	16	17	18	19	20	21	22	23	24
Total Population 0.149 mm	147	20	95	25	80	6	110	27	28	13	61	15	12	7	11	5	26	2	15	14	47	34	162	11
Ammoastuta inepta				4			14																	
Ammotium dilatatum	8		1				3				3													
Ammotium salsum	2		2		3				4		8						4							
Arenoparella mexicana	5	35	3	13			26	4	27		5	13								7				
Bucella hannai	5						.9																	
Cornuspira sp.							.9																	
Elphidium advenum																					2			
E. discoidale					3																			
E. gunteri	42	20	39		23	33	12	33	8	54	31	20	50	14	36	20	39		33	7	19	27	18	9
E. mexicanum			1		1		2											50			2		4	
E. matagordanum					3		4			15	7	7	33		27				27	7	26	18	26	18
E. poeyanum			1			17					16		17				8		13	29	15	15	29	46
Haplophragmoides wilberti					1		2		4															
Miliammina fusca		15									3	13											.6	
Nonionella atlantica																				21	2			27
N. opima																						9		
Proteonina lagenaria			1																					
Quinqueloculina compta	.7																4			14		18	1	
Q. funafutiensis																						9		
Q. lamarckiana									4												6		8	
Q. poeyana														14										
Q. rhodiensis																	8			14	2		1	
Q. seminulum																					2		1	
Q. wiesneri																							4	
Recurvoides sp.			1																					
Ammonia beccarii var. A	30	10	41	4	43		14	19	12	23	16	13		29			15		20		9		.6	
Ammonia beccarii var. B			5		7		6	4	4		7	7		43	36	80	19	50	7		13	6	8	
Trochammina comprimata	1		1	4	4												4							
T. inflata	.7	20	4	74	13	50	16	41	39	8	3	27									2			
T. laevigata	5																							
T. macrescens							.9																	

TABLE 7

DISTRIBUTION OF FORAMINIFERA IN LITTORAL SEDIMENT SAMPLES FROM TIMBALIER BAY, MAY 1973

Station	MES-1						MCS-3						SES-2						SCS-4					
Sample	1	2	3	4	5	6	7	8	9	10	11	12	13	14	15	16	17	18	19	20	21	22	23	24
Total Population 0.149 mm	60	15	72	21	127	7	75	26	60	8	18	6	18	11	16	10	30	12	11	9	32	29	84	21
Ammoastuta inepta	15						8		7															
Ammotium dilatatum			1		.8		7				11										3			
Ammotium salsum	8				6		11		12		17					10						3	2	
Arenoparella mexicana	23	20	7	24	6	14	11	27	33	25														
Elphidium delicatulum																							1	
E. discoidale	3																							
E. gunteri	17	13	43	10	54	29	1	19	7		28	17	33	27	24	30	27	17	36	22	28	21	14	14
E. mexicanum								4																
E. matagordanum	2		3		.8		1	15			6	17		18	14		7		18	33	13	17	23	5
E. poeyanum			6		6		4						11				10	8	9	22	19	31	28	38
Haplophragmoides wilberti		7					3																	
H. manilaensis							1																	
Miliammina fusca					.8						6						3							
Nonionella atlantica													6					8				7		
N. opima																20					9		8	5
Quinqueloculina compta															5		10		9			7		14
Q. funafutiensis																							4	
Q. lamarckiana																							4	
Q. rhodiensis														9						11				
Q. wiesneri															5						6			
Ammonia beccarii var. A	12	7	24		13	14	16		3	13	28	33	17	9	19	20	40	33	27		9	3	11	10
Ammonia beccarii var. B			1		4		3		3		6		33	36	33	20	3	33	9	11	13	11	5	14
Trochammina comprimata	2	7	3	19	4	14	5		2														1	
T. inflata	18	33	11	29	6	29	29	35	32	38		33												
T. laevigata		13		19																				
T. macrescens			1						2															

TABLE 8

Monthly analyses of the number of species, number of individuals, and calculated species diversity index of each sampling site.

Station Number	Date	Number of Species	Number of Individuals	Species Diversity Index
MES-1	9-19-72	19	382	2.43
SES-2	9-20-72	18	347	2.22
MCS-3	9-19-72	18	465	2.67
SCS-4	9-20-72	21	529	2.54
MES-1	1-31-73	21	371	2.19
SES-2	2-1-73	11	63	2.21
MCS-3	1-31-73	17	246	2.66
SCS-4	2-2-73	18	283	2.51
MES-1	5-18-73	15	300	2.35
SES-2	5-19-73	12	102	2.28
MCS-3	5-18-73	16	173	2.70
SCS-4	5-19-73	16	187	2.52

sediment samples from Timbalier Bay (tables 5-7). Waldron listed 17 of these species in his study of this bay.

Heaviest populations were recorded in September, with decreasing numbers in January and May (figure 2). This seasonal cycle differs from the one Waldron found, where minimum populations were counted in January and increasing numbers in May and September. The large number of living Foraminifera found in September may be attributed to the increased food supply during the summer months, with the decreased number in January correlated with a drop in the food supply. The minimal populations of May may be accounted for by such factors as the spring reproductive period and its subsequent reduction in adult

Foraminifera. The excessive precipitation of the spring of 1973, causing record Mississippi River runoff, which decreased the salinity in the bay, was perhaps the most important reason for the population decrease in May.

The distribution of Foraminifera also appeared correlated with the seasonal cycle of temperature. Highest temperatures were recorded in September and lowest temperatures in January. The density of Foraminifera exhibited a similar seasonal cycle. Temperature changes affect food supplies and the rate of reproduction. Foraminifera reproduce predominantly during the warmer spring and summer months (Myers 1943).

The seasonal distribution of Foraminifera and that of salinity exhibited similar cycles. Salinity also appears related to the geographic variation throughout Timbalier Bay. Species predominantly found at the northern mud stations (MES-1 and MCS-3) were collected from a salinity range of 16 ppt to 30 ppt; those collected primarily from the southern sand stations (SES-2 and SCS-4) were associated with a salinity range of 28 ppt to 39 ppt.

It is imposible with the data available to determine whether or not dissolved nutrient substances such as orthophosphate, nitrate-nitrite, and sulfate levels had any direct effect upon the populations of Foraminifera. These nutrients, however, influence the production of photosynthetic organisms, which constitute the principal food for the benthic Foraminifera, and periods of growth and reproduction in Foraminifera generally coincide with periods when nutrients are most abundant. It is, therefore, believed that by directly limiting the food supply of the Foraminifera, nutrients indirectly determine the density of foraminiferal populations.

The inorganic and organic carbon levels decreased from September to January and increased slightly in May (see table 2). The seasonal cycle for the Foraminifera was similar except for the continued decrease in numbers in May (figure 2). It is felt that inorganic or organic carbon levels had little direct effect on the seasonal distribution of Foraminifera, because these organisms are holozoic and low pH levels of mud habitats cause breakdown of calcareous test (Eltringham 1971).

The vertical distribution of Foraminifera seems attributable to scarcity of food and the presence of toxic substances in the deeper sediment (Myers 1943). Foraminifera have symbiotic algae to supply oxygen (Boltovskoy 1966). More Foraminifera were collected at the 15-30 cm depth range in sand beaches than in mud beaches. Foraminifera penetrate sandy sediments more readily because spaces between particles allow better aeration, greater flushing action, greater penetration of food supplies, and easier movement.

The Foraminifera throughout Timbalier Bay form two distinct biofacies: those primarily found in the northern mud stations (MES-1 and MCS-3) and those predominantly found at the southern sand stations (SES-2 and SCS-4). A third group of organisms is common to both mud and sand beaches.

The mud biofacies comprises the following species: *Ammoastuta salsa*, *Ammotium salsum*, *A. dilatatum*, *Arenoparella mexicana*, *Haplophragmoides manilaensis*, *H. wilberti*, *Trochammina comprimata*, *T. inflata*, and *T. macrescens*.

Ammotium salsum and *Ammotium dilatatum* were also found in high numbers at the two sand stations (SES-2 and SCS-4) in September; however, in January and May they were recorded primarily from the mud stations. All of those species that characterize the mud biofacies have arenaceous, or agglutinated, tests.

The sand biofaces comprises *Elphidium matagordanum*, *E. poeyanum*, *Nonionella atlantica*, *N. opima*, *Quinqueloculina compta*, *Q. lamarckiana*, *Q. rhodiensis*, and *Q. wiesneri*.

All of these species have calcareous tests. Foraminifera having calcareous tests are rarely found in mud habitats (Lankford 1959). The following calcareous species occurred at all stations, however, and were generally major species in every population: *Elphidium gunteri*, *Streblus beccarii* var. *parkinsoniana*, and *S. beccarii* var. *tepida*. Phleger (1954) found *Elphidium* spp. and *Streblus beccarii* vars. confined to high-salinity open-Gulf water in the Mississippi Sound area; whereas Parker et al. (1953) found these species not only in the open Gulf but also in San Antonio Bay, Texas, whose waters originate in the open Gulf. The distribution of these organisms in the relatively saline waters of Timbalier Bay verifies the conclusions of Phleger and Parker.

Foraminiferal distribution along the beach face varied between mud beaches and sand beaches. Higher numbers of living Foraminifera were collected from the high-water mark of the mud beaches, whereas the most dense populations from sand beaches were always recorded from the low-water mark. The larger size of the sand grains in sand beaches results in rapid drying of the upper reaches of these beaches during low tide and subsequent sharper variations of ecological parameters, whereas the capillary attraction of closely packed particles in mud beaches prevents rapid drying out (Eltringham 1971).

The two stations located near active oil production sites consistently yielded lower calculated species diversity indices than control sites (see table 8). Warren (1956) stated that a marked environmental change, without the passage of sufficient time for the evolution of a new community, tends to lead to a reduction of species diversity at the affected location.

It appears that Foraminifera in closer proximity to oil industry activities are experiencing some environmental stress; because the indices are relatively high at both experimental and control stations, however, it is believed the stress is only minimal and the presence of the oil industry in Timbalier Bay has shown little long-term effect on the distribution and abundance of littoral Foraminifera inhabiting the area.

ACKNOWLEDGMENTS

This research was part of grant OB 64 HJM, The Study of Interstitial Organisms, of the Gulf Universities Research Consortium's Offshore Ecology Investigation, and resulted in a Master of Science Degree at the University of Southern Mississippi for Geoffrey Fish. Special thanks are accorded to all personnel who assisted in this research.

REFERENCES CITED

Bandy, O. L., J. C. Ingle, and J. M. Resig. 1964. Foraminifera, Los Angeles County outfall. Limnology and Oceanography **9**:124-137.

Boltovskoy, E. 1966. Depth at which Foraminifera can survive in sediments. Contributions, Cushman Foundation for Foraminiferal Research **17**:43-45.

Bradshaw, J. S. 1968. Environmental parameters and marsh Foraminifera. Limnology and Oceanography **13**:26-38.

Buzas, M. A. 1969. Foraminiferal species densities and environmental variables in an estuary. Limnology and Oceanography **14**:411-422.

Eltringham, S. K. 1971. Life in Mud and Sand. New York: Crane, Russak, and Company.

Hulings, N. C. and U. S. Gray. 1971. A Manual for the Study of Meiofauna. Smithsonian Contributions to Zoology No. 78.

Kornfeld, M. M. 1931. Recent littoral Foraminifera from Texas and Louisiana. Contributions, Department of Geology, Stanford University **1**:77-101.

Lankford, R. R. 1959. Distribution and ecology of Foraminifera from east Mississippi delta margin. American Association of Petroleum Geologists Bulletin **43**:2068-2099.

Lehmann, E. P. 1957. Statistical study of Texas coast recent foraminiferal facies. Micropaleontology **3**:325-356.

Margalef, R. 1956. Informacion y diversidad especifica en las comunidades de organismos. Investigacion Pesquera **3**:99-106.

Moore, H. B. 1958. Marine Ecology. New York: John Wiley.

Myers, E. H. 1943. Life activities of Foraminifera in relation to marine geology. Proceedings of the American Philosophical Society **86**:439-458.

Oetking, P., R. Back, R. Watson, and C. Merks. 1974. Surface and shallow subsurface sediments on the nearshore continental shelf of South Central Louisiana. Gulf Universities Research Consortium. Report No. 138.

Parker, F. L., F. B. Phleger, and J. F. Peirson. 1953. Ecology of Foraminifera from San Antonio Bay and environs, southwest Texas. Cushman Foundation for Foraminiferal Research, Special Publication No. 2.

Phleger, F. B. 1954. Ecology of Foraminifera and associated microorganisms from Mississippi Sound and environs. American Association of Petroleum Geologists Bulletin **38**:584-647.

Seiglie, G. A. 1968. Foraminiferal assemblages as indicators of high organic carbon content in sediments and of polluted waters. American Association of Petroleum Geologists Bulletin **52**:2231-2241.

Waldron, R. P. 1963. A seasonal ecological study of Foraminifera from Timbalier Bay, Louisiana. Gulf Research Report **1**:131-188.

Walton, W. R. 1952. Techniques for the recognition of living Foraminifera. Contributions, Cushman Foundation for Foraminiferal Research **3**:56-60.

Warren, A. E. 1956. Ecology of Foraminifera of the Buras-Scofield Bayou region, southeast Louisiana. Gulf Coast Association Geological Society Transactions **6**:131-152.

Warren, C. E. 1971. Biology and Water Pollution Control. Philadelphia: W. B. Saunders Co.

Watkins, J. G. 1961. Foraminiferal ecology around the Orange County, California, ocean sewer outfall. Micropaleontology **7**:199-206.

Wentworth, C. K. 1922. A scale of grade and class terms for clastic sediments. Journal of Geology **30**(5):377-392.

THE ECOLOGY OF THE LITTORAL MARINE POLYCHAETES OF TIMBALIER BAY

by Philip L. Lewis and A. Geoffrey Fish

ABSTRACT

The marine polychaetes were studied for community structure, diversity, biomass, and abundance, and to determine the relationship of chemical and physical factors with their distribution in beaches located near petroleum production and control areas in Timbalier Bay, Louisiana.

Two northern mud beaches and two southern sand beaches were studied during July and October 1973 and January 1974. At each locality one beach was close to an oil production site; the control beach was remote from such activities. At the high-, mid-, and low-tide region of each beach, biological, chemical, and physical samples were collected. Total numbers and wet-weight biomass were determined and the biological data were analyzed by the Shannon and Weaver Diversity Index, Morisita-Ono Index of Faunal Affinity, and Analysis of Variance.

Twenty-one species were identified. Average numbers increased from July to January. The sand control beach was the richest area, followed by the mud production beach. The sand production beach was the poorest. Wet-weight biomasses were greatest in mud beaches and least in sand beaches. Faunal resemblances were greatest between the two mud beaches, the two sand beaches, and between the mud and sand production beaches. Sand beaches had higher diversities than mud beaches in July; the reverse was true in October. In January, production

---

Philip Lewis is with the Mississippi Marine Research Council, University of Southern Mississippi, Long Beach, MS; Geoffrey Fish is in the Department of Biology, University of Southern Mississippi, Hattiesburg, MS.

beaches had lower diversities than control beaches. The analysis of variance demonstrated that there were three distinct species-substrate associations. Some species were found only in mud beaches, a second group in sand beaches, and a third group in both mud and sand beaches. The mud production beach with an absence of petroleum had a larger population than its control; therefore, other factors seem to influence polychaete abundance in this beach. The sand production beach with visible hydrocarbons had lower numbers than its control. Other parameters were similar. Hydrocarbons may be associated with distribution in this beach, or other factors may have been responsible.

INTRODUCTION

In 1972, the Gulf Universities Research Consortium, in conjunction with affiliates from the petroleum industry, established the Offshore Ecology Investigation (OEI). The purpose was to determine whether petroleum production and related activities had any effects on marine organisms over an extended period of time.

The objectives of our research were to (1) evaluate the polychaete community structure, (2) compare the diversity, biomass, and abundance of polychaetes, and (3) determine the chemical and physical factors most important to the distribution of polychaetes in beaches located near petroleum-producing activities and control areas in Timbalier Bay, Louisiana.

Polychaetes are one of the most abundant groups of metazoans in marine sediments (Day 1967). Wass (1967) cited several studies of polychaetes known to tolerate oil pollution. Gilet (1960) reported that the polychaete *Capitella capitata* is one of the few animals to occur in large numbers on the bottom of the heavily oil-polluted port of Marseilles, and Reish (1960) cited *C. capitata* as an indicator of pollution in Japan and California. Nelson-Smith (1973) cited other studies of polychaetes in polluted areas. Orton (1925) observed large numbers of polychaetes burrowing in a lump of weathered oil, and George (1971) reported that a fresh fuel-oil spill at Southampton had no detectable effects on the mortality, growth, or reproduction of *Cirriformia tentaculata* or *Cirratulus cirratus*.

The Louisiana coastal area is well known for sustained long-term petroleum production and its associated activities. Special efforts have been made to study the role of marine organisms in this area in relation to petroleum-producing activities (Galtsoff et al. 1935; Mackin and Hopkins 1961; St. Amant 1971, 1972). These studies were directed toward selected communities, primarily oyster biocoenoses. Little attention has been given to the marine polychaetes of the coastal waters of Louisiana. Warren (1942) surveyed the marine annelids of Grand Isle and Hartman (1951) later described several littoral marine annelids collected by Dr. E. H. Behre and Dr. J. H. Roberts from the same locality.

MATERIALS AND METHODS

Description of the study area and station sites

Timbalier Bay is a segment of the system of large, shallow bays that are typical of the Louisiana coast. The bay area extends from latitudes 29°03′ to 29°19′ N and longitudes 90°13′ to 90°29′ W and is bounded on the east by the Lafourche Bayou region and on the west by Terrebonne Bay (figure 1). The maximum width is approximately 23.2 km and the length is 27.2 km. The total surface area is approximately 362.6 square km. The bay averages 1.8 m in depth. Access to the bay is through Grand Pass Timbalier, which connects the bay to the Gulf of Mexico. There is one tidal cycle each twenty-four hours with an average tidal range of about 0.3 m (Waldron 1963). Many of the islands within the bay are remnants of natural levees and consist of low-lying areas and marshes of muddy-sand sediments. The southern boundary of the bay is composed of natural sand barrier islands (Waldron 1963).

Four stations were established within the bay: two from the mud bottomed north-central region and two from the sand bottomed southern region. At both of these localities one station was located in major petroleum producing areas and designated the experimental station and the other was located at some distance from oil industry related activities and designated the control station. Descriptions of the stations are as follows: Station 1 (MES-1, figure 1) was located on the southeastern tip of Northwest Island at latitude 29°11.73′ N and longitude 90°20.93′ W. The island is located in the north central portion of Timbalier Bay. The study area was a narrow beach characterized by sediments containing a high silt-clay content and a high organic carbon content, which was sporadically covered with crushed shell material. This site was located within an oil producing area.

Station 2 (SES-2) was located on the northern side of East Timbalier Island, approximately 100 m northwest of a Gulf Oil Company installation at latitude 29°04.8′ N and longitude 90°17.20′ W. Sediments at this station had a low silt-clay content and a low organic carbon content.

Station 3 (MCS-3) was situated on the southwestern tip of a small, low-lying island located in the northeastern central portion of Timbalier

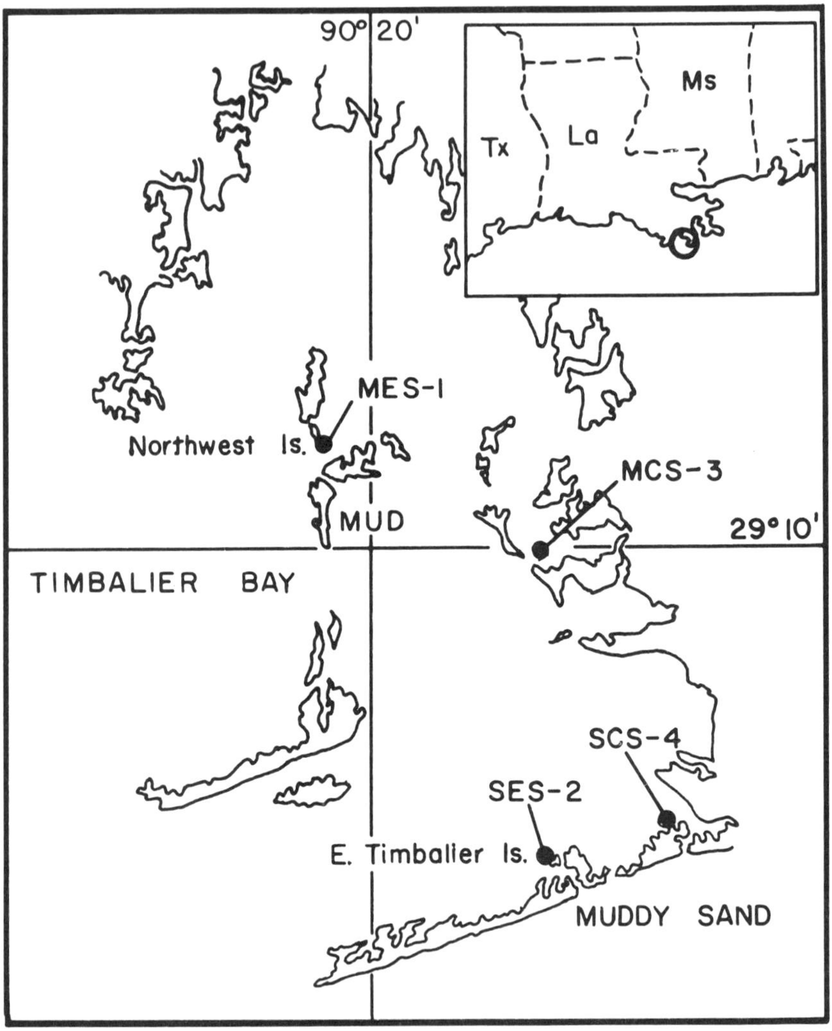

FIG. 1. LOCATION OF SAMPLING STATIONS WITHIN TIMBALIER BAY, LA.

Bay at latitude 29°09.80′ N and longitude 90°17.18′ W. The sample site was an isolated beach bound on both sides by *Spartina* and having sediments with a high silt-clay content and a high organic carbon content. This site was situated a suitable distance from oil producing areas and separated from these areas by small islands.

Station 4 (SCS-4) was established on a narrow sand island in the southeastern corner of Timbalier Bay at latitude 29°05.65′ N and longitude 90°14.59′ W. The beach was characterized by sediments with a low silt-clay and organic carbon content. This station was also located as far as possible from oil production activities and was designated the control station for Station 2.

Sampling techniques

A transect of the littoral region was established at the four selected sites in order to assess the variability of the polychaete community. A stratified sampling design was used (Hulings and Gray 1971). The beaches were divided according to tidal level into three sampling areas along the transect. Three replicate samples were obtained from each tidal zone with a hand operated corer, which was inserted to 10 cm in the substrate and circumscribed an area of 14 cm^2. The core was removed, stored in a glass jar, and preserved in 10% formalin. The number of replicates was increased to five following analyses of the July samples. Sediment samples for total organic and inorganic carbon analysis, hydrocarbon analysis, salinity, and pH were taken at each tide region at each station. Methods for the analysis of these factors followed the procedures in Standard Methods (1971) and Hulings and Gray. Hydrocarbons were determined by Dr. J. L. Laseter, University of New Orleans.

In the laboratory the faunal samples were washed through a set of two sieves (McIntyre 1964). The top sieve had a mesh opening of 2.0 mm and the bottom sieve had a mesh opening of 0.3 mm. Material in the top sieve was examined for polychaetes; material in the bottom sieve was washed into a vial and stained with a mixture of rose Bengal and 70% isopropyl alcohol. The polychaetes were removed from the stain after 24 hours, sorted, identified, and stored.

The Morisita-Ono Index of Faunal Affinity (Morisita 1959; Ono 1961) was used to determine similarity between stations. The diversity index as defined by Shannon and Weaver (1963) was used to measure the faunal similarity between the production and control station sites.

RESULTS

Environmental

The data collected from the physical and chemical measurements made during the survey are summarized in tables 1-4.

A comparison of sediment temperatures (table 1) indicated only slight variations between tide levels for all stations. Temperatures at stations SES-2 and SCS-4, in the southern bay region, were found to be higher than those at stations MES-1 and MCS-3 in the northern bay region. The variation may be attributed to differences in the slope and exposure of the beaches.

Table 1

Temperature (^{o}C) observed at the high, mid, and low-tide regions in Timbalier Bay, from July 1973 to January 1974.

Station Number	Tidal Zone	Temperature ^{o}C		
		July 1972	October 1972	January 1973
MES-1				
Sediment	H	27	22	17
	M	27	22	17
	L	27	22	17
Water		27	22	17
SES-2				
Sediment	H	31	24	16
	M	31	23	16
	L	31	23	16
Water		31	24	16
MCS-3				
Sediment	H	27	23	16
	M	27	23	16
	L	27	23	17
Water		28	22	16
SCS-4				
Sediment	H	30	24	16
	M	29	24	16
	L	29	24	16
Water		30	25	16

Interstitial water was found to have a greater salinity range, 17 ppt to 34 ppt, than the adjacent seawater with a range 15 ppt to 27 ppt (table 2). The salinities at stations SES-2 and SCS-4 were, in general, higher and had a greater range, 21 ppt to 34 ppt, than those at stations MES-1 and MCS-3 with a range of 17 ppt to 27 ppt. The higher values at stations SES-2 and SCS-4, in the southern bay region, were probably due to the proximity of those stations to high salinity waters of the Gulf of Mexico.

A comparison of the organic content between stations (table 3) indicated similarity of readings between the two northern mud stations, MES-1 and MCS-3, and between the two southern sand stations, SES-2 and SCS-4, but little similarity between northern and southern stations. The higher organic content of MES-1 and MCS-3 was probably the result of dense aquatic vegetation, which extended to the water's edge. Aquatic

Table 2

The salinity (^{o}/oo) of the interstitial water and adjacent sea water of Timbalier Bay from July 1973 to January 1974.

Station Number	July 1973	October 1973	January 1974
MES-1	17	27	22
sea	15	27	19
SES-2	30	34	26
sea	21	25	21
MCS-3	19	22	23
sea	17	19	19
SCS-4	30	23	21
sea	21	23	20

Table 3

Organic carbon content (^{o}/oo) at the high, mid, and low-tide regions in Timbalier Bay from July 1973 to January 1974.

Station Number	Tidal Region	July 1973	October 1973	January 1974
MES-1	H	13.52	13.34	14.63
	M	14.35	10.91	24.49
	L	35.66	9.69	28.15
SES-2	H	8.00	3.20	3.93
	M	0.83	1.66	1.72
	L	3.12	2.63	2.57
MCS-3	H	12.19	11.00	16.39
	M	13.45	40.17	36.27
	L	9.44	22.39	27.31
SCS-4	H	1.25	6.31	8.16
	M	3.29	1.16	3.26
	L	1.49	3.92	4.89

Table 4

Sediment Analysis of samples taken at the high, mid, and low-tide regions in Timbalier Bay, October 1973

Station Number	Tidal Region	Percent Mud	Percent Sand and Gravel	Sorting Coefficient	Mean Grain Size (mm)
MES-1	H	85.14	14.86	-	-
	M	87.26	12.74	-	-
	L	93.26	6.74	-	-
SES-2	H	4.39	95.61	0.42	0.12
	M	4.39	95.37	0.37	0.11
	L	1.74	98.26	0.35	0.12
MCS-3	H	89.20	10.80	-	-
	M	94.10	5.89	-	-
	L	86.92	13.08	-	-
SCS-4	H	37.60	62.40	0.55	0.12
	M	21.38	78.62	0.56	0.13
	L	61.93	38.07	0.53	0.11

vegetation was lacking at SES-2 and SCS-4. The organic carbon content in general was higher in January than in July, perhaps because of the incorporation of dead material from salt marsh and terrestrial vegetation into the sediments from the past season. Considerable variation was encountered between levels within a beach and no level was consistently higher in organic carbon than another. Measurements tended to be higher in the northern mud beaches than in the southern sand beaches. The results may indicate the effects of particle size upon the distribution of organic carbon within a beach. Fine-grained particles retain organic carbon particles, and because of poor percolation along the beach, greater variation can be expected. In beaches with better drainage, SES-2 and SCS-4, the variation between beach levels was somewhat reduced.

A mechanical analysis was carried out on sediment samples collected on the October 1973 survey (table 4). The highest values for percentage of silt-clay (mud) were recorded for MES-1 and MCS-3. Sand fractions were small, and the mean grain size and sorting coefficients could not be determined. A high percentage of the particles in the sand fractions at MES-1 and MCS-3 was shell fragments. The sediments at SES-2 consisted of very fine sand, which was well sorted. Sediments at SCS-4 were also composed of very fine sand, but the sorting was not as well defined.

Biological

Twenty-one species of polychaetes were collected from the littoral region of four selected beaches in Timbalier Bay. The species composition, total numbers, and mean number of polychaetes per core are shown in table 5 (pp. 520-521). Of the 21 species collected, six were common to MES-1 and MCS-3 but absent at SES-2 and SCS-4. Many species collected in the study area were rare. Their absence in some samples may be due to inherent error associated with random sampling.

The seasonal change in abundance indicates that polychaetes are generally more abundant in January than in October (figure 2, p. 522). The lowest number of polychaetes was recorded for July.

There was little consistency to the pattern of distribution of organisms with respect to tidal level at MES-1 and MCS-3. During October at MES-1, mean numbers for all tide levels were greater than at MCS-3 and the reverse was somewhat true in July and January. A clearer distribution occurred at SES-2 and SCS-4. The mean numbers of organisms per tidal level were always greater at SCS-4 than at SES-2, with the exception of July for the high- and mid-tide levels. A mean was calculated for the three tide levels per collection to reduce variation with respect to tide level. These data indicate that mean numbers of organisms were greater at the control stations (MCS-3 and SCS-4) than at the experimental stations (MES-1 and SES-2) with an exception during October at MES-1. The two southern stations also showed the greatest fluctuation in mean numbers. The richest area quantitatively was SCS-4, whereas SES-2 had the lowest mean numbers. The two northern bay stations (MES-1 and MCS-3) showed more uniformity in means. At all stations mean numbers tended to rise from July to January with maximum means in January.

The wet-weight biomass of the polychaetes is shown in table 6. In general, the highest values were reported in January at all stations. Biomass means at MES-1 and MCS-3 were higher than those at SES-2 and SCS-4. The higher values at MES-1 and MCS-3 may be attributed to the large populations of *Neanthes succinea* at these stations (table 5).

The Morisita-Ono Index of Faunal Affinity between stations is shown in table 7. A $C\lambda$ value of 1.0 indicates complete faunal similarity and a $C\lambda$ value of 0 indicates no faunal similarity. The faunal resemblance was greatest between MES-1 and MCS-3 in the northern bay region. The second highest faunal resemblance was between SES-2 and SCS-4 in the southern bay region. Faunal resemblances were least between northern and southern bay stations; the faunal resemblance was greater between production stations (MES-1 and SES-2) than between control stations (MCS-3 and SCS-4).

The Shannon-Weaver species diversity index (H ′) was calculated for each station to determine the community diversity (table 8). Diversities at SES-2 and SCS-4 were found to be higher than those at MES-1 and MCS-3 for July. In October, only slight variations between stations occurred; however, the values were higher at MES-1 and MCS-3.

TABLE 5

THE SPECIES COMPOSITION OF POLYCHAETES, TOTAL NUMBERS PER TIDE LEVEL, MEAN NUMBER OF ORGANISMS PER CORE, AND MEAN OF THREE TIDE LEVELS PER MONTH FOR JULY, OCTOBER, AND JANUARY 1973-74

STATION NUMBER	MES-1									SES-2								
DATE	JULY			OCT			JAN			JULY			OCT			JAN		
TIDAL LEVEL	H	M	L	H	M	L	H	M	L	H	M	L	H	M	L	H	M	L
SPECIES COMPOSITION																		
Etenoe heteropoda	0	0	0	0	0	1	0	0	0	0	0	0	1	0	0	0	0	0
Loandalia americana	0	0	0	0	0	0	1	0	0	0	0	0	1	0	0	0	0	0
Streblospio benedicti	0	0	0	2	2	0	0	1	2	2	0	0	1	3	3	0	3	9
Polydora socialis	0	0	0	0	1	0	0	0	0	0	0	0	0	1	0	0	0	1
Notomastus hemipodus	0	0	1	0	6	1	0	0	0	0	0	0	0	6	1	0	0	0
Heteromastus filiformis	0	0	0	6	0	1	0	0	0	0	0	0	0	0	0	0	0	0
Capitella capitata	0	0	0	2	35	22	11	33	55	2	2	0	0	0	9	4	19	3
Malanid sp.	0	0	0	0	1	0	0	0	0	0	0	0	0	0	0	0	0	0
Hesionid sp.	0	0	0	0	1	4	0	0	2	0	0	0	0	0	0	0	0	0
Neanthes succinea	13	4	9	110	56	27	46	46	135	1	0	0	0	0	2	0	4	0
Spiochaetopterus oculatus	0	0	0	0	0	11	0	0	0	0	0	0	0	1	0	0	0	0
Paraeurythoe americana	0	0	0	0	1	0	0	0	0	0	0	1	0	0	0	0	0	0
Megalomma bioculatum	0	0	0	3	17	6	2	0	9	0	0	0	0	0	0	0	0	0
Dorvillea rudolphi	0	0	0	0	0	1	1	4	0	0	0	0	0	0	0	0	0	0
Sphaerosyllis sp. a	0	0	0	5	26	12	1	2	1	0	0	0	0	0	0	0	0	0
Sphaerosyllis sp. b	0	0	0	0	0	0	1	0	0	0	0	0	0	0	0	0	0	0
Autolytus fasciatus	0	0	0	0	8	1	0	2	6	0	0	0	0	0	0	0	0	0
Exogone dispar	0	0	0	0	0	0	0	1	12	0	0	0	0	0	0	0	0	0
Tharyx monilaris	0	0	0	0	0	0	0	0	0	0	0	2	0	0	0	0	0	0
Sthenelais boa	0	0	0	0	0	1	0	0	0	0	0	0	0	0	0	0	0	0
Paranoid sp.	0	0	0	0	0	0	0	0	0	0	0	0	0	0	0	0	0	0
MEAN NO. ORGANISMS PER CORE	4.3	1.3	3.3	25.6	30.8	17.6	12.6	17.8	44.4	1.6	0.6	1.0	0.6	2.2	3.0	0.8	5.2	2.6
MEAN OF 3 TIDE LEVELS PER MONTH	3.0			24.7			24.9			1.1			1.9			2.9		

MCS-3									SCS-4								
JULY			OCT			JAN			JULY			OCT			JAN		
H	M	L	H	M	L	H	M	L	H	M	L	H	M	L	H	M	L
0	0	0	0	0	0	0	0	0	0	0	3	0	0	0	0	0	3
0	0	0	0	0	0	0	0	0	0	0	0	0	0	0	0	0	0
0	0	0	0	0	0	4	2	4	0	0	45	16	58	6	9	238	167
0	0	0	0	0	0	3	10	6	0	0	0	0	0	0	0	32	1
0	0	1	0	0	0	101	30	5	0	0	4	0	2	3	0	0	0
0	0	1	0	0	0	0	0	0	0	0	3	0	0	0	0	0	0
5	1	0	5	5	3	19	8	11	0	0	0	18	4	6	16	7	42
0	0	0	0	0	0	0	0	0	0	0	0	0	0	0	0	0	0
0	0	0	0	3	0	0	1	0	0	0	0	0	0	0	0	0	0
6	51	9	36	8	0	47	51	47	0	0	4	2	2	3	0	4	0
0	0	0	0	15	12	0	0	0	0	0	0	0	0	0	0	0	0
0	0	0	0	0	0	0	0	0	0	1	13	0	0	0	0	0	20
0	0	1	7	4	2	10	22	14	0	0	0	0	0	2	0	1	1
0	0	0	0	0	0	0	2	0	0	0	0	0	0	0	0	0	0
0	0	0	7	18	4	1	0	0	0	0	0	0	0	0	0	0	0
0	0	0	0	0	0	0	14	9	0	0	0	0	0	0	0	0	0
0	0	0	1	0	0	0	1	0	0	0	0	0	0	0	0	0	0
0	0	0	0	0	0	19	9	0	0	0	0	0	0	0	0	0	0
0	0	0	0	0	0	0	0	1	0	0	5	5	3	0	0	0	23
0	0	0	0	1	0	0	0	0	0	0	0	0	0	0	0	0	0
0	0	0	0	0	0	0	0	0	0	0	5	6	0	0	0	0	0
	17.3		11.2		4.2		30.0			0.3		9.4		4.0		56.3	
3.6		4.0		10.8		40.8		19.4	0.0		27.3		13.8		5.0		53.2
8.3			8.7			30.1			9.2			9.1			38.2		

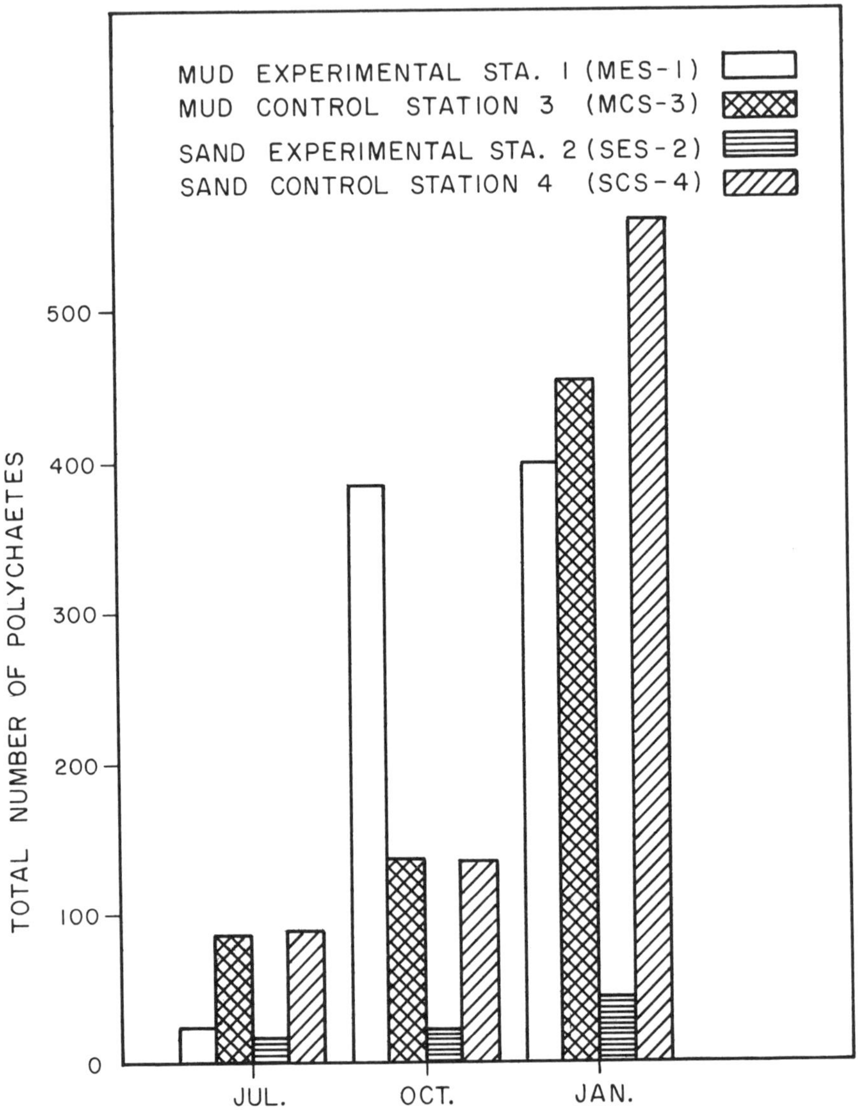

FIG. 2. TOTAL NUMBER OF POLYCHAETES COLLECTED from MES-1, MCS-3, SES-2, and SCS-4 during July, October, and January.

Table 6

The mean biomass per core (mg/cm^3) for each tide region at each station site from July 1973 to January 1974.

Station Number	Tidal Region	July 1973	October 1973	January 1974
MES-1	H	0.010	0.159	0.102
	M	0.001	0.153	0.816
	L	0.005	0.081	1.370
	Mean	0.005	0.131	0.763
SES-2	H	0.011	0.001	0.007
	M	0.005	0.022	0.022
	L	0.013	0.021	0.018
	Mean	0.009	0.015	0.016
MCS-3	H	0.006	0.123	0.821
	M	0.202	0.036	0.827
	L	0.214	0.004	1.060
	Mean	0.140	0.057	0.902
SCS-4	H	0.001	0.104	0.013
	M	0.001	0.102	0.158
	L	0.438	0.059	0.978
	Mean	0.146	0.088	0.383

Table 7

The Morisita-Ono Index of Faunal Affinity between Stations of Timbalier Bay from July 1973 to January 1974.

	MES-1	SES-2	MCS-3	SCS-4
MES-1	+	0.44	0.84	0.10
SES-2		+	0.37	0.61
MCS-3			+	0.08
SCS-4				+

Table 8

The number of species, number of individuals, and species diversity for all stations from July 1973 to January 1974.

Station Number	Date	Number of Species	Number of Individuals	Species Diversity Index (H')
MES-1	July	2	27	0.23
	October	16	370	2.35
	January	11	374	1.67
SES-2	July	5	10	2.12
	October	7	28	2.13
	January	4	43	1.39
MCS-3	July	5	75	0.69
	October	9	132	2.43
	January	13	451	2.74
SCS-4	July	8	83	1.73
	October	8	137	2.11
	January	9	572	1.48

Diversity indices at MES-1 and SES-2 were noticeably lower than those at MCS-3 and SCS-4 in January.

DISCUSSION

Three analytical methods were used in this study to establish whether or not changes in the polychaete community were associated with petroleum producing activities. Several physical and chemical measurements were also taken in order to determine which environmental factors influenced the distribution of organisms.

It has been previously emphasized that biological samples were taken from petroleum production areas and control areas located maximum distances from petroleum producing activities. Although the control areas were not isolated from potential petroleum pollutants, we assumed that they were not affected by petroleum producing activities.

McIntyre (1969) suggested that temperature and salinity operating directly, and grain size operating indirectly, control the distribution of benthic organisms. Pettibone (1967) indicated that polychaetes are highly euryhaline organisms adapted to living in varying salinities. In the present study the northern stations had salinities similar to those of the southern stations, but the polychaete composition was different. Thus, there appears little evidence to indicate that salinity was an important factor in the distribution of polychaetes in Timbalier Bay.

McIntyre reported that the polychaetes were fewer in the spring and more numerous in late autumn and winter. A similar seasonal cycle was found in the present study. Pettibone indicated that sexually mature polychaetes may be found swimming at the surface during the spawning season between early spring and late summer. Therefore, fewer polychaetes are found in the benthic environment in July, the time at which the higher water temperatures occurred.

The abundance of polychaetes in the present study increased in October and January. Tietjen (1969) also found significant peaks in total numbers of polychaetes during late summer and early fall. The increase in total numbers might be due to settlement of the metamorphosed larvae.

The effects of grain size are of fundamental importance in the ecology of benthic invertebrates. Larger grain sizes allow more percolation of tidal waters, which removes trapped organic material. Sandy beaches are characteristically low in organic detritus. In sediments containing a large percentage of clay, the percolation of tidal waters is decreased and the organic material is not readily removed; therefore, mud beaches are characteristically higher in organic detritus.

In the present study the data indicate significant differences in polychaete abundances throughout Timbalier Bay, suggesting the occurrence of two distinct biofacies. There are those primarily found in the northern mud beaches, and those predominantly found in the southern sand beaches. A third group of species occurred in both mud and sand beaches.

The mud biofacies was characterized by the occurrence of *Neanthes succinea*, *Megalomma bioculatum*, *Exogone dispar*, *Sphaerosyllis* species *a*, and *Sphaerosyllis* species *b*.

All samples of polychaetes from MES-1 and MCS-3 were dominated by *Neanthes succinea* and *Megalomma bioculatum*. *Sphaerosyllis* species *a* was prominent in October, but decreased in January; the reverse relationship was true for *Sphaerosyllis* species *b*. *Exogone dispar* was not found in samples collected in October, but was common in most samples collected in January. The northern stations (MES-1 and MCS-3) had a high silt-clay content and were rich in organic detritus. These stations supported large populations of deposit feeders such as *Neanthes succinea* and *Exogone dispar*.

The sand biofacies was characterized by the dominance of *Streblospio benedicti*, *Polydora socialis*, and *Tharyx monilaris*.

Streblospio benedicti was the dominant species found in samples collected at SES-2 and SCS-4. *Polydora socialis* and *Tharyx monilaris* were negligible in October, but prominent in January. The southern stations (SES-2 and SCS-4) were composed of a very fine sand and little organic detritus was present. Here suspension feeders, such as *Streblospio benedicti*, replaced the deposit feeders, a replacement expected in view of the lack of detritus.

Notomastus hemipodus and *Capitella capitata* occurred at all stations and were generally abundant in Timbalier Bay.

Dr. J. L. Laseter (personal communication) was not able to support or disprove the hypothesis that petroleum hydrocarbons exist at any station because hydrocarbon samples taken during the present study produced insufficient results. He did comment that the hydrocarbons present seem to be in an advanced state of degradation.

The results (table 5) indicate that MES-1, in a production area, supported larger populations of *Neanthes succinea*, *Capitella capitata*, and *Sphaerosyllis* species *a* than its control MCS-3. Four other species, *Heteromastus filiformis, Hesionid* sp., *Dorvillea rudolphi,* and *Autolytus fasciatus,* although present in very small numbers, also occurred in greater numbers at MES-1 than at MCS-3. Four other species, *Eteone heteropoda, Loandalia americana, Malanid* sp., and *Paraeurythoe americana* were represented only at MES-1.

Spiochaetopterus oculatus, *Megalomma bioculatum*, *Streblospio benedicti*, *Polydora socialis*, *Notomastus hemipodus*, *Sphaerosyllis* species *b*, and *Exogone dispar* were more common at MCS-3 than at MES-1, but only one species, *Tharyx monilaris*, was found at MCS-3 and not at MES-1.

The faunal affinity between MES-1 and MCS-3 (table 8) was greater than between any other combination of two stations, and the physical and chemical parameters at both MES-1 and MCS-3 were quite similar (tables 1-4). There was also no indication of the physical presence of petroleum at either station. It is concluded that factors other than oil pollution were controlling the abundance of these polychaetes at MES-1.

The lowest number of polychaetes was always recorded from SES-2 (table 5). This station was located close to workover rigs and just northwest of the channel leading to a Gulf Oil Company installation. Although hydrocarbons from petroleum could not be identified by Dr. J. L. Laseter, tarry balls were found in sediment samples taken in October and January, and an oil film was always present in the holes dug for the collection of interstitial water at all tide levels. On the other hand, SCS-4, which had similar physical and chemical parameters and no indication of petroleum hydrocarbons, supported large populations of polychaetes, in particular *Streblospio benedicti* and *Capitella capitata, Eteone heteropoda*, *Polydora socialis*, *Notomastus hemipodus*, *Neanthes succinea, Paraeurythoe americana, Tharyx monilaris,* and *Paranoid* sp. Only one species, *Loandalia americana*, represented by one specimen, was collected at SES-2 and not at SCS-4. The faunal affinity index (table 8) was also relatively low between these stations. The sparsity of polychaetes at SES-2 may bear some relationship to such factors as boat traffic in and out of the Gulf Oil Company installation, the presence of petroleum hydrocarbons present within the substrate, or some other unknown factor.

ACKNOWLEDGMENTS

This research was part of grant OB 64 HJM, The Study of Interstitial Organisms, of the Gulf Universities Research Consortium's Offshore Ecology Investigation.

REFERENCES CITED

Day, J. H. 1967. A monograph on the Polychaeta of Southern Africa. Part I. Errantia; Part 2. Sedentaria. Trustees of the British Museum (Natural History) Publications.

Galtsoff, P. S., H. F. Prytherch, R. D. Smith, and V. Koehring. 1935. Effects of crude oil pollution on oysters in Louisiana waters. Fisheries Bulletin 18, U.S. Fish and Wildlife Service **48**:143-210.

George S. S. 1971. On the importance of marine meiofauna for benthos communities. Oecologia (Berlin) **6**:176-190.

Gilet, R. 1960. Water pollution in Marseille and its relation with flora and fauna. Proceedings of the International Conference, Waste Disposal in the Marine Environment **1**:39-56.

Hartman, O. 1951. The littoral marine annelids of the Gulf of Mexico. Publications of the Institute of Marine Science, The University of Texas **2**(1):7-124.

Hulings, N. C. and J. S. Gray. 1971. A Manual for the Study of Meiofauna. Washington: Smithsonian Institution Press.

Mackin, J. and S. Hopkins. 1961. Studies on oyster mortality in relation to natural environments and to oil fields in Louisiana. Publications of the Institute of Marine Science, The University of Texas **7**:1-126.

McIntyre, A. A. 1964. The meiobenthos of sub-littoral muds. Marine Biological Association of the United Kingdom, Journal **44**:665-674.

———. 1969. Ecology of the marine meiobenthos. Biological Reviews **44**:245-290.

Morisita, M. 1959. Measuring of interspecific association and similarity between communities. Memoirs, Faculty of Science, Kyushu University, Series E **2**:215-235.

Nelson-Smith, A. 1973. Oil Pollution and Marine Ecology. New York: Plenum Press.

Ono, Y. 1961. An ecological study of the Brachyuran community on Tomioka Bay, Amakusa, Kyushu. Records of Oceanographic Works in Japan **5**:199-209.

Orton, J. H. 1925. Possible effects on marine organisms of oil discharge at sea. Nature **115**:910-911.

Pettibone, M. H. 1967. Marine polychaete worms of the New England region, Families Aphroditidae through Trochochaetidae. U.S. National Museum Bulletin **227**(1):1-356.

Reish, D. J. 1960. The use of marine invertebrates as indicators of water quality. *In* E. A. Pearson, ed., Waste Disposal in the Marine Environment. Oxford: Pergamon Press, pp. 92-103.

St. Amant, L. S. 1971. Impact of oil on the Gulf Coast. Transactions of the Thirty-sixth North American Wildlife and Natural Resources Conference, pp. 206-219.

———. 1972. The petroleum industry as it affects marine and estuarine ecology. Journal of Petroleum Technology **24**(4):385-392.

Shannon, H. L. and L. Weaver. 1963. Science and Information Theory. 2nd ed. New York: Academic Press.

Standard Methods for the Examination of Water and Wastewater, 13th ed., 1971. Washington: American Public Health Association.

Tietjen, J. H. 1969. The ecology of shallow water meiofauna in two New England estuaries. Oecologia (Berlin) **2**:251-290.

Waldron, R.P. 1963. A seasonal ecological study of Foraminifera from Timbalier Bay, Louisiana. Gulf Research Reports **4**:132-188.

Warren, W. S. 1942. A survey of annelid worms of the Grand Isle region. Bioscience **13**(1):39-46.

Wass, M. L. 1967. Biological and physiological basis of indicator organisms. *In* T. A. Olson and F. J. Burgess, eds., Pollution and Marine Ecology. New York: Interscience Publishers.

PELAGIC, EPIBENTHIC, AND INFAUNAL INVERTEBRATES OF TIMBALIER BAY AND OFFSHORE ENVIRONMENT

by Richard S. Waller

ABSTRACT

Offshore and Timbalier Bay, Louisiana, sites were sampled with trawls and benthic grabs to determine if there have been any effects of petroleum drilling and production activities on the invertebrate fauna of the area. Collections were made at a drilling site, a production platform, and a control site at both inshore and offshore localities. The dominant invertebrate fauna was similar at all stations within a locality. The catch per unit effort was highest at the control station in two instances and at the upstream drilling station in two sampling series. The biomass was highest at the control sites for both benthic grab series of samplings. On the basis of these data no major differences were noted between the petroleum activity sites and control area.

INTRODUCTION

The "Oil Patch" around the Grand Isle, Louisiana, area has long been regarded as a highly productive fisheries area particularly with reference to the shrimp, blue crab, and oyster fisheries.

Some insight into the biological nature of this area has been provided, but years after petroleum production was initiated. Perret et al. (1971) noted that only 13 invertebrate species were taken from the Terrebonne-Timbalier Bay area in 168 seine and trawl samples. Adkins (1972) sampled this same area for blue crabs parasitized by the barnacle *Loxothylacus*

Richard Waller is at the Gulf Coast Research Lab, Ocean Springs, MS.

texanus and reported only on the catch of blue crabs. Gaidry and White (1973) reported on brown and white shrimp taken from five trawl stations. St. Amant et al. (1965) worked on brown shrimp in Barataria Bay and Perret (1966, 1967) studied the fauna of Vermilion Bay. Gunter (1952) reported on the changes in the salinity regime of the general southwestern area of Louisiana due to leveeing of the Mississippi River and its tributaries. Behre (1950) and Dawson (1966) provided checklists of the fauna of the Grand Isle area. Felder (1973) presented a key to the reptantian decapods of the northwestern Gulf of Mexico. Comprehensive biological surveys of this area of the continental shelf along the Louisiana coast were not made prior to oil exploration and development. In the absence of previous data it is difficult to ascertain whether or not 35 years of oil production activities have led to any adverse effect on this bay and offshore environment. This study was conducted to determine what effects these activities may have had on the benthic invertebrate populations in Timbalier Bay and the adjacent offshore area.

MATERIALS AND METHODS

Samples were collected from the offshore sites for epifauna with a 9.4 m trawl provided with 3 cm mesh and for infauna with a 0.2 m^2 Van Veen type bottom grab. Surface trawls were taken for pelagic species with a 1.2 × 2.4 m opening and 0.6 cm mesh webbing along the sides and 0.3 cm mesh webbing at the end.

Samples in Timbalier Bay were collected with a 4.9 m trawl provided with 1.9 cm mesh netting along the sides and 0.3 cm mesh in the end. Benthic grabs were taken with a 0.1 m^2 Petersen-type sampler. Surface plankton tows were taken with a 0.9 × 1.2 m framed mouth and 0.6 cm mesh webbing in the body and 0.3 cm mesh webbing in the end.

All nets were towed 20 minutes. Samples were fixed in 10% formalin on the boat and transported to the Gulf Coast Research Laboratory for analysis. Samples were sorted to lowest taxa possible and aliquot samples of up to 100 individuals from each group were taken, weighed to the nearest 0.1 g, and the length measured to the nearest mm. All specimens were counted and weighed.

Bottom grab samples were washed through a 0.5 mm sieve and fixed in 5% buffered formalin for transfer to GCRL. The material was identified to the lowest taxon possible. Total count and weights were made for each group and species.

Species diversity values were computed using the equation of Wilhm (1968), where species diversity $H^1 = \sum_{I}^{S} pi \log 2pi$; with $pi = ni/N$, ni = number of species, and N = total number of organisms.

Samples were taken upstream and downstream from an active drilling rig and an active production rig, and at a control site that was located as far as possible from any petroleum production activities. The offshore drilling sites were occupied only twice because drilling ceased. Production and control sites were visited on each cruise.

RESULTS

The data for the offshore and Timbalier Bay trawls and grabs are summarized in tables 1-4. Comparisons of data for the most part cannot be made because of the different number of samples taken at the four sites. However, some comparisons can be made by comparing catch per unit effort between stations and for the dominant species. The catch per unit effort was highest at the control site for the offshore trawling samples and the Timbalier Bay grab samples. The upstream drilling site had the highest values for the offshore grab samples and Timbalier Bay trawling samples. The dominant species at the offshore drilling sites were *Calliactis tricolor*, *Trachypeneus similis*, *Pagurus pollicaris*, and *Lolliguncula brevis*, with the greatest number of specimens per unit effort taken from the control site (table 1). The dominant species from the offshore benthic grabs were counted but not identified; the highest number per sample were taken from the upstream drilling site (table 2). The dominant species encountered in the Timbalier Bay trawling samples were *Penaeus aztecus, P. setiferus, Callinectes sapidus, C. similis,* and *Lolliguncula brevis*, with the greatest number of specimens per unit effort taken from the downstream drilling site (table 3). The polychaete *Spiochaetopterus costarum* was the dominant species taken in the benthic grabs in Timbalier Bay, but the number of specimens collected was not determined (table 4).

The biomass per sample was calculated for the benthic grabs for offshore and Timbalier Bay. In both instances the biomass was considerably higher in the control than those values for the remaining three samples within a series.

DISCUSSION

On the basis of these data it is impossible to state whether or not there is any statistically significant difference in the biota of one site over the others. The number of samples, which were as few as one in some instances, varied from site to site, and not all the specimens were identified from the benthic samples. On the qualitative comparison of the dominant species present, the four stations within a site were similar. The

Table 1

Number of Species and Specimens Collected in Offshore Trawls

Species	Number of Specimens			
	Upstream Drilling	Downstream Drilling	Production	Control
Coelenterata				
Chrysaora quinquecirtha			1	
Pelagia noctiluca		abundant		common
Anthozoa			1	
Calliactis tricolor	87	98	53	347
Crustacea				
Lironeca texana		1		
Neroclia lanceolata	1			
Hyperiidae			1	
Decapoda				2
Penaeus aztecus			10	29
P. setiferus	12	27	1	14
Trachypeneus similis	12	18	14	199
Sicyonia dorsalis		1		5
Acetes americanus			1	
Latreutes fucorum				6
Leander tenuicornis				2
Calappa sulcata		3	2	2
Hepatus epheliticus	2	2		1
Persephonia crinita	2			2
Libinia dubia	2			1
Callinectes sapidus	2	1	2	
C. similis	4	4		13
Metoporhaphis calcarata				1
Portunus gibbesii	2	1		24
P. spinimanus	4	6	2	7
Albunea paretii				6
Paqurus pollicaris	36	42	22	253
Petrochirus diogenes	3	8		2
Porcellana sayana	2	9	1	16
Porcellanidae				1
Squilla empusa				12
Stomatopoda	7	3	1	12
Mollusca				
Loligo peali			10	5
Lolliguncula brevis	18	24	44	229
Octopus vulgaris		1		
Busycon contrarium	1			
Cantharus cancellarius	1	1		14
Crepidula fornicata	1	3		15
Murex fulvescens			1	
Sinum perspectivum			1	
Noetia ponderosa				2
Nudibranchia	3			1
Total Number of Species	20	20	18	30
Total Number of Specimens	202	253+	167	1223+
Species Diversity	4.3	5.0	5.1	5.4
Catch/Unit Effort	101	125.5	23.9	152.9
Number of Trawls	2	2	7	8

Table 2

Number of Specimens of Infaunal Species
Collected in Offshore Grab Samples

Species	Number of Specimens			
	Upstream Drilling	Downstream Drilling	Production	Control
Coelenterata				
Actiniaria			1	
Nemertea				
Cerebratulus lacteus				1
Polychaeta				
Spiochaetopterus costarum				abundant
unidentified	38	17	41	411
Crustacea				
Amphipoda	1		3	10
Mysidacea	1	1		1
Oxyurostylis smithi	1		1	
Leptochela serratorbita				1
Ambidexter symmetricus				1
Calappa sp.				1
Squilla empusa				1
Mollusca				
Canthurus cancellarius	1		1	
Corbula krebsiana	20		9	
C. sp.	17	2	6	
Abra aequalis	2	2		2
Nucula concentrica	2		3	7
N. acuta			5	
N. proxima	2			2
Solen veridis			1	
Lucina amianthus			2	
Tellina iris				3
T. sp.	2		1	
Cumingia tellionides				3
Callocardia texasiana				1
Chione latilirata				1
Phacoides filosus				2
Mercenaria campechiensis				1
Vitrinellidae				2
pelecypods, unidentified		1	1	
Oliva sayana			1	
Ophiuroidea, unidentified	10	5	4	40
unidentified specimens	12	14	35	34
Total Number of Species	13	7	16	21
Total Number of Specimens	109	42	110	232+
Biomass (g)/Sample	3.9	0.9	6.2	15.8+
Number of Samples	1	1	4	4
Catch/Unit Effort	109	42	28	58+

Table 3

Number of Species and Specimens Collected in Timbalier Bay Trawls

	Number of Specimens			
Species	Upstream Drilling	Downstream Drilling	Production	Control
Polychaeta, unidentified			1	
Crustacea				
Aegathoa oculata	1		1	
Cymothoa excisa	3	7		
Lironeca ovalis			1	
Penaeus aztecus	102	34	81	109
P. setiferus	29	43	112	175
P. duorarum				4
Trachypenaeus similis			14	4
Palaemonetes vulgaris				2
Xiphopeneus kroyeri			1	
Callinectes sapidus	32	8	31	46
C. similis	7	7	43	26
Portunus gibbesii				1
Xanthidae	2		1	1
Clibanarius vittatus	1		1	3
Panopeus herbstii				1
Pagurus annulipes				1
P. longicarpus	1	1	7	9
P. pollicaris	5	1	2	8
Squilla empusa	1	1	10	44
Mollusca				
Lolliguncula brevis	108	147	83	57
Nassarius vibex	1		1	1
Anchis avara				1
Polinices duplicatus			1	
Thais laemastoma	2		1	5
Crepidula fornicata				1
Chione cancellata			1	
Mulinia lateralis			2	6
Nuculana concentrica				1
Ectoprocta				
Bugula neritina	present		common	abundant
Number of Species	15	9	21	23
Number of Specimens	375+	249	395+	507+
Species Diversity	3.0	2.5	5.5	5.1
Catch/Unit Effort	46.9	35.8	35.9	46.1
Number of Trawls	6	5	9	9

sediment characteristics were more closely compacted around the drilling and production platforms, possibly as a result of the discharge of drilling mud, which may account for the higher biomass values measured from the benthic grab samples at the control site.

Trawls were taken out from two production rigs to determine whether or not any difference in number of species and species diversity occurred. No relationship was found to exist between the species composition or species diversity with respect to distance from the rig.

Table 4

Number of Specimens of Infaunal Species
Collected in Timbalier Bay Grab Samples

Species	Number of Specimens			
	Upstream Drilling	Downstream Drilling	Production	Control
Polychaeta				
Spiochaetopterus costarum	+	+	+	abundant
Lepidonotus sp.				7
Amphinomidae				1
Drilonereis sp.				4
Lumbrineris sp.				3
Branchioasychis americana				4
Magelona sp.				1
Onuphidae				1
unidentified	8	4	40	115
Crustacea				
Amphipoda		1		2
Ogyrides limicola	4		7	
Pagurus annulipes				1
P. sp.				2
Callinectes sp.				1
Mollusca				
Mulinia lateralis		2	8	7
Nuculana concentrica			1	6
Tellina versicolor				2
Corbula contracta				4
Tagelus divisus				1
Ensis minor				8
Anachis avara				4
Nassarius acutus			1	10
Ophiuroidea, unidentified			23	75
Pisces				
Anchoa mitchilli	3			1
Myrophis punctatus				2
Ophidion welshi	1			
unidentified specimens	5	5	9	21
Number of Species	6	5	8	25
Number of Specimens	21+	12+	89+	283+
Biomass (g)/Sample	0.4+	0.1+	1.3+	5.3+
Number of Samples	2	2	4	4
Catch/Unit Effort	10.5+	6+	22.3+	71

CONCLUSIONS

Any definitive conclusions concerning the impact of oil production activities on the macroinvertebrate fauna in this area are difficult to draw primarily because of the small number of samples collected and the lack of historical data. Higher biomass values at the control site may be accounted for by the presence of compact drilling muds at the drilling and production sites.

ACKNOWLEDGMENT

This research was supported by contract number OB-74-HJM to the Gulf Coast Research Laboratory.

REFERENCES CITED

Adkins, Gerald. 1972. Notes on the occurrence and distribution of the rhizocephalon parasite (*Loxothylacus texanus* Boschma) of blue crabs (*Callinectes sapidus* Rathbun) in Louisiana estuaries. Louisiana Wildlife and Fisheries Commission, Technical Bulletin No. 2.

Behre, E. H. 1950. Annotated list of the fauna of the Grand Isle region. Occasional Papers of the Marine Laboratory, Louisiana State University. No. 6.

Dawson, C. E. 1966. Additions to the known marine fauna of Grand Isle, Louisiana. Proceedings, Louisiana Academy of Science **11**:175-180.

Felder, D. L. 1973. An annotated key to crabs and lobsters (Decapoda, reptantia) from coastal waters of the Northwestern Gulf of Mexico. Center for Wetland Research, Louisiana State University, Publication No. LSU-SG-73-02.

Gaidry, W. J. and C. J. White. 1973. Investigations of commercially important penaeid shrimp in Louisiana estuaries. Louisiana Wildlife and Fisheries Commission, Technical Bulletin No. 8.

Gunter, Gordon. 1952. Historical changes in the Mississippi River and the adjacent marine environment. Publications of the Institute of Marine Science, The University of Texas **2**:119-140.

Perret, William S. 1966. Occurrence, abundance, and size distribution of fishes and crustaceans collected with otter trawl in Vermilion Bay, Louisiana. M.S. thesis, University of S.W. Louisiana.

———. 1967. Occurrence, abundance, and size distribution of blue crab, *Callinectes sapidus*, taken with otter trawl in Vermilion Bay, Louisiana, 1964-65. Proceedings, Louisiana Academy of Science **20**:63-69.

———, W. R. Latapie, J. F. Pollard, W. R. Mock, G. B. Adkins, W. J. Gaidry, and C. J. White. 1971. Section 1. Fishes and invertebrates collected in trawl and seine samples in Louisiana estuaries. Cooperative Gulf of Mexico Estuarine Inventory and Study, Louisiana. Phase IV, Biology. Louisiana Wildlife and Fisheries Commission, pp. 41-105.

St. Amant, L. S., J. G. Brown, and T. B. Ford. 1965. Studies of the brown shrimp, *Penaeus aztecus,* in Barataria Bay, Louisiana, 1962-1965. Bulletin Marine Science, Gulf and Caribbean **18**:1-16.

Wilhm, J. L. 1968. Use of biomass units in Shannon's Formula. Ecology **49**:153-156.

FISH OF TIMBALIER BAY AND OFFSHORE LOUISIANA ENVIRONMENTS COLLECTED BY TRAWLING

by Allison Perry

ABSTRACT

Trawlings were made for fish in the vicinity of a drilling rig and production platform, and at a control site in Timbalier Bay and off Grand Isle, Louisiana, to determine whether or not oil activities have had any significant effect on fish populations. Species diversity was highest at the control site and lowest at the drilling site in Timbalier Bay; however, biomass was highest at the drilling site and lowest at the production platform. At the offshore area the drilling site had the highest species diversity and the production site had the lowest. The highest biomass was measured at the control site and the lowest at the production site.

Earlier studies of fish populations pre-dating oil activities were similar to these findings, indicating that oil production has not had an adverse effect on fish populations.

INTRODUCTION

The effect of oil drilling and production activities on fish populations in the Gulf of Mexico has been a topic of subjective discussion for many years. To date there have been no published accounts of objective survey work using bottom nets to sample the fish populations in oil production tracts. Therefore, this investigation was designed to determine by trawl survey whether normal offshore drilling

---

Allison Perry is with the Gulf Coast Research Laboratory, Ocean Springs, MS.

and production operations have had any significant effect on the ecology of estuaries or continental shelf areas as related to the various bottom fish populations in the Grand Isle, Louisiana, petroleum activity regions.

The first studies of trawl catches of fishes in the Gulf of Mexico were made by Gunter from 1931 to 1933 (Gunter 1936, 1938a, 1938b). His collections were made in Barataria Bay and the adjacent Gulf. Fish counts were made monthly over the seasons, and the data were used in various analyses. They were also used to compare with similar data collected later in Texas waters where total lengths, as well as counts, were made of all fish. The data from the two different areas enabled Gunter to compare the fish populations of Louisiana and Texas shallow Gulf waters (Gunter 1941, 1945). These original studies were located near Timbalier Bay and covered a salinity gradient similar to that of the present study. These previous studies by Gunter enabled us to compare present fish populations with those collected in the 1930s to determine whether or not there has been an adverse effect from oil activities in the ensuing years.

MATERIALS AND METHODS

Samples of the bottom-dwelling fish were collected in Timbalier Bay with a 4.9-meter trawl provided with 2.9 cm webbing and a 0.6 cm mesh tail liner. Samples off Grand Isle were taken with a 9.1 m trawl provided with a 2.9 cm mesh net. Nektonic fish within the bay were collected with a 0.9 × 1.2 m frame and offshore with a 1.2 × 2.4 m frame. The nets in each case consisted of 0.6 cm mesh in the body and 0.3 cm mesh at the tail.

All nets were towed for a period of 20 minutes. The fish were fixed in 10% formalin and transported to the laboratory. Specimens were identified to species and counted, and lengths and weights were measured for up to 100 specimens per species per tow. Collections were made over an 18-month period from August 1972 through January 1974, with 26 tows taken in Timbalier Bay and 14 from offshore waters (table 1). Data were analyzed according to relative apparent abundance (Marr 1951), species diversity (Wilhm 1968), species redundancy (Kohn 1966), and biomass.

Samples were taken upstream and downstream from a drilling site and a production platform, and at a selected control site. Drilling activities within the bay were rarely at the same site from one collection period to the next, necessitating changing locations from time to time. The Philo Brice Island platform was used for the bay production station, and the control station was located at Wrong Name Pass. There were three sites used in the offshore sampling program. The drilling rig, HOR-ST-

TABLE 1

Number of Species and Specimens of Fish from

Timbalier Bay and Offshore Waters of Louisiana

Species	Timbalier Bay				Offshore Waters				Totals of all Stations
	Drilling Upcurrent	Drilling Downcurrent	Production	Control	Drilling Upcurrent	Drilling Downcurrent	Production	Control	
Anguilliformes, larvae		1	6	9			1	3	20
Anchoa hepsetus	2	5	35	23		8	17	117	207
A. lyolepis				8					8
A. mitchilli	9620	7463	8728	3808	2			12	29,633
A. sp., larvae								5	5
Arius felis	18	9	26	28	72	69	225	14	461
Antennarius radiosus					2	4		16	22
Bagre marinus	2		71	9					82
Bairdiella chrysura	6		2	22					30
Balistes capriscus					1		1		2
Brevoortia patronus	12	4	7	35					58
Caranx hippos			1			1		1	3
Centropristis philadelphicus					7	7	9	1	24
Chaetodipterus faber	1		14	10	11	12	111	5	164
Chloroscombrus chrysurus	2	5	43	14			2	20	86
Citharichthys spilopterus	1	1	9	8	1		1	3	24
Clupeidae, larvae	2		155	25					182
Cynoscion arenarius	10	9	84	13	82	67	8	40	313
C. nebulosus			1						1
C. nothus	1				56	77	14	226	374
Dasyatis sabina	1								1
Dorsoma cepedianum			1						1
Etropus crossotus	4	2	29	22	77	76	26	157	393
Halieutichthys aculeatus					3	5	3	6	17
Harengula pensacolae	1	1	5					1	8
Lagodon rhomboides						2	5		7
Larimus fasciatus			2		3	5	6	9	25
Leiostomus xanthurus	3		12	28	26	56	92	42	259
Lutjanus campechanus					2		19		21
Membras martinica	2	1		1					4

TABLE 1 (continued)

Species	Timbalier Bay				Offshore Waters				Totals of all Stations
	Drilling Upcurrent	Drilling Downcurrent	Production	Control	Drilling Upcurrent	Drilling Downcurrent	Production	Control	
Menticirrhus americanus				2	1	2	1	3	9
Micropogon undulatus	4	3	251	617	42	78	122	764	1881
Mugil curema		1							1
Ogcocephalus parvus						1			1
Ophisthonema oglinum	1		2	2			1		6
Orthopristis chrysoptera						2	2		4
Paralichthys lethostigma					2	3	1		6
Peprilus alepidotus				2	1				3
P. burti		1	67	30		1	9	68	176
Polydactylus octonemus			24	4			12	13	53
Pomatomus saltatrix				1			1	1	3
Pontinus longispinis					1			1	2
Porichthys porosissimus					1	1		1	3
Prionotus roseus						1			1
P. rubio			1	2	3	8	7	32	53
P. scitulus			1	1					2
P. sp., larvae				1				2	3
P. tribulus		1	5	4				2	12
Rhizoprionodon terraenovae			1						1
Sciaenidae, larvae	1								1
Scomberomorus maculatus	1			5					6
Sphoeroides parvus	3	5	45	20	7	6	12	40	138
Stellifer lanceolatus			1		2	8		35	46
Stenotomus caprinus							3		3
Syacium gunteri					2	8	2		12
Symphurus civitatus					10	5		36	51
S. plagiusa	1	1	3	1	27	17	4	30	84
Syngnathus louisianae				1					1
Synodus foetens	2	3	2	2		4	1	6	20

TABLE 1 (continued)

Species	Timbalier Bay				Offshore Waters				Totals of all Stations
	Drilling Upcurrent	Drilling Downcurrent	Production	Control	Drilling Upcurrent	Drilling Downcurrent	Production	Control	
Trachurus lathami					1			2	3
Trichiurus lepturus	42	6	45	29	22	23	47	529	743
Urophycis floridanus					1	1		1	3
Vomer setapinnis	9	9	3	18	11	12	7	14	83
Total Number of Species	26	20	33	34	30	31	32	37	63
Total Number of Specimens	9752	7531	9682	4805	479	570	772	2258	35,849
Number of Trawls	4	4	9	9	2	2	5	5	40
Relative Apparent Abundance	2438	1883	1076	534	240	285	154	452	896
Species Diversity	1.62	1.84	2.18	2.90	6.33	8.01	3.81	3.95	-
Species Redundancy	0.69	0.77	0.65	0.55	0.29	0.23	0.28	0.50	-
Biomass (g/ha)	3019.	1922	1577	1795	7377	8876	5772	11,565	-

54E, was used twice for the drilling station before it ceased operation. The platform HOR-ST-54A (called Exxon 54A in other papers in this volume) was used for the production site, and the control station was located 5.3 nautical miles northeast of 54A in an undisturbed area.

RESULTS

The data for the fish collected in the 40 trawls are summarized in table 1 according to species, number of specimens, relative apparent abundance, species diversity, species redundancy, and biomass. Table 2 ranks the most commonly encountered fish according to the type of petroleum activity in Timbalier Bay and offshore waters. A total of 35,849 specimens in 63 species was collected from both areas during the 18-month period. *Anchoa mitchilli* accounted for nearly 83% of the catch, most of which were taken from Timbalier Bay. If this species is excluded from the total fish catch, then only 2,131 specimens were collected from the bay as compared to 4,055 from offshore. Forty-seven species were taken in offshore waters and 43 from Timbalier Bay; 32 species were common to both areas. The number of species was largest from both control stations followed by the production site; however, none of the species unique to the control site was present in large numbers. If the number of specimens of *Anchoa mitchilli* is excluded, then the numbers of specimens collected from Timbalier Bay sites become 111 and 106 per trawl at the control and production site, respectively, as compared to 17 and 33 per trawl at the drilling platform. Nearly twice as many specimens were taken from the control site at the offshore locality as from the drilling and production sites. The species diversity and biomass were considerably higher at the oil offshore stations compared to the bay.

Anchoa mitchilli was the dominant fish at all localities within the bay (table 2). *Micropogon undulatus* was the second most common species, at both the production and control stations.

There was a greater degree of species variability among production platform, control stations, and bay stations than between the offshore stations. The greatest variability at the offshore stations was within the dominant species, *Micropogon undulatus*.

DISCUSSION

The bay anchovy, *Anchoa mitchilli*, was the most abundant species in the bay and was the largest contributor to the biomass. The croaker, *Micropogon undulatus*, cutlass fish, *Trichiurus lepturus*, and the white

TABLE 2

Ranking of Most Abundant Species by Station*

Species	Timbalier Bay				Offshore Waters			
	Drilling Upcurrent	Drilling Downcurrent	Production	Control	Drilling Upcurrent	Drilling Downcurrent	Production	Control
Anchoa hepsetus			10	9			8	5
A. mitchilli	1	1	1	1				
Arius felis	3			6	3	4	1	
Bagre marinus			5					
Bairdiella chrysura				10				
Brevoortia patronus	4			3				
Chaetodipterus faber					9	9	3	
Chloroscombrus chrysurus			9					
Clupeidae, larvae			3	8				
Cynoscion arenarius	5		4		1	5		
C. nothus					4	2	9	3
Etropus. crossotus				10	2	3	6	4
Leiostomus xanthurus				6	7	6.	4	
Lutjanus campechanus							7	
Micropogon undulatus			2	2	5	1	2	1
Peprilus burti			6	4				6
Polydactylus octonemus							10	
Prionotus rubio								10
Sphoeroides parvus			7				10	7
Stellifer lanceolata								9
Symphurus civitatus								8
S. plagiusa •					6	8		
Trichiurus lepturus	2		7	5	8	7	5	2
Vomer setapinnis					9	9		

*Based on a minimum of 10 specimens per station

trout, *Cynoscion arenarius*, were the next important species. These fish were the most important fish in Caminada Bay, Louisiana (Wagner, in Stone 1972), Barataria Bay, Louisiana (Gunter 1938b; Behre 1950), Terrebonne and Timbalier Bays, Louisiana (Perret et al. 1971), Copano and Aransas Bays, Texas (Gunter 1945), Alabama bays (Boschung 1957; Swingle 1971), and Mississippi bays (Christmas and Waller 1973). Drilling activities seemed to affect the fish populations in shallow Timbalier Bay as indicated by the lower catches and lower species diversities at these stations; however, these lower figures are apparently only temporary during drilling operations, as indicated by data from the production site.

The offshore fish most frequently caught was the croaker, *Micropogon undulatus*, followed by the cutlassfish, *Trichiurus lepturus*, hard-head catfish, *Arius felis*, the silver seatrout, *Cynoscion nothus*, and the white trout, *Cynoscion arenarius*. Hildebrand (1954) analyzed the catch of shrimp boats that had trawled in the same general area, and he reported that the croakers and cutlass fish were the dominant species, followed by the two species of sea trout. He caught many more specimens of the blackfin searobin, *Prionotus rubio*, but this was probably because his trawls were from deeper waters. Moore et al. (1970) found the croaker to be the most abundant species along the Louisiana coast. Gunter obtained similar results in trawling studies south of Grand Isle, Louisiana.

Several species of fish were more common downcurrent from the rig in the path of overboard discharge than upcurrent in cleaner waters. These fish include the silver seatrout, spot, croaker, and the moonfish, *Vomer setapinnis*. No explanation is offered to account for this difference.

CONCLUSIONS

Comparisons of this 18-month fish trawling study in Timbalier Bay and offshore waters to present work in similar habitats of the northern Gulf region are in agreement. While there seems to be some disruptive effect on fish populations as a result of drilling, especially within Timbalier Bay, this adverse effect was localized and temporary, as shown by the similarity of the present fish populations to those of 20-40 years ago and the greater number of fish present around production platforms. Lower biomass values measured around both the bay and offshore production platforms may be due in part to the compact sediment present around rigs, from which normal infauna of soft bottom substrata would be excluded. Such an exclusion would result in lower benthic biomass values, which in turn would result in lower food availability for fish. High barium concentrations, which are indicative of oil drilling, were reported by Chan and Hanor (1973) near production platforms.

REFERENCES CITED

Behre, E. H. 1950. Annotated list of the fauna of the Grand Isle region. Occasional Papers of the Marine Laboratory, Louisiana State University, No. 6.

Boschung, H. T. 1957. The fishes of Mobile Bay and the Gulf coast of Alabama. Ph.D. dissertation, University of Alabama.

Chan, L. H. and J. S. Hanor. 1973. Input of barium into the marine environment as a result of offshore drilling operations. First Annual Report to the Gulf Universities Research Consortium. Department of Oceanography, Louisiana State University.

Christmas, J. Y. and R. S. Waller. 1973. Cooperative Gulf of Mexico estuarine inventory and study, Mississippi: Phase IV-Biology. Gulf Coast Research Laboratory, pp. 320-434.

Gunter, G. 1936. Studies of the destruction of marine fish by shrimp trawlers in Louisiana. Louisiana Conservation Review **5**:18-46.

———. 1938a. The relative numbers of species of marine fish on the Louisiana coast. American Naturalist **72**:77-83.

———. 1938b. Seasonal variations in abundance of certain estuarine and marine fishes in Louisiana, with particular reference to life histories. Ecological Monographs **8**:313-346.

———. 1941. Relative number of shallow water fishes of the Northern Gulf of Mexico with some records of rare fishes from the Texas coast. American Midland Naturalist **26**:194-200.

———. 1945. Studies on marine fishes of Texas. Publications of the Institute of Marine Science, The University of Texas **1**:1-190.

Hildebrand, H. H. 1954. A study of the fauna of the brown shrimp (*Penaeus aztecus*, Ives) grounds in the western Gulf of Mexico. Publications of the Institute of Marine Science, The University of Texas **3**:232-366.

Kohn, A. J. 1966. Food specialization in *Conus* in Hawaii and California. Ecology **47**:1041-1043.

Marr, J. C. 1951. On the use of the terms *abundance*, *availability*, and *apparent abundance* in fishery biology. Copeia (1951) **2**:163-169.

Moore, D., H. A. Brusher, and L. Trent. 1970. Relative abundance, seasonal distribution, and species comparison of demersal fishes off Louisiana and Texas, 1962-64. Contributions in Marine Science, The University of Texas **15**:45-70.

Perret, W. S., W. R. Latapie, J. F. Pollard, W. R. Mock, G. B. Adkins, W. J. Gaidry, and C. J. White. 1971. Section I. Fishes and invertebrates collected in trawls and seine samples in Louisiana estuaries. Cooperative Gulf of Mexico Estuarine Inventory and Study, Louisiana, Phase IV-Biology. Louisiana Wildlife and Fisheries Commission, pp. 41-105.

Stone, J. H. 1972. Louisiana superport studies: Report 2—Preliminary assessments of the environmental impact of a superport on the southeastern coastal area of Louisiana. Center for Wetland Resources, Louisiana State University. LSU-SG-72-05.

Swingle, H. A. 1971. Biology of Alabama estuarine areas—Cooperative Gulf of Mexico estuarine inventory. Alabama Marine Resources Bulletin **5**:1-123.

Wilhm, J. L. 1968. Use of biomass units in Shannon's formula. Ecology **49**:153-156.

A STUDY OF THE TEMPORAL CHANGES IN OFFSHORE MACROFAUNA IN THE NORTHERN GULF OF MEXICO DURING THE DEVELOPMENT OF THE OFFSHORE OIL INDUSTRY

by John R. Thompson

ABSTRACT

To supplement the relatively short-term trawl studies conducted by other investigators in the OEI, a longer range study was conducted using existing data available from the Bureau of Commercial Fisheries (now the National Marine Fisheries Service). Data taken in the Northern Gulf of Mexico by the M/V *Oregon* and chartered vessels during the period in which the offshore oil industry was developed in Southern Louisiana were used in an effort to detect changes in macrofauna attributable to this buildup of estuarine and offshore oil drilling and production.

INTRODUCTION

The fish in coastal waters of Louisiana have been studied since the first half of this century (Raymond 1905; Gunter 1935). Such studies became much more intensive during the 1950s. These primarily individual studies provide much useful information on the taxonomy, life history, and ecology of fish populations in this area.

The most consistent and systematic set of data obtained during the period of oil industry growth in coastal and offshore Louisiana waters was developed by the Bureau of Commercial Fisheries (now the National Marine Fisheries Service and referred to hereafter as BCF/NMFS). These

At the time of this study, John R. Thompson was at the Department of Biology, University of Southern Mississippi, Hattiesburg, MS.

data, covering the 1950s and 1960s, were collected in order to assess the living marine resources in the Northern Gulf of Mexico. Chartered vessels, primarily the M/V *Oregon*, were used in these surveys. The data collected are reported in two "collection reports" (Springer and Bullis 1956; Bullis and Thompson 1965).

METHODS AND MATERIALS

For this paper, only data obtained from the *Oregon* cruises off the Louisiana coastline between longitudes 89°30′W and 92°30′W were used. This area includes the outflow areas of the Mississippi and Atchafalaya River systems and, most importantly, the areas off Barataria and Timbalier Bays designated as the primary study area of the OEI.

Records covering the period 1950 to 1965 taken at 523 stations in this area by the BCF/NMFS were used. A "station" according to BCF/NMFS designation consists of a specific set of operations—trawl haul, dredge haul, longline set or similar sampling/observation unit—at a precise, but not necessarily reoccupied, position fixed by Loran.The gear most commonly used for acquisition of the data in this study were either 40-foot (opening across the mouth) shrimp trawls as described by Bullis (1951) and later workers, or fish trawls rigged New England style with rollers (Captiva and Rivers 1960).

Computer facilities (UNIVAC 9200) at the BCF/NMFS Pascagoula, Mississippi, Laboratory were used in the compilation and analysis of these data. Computer analyses covered only the dominant (by abundance or percentage of occurrence) fishes and shellfishes. Distribution densities and availability densities of these dominant species were used to identify any trends in the populations and their distributions that might be associated with oil drilling and production activity.

RESULTS

The total species complement collected and used for analysis in this study was 630 fish, 146 mollusk, and 234 crustaceans. Computer analyses revealed that eight species of fish were most abundant, with Sciaenids being the dominant group present (table 1). Offshore, blue crabs, brown shrimp, and rock shrimp were found to be the dominant crustaceans taken.

Studies of annual populations and distributions revealed no significant *trends* over the period 1960-1965. Variations in the data were well within those expected in trawl operations, and no consistent changes were evident over this time span.

Table 1. Predominant Fishes and Invertebrates in Catches from M/V OREGON Trawl Work -- 1950-65.

Scientific Name	Common Name
Fishes	
Synodus foetens	Lizardfish
Arius felis	Sea Catfish
Leiostomus xanthurus	Spot
Micropogon undulatus	Croaker
Menticirrhus americanus	Southern Kingfish
Cynoscion nothus	Silver Sea-Trout
Stenotomus caprinus	Longspine Porgy
Trichiurus lepturus	Atlantic Cutlassfish
Invertebrates	
Callinectes danae	Offshore blue crab
Penaeus aztecus	Brown shrimp
Sicyonia brevirostris	Rock shrimp

The seasonal variations apparent from study of the data were those well known to commercial fisherman and as reported by Russian fleets conducting exploratory operations in the area (Sal'nikov 1969). Lowest catches occurred in June and early July (though least effort was apparently expended during these months) and highest catches occurred in winter (November-March).

DISCUSSION

The BCF/NMFS data used in this study were taken in an area offshore described by Gunter (1963) as the "Fertile Fisheries Crescent." The "Crescent" extends from Pascagoula, Mississippi, to Port Arthur, Texas. It is reputed to be among the most productive fisheries areas in the world (Roithmayr 1965; Gunter 1967; Perrett et al. 1971; Juhl et al. 1973). It is estimated that 90%, by weight, of the fish species commercially harvested from the Gulf of Mexico are estuarine dependent (Gunter 1938a,

1938b). A similar estuarine dependence exists for most crustaceans including commercially important blue crabs and penaeid shrimp. Since the most sensitive period in these organisms' life cycle (eggs and larvae) is associated with the bays and sounds, any adverse effects on these environments resulting from oil drilling and production should have maximum effect on abundance and community structure.

In addition to the possible introduction of effluents from oil drilling and production (drilling mud, crude oil, etc.), these activities also introduce a physical change in the environment. It is generally accepted that artificial structures, such as oil platforms, attract and concentrate fish populations. Unless the effect of effluents is more significant, this effect on populations should be observable. Trawls studies directed to investigating this possibility were conducted under another OEI investigation by Mr. Allison Perry. Extensive studies by the Louisiana Wildlife and Fisheries Commission (Duffy 1968) concluded that sport fishing was enhanced by the artificial reef effect of the numerous oil production platforms in the coastal and offshore waters.

CONCLUSIONS

No correlations could be demonstrated between macrofauna populations and oil related activities in the time period and area covered in this study. Seasonal or "natural variability" best correlates with the data.

An offshore groundfish fishery was developed during the period of data collection (Roithmayr 1965). The dominant group of fishes supporting this industry (Sciaenids) is the same group shown to be dominant in the BCS/NMFS data. These data and groundfish catch history show that the populations of fishes offshore of the OEI study are sufficient to sustain a large-scale groundfish industry.

The value of the BCF/NMFS data for the correlations made in this study could have been improved through greater standardization and an increase in sampling density. This would enhance the statistical validity of the studies as well as provide a more comprehensive basis for the detection of either transient or consistent trends in populations and their distributions.

ACKNOWLEDGMENTS

Three senior graduate students at the University of Southern Mississippi are responsible for much of the background work upon which the results and discussion herein rest:

Nicholas Funicelli, Larry L. Massey, and Thomas McElwain. Students Louis Eaton and Vincent Guido also contributed in the early stages of the study, and many others contributed to stimulation of thought and aid in various facets. The National Marine Fisheries Service, Southeast Fisheries Center, supplied data printouts and computer usage. Richard B. Roe of NMFS, Washington, D.C., was particularly helpful in this, as was his technician, Judith Gatlin.

REFERENCES CITED

Bullis, J. R., Jr. 1951. Gulf of Mexico shrimp trawl designs. U.S. Department of the Interior, Fishery Leaflet 394.

——— and John R. Thompson. 1965. Collections by the Exploratory Fishing Vessels *Oregon, Silver Bay, Combat,* and *Pelican* Made During 1956-1960 in the Southwestern North Atlantic. U.S. Department of the Interior, Special Scientific Report, Fisheries No. 510.

Captiva, Francis J. and Joaquim B. Rivers. 1960. Development and use of the otter-trawling gear for red snapper fishing in the Gulf of Mexico, June 1957-May 1959. Commercial Fisheries Review **22**(10):1-14.

Duffy, McFadden. 1968. Sport fishing along the Gulf Coast. Louisiana Conservationist, Sept.-Oct., pp. 15-18.

Gunter, G. 1935. Records of fishes rarely caught in shrimp trawls in Louisiana. Copeia **1**:39-45.

——— 1938a. Seasonal variations in abundance of certain estuarine and marine fishes in Louisiana, with particular references to life histories. Ecological Monographs **8**(3):313-346.

——— 1938b. The relative numbers of marine fishes on the Louisiana coast. American Naturalist **72**:77-83.

——— 1963. The fertile fisheries crescent. Mississippi Academy of Sciences Journal **9**:286-290.

——— 1967. Some relationships of estuaries to the fisheries of the Gulf of Mexico. *In* Estuaries. American Association for the Advancement of Science **83**:621-638.

Juhl, Rolf, Elmer J. Gutherz, Charles M. Roithmayr, and Gary M. Russell. 1973. Southeast groundfish program status report to the Gulf States Marine Fisheries Commission, October 1973.

Perrett, William S., Barney B. Barrett, Walter R. Latapie, Judd F. Pollard, Woodrow R. Mock, Gerald B. Adkins, Wilson J. Gaidry, and Charles J. White. 1971. Cooperative Gulf of Mexico Estuarine Inventory and Study, Louisiana. Phase I. Area Description and Phase IV. Biology. Louisiana Wildlife and Fisheries Commission.

Raymond, R. I. 1905. The Fishes of Louisiana. M.S. thesis, Tulane University.

Roithmayr, Charles M. 1965. Industrial Bottomfish Fishery of the Northern Gulf of Mexico. 1959-1963. U.S. Department of the Interior, Special Scientific Report, Fisheries No. 518.

Sal'nikov, N. E. 1969. Fishery research in the Gulf of Mexico and the Caribbean Sea. *In* A. S. Bogdanov, ed., Society-Cuban Fishery Research, translation by U.S. Department of the Interior, NSF, pp. 78-171, Israel Program for Scientific Translations.

Springer, Stewart and Harvey R. Bullis, Jr. 1956. Collections by the *Oregon* in the Gulf of Mexico. U.S. Department of the Interior, Special Scientific Report, Fisheries No. 196.

BIOFOULING COMMUNITY DYNAMICS IN LOUISIANA SHELF OIL PLATFORMS IN THE GULF OF MEXICO

by Robert Y. George and Philip J. Thomas

ABSTRACT

This paper presents the quantitative results on the standing stock and population densities of the fouling community on the submerged oil platform structures in the area of the Offshore Ecology Investigations on the Louisiana Shelf. Climax fouling community was investigated on the basis of quantitative samples and *in situ* photographs. Patterns in biomass and density of seasonal fouling settlement were explained on the basis of periodic examination of *in situ* test panels. Oil platforms act as artificial reefs, offering large submerged surfaces for promoting biofouling growth in this environment. Comparisons of conditions today with baseline information obtained two decades ago from platforms point out some faunal changes. Vertical zonation in types of sessile foulers in Exxon Platform 54A suggest the dominance of the barnacle-*Enteromorpha* community in the tidal and surge zone; the hydroid-barnacle community in the submerged zone, and the hydroid *Bougainvillia*-serpulid community in the near-bottom turbid zone.

INTRODUCTION

In the photic zone of the continental shelf environment, submerged hard surfaces have been known to form an ideal base for the develop-

---

Robert George and Philip Thomas are with the Institute of Marine Biomedical Research, University of North Carolina at Wilmington.

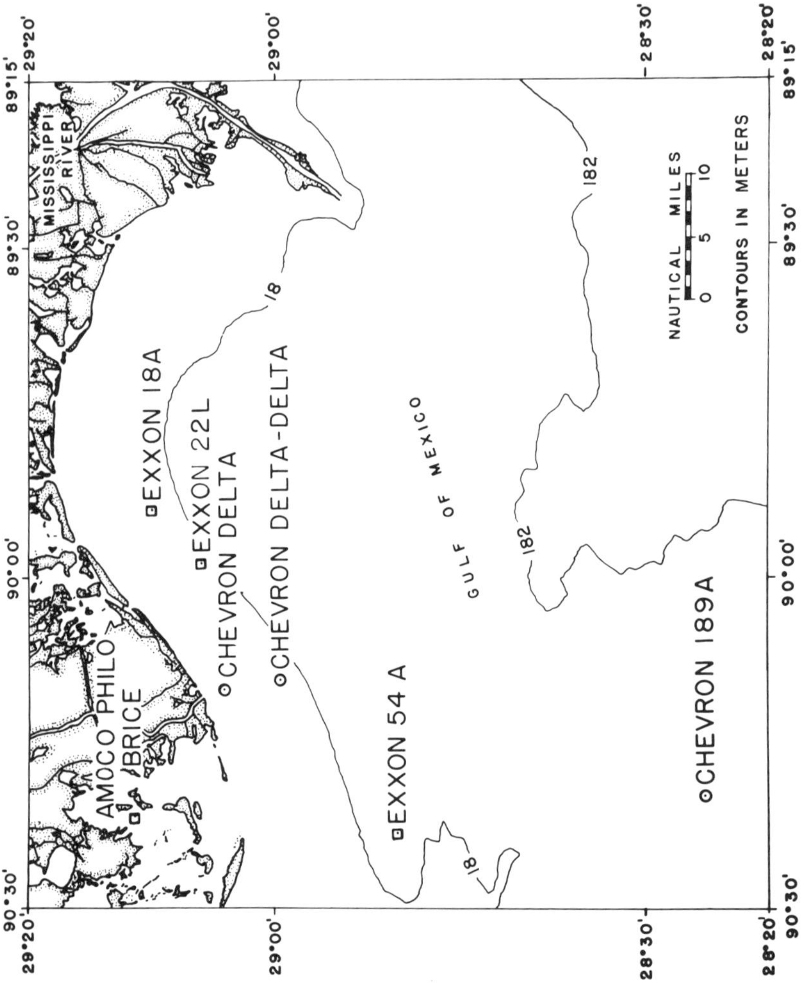

FIG. 1. MAP OF STUDY AREA.

ment and establishment of the fouling community. Therefore, the submerged structures of oil platforms present a suitable substratum, as an artificial reef, offering surface for successful settlement for all types of sessile organisms. This settlement eventually constitutes a mat of sessile plants and animals, which in turn offers shelter and food for a variety of motile organisms such as crabs, amphipods, isopods, polychaetes, and fish. The chlorophyll-bearing phytobenthos, including the filamentous algae *Enteromorpha* spp. and sessile diatoms, are the primary components of this fouling community in the tidal-splash zone of the platform structures. The animal components include both herbivorous and carnivorous organisms. The biofouling growth seems to prevail from the surface to the bottom as deep as 43 m in Gulf of Mexico platform structures. The growth attracts grazing fishes, which provide a renewable food supply for carnivorous fishes that are of recreational and commercial importance.

Biofouling studies on oil platform structures are limited. In the Gulf of Mexico two studies have been conducted on offshore fouling communities. Gunter and Geyer (1955) investigated the composition and abundance of these organisms in near-shore platforms off Grand Isle, Louisiana. Pequegnat and Pequegnat (1968) carried out studies on offshore fouling off Panama City, Florida.

The fouling community on seven different producing oil platforms, including one of those sampled by Gunter and Geyer, was examined in this study (figure 1). We investigated the effects of environmental changes such as the 1973 flood, the role of platforms as artificial reefs, the standing stock and seasonal trends, and the vertical and horizontal patterns of biofouling growth. The fouling community on oil platforms is an integral part of the Louisiana shelf biota, a status that developed after the introduction of these man-made structures, which offered large surfaces for the settlement of organisms. Sessile foulers can be used to estimate the possible impact of oil-production activities. Our study is therefore an important reference base for future evaluation of possible environmental impact.

METHODS AND MATERIALS

The quantitative data on the biofouling growth on oil platorms were obtained by three methods. Quantitative samples were taken of a known area of climax fouling growth at various depth levels from the submerged oil platform legs. A diver-operated scraping device (figure 2) was strapped around the platform leg and the biofouling growth was scraped with the movable blades and collected into an attached bag. The sample

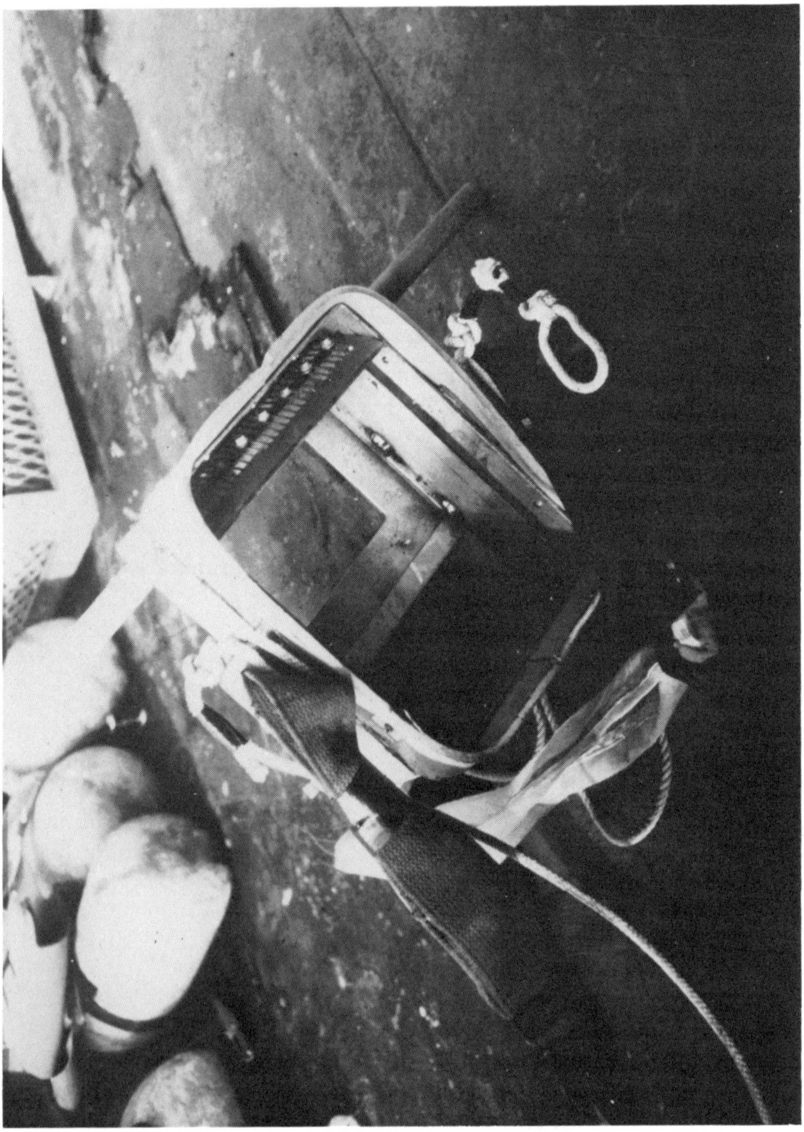

FIG. 2. THE DIVER-OPERATED SCRAPING DEVICE used for obtaining samples of fouling growth from submerged platform legs.

was sorted and estimations made on species composition, biomass, and density. The scraping sampling was done in all four seasons from the same platforms at similar depths.

An *in situ* photographic survey of biofouling growth on platform legs was made by divers using a Nikonas camera with a special close-up lens. The photographs were enlarged and analyzed for types of fouling organisms on platform legs from various depths.

For the third type of sampling, Plexiglas test panels (15 cm × 5 cm × 0.6 cm = total surface area of 174 cm^2) were suspended from surface to bottom beneath four different platforms at intervals of one meter. The test panels were located on an Amoco platform in the bay; on the near-shore Chevron Platform Delta; on Chevron Platform Delta Delta in the Bay Marchand area; and offshore on Chevron Platform 189A. The test panels were exposed for fouling growth during spring, summer, fall, and winter of the year 1972-73 for seasonal estimations of kinds and quantities of biofouling growth. The hydrographic data in the vicinity of platforms come primarily from Oetking (1973).

RESULTS

Fouling communities on platforms

Samples were analyzed by evaluating biomass and population density of the community and the member species. The summer sample was taken on July 11, 1973, when the water temperature had a mean value of 23° C at the bottom at 16.8 m and 30° C at the surface. The winter samples were collected on January 17, 1974, with the water temperature ranging between 18° and 19.5° C. The salinity of the water showed significant dilution in the summer of 1973, with a low of 17.8 ppt at the surface and a mean bottom salinity of 36.2 ppt. Salinity in the winter months ranged from surface to bottom between 31 and 34 ppt (Oetking 1973). The biological changes in the fouling community were examined in light of the fluctuations in the physical environment. Seasonal fluctuations in the climax community were subtle for the sessile foulers but well-marked in motile organisms and algae. Sessile acorn barnacles were the dominant organism at all depths (tables 1 and 2). Among the motile animals, the dominant amphipods in the climax community during the summer were mainly *Corophium acherusicum* and *Stenothoe* sp., but in the winter months caprellids and gammarids were the dominant forms. Algal biomass in the surface diminished during the winter months from 85.6 g/m^2 in summer to 24.5 g/m^2 in the winter. The biomass and density of the fouling species in the summer and winter period are tabulated for each group in tables 1 and 2.

Table 1

Biomass and Density of Fouling Organisms at Exxon 18 "A" on February 17, 1974

Type of Organism	2.4m		9.1m		12.8m	
	No./m^2	g/m^2	No./m^2	g/m^2	No./m^2	g/m^2
Barnacles	1390	5233.3	900	1525.1	50	30.0
Algae	++	24.5	--	--	--	--
Anemones	5120	49.7	10890	102.7	--	--
Amphipods	3440	16.5	35880	37.8	--	--
Polychaetes	3560	25.2	2800	36.1	475	0.5
Crabs	240	53.7	180	57.1	--	--
Copepods	--	--	750	7.4	--	--
Bivalves	40	5.9	12	2.1	--	--
Hydroids	--	86.5	--	18.9	--	3.5
Sponge	--	163.1	--	81.6	--	--
Bryozoa	--	2.8	--	1.1	--	4.0
Total Biomass		5661.2		1869.9		38.0
Sediment		289.8		202.4		24.5

Bathymetric Zonation of Climax Fouling Community on Exxon Platform 54A. The vertical zonation patterns of the fouling mat in all of the platform samples appeared to show similar qualitative distribution. A generalized picture of the major fouling assemblages of a platform leg from Exxon 54A is as follows: At the surface, just below mean high tide level, the outer covering of the fouling mat is dominated by the green algae *Enteromorpha*. The thickest concentration is usually in the upper 5 cm, with the algae rapidly diminishing below one meter, and only found in patches down to 4.6 m. Interspersed within the algal growth are the hydroids *Syncoryne*, and small motile animals, such as the amphipods *Corophium*, xanthid crabs *Neopanope texana*, and pycnogonids. The barnacles, underneath the filamentous growth, are *Balanus reticulatus* and *B. improvisus* in a 55-45 ratio. The barnacles are usually somewhat smaller (approximately 1-5 mm) than those found deeper.

Table 2

Biomass and Density of Fouling Organisms
at Exxon 18 "A" on July 11, 1974

Type of Organism	2.4m		9.1m		12.8m	
	No./m^2	g/m^2	No./m^2	g/m^2	No./m^2	g/m^2
Barnacles	2600	5069.3	925	2769.2	50	43.2
Anemones	2010	18.8	36925	42.4	50	0.1
Amphipods	21700	8.8	10375	2.5	--	--
Polychaetes	825	1.5	200	0.2	550	0.5
Crabs	25	1.3	325	3.0	--	--
Flatworms	--	--	100	0.1	--	--
Isopods	--	--	--	--	50	0.5
Hydroids	--	11.5	--	13.4	--	3.5
Sponge	--	10.9	--	24.8	--	--
Bryozoa	--	18.4	--	12.4	--	4.0
Total Biomass		5140.5		2868.0		48.8
Sediment		362.3		162.2		30.2

The algae-hydroid fouling mat is replaced by the sea anemones, *Aiptasia,* at depths of 2.4 to 6 m. The barnacles are larger, with the ratio of *B. reticulatus* to *B. improvisus* increasing to 70-30. *B. eburneus* is occasionally found intermixed with these barnacles but composes less than 1% of the total population. The amphipods *Corophium, Stenothoe,* and *Caprella,* xanthid crabs, and the blemmy *Hypleurochilus* dominate the motile fauna at this depth.

Hydroids dominate the outer covering of the barnacles below 7.5 m. The barnacles are found in diminishing patches from 7.5 to 10.7 m. *Balanus reticulatus* completely dominates the barnacle fauna, with only a few *B. improvisus*; *B. tintinnabulum* is present in patches and usually as empty shells.

The fouling fauna is completely dominated by hydroids from 12.2 m to the bottom. A few scattered, empty shells of *B. reticulatus* are present, but no living barnacles were recorded. A few serpulid worm

Table 3

Biomass (g) of Fouling Organisms Expressed in Terms of Each Quantitative Scraping Sample of 400 cm^2. Humble Platform 54A.

Species	Surface		2.4m		6.1m		12.2m		Total	
	Biomass	No.	Biomass	No.	Biomass	No.	Biomass	No.	Biomass	No.
Algae										
Polysiphonia sp.	+	+	+	+					+	+
Enteromorpha sp.	+	+	+	+	+	+			+	+
Coelenterata										
Syncoryne sp.			+	+					+	+
Bougainvillia tennella							+	+	+	+
Bundosoma cavernata							.05	7	.05	7
Ectoprocta										
Bugula neritina			+	+					+	+
encrusting sp. 1							8		8	
Sipunculida										
Sipuniculid sp. 1	.07	1			.04	1			.11	2
Annelida										
larval polychaetes			.04	13	.01	1	.01	3	.06	17
Mollusca										
Lithophaga aristata							.56	2	.56	2
Arthropoda										
Balanus reticulatus	140	490	120	420	100	350	29	101	389	1361
Neopanope sp.	.47	8	.03	3	.03	3	.1	1	.66	15
Synalpheus townsendi			.36	4	.13	3	.06	2	.55	9
Chordata										
Tunicate sp. 1					19	+			19	+
Hypleurochilus geminatus			.25	2					.25	2
Total Biomass	140.47		120.68		119.21		37.78		399.69	

tubes were found in the near-bottom zone, which is apparently under the intermittent influence of the turbid layer observed during the OEI investigations (Oetking 1973).

The biomass distribution of the total fouling growth shows a depth-related trend with abrupt decrease beyond 12.2 m (table 3). Such a sharp fall in biomass with increase in depth, which is correlated with the population of *B. reticulatus*, is also seen in the fouling biomass in two other adjacent Exxon oil platforms, 63A and 22L (figure 3).

Acorn Barnacles on Louisiana Oil Platforms. Four species of barnacles were encountered in the fouling studies, all of which belong to the genus *Balanus: B. reticulatus*, *B. improvisus*, *B. eburneus*, and *B. tintinnabulum*. The dominant species on the oil platform legs was *B. reticulatus*; this is the first record of it from the Louisiana coast. This species was absent from the barnacle faunal lists of Henry (1954), Gunter and Geyer (1955), Wells (1966), and Pequegnat and Pequegnat (1968). *B. reticulatus* was found at all stations and at all depths (0-18.3 m), predominantly above the 10.7 m depth on the Louisiana offshore platforms. It ranged in size from 2 to 30 mm in diameter. *B. reticulatus* dominated the growth on the test panels during the summer months. It was rarely found in January and completely absent in the spring. These data suggest a breeding and settlement cycle between May and December that corresponded to Utinomi's (1967) records. The height was between 3 and 20 mm with specimens ranging from 10 to 17 mm in diameter.

The second most abundant species of barnacles was *B. improvisus*, which is well known in the Gulf (Henry 1954). This barnacle was present at all stations, but restricted to the upper 9.1 m on the pilings. It grew on the shells of *B. reticulatus* and was rarely observed directly attached to the platform leg. Apparently, off the Louisiana coast *B. improvisus* is restricted by competition from *B. reticulatus*, a situation similar to the interaction of *B. improvisus* and *B. balanoides* in the Massachusetts coast, where *B. balanoides* is dominant over *B. improvisus* (Zullo 1963, 1966). In the test panel studies, *B. improvisus* was found to dominate only in the January and March samples, reflecting a winter breeding and settlement cycle. This cycle corresponded with the breeding cycle observed by McDougall (1943), but conflicted with the data reported by Vissicher and Luce (1928) from North Carolina. Gunter and Geyer reported larval settlement in all seasons of the year, and Pequegnat and Pequegnat similarly found the species settling throughout the year. Therefore, *B. improvisus* in the Gulf of Mexico is capable of seasonal breeding, but settlement of larvae is restricted to the winter months because of competition from *B. reticulatus*.

B. eburneus was the third species encountered occasionally on the

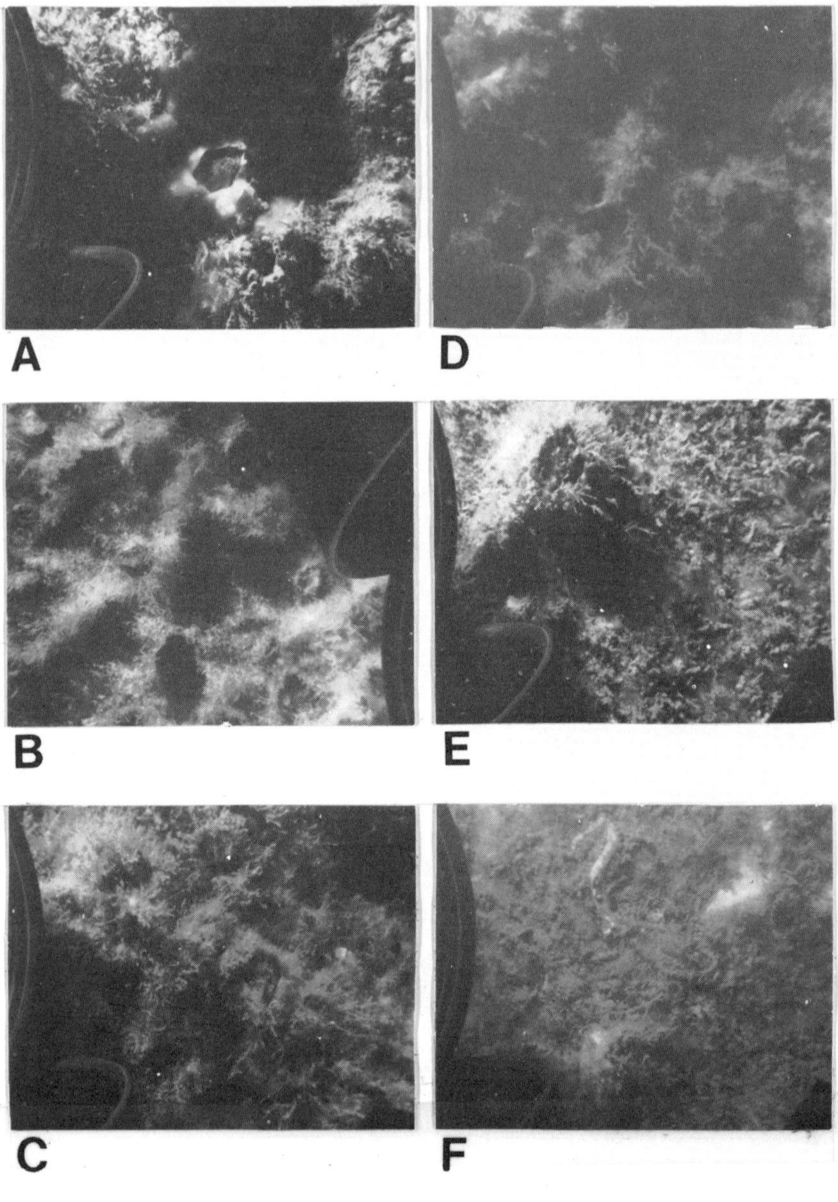

FIG. 3. CLOSEUP *IN SITU* PHOTOGRAPHS OF FOULING GROWTH ON PLATFORM LEGS. Platform CAGC-G1-63A, April 3, 1973: (a) 1.5 m; (b) 7.6 m; (c) 10 m; (d) 13.4 m; (e) 30.5 m; (f) 27.1 m.

platform legs in depths from 3 to 9 m. The few specimens recorded were 18-25 mm in size and were usually growing in between *B. reticulatus*. A few specimens of *B. tintinnabulum* were also observed on Exxon Platform 54A, but all specimens were dead; the empty shells measured 40-60 mm. All four barnacle species are shown in figure 4.

This analysis of the distribution of the four acorn barnacle species indicates an invasion of *B. reticulatus* as a new colonizer to the Louisiana shelf environment. The dominant nature of *B. improvisus* in certain seasons is also evident. *B. eburneus* and *B. tintinnabulum* are much less frequently found on offshore platform legs.

Seasonal Impact of Spring Flood on Settling Fouling Community on Oil Platform Delta. The fouling community was examined quantitatively before and after the 1973 spring flood at the near-shore Platform Delta and the Timbalier Bay platform. Significant changes in both density and biomass occurred during the post-flood periods when the waters were diluted down to 15-25 ppt. This condition facilitated the migration of euryhaline bay populations into the inner shelf environment. This trend was clearly noted in the pronounced increase in biomass and density of populations in a comparison between the pre-flood and post-flood conditions in 1973 at the near-shore Platform Delta. The total biomass of settling fouling community ranged between 100 and 550 g/m^2 in spring 1973 and increased to 2500-6500 g/m^2 in the summer. The summer development of fouling organisms on this near-shore platform in the warm and less saline period of summer 1973 was attributed to the influence of the flood and subsequent introduction of euryhaline fouling species into the inner shelf environment. A marked decrease in both biomass and density was encountered subsequently in fall 1973 when the biomass ranged between 100 and 900 g/m^2. These changes suggest that seasonal variation in the standing stock of the fouling community in the Louisiana shelf oil platform is hydrographically regulated by the flux of water from the coastal bays and Mississippi River, or by the increase in water, or by a combination of these two factors.

Fouling communities on test panels

Bathymetric Zonation of Settling Fouling Community. The fouling community structure on Platform Delta was based on quantitative analysis of biomass at various depth levels (figure 5). It is evident that the fouling community is strikingly different in composition between surface and bottom. Barnacles compose about 25% of the biomass at the surface and increase to about 55% between 1.2 and 4.9 m, followed by a decrease to zero at the bottom. However, empty barnacles were encountered in greater density at mid-depths and at the bottom. Test panels

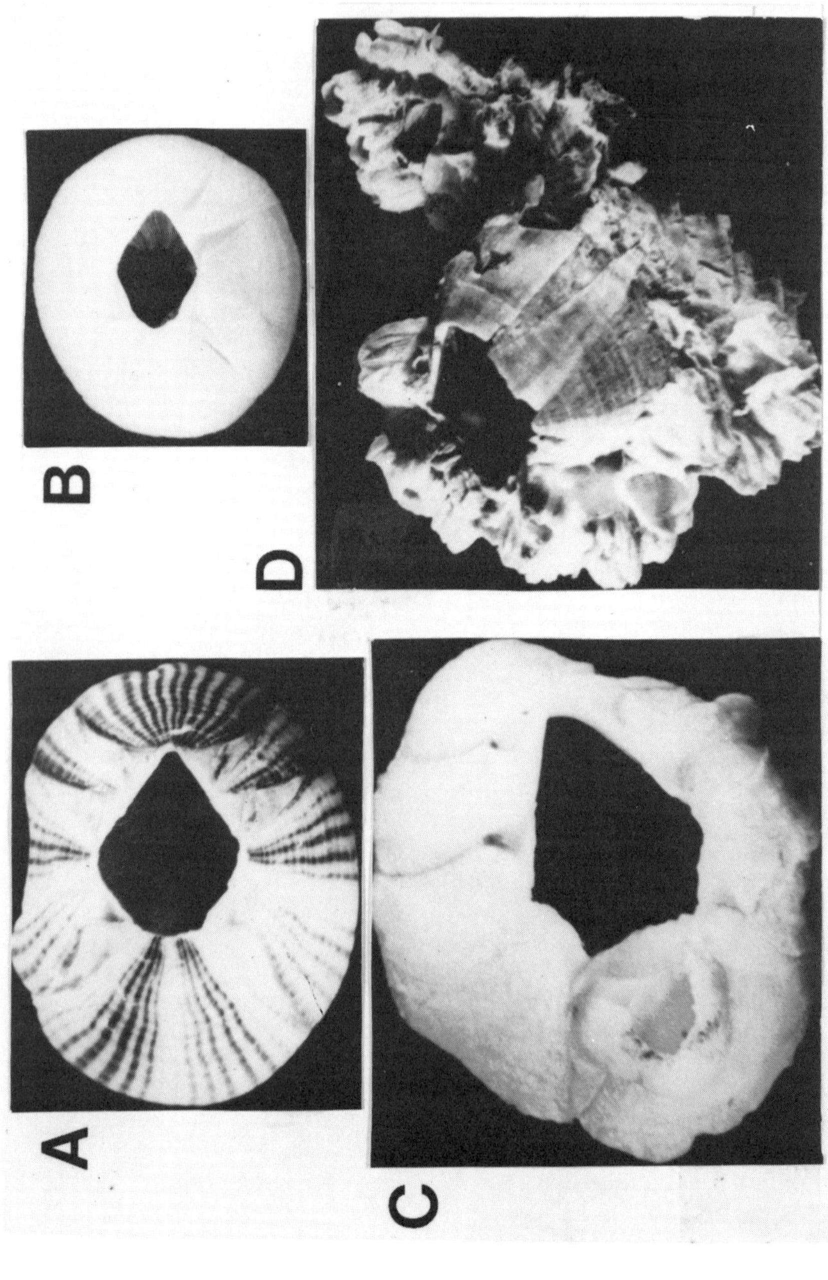

FIG. 4. THE FOUR BARNACLE SPECIES ENCOUNTERED IN THE BIOFOULING GROWTH ON LOUISIANA OFFSHORE OIL PLATFORMS: (a) *Balanus reticulatus*; (b) *B. improvisus*; (c) *B. eburneus*; (d) *B. tintinnabulum*.

submerged for 60 days contained such empty shells of about 2.5 cm in diameter, suggesting an initial rapid growth of barnacles following settlement. The ratio of living to dead barnacles showed a relationship with depth and water characteristics. A three-layered vertical water structure was recognized with murky conditions occupying the surface layer, clear water in the middle zone, and a turbid layer on the bottom. The decrease in biomass with increase of depth seemed related to water characteristics, with evidence of diminished growth in the turbid layer zone. Encrusting bryozoans became more numerous as the population of barnacles declined.

The distribution of the two common amphipods at the Chevron Platform Delta Delta seemed to be depth-related. *Corophium* sp. reached its density peak in the upper 3 m and *Caprella* sp. was less common. In the mid-depths, both species were equally represented, and at depths beyond 10.6 m *Caprella* sp. reached its dominance and *Corophium* sp. declined. Both species are evidently competitors and separation of them by depths suggests a biological equilibrium dependent upon food or habitat.

Comparison of Inshore and Offshore Fouling Community. The relationship of distance from the shore and standing stock of the fouling community on platforms was examined on the basis of biomass of settlement. The data for the inshore oil platform were obtained from an Amoco platform located in the northern part of Timbalier Bay near Philo Brice Island. The water is 2.4-3 m deep; salinity ranged from 20 to 28 ppt in 1972-1974, and temperature ranged between 8° and 29° C. The biomass averaged 4175 g/m^2 at the surface, 3108 g/m^2 at 1 m, 2336 g/m^2 at 1.8 m, and 84 g/m^2 below 2 m (figure 6). The water in the bay was turbid and the low biomass in the bottom depth was probably the result of the suspended sediments and high turbidity. The total biomass was high at the surface, with green algae *Enteromorpha* constituting 20% of the biomass. Barnacles were the dominant member of the inshore fouling community with a population of 60,402/m^2, followed by polychaetes (56,781/m^2) and the flatworm *Leptoplana* sp. with 20,747/m^2. This flatworm occurred only in the Timbalier Bay fouling community and was not encountered offshore.

Three platforms were compared to determine the function of the distance from shore on the settlement of organisms on test panels. The platform near Philo Brice Island was used for the Timbalier Bay station. Bay Marchand oil platform is located four miles from shore and is within the influence of the Mississippi River. Chevron Platform 189 (see figure 1) is located 45 miles from shore in the blue water zone of the Gulf and is probably outside the influence of the Mississippi River

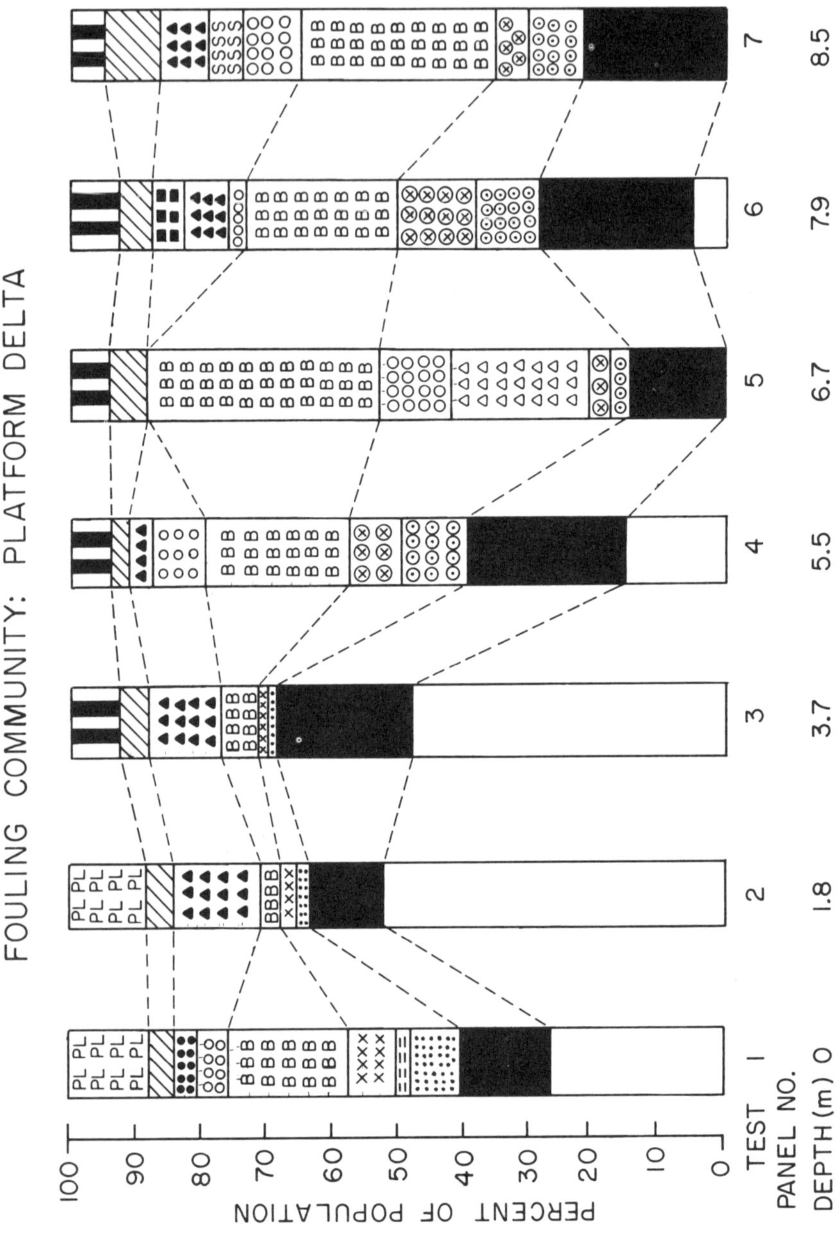
FOULING COMMUNITY: PLATFORM DELTA
PERCENT OF POPULATION
100
90
80
70
60
50
40
30
20
10
0
TEST I
PANEL NO. 1 2 3 4 5 6 7
DEPTH (m) 0 1.8 3.7 5.5 6.7 7.9 8.5

FIG. 5. BIOMASS STRUCTURE OF FOULING COMMUNITY based on quantitative analysis of fouling growth after 60 days exposure at 7 different depth levels beneath Chevron Platform Delta. Percentage representation of various taxa is given for each test panel. Code for symbols used is as follows:

- Living barnacle *Balanus reticulatus*
- Dead barnacle *Balanus reticulatus*
- Bryozoan *Bugula neritina*
- Chlorophyta - filamentous alga *Enteromorpha* sp.
- Hydroid *Bougainvillia tenella*
- Hydroid *Syncoryne* sp.
- Sea anemone sp. #2
- Sea anemone sp. #3
- Polychaete
- Decapod Xanthid crabs
- Crustacean larvae
- Amphipod *Caprella* sp.
- Tunicate unidentified
- Appendages, Mucus and Moults
- Sediment and accumulated debris from suspended matter

Total biomass ranged from 4150 g/m^2 at the surface to 1740g/m^2 at the bottom.

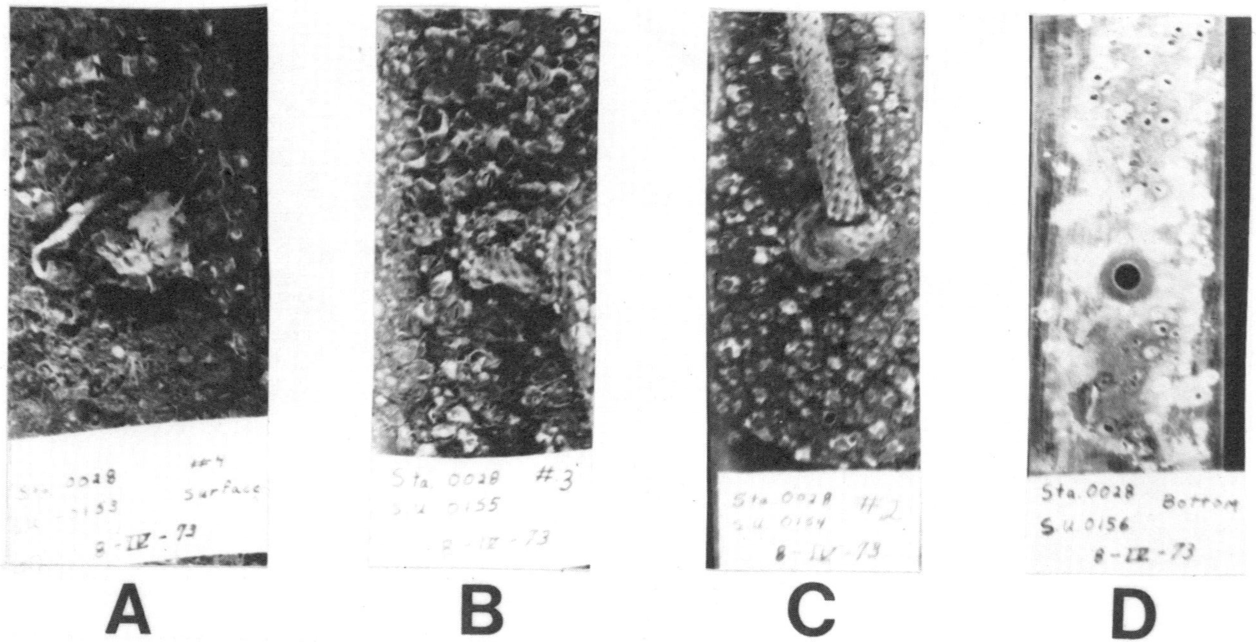

FIG. 6. ASSEMBLAGE OF PHOTOGRAHS OF *IN SITU* TEST PANELS showing fouling growth in summer 1973 at the Amoco oil platforms at Timbalier Bay. Note the reduced fouling settlement near the mud bottom: (a) surface; (b) 0.9 m; (c) 1.8 m; (d) 2.7 m.

plume. The species distribution from the three platforms is given in table 4. The greatest diversity of life was present at the near-shore platform, where nearly twice the number of species were present as at the other two sites. The increase in number of species at the near-shore station was equally divided between sessile and motile organisms. The biomass, in contrast, was greatest at the offshore platform. Three species of grazing fish, the sheepshead *Archosargus probatocephalus*, spade fish *Chaetodipterus faber*, and trigger fish *Balistus capriscus* were commonly found at the near-shore platform, and the red snapper *Lutjanus campechanus* was observed at the offshore station.

DISCUSSION

These studies of the fouling communities of the oil platforms clearly demonstrate the large amount of biomass present on these structures. Since each platform averages about 0.6 hectares of surface area, and since there are over 1500 of these structures in the OEI area, these represent a total of 900 hectares of surface area available for settlement of fouling organisms. If the minimum and maximum values of biomass of 100 to 8000 g/m^2 are used to calculate the total amount of biomass, then the quantity of fouling organisms present on the platforms of the OEI area ranges from 9×10^8 to 7.2×10^9 g total for all structures. Therefore, this extensive surface area with luxuriant fouling growth out to at least 50 miles offshore acts as a reef, resulting in the aggregation of fishes in the vicinity of these structures. The common grazing fishes present around platforms included the red snapper *Lutjanus campechanus*, mangrove snapper *Lutjanus griseus*, a grouper, blue runner *Caranx crysos*, bull shark *Carcharhinus leucas*, blenny *Hypleurophilus geminatus*, ocean trigger fish *Canthidermis sufflamen*, Atlantic spade fish *Chaetodipterus faber*, and the striped sheepshead *Archosargus probatocephalus*. Fishes were not commonly seen grazing on the fouling growth within the bottom turbid (= nepheloid) layer.

The reef effect possibly promotes the standing stock of fish populations or perhaps merely dislocates the fish concentrations. Added studies are needed to characterize the reef effect of platforms and to determine the long-term biological impact of attracting the colonization of species from other adjacent or distant provinces.

Gunter and Geyer studied the fouling community on four oil platforms off the Louisiana coast in 1951-1953, two of which are within the present study area of the OEI. These two Exxon oil platforms within the same area were chosen for a comparative study. We have interpreted Gunter and Geyer's data in our tables 5 and 6.

Table 4

Qualitative Occurrence of Fouling Organisms Present
On Oil Platforms in the O.E.I. Area, Louisiana
June 1972 - July 1973

m = motile s = sessile

Species	Timbalier Bay	Nearshore	Offshore
Phylum Chlorophyta			
Enteromorpha sp. (s)	X	X	X
Phylum Rhodophyta			
Polysiphonia sp. (s)	X	X	
Phylum Porifera			
sponge, unidentified (s)			X
Phylum Coelenterata			
Bougainvillia tennella (s)		X	
Bundosoma cavernata (s)		X	
Astrangia astreiformis (s)			X
Phylum Platyhelminthes			
Leptoplana sp. (m)	X		
Phylum Annelida			
Nereis sp. (m)	X		
Neanthes succinea (m)	X	X	X
Eunice subra (m)	X		
Pomatoceros sp. (s)		X	
Phylum Mollusca			
Ostrea equestris (s)		X	
Lithophaga aristata (s)		X	
Crassostrea virginica (s)	X	X	
Mitrella sp. (m)		X	
Phylum Arthropoda-Crustacea			
Copepoda			
Lapidocera aestiva (m)		X	
Acartia tonsa (m)		X	
Cirripeda			
Balanus reticulatus (s)	X	X	X
Balanus eberneus (s)		X	X
Lepas sp. (s)			X
Amphipoda			
Hemiaegina sp. (m)		X	
Stenothoe sp. (m)		X	
Corophium sp. (m)	X	X	
Caprella sp. (m)	X	X	
Tanaidacea			
Tanaid, unidentified (m)		X	
Isopoda			
Sphaeroma quadridentatum (m)	X		
Limnoria tripunctata (m)	X		
Decapoda			
Synalphaus townsendi (m)		X	
Neopanope texana (m)		X	X
Pilumus pannosus (m)		X	X
Eurypanopeus abbreviatus (m)	X	X	X
Phylum Ectoprocta			
Bugula neritina (s)		X	
Acanthodesia sp. (s)			X
Phylum Echinodermata			
Arbacia punctulata (m)		X	X
Phylum Chordata			
ascidian, unidentified (s)			X
Number of Species	13	25	13

Table 5

Faunal List of Fouling Assemblage off the Louisiana Coast
Tabulated from Gunter and Geyer (1955)

Species	Relative Abundance	Optimum Depth (meters below surface)	Depth Range (meters below surface)
Algae	++	0-0.6	0-0.6
Phylum Coelenterata			
Astrangia asteriformis	+	9.1	5.2-14.6
Anthopleura krebsi	+++	9.4-14	surface-bottom
Aiptasia pallida	+++	0.6-5.2	surface-bottom
hydroids	+++	0.6-5.2	surface-bottom
Phylum Annelida			
Eupomatus	+	bottom	bottom
Phylum Mollusca			
Crepidula	rare	bottom	9.4-14
Pteria eolymbus	rare	9.4	9.4
Arca transversa	rare	14	14
Ostrea equestris	rare	14.6	9.4-14.6
Crassostrea virginica	rare	0.6	0.6
Thais haemostoma	rare	5.2	0.6- 7.9
Phylum Arthropoda-Crustacea			
Cirripedia			
Balanus improvisus	+++	0.6-9.4	surface-bottom
Balanus amphitrite	rare	1.2	1.2
Amphipoda			
Corophium	+++	0.6-5.2	surface-bottom
Decapoda			
Menippe	rare	1.2	1.2
Phylum Ectoprocta			
Aeanthodesia	+	bottom	surface-bottom

+++ = Abundant, ++ = Common, + = Present, rare = Less than 10 individuals observed.

These data reflect three apparent differences between the climax fouling community now and two decades ago: (1) the composition of the motile components, (2) the presence of sponges in the 1973-74 study, and (3) the presence of the barnacle *Balanus reticulatus*. The first two differences are possibly due to sampling technique, and the third due to the introduction of the barnacle *B. reticulatus* into the Gulf of Mexico, perhaps as a consequence of increased shipping operations in the past two decades.

The fouling samples in the present study were collected with an *in situ* scraping device that collected both sessile and motile species. Gunter and Geyer suspended steel cylinders from a platform and retrieved them from the surface, allowing the escape of the motile animals such as

Table 6

Faunal List of Fouling Assemblage off the Louisiana Coast
From Exxon 22 "L" and Exxon 18 "A"

Species	Relative Abundance	Optimum Depth (meters below surface)	Depth Range (meters below surface)
Algae			
Enteromorpha	++	0- 0.3	0- 4.6
Phylum Porifera			
Cliona	+++	4.6- 7.6	3-10.7
Lissodendoryx	+	4.6- 7.6	3-10.7
Phylum Coelenterata			
Aiptasia pallida	+++	3- 7.6	5-10.7
Astrangia asteriformis	rare	12.8	12.8
hydroids	+++	surface	surface-bottom
Phylum Annelida	+++	6.1- 9.1	surface-bottom
Phylum Mollusca			
Crassostrea virginica	rare	0- 3	0- 4.6
Ostrea equestris	rare	7.6-10.7	6.1-10.7
Thais haemostoma	rare	9.1	7.6-10.7
Phylum Arthropoda-Crustacea			
Copepoda			
Labidocera aestiva	rare	0- 3	0- 9.1
Cirripedia			
Balanus reticulatus	+++	5- 9.1	surface-bottom
B. improvisus	++	5- 6.1	above 9.1
B. eburneus	rare	4.6	3- 9.1
B. tintinnabulum	rare	9.1	7.6-10.7
Isopoda	rare	12.8	10.7-13.7
Amphipoda			
Corophrium	+++	0- 6.1	0-10.7
Stenothoe	+++	0- 6.1	0- 9.1
Caprella	++	3- 9.1	0- 9.1
Decapoda			
Neopanope texana	+	3- 9.1	0-10.7
Eurypanopeus	+	3- 9.1	0-10.7
Phylum Ectoprocta			
Bugula neritina	+	2.4	3- 6.1
Membranipora	+	10.7	surface-bottom
Chordata			
Ascidian	+	6.1	3-10.7
Hypleurochilus	+	7.6	3-10.7

+++ = Abundant, ++ = Common, + = Present, rare = Less than 10 individuals observed.

amphipods. The difference in collecting techniques may explain the near absence of motile crustacea in the study by Gunter and Geyer (see table 5).

The absence of the sponge, *Cliona*, from Gunter and Geyer's work could be attributed to the duration of their experiment. The longest exposure period was 14 months from November 1950 to January 1952. Platform legs that had been exposed for 14 months (June 1972 to August 1973) had a significantly smaller amount of *Cliona* growth than was present on the older portion of the structure from which scraping samples were taken.

These results point out colonization of new species on the same platform, suggesting faunal change in the ensuing 20 years. The results also revealed the obvious influence of 1973 spring flood by promoting the invasion of less saline shore species into the near-shore and offshore environment. Evidently the shelf ecosystem in Louisiana coast is strongly influenced by the Mississippi drainage into the shelf.

Comparison of the results of this study and the Northeastern Gulf studies conducted by Pequegnat and Pequegnat is difficult because of geographic distinctions between the two regions. The fouling community in the Northeastern Gulf was found to be under the influence of East Gulf Loop current and apparently unaffected by the Mississippi River plume. A benthic "turbid layer" was not observed off the coast of Florida. One readily discernible difference is that the dominant biomass on the Louisiana platforms is made up of barnacles and hydroids, whereas in the Florida offshore locations the dominant organisms were sponges, corals, and tunicates. A recent photo-survey of offshore structures off Panama City also confirmed this distinction.

CONCLUSIONS

1. High standing stock and density of fouling on submerged oil platform structures offer feeding grounds for grazing fishes. Therefore, the oil platforms compare ecologically to artificial reefs.

2. The record spring flood in 1973 diluted the water, leading to invasion of shore fouling species into near-shore and offshore platforms and significantly enhancing the total standing stock of the fouling community.

3. Comparison of the climax community of fouling growth in the same Exxon platforms today and that of 20 years ago revealed faunal changes in association with environmental changes of ecological consequence. This points out the dynamic, not static, nature of the ecosystem.

4. Trends in both horizontal (distance from shore) and vertical (depth-related) variations in biomass and density distribution of the fouling community are evident.

5. The fouling in Louisiana oil platforms differs from known off-

shore fouling in Florida coast waters primarily because of hydrological conditions.

ACKNOWLEDGMENTS

We acknowledge with gratitude the cooperation and support provided by the Gulf Universities Research Consortium. The use of helicopters offered by the Chevron Oil Company facilitated recovery of test panels from several offshore oil platforms.

REFERENCES CITED

Gunter, G. and R. A. Geyer. 1955. Studies on fouling organisms of the northwest Gulf of Mexico. Publications of the Institute of Marine Science, The University of Texas **4**(1):37-87.

Henry, D. P. 1954. Cirripedia: The barnacles of the Gulf of Mexico. *In* P. S. Galtsoff, ed., Gulf of Mexico: Its Origin, Waters, and Marine Life. Fisheries Bulletin 89, U.S. Fish and Wildlife Service **55**:443-446.

McDougall, K. D. 1943. Sessile marine invertebrates of Beaufort, North Carolina. Ecological Monographs **13**(3):321-374.

Oetking, P. 1973. Hydrography on the nearshore continental shelf of south central Louisiana. Report submitted to the September Meeting of the Offshore Ecology Investigation, Tampa, Florida.

Pequegnat, W. E. and L. H. Pequegnat. 1968. Ecological aspects of marine fouling in the northeastern Gulf of Mexico. Technical Report to the Office of Naval Research.

Utinomi, H. 1967. Comments on some new and already known cirripeds with emended taxa, with special reference to the parietal structure. Publications of the Seto Marine Biological Laboratory **15**(3):199-237.

Vissicher, J. P. and R. H. Luce. 1928. Reactions of the cyprid larvae of barnacles to spectral colors. Biological Bulletin **54**:336-350.

Wells, H. W. 1966. Barnacles of the northeastern Gulf of Mexico. Quarterly Journal of the Florida Academy of Science **29**(2):81-95.

Zullo, V. A. 1963. A preliminary report on systematics and distribution of barnacles (Cirripedia) of the Cape Cod Region. Marine Biological Laboratory, Woods Hole, Massachusetts.

———. 1966. Thoracic Cirripedia from the continental shelf off South Carolina, U.S.A. Crustaceana **11**(3):229-244.

RECENT GEOLOGICAL HISTORY OF THE TIMBALIER BAY AREA AND ADJACENT CONTINENTAL SHELF

by James P. Morgan

ABSTRACT

The basic outlines of Timbalier Bay and the adjacent continental shelf were sketched at the close of the last Ice Age, approximately 3600 years ago. Since then, the area has been shaped and changed by three major processes: river deposition, sea level fluctuation, and land variation. Sedimentation from the Mississippi River has caused the deltaic plain to prograde seaward, while a rise in sea level and land subsidence have caused a general inundation of areas that were, until very recently, dry land.

INTRODUCTION

To comprehend the complex and changing environment of the central Louisiana continental shelf and adjacent beaches, bays, and marshlands, it is necessary to consider the geological development of this region during the Holocene, or Recent epoch. This latest geological episode is generally thought to span the past 16,000 to 18,000 years, an interval of ameliorating climate following the last of the glacial ages.

During the last few thousand years of the Holocene, approximately one-third of Louisiana's present land area was fashioned by a series of highly dynamic geological processes. Although the processes are numerous and complex, there are three dominant and interrelated aspects to be considered: (a) deltaic sedimentation by the Mississippi River, (b)

James Morgan is Professor Emeritus of Geology at Louisiana State University.

sea-level fluctuations (mainly rise), and (c) land-level variations (principally subsidence). The Gulf of Mexico surface establishes the base level of the Mississippi River, as well as that of other streams. Base level also controls deltaic sedimentation. Therefore, fluctuations of land level relative to water level significantly influence sedimentation and must be considered in addition to the depositional processes themselves.

SEA-LEVEL VARIATIONS

At the culmination of the last ice age, glacial ice covered approximately 15,000,000 square miles of the earth's polar and mountainous continental land areas. During the subsequent 17,000 or so years, most of this ice melted, until at present less than 6,000,000 square miles of land are ice covered. The melt waters returned to their source, and sea level rose some 400 feet during the same period (see figure 1). Glacial melting was rapid at first and produced rapid sea-level rise, but the rates diminished significantly during the past 7000 years (Curray 1965).

Scientists disagree on the details of sea-level fluctuations during these last few thousand years, and definitive work remains to be done. In general, however, three different interpretations have been presented: (a) a continuous, slow rise of sea level for the past several thousand years; (b) a rise until approximately the present level was reached some 3600 to 5000 years ago, with essential stability since that time; and (c) continuing rise until about 6000 years ago to a few feet higher than at present, followed by a minor fall to the present level. Though I will not evaluate the details here, the second of the concepts above seems best to fit the facts known for Louisiana (Coleman 1966). Sea-level rise seems to have

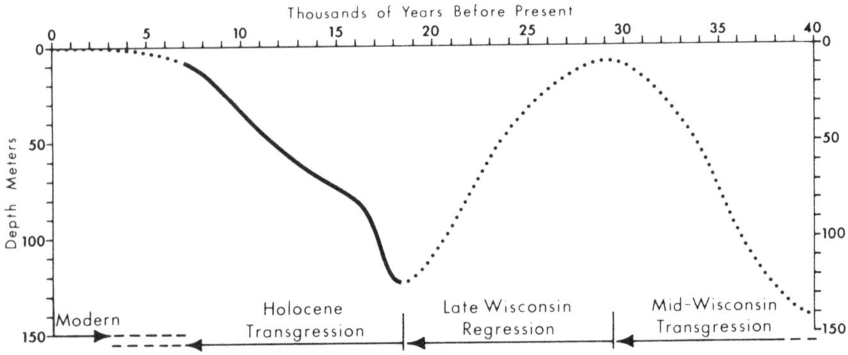

FIG. 1. GENERALIZED CURVE OF LATE QUATERNARY SEA-LEVEL FLUCTUATIONS. Modified from Curray (1965).

continued at a decelerating rate until about 3600 years ago, at which time it reached its present level. Subsequently, there seem to have been only minor fluctuations of a foot or so.

Precise data on sea-level variations are furnished by continuously recording tide gauges that are maintained at many seaports around the world. Since accurate gauging has been technically possible for less than a hundred years, reliable data are available only for a relatively short interval. For example, the first tide gauge established in the Gulf of Mexico (at Galveston, Texas) has been in operation only since 1909 (figure 2). In this relatively short time, the gauge reveals an overall sea-level rise of more than one foot. The curve shows a considerable variation in level from year to year and the "average" or "mean sea level" curve can be drawn and interpreted in various ways.

Furthermore, such a gauge, mounted on pilings, records not only rise of sea level but also subsidence of the underlying land. It is very difficult to separate these two processes that have the same end result. Whatever its cause, the "relative" sea-level rise at the Galveston gauge means that there has been a resulting inundation of low-lying lands by the sea, an event that geologists call "marine transgression." Such transgressions result in ecological changes: sea water replaces fresh water, with resulting effects on the fauna and flora that inhabit the area.

Although a one-foot sea-level rise hardly constitutes a major marine transgression, in the low marshlands of Terrebonne and Lafourche

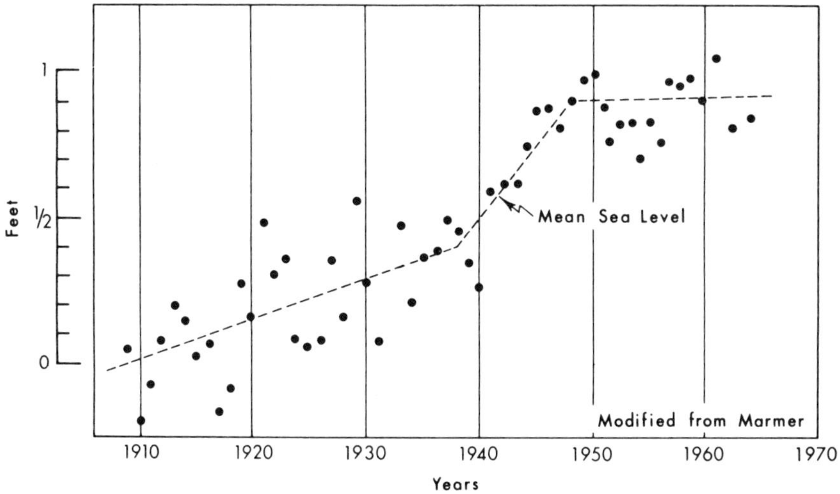

FIG. 2. SEA-LEVEL VARIATIONS AT GALVESTON, TEXAS. Modified from Marmer (1952).

parishes, many hundreds of square miles have been influenced. These natural changes have taken place concurrently with a growing human use of the marshlands for trapping, commercial and pleasure boating, and especially the search for and exploitation of petroleum, natural gas, sulfur, and salt. It becomes very difficult to separate the changes caused by the minor marine transgression from those related to man's alteration of the marshlands.

In contrast to the minor sea-level rise recorded during the past 75 to 100 years, there has been a major marine transgression since the end of the last ice age. The rise, more than 400 feet, caused the shoreline to move from approximately the continental shelf edge to its present position, a distance of as much as 140 miles in western Louisiana (figure 3). Numerous transgressive shoreline features were left on the inundated continental shelf as sea level rose. Elongate, sandy deposits, such as Ship Shoal, 10 to 15 miles seaward of Isles Dernieres (figure 3) still await study, but they probably represent beach deposits formed during minor cessations or changes in rate of sea-level rise (Fisk 1955).

Shallow borings in southwest coastal Louisiana reveal a series of peat deposits related to early Mississippi River deltaic deposition (Coleman 1966). These peat deposits have been radiocarbon dated and reveal various ages back to some 7200 years ago, when sea level stood about 25 feet lower than at present. From data such as these, it can be argued that the Mississippi deposited deltaic sediments at various places across the continental shelf during the post-glacial transgression. Details of these events are unknown, but it is quite possible that the Louisiana shelf is covered by a complex of deltaic deposits, as well as by reworked beach and near-shore features.

LAND-LEVEL VARIATIONS

The east-west-trending Louisiana coastline generally follows the axis of a downwarped section of the continental margin known as the "Gulf Coast geosyncline." Tens of thousands of feet of deltaic and shallow-marine sediments have accumulated in this subsiding trough during the past few tens of millions of years. (These Cenozoic-age sediments are the source beds for south Louisiana's rich petroleum accumulations.) The geosynclinal sediments reveal a nearly continuous record of basin subsidence contemporaneous with sediment deposition. Such subsidence continues to affect the Gulf Coastal Plain, but at rates that become significant only when considered in spans of many thousands of years.

A more significant, short-term factor is the localized subsidence that accompanies compaction and water expulsion from newly deposited

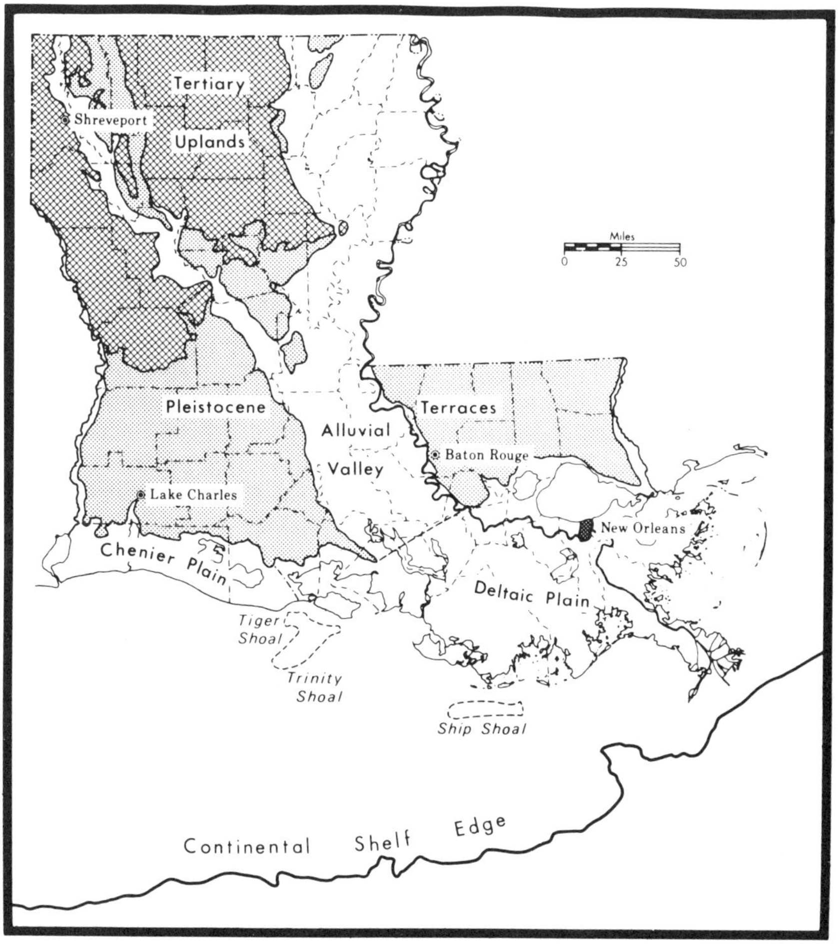

FIG. 3. PHYSIOGRAPHY AND GENERALIZED GEOLOGY OF THE LOUISIANA COASTAL PLAIN AND SHELF.

sediments. The significance of this compaction can be seen in its influence on the materials dumped rapidly by the Mississippi in fashioning its delta. Precise measurements are difficult, but sediment compaction in the delta vicinity has lowered the surface at rates ranging from 1.3 to 4.3 cm per year (Swanson and Thurlow 1973). Such subsidence rates reflect compaction and water expulsion from the recently deposited sediment. Compaction, in turn, is determined by the nature of the sedimentary particles and by the amount of water contained in the spaces between the particles.

The Mississippi delta sediments, like those of most deltas, vary greatly both horizontally and vertically in particle size. They are composed dominantly of clay and silt-sized particles, with only a small percentage of fine and very fine sand. The clays contain a large quantity of water, both attached to and trapped between the flake- or sheet-like clay particles. Clay deposits yield this water slowly but continuously under compaction. Silts and sands behave differently from clays in that individual particles are more nearly spherical, with water confined to the voids between particles. Under compaction, this water is expelled until the individual sediment grains are in physical contact, at which time compaction virtually ceases. Hence, deposits containing significant fractions of clay subside under compaction for longer periods, but sandy strata compact rapidly. Thus, variable subsidence rates reflect, among other factors, local variation in sediment type deposited by the river.

DELTAIC SEDIMENTATION—MISSISSIPPI RIVER

Depositional processes

Rapid sea-level rise and resultant marine transgression typified the early part of the Holocene. Some 7000 years ago, the rate of rise began to slow, and depositional processes by the Mississippi River, previously overwhelmed, became more prevalent. A number of deltaic depositional sites developed on the continental shelf, only to be abandoned and subsequently inundated by the rising Gulf of Mexico. About 5000 years ago, the river discharged its sedimentary load in central coastal Louisiana and fashioned a broad, delta-shaped mass, known as the "Sale-Cypremort delta" (figure 4). Gulf level at that time was apparently still some 8 to 12 feet lower than at present. Following abandonment of the Sale-Cypremort delta in favor of a depositional center farther east, the Sale-Cypremort sedimentary mass was reworked by the still-rising sea, leaving Tiger and Trinity shoals as remnants (see figure 3).

About 3600 years ago, as the Gulf reached its still-stand position, the episode of eustatic sea-level rise virtually ceased. River deposition subsequently became the dominant process and has fashioned a broad deltaic plain in southeastern Louisiana (see figure 4).

Cyclical deltaic sedimentation by the Mississippi River has been rather thoroughly described (Scruton 1960; Coleman and Gagliano 1964; Morgan 1970) and can be briefly summarized for the central coastal area as follows: On reaching the Gulf (base level), an alluviating river, such as the Mississippi, is forced to deposit its entrained sedimentary load because its flow-velocity diminishes. Resulting sedimentary deposits accumulate,

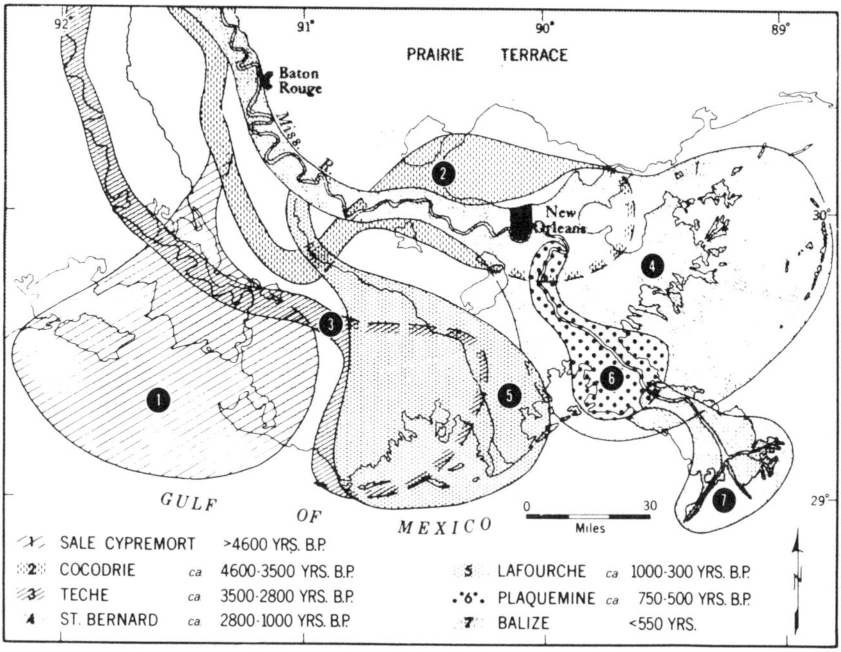

FIG. 4. CHRONOLOGY OF DELTAS THAT COMPOSE THE HOLOCENE MISSISSIPPI RIVER DELTAIC PLAIN. Modified from Kolb and Van Lopik (1958).

forming a delta and gradually building seaward (prograding). The river branches time and again to form an anastomosing, radiating pattern typical of deltaic distributaries. As each distributary lengthens, its gradient diminishes, and the channel becomes increasingly inefficient. The deltaic distributaries ultimately cannot handle their water and sediment loads, and the river is forced to divert through a new distributary that has a shorter route (and steeper gradient) to base level. During the past several thousand years, this sequence has recurred several times, with the result that the river has fashioned its deltaic plain from a number of overlapping deltaic lobes (see figure 4). Subsidence, mainly compaction of soft, newly deposited sediments, has allowed later delta lobes to prograde across and over older sequences.

Details are increasingly vague for the progressively older delta lobes; however, a study of the modern delta reveals a series of smaller subdeltas surrounding and peripheral to the major delta framework (figure 5). These details of subdeltaic development during the past three centuries are revealed by comparisons of old maps. It can be assumed with consid-

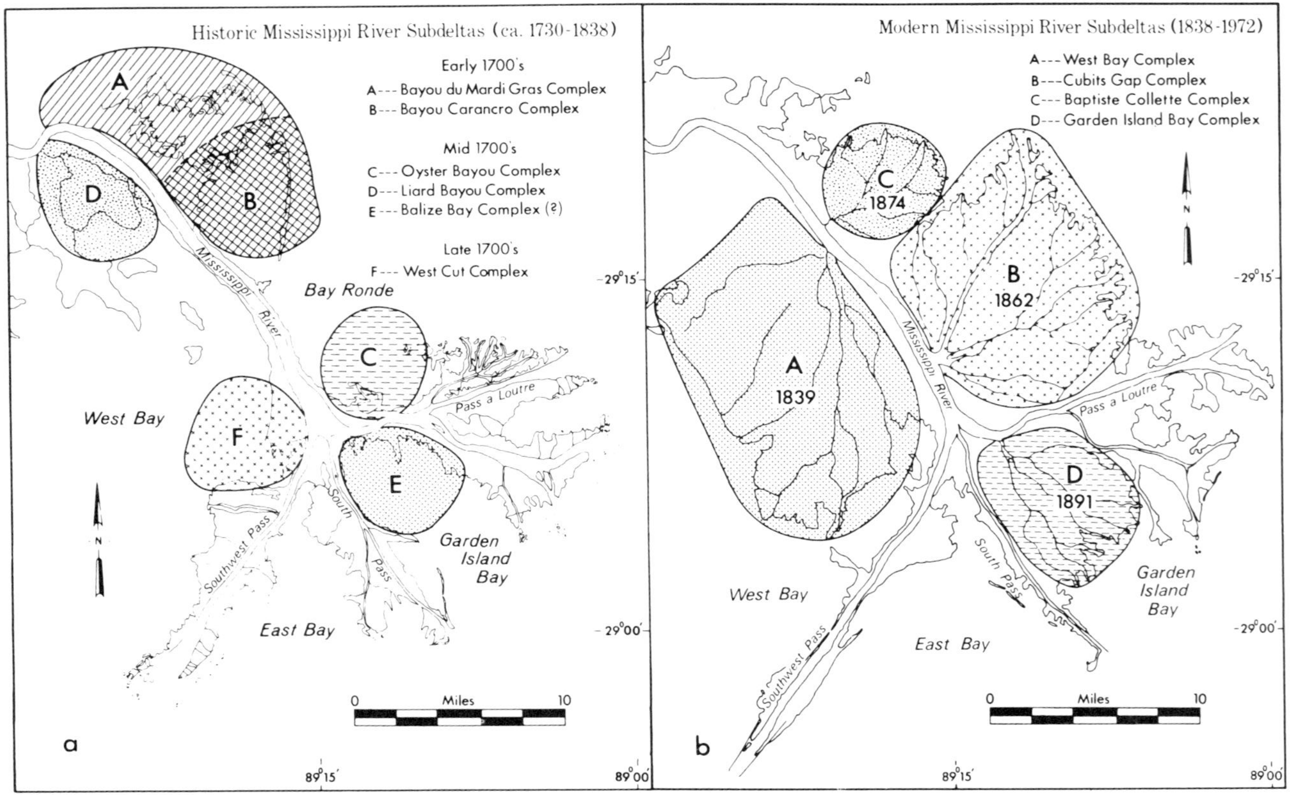

FIG. 5. SUBDELTAS OF THE MISSISSIPPI RIVER. (a) approximate ages of earlier subdeltas. (b) modern subdeltas, showing dates of initial crevasses. After D. J. Morgan (1977).

erable certainty that similar subdeltaic development occurred during deposition of the earlier delta masses.

Early Lafourche Delta

The central Louisiana coast of Terrebonne and Lafourche parishes displays a distinctive pattern of deltaic distributary levees and flanking marshes, numerous ponds, lakes, and bays lying between their radiating channels (figure 6). Radiocarbon dating suggests that this delta formed during the period from 1000 to 300 years ago, although some distributaries no doubt had entered the area a few hundred years earlier, and the last active channel, Bayou Lafourche, carried flow until it was dammed off in 1906. The Lafourche system formed a very extensive delta in a relatively short period of time because it built seaward into quite shallow water left by the subsidence of the older Teche-Mississippi delta (see figure 4).

Insufficient data are available to reconstruct a detailed history of the Lafourche delta, but it certainly formed in at least two episodes, referred to for simplicity as "Early" and "Late." Distributaries of the Early Lafourche subdelta extended in a southerly direction from a point of breakup (analogous to present-day Head of Passes) at the site of the present town of Thibodaux (see figure 6). Bayous Little Black, Terrebonne, and Blue were the principal distributaries, with lower Bayou Lafourche at that time being either of lesser importance or nonexistent. These distributaries branched again near the present town of Houma to form a rather densely radiating network consisting of Bayous Black, Mauvais Bois, du Large, Grand Caillou, Petit Caillou, Terrebonne, Pointe au Chien, and Blue. Older Bayou Teche levees in the vicinity of Houma probably created a topographic barrier that, when finally breached, determined the positions of the numerous radiating deltaic distributaries.

A half dozen of the major distributaries branched again and again as they prograded seaward, forming a truly Δ-shaped land mass that extended a mile or so seaward of the trend of present-day Isles Dernieres. These distributaries apparently became so overextended that inefficient gradients resulted in the gradual diversion of the majority of the Lafourche-Mississippi's flow to the present course of Bayou Lafourche. As the numerous distributaries of the Early Lafourche delta were gradually deprived of their flow and sediment load by this diversion, marine processes became increasingly dominant. Attack by waves and marine currents winnowed the silts and clays from the deltaic deposits, concentrating the coarser materials (fine sands) in an elongated barrier-beach complex including Isles Dernieres, Wine, Caillou, Brush, and Casse Tete islands (see figure 6).

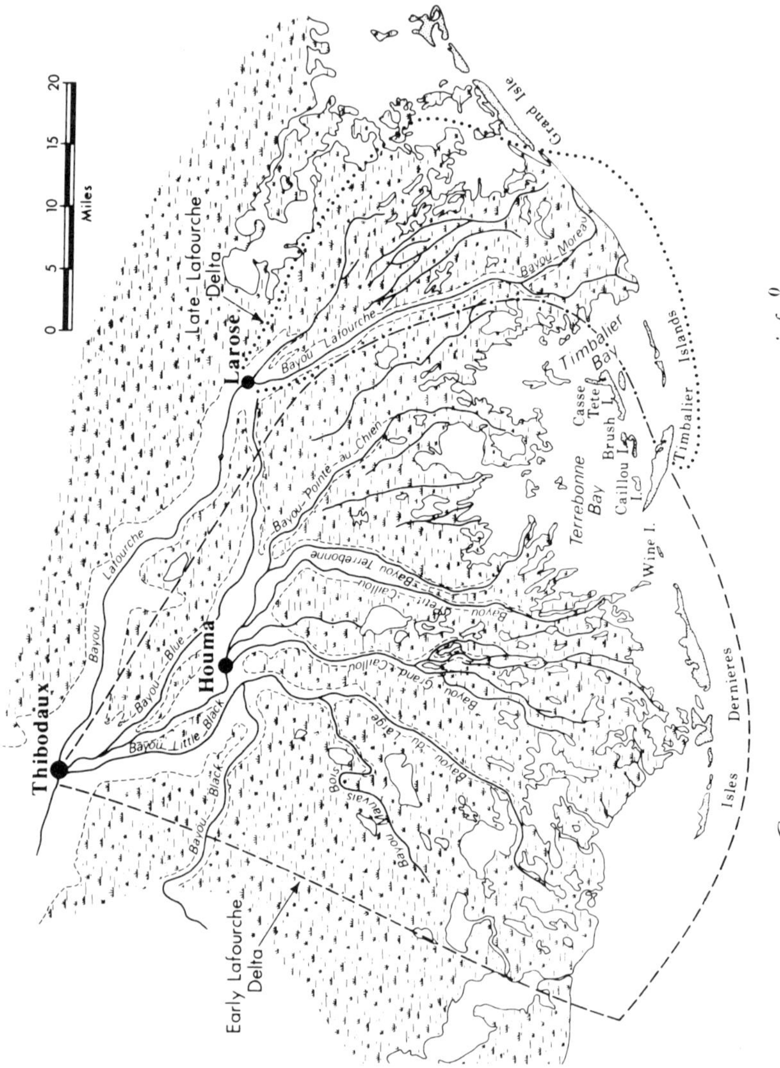

FIG. 6. THE LAFOURCHE-MISSISSIPPI RIVER DELTA, Terrebonne and Lafourche parishes, Louisiana.

With abandonment, sediment deposition diminished but subsidence through compaction continued. For a while, marsh growth was able to keep pace with subsidence, but interdistributary marshes were ultimately destroyed by enlarging ponds, lakes, and bays during this "destructional phase" of the deltaic cycle (Morgan 1973).

Late Lafourche Delta

Diversion from the Early to the Late Lafourche system probably was not a sudden event but one that required a hundred years or so (Fisk 1952). As Bayou Lafourche received an increasing volume of water and sediment, it prograded rapidly and built at least one major subdelta to the southeast of the present town of Larose (see figure 6). In its final phase of progradation, Bayou Lafourche and its principal distributary, Bayou Moreau, extended seaward, well beyond the destructional barrier island shoreline of the Early Lafourche system.

Diversion of the Mississippi from the Late Lafourche subdelta to its modern site took place several hundred years ago, and by the time of earliest historical records and maps, the Late Lafourche system was well into its destructional phase. Wave attack on the distributaries and marshes of Bayous Moreau and Lafourche caused rapid shoreline retreat, and the coarser fractions (fine and very fine sands) were concentrated to form the forerunners of Timbalier Island. Map comparison reveals that shorelines near the mouth of Bayou Lafourche were retreating rapidly during the last half of the nineteenth century. Retreat rates of more than 50 feet per year reflect the rapidity of subsidence and the consequent marine transgression (Morgan and Larimore 1957). The process has continued during the twentieth century, a result not only of continuing compaction and subsidence but also of sea-level rise. While Timbalier Island has retreated landward under wave attack, it has also migrated westward under the influence of the prevailing littoral currents. During a period of 116 years, the west end of Timbalier Island has moved northwestward about 4.25 miles, establishing a new shoreline trend seaward of the Caillou-Brush-Casse Tete island part of the Early Lafourche delta.

The earliest reasonably accurate maps of Terrebonne and Lafourche parishes were the township plats made during the 1830s and 1840s. These rather crude township surveys, augmented by other data, were compiled and included as part of the "Map of the Alluvial Valley of the Mississippi River" published in 1887 by the U.S. Army Corps of Engineers (figure 7). Configurations shown on this map indicate more the form of the coast during the earlier part of the nineteenth century than that of the map publication date.

In the intervening century or so, not only has the Timbalier-

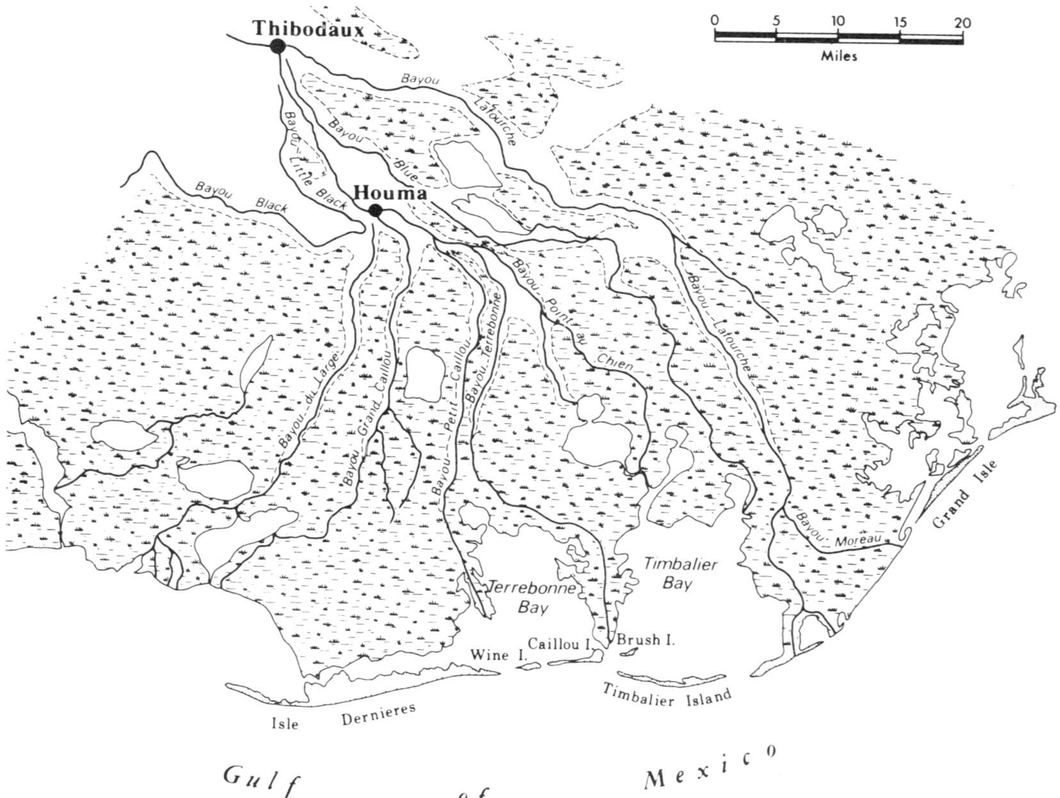

FIG. 7. THE LAFOURCHE-MISSISSIPPI DELTA as depicted on the "Map of the Alluvial Valley of the Mississippi River, 1887," U.S. Army Corps of Engineers, which was based on surveys made several decades earlier.

Dernieres Islands complex migrated northwestward in dramatic fashion, but there have been equally significant changes within the coastal bays (figures 6 and 7). Subsidence and erosion have removed surface expression of the seaward extension of Bayou Terrebonne, thereby allowing Timbalier and Terrebonne Bays to merge. Formerly extensive marshlands have deteriorated, to leave only scattered small islands throughout the rapidly enlarging bay.

Post-Deltaic history

Mississippi floods through the Bayou Lafourche distributary ended finally in 1906 with the construction of a permanent dam at its head at Donaldsonville. A small amount of river water has subsequently been pumped into the distributary to allow some flushing action. Following the disastrous flood of 1927, artificial-levee improvement on the lower Mississippi precluded further overbank flooding. Since that time, the Lafourche delta complex has been completely deprived of fresh water, except for local rainfall. Consequently, destructional processes have dominated the Lafourche delta for the past fifty years. Continuing subsidence has caused the radiating framework of distributary levees gradually to become reduced in height, length, and width. Interdistributary marshes have opened up, becoming ponds, lakes, and bays. As water bodies enlarged under wave attack, they also deepened.

Near the coast, a slowly growing network of tidal channels allows increasing brackishness in the interdistributary basins; flora and fauna change accordingly. The brackish-water clam *(Rangia cuneata)* and brackish-saline oyster *(Crassostrea virginica)* thrive in the delta-margin bays, and their shells become a major component in the bay-bottom sediments. To a lesser degree, other brackish-to-saline organisms also contribute to the sediments in the bays and lakes.

During the Lafourche-Mississippi's destructional phase, its deltaic sediments are rapidly being reworked and modified. Much of the reworking results naturally from more effective wave erosion, stemming from increased fetch and greater wave size, as well as the influence of tidal currents. Interdistributary marshes, attacked by these waves and currents, add their highly organic clays to the materials flooring the lakes and bays. In addition, some of the coarser-grained deltaic sediments, reworked by Gulf waves, are transported by currents into the more seaward parts of the bays.

HUMAN INFLUENCE IN THE LAFOURCHE DELTA

In addition to the essentially natural deltaic destructional processes previously described, man plays an increasingly dominant role in causing changes within the abandoned Lafourche delta. Most environmentally significant human activities within this marginal environment result in changes that parallel or accelerate natural processes. Hence, they are exceedingly difficult to document and, especially, to quantify.

For example, canalization increases water-exchange rates, which almost invariably lead to some degree of salt water intrusion toward the interior, with resultant changes in the flora and fauna. These modifications parallel those resulting from natural marine transgression, and it is very difficult to determine the proportional influence of each set of factors.

Boat traffic and shrimp trawling likewise play definite but poorly assessed roles in increasing water turbidity, with resulting detrimental effects on much of the flora and fauna. The results cannot be easily differentiated from discharge of oil-well cuttings or, for that matter, the bottom-stirring action of storm waves or abnormal tidal currents.

Detrimental influences of these activities—dredging, boat traffic, shrimp trawling, and discharge of cuttings or effluents—are greater within the bays than offshore. Shallow water depths and restricted circulation within the shallow bays and estuaries make those areas more susceptible to human influences than is the adjacent continental shelf, where more effective circulation patterns prevail.

To prevent or reduce adverse effects of human activities within the delta plain and adjacent shelf, it is necessary to understand in detail the physical, chemical, and biological characteristics influencing the two areas. Only when adequate scientific data are accumulated and evaluated will it become possible to assign environmental changes to human or natural influences.

NOTE

This article originally appeared in *Mêlanges* no. 9 (November 15, 1974), Museum of Geoscience, Louisiana State University. Reprinted with permission. A few minor changes to the original text have been made.

REFERENCES CITED

Coleman, J. M. 1966. Recent Coastal Sedimentation, Central Louisiana Coast. Louisiana State University, Coastal Studies Institute, Technical Report no. 29.

——— and S. M. Gagliano. 1964. Cyclic sedimentation in the Mississippi River deltaic plain. Gulf Coast Association of Geologists Society, Transactions **14**:67-80.

Curray, J. R. 1965. Late Quaternary history, continental shelves of the United States. *In* H. E. Wright, Jr., and D. G. Frey, eds., The Quaternary of the United States. Princeton University Press, pp. 723-735.

Fisk, H. N. 1952. Geological Investigations of the Atchafalaya Basin and the Problem of Mississippi River Diversion. U.S. Army Corps of Engineers, Mississippi River Comm., Vicksburg, Mississippi.

———. 1955. Sand facies of Recent Mississippi delta deposits. *In* Proceedings, 4th World Petroleum Congress, Rome, Sec. 1-C, pp. 377-398.

Kolb, C. R. and J. R. Van Lopik. 1958. Geology of the Mississippi River Deltaic Plain, Southeastern Louisiana. U.S. Army Corps of Engineers, Waterways Experimental Station, Vicksburg, Mississippi. Technical Report 3-483 and 3-484.

Marmer, H. A. 1952. Changes in sea level determined from tide observations. *In* Proceedings, 2d Conference on Coastal Engineering, Council on Wave Research, University of California, Berkeley, pp. 62-67.

Morgan, D. J. 1977. The Mississippi River delta: Legal-geomorphic evaluation of historic shoreline changes. Geoscience and Man **16**:1-196.

Morgan, J. P. 1970. Deltas, A resumé. Journal of Geological Education **18**:107-117.

———. 1973. Impact of subsidence and erosion on Louisiana coastal marshes and estuaries. Coastal Marsh and Estuary Symposium, Baton Rouge, Louisiana.

——— and P. B. Larimore. 1957. Changes in the Louisiana Coastline. Gulf Coast Association of Geologists Society, Transactions **7**:303-310.

Scruton, P. C. 1960. Delta building and the deltaic sequence. *In* Recent Sediments, Northwest Gulf of Mexico. Tulsa, Oklahoma: American Association of Petroleum Geologists, pp. 82-102.

Swanson, R. L. and C. I. Thurlow. 1973. Recent subsidence rates along the Texas and Louisiana coasts as determined from tide measurements. Journal of Geophysical Research **78**:2665-2671.